Proceedings
of the
STEKLOV INSTITUTE
OF MATHEMATICS

1985, ISSUE 2

Orthogonal Series and Approximation of Functions

Collection of Papers Dedicated to the One-hundredth Birthday of Academician

N. N. LUZIN

Translation of
ТРУДЫ
ордена Ленина
МАТЕМАТИЧЕСКОГО ИНСТИТУТА
имени В. А. СТЕКЛОВА

Том 164 (1983)

NIKOLAĬ NIKOLAEVICH
LUZIN
(1883–1950)

АКАДЕМИЯ НАУК

СОЮЗА СОВЕТСКИХ СОЦИАЛИСТИЧЕСКИХ РЕСПУБЛИК

ТРУДЫ

ордена Ленина

МАТЕМАТИЧЕСКОГО ИНСТИТУТА

имени В. А. СТЕКЛОВА

CLXIV

ОРТОГОНАЛЬНЫЕ РЯДЫ И ПРИБЛИЖЕНИЕ ФУНКЦИЙ

СБОРНИК СТАТЕЙ

Посвящается
100-летию со дня рождения
академика
Н. Н. Лузина

издательство "наука"

МОСКВА 1983

Library of Congress Cataloging-in-Publication Data

Ortogonal'nye rīady i priblizhenie funkfsiī. English.
Orthogonal series and approximation of functions.
(Proceedings of the Steklov Institute of Mathematics, ISSN 0081-5438; 1985, issue 2)
Translation of: Ortogonal'nye rīady i priblizhenie funkfsiī.
Number in Russian series statement: tom. 164 (1983).
1. Series, Orthogonal—Addresses, essays, lectures. 2. Approximation theory—Addresses, essays, lectures. 3. Luzin, N. N. (Nikolai Nikolaevich), 1883–1950. 4. Mathematicians—Soviet Union—Biography. I. Title. II. Luzin, N. N. (Nikolai Nikolaevich), 1883–1950. III. Nikol'skiĭ, S. M. (Sergeĭ Mikhaĭlovich), 1905– . IV. Kashin, B. S. V. Stechin, S. B. VI. Series: Trudy Ordena Lenina Matematicheskogo instituta imeni V. A. Steklova. English; 1985, issue 2.
QA1.A413 no. 164 [QA404.5] 510s [515'.55] 85-20001
ISBN 0-8218-3086-4

September 1985

ABSTRACT. This collection of papers is dedicated to the one hundredth birthday of one of the founders of the Moscow mathematical school, Academician Nikolaĭ Nikolaevich Luzin (1883–1950). It is prefaced by brief recollection of D. E. Men′shov, Corresponding Member of the Academy of Sciences of the USSR, on Luzin's scientific school. There are original papers on various aspects of the theory of functions.

1980 *Mathematics Subject Classification*. Primary 01A70, 20-02, 26D05, 26D10, 33A65, 41A05, 41A15, 42A10, 42A24, 42A50, 42B05, 42B15, 42B99, 42C05, 42C10, 42C15, 42C30, 46B15, 46E15, 46E35, 47E05, 58C05; Secondary 30E05, 31A15, 39A10, 41A15, 41A46, 41A50, 41A63, 42A50, 42B15, 42C20, 49D25, 46E30.

PROCEEDINGS OF THE STEKLOV INSTITUTE OF MATHEMATICS IN THE ACADEMY OF SCIENCES OF THE USSR

TABLE OF CONTENTS

RUSSIAN TABLE OF CONTENTS*

*The American Mathematical Society scheme for transliteration of Cyrillic may be found at the end of index issues of *Mathematical Reviews.*

Correspondence between *Trudy Mat. Inst. Steklov.* and *Proc. Steklov Inst. Math.*

Russian imprint	Vol.	English imprint	Russian imprint	Vol.	English imprint	Russian imprint	Vol.	English imprint Year	English imprint Issue
1966	74	1967	1971	107	1976	1980	139	1982	1
1965	75	1967	1968	108	1971	1976	140	1979	1
1965	76	1967	1971	109	1974	1976	141	1979	2
1965	77	1967	1970	110	1972	1976	142	1979	3
1965	78	1967	1970	111	1972	1977	143	1980	1
1965	79	1966	1971	112	1973	1979	144	1980	2
1965	80	1968	1970	113	1972	1980	145	1981	1
1966	81	1968	1970	114	1974	1978	146	1980	3
1966	82	1967	1971	115	1974	1980	147	1981	2
1965	83	1967	1971	116	1973	1978	148	1980	4
1965	84	1968	1972	117	1974	1978	149	1981	3
1966	85	1967	1972	118	1976	1979	150	1981	4
1965	86	1967	1973	119	1976	1980	151	1982	2
1966	87	1967	1974	120	1976	1980	152	1982	3
1967	88	1969	1972	121	1974	1981	153	1982	4
1967	89	1968	1973	122	1975	1983	154	1984	4
1967	90	1969	1973	123	1975	1981	155	1983	1
1967	91	1969	1976	124	1978, Issue 3	1980	156	1983	2
1966	92	1968	1973	125	1975	1981	157	1983	3
1967	93	1970	1973	126	1975	1981	158	1983	4
1968	94	1969	1975	127	1977	1983	159	1984	2
1968	95	1971	1972	128	1974	1982	160	1984	1
1968	96	1970	1973	129	1976	1983	161	1984	3
1968	97	1969	1978	130	1979, Issue 4	1983	162	1985	1
1968	98	1971	1974	131	1975	1984	163	1985	4
1967	99	1968	1973	132	1975	1983	164	1985	2
1971	100	1974	1973	133	1977	1984	165	1985	3
1972	101	1975	1975	134	1977				
1967	102	1970	1975	135	1978, Issue 1				
1968	103	1970	1975	136	1978, Issue 2				
1968	104	1971	1976	137	1978, Issue 4				
1969	105	1971	1975	138	1977				
1969	106	1972							

Труды Матем. Инст. Стеклов.
164(1983)

Proceedings of the Steklov
Institute of Mathematics
1985, Issue 2

ON THE FIRST STEPS OF N. N. LUZIN'S SCHOOL

D. E. MEN'SHOV

Nikolaĭ Nikolaevich Luzin is the acknowledged founder of the Moscow school in the theory of functions of a real variable. But the significance of his activity is far broader. The enormous role he played in the development of Soviet mathematics as a whole is becoming more and more obvious now, as we mark the 100th anniversity of his birth. Many independent schools of research have branched off from Luzin's school. Examples are the topological school founded by his students P. S. Aleksandrov and P. S. Urysohn, and the school in probability theory that grew from the work of A. N. Kolmogorov and A. Ya. Khintchine, also students of Luzin. In addition, Khintchine developed function-theoretic methods in number theory. Another student of Luzin who obtained deep results in number theory was L. G. Schnirelmann.

Adding to this the contributions of Luzin's school in the evolution of the theory of analytic functions (M. A. Lavrent'ev), functional analysis and differential equations (L. A. Lyusternik), mathematical logic (P. S. Novikov), and differential geometry, we see that the outstanding pedagogical and scientific activities of Luzin exhibit a deep influence on the mathematical life in our country up to the present time.

In these brief notes about Luzin I have dwelt only on the first years of his fruitful activity at Moscow University, when his school had just come into being.

Luzin began his systematic teaching at the university in the fall of 1914. Before that he had spent several years abroad in Göttingen and Paris. There he acquainted himself in depth with the latest investigations of the greatest European mathematicians, was on personal terms with many of them (Landau, Hadamard, Picard, Borel, Lebesgue, and Denjoy), and also carried out intensive research which was summed up in his famous dissertation, "The integral and trigonometric series".

Luzin thus began his work at the university as an established scholar having independent views on the path of development of mathematics.

It should be said that, by the time he began his systematic work at the university, student interest in the theory of functions had been prepared to a considerable extent by the seminar of D. F. Egorov, under whom Luzin studied. Because of this seminar, many students were well acquainted with the foundations of the theory of functions. Therefore, Luzin could at once give a special advanced course in the theory of functions and attract some of his listeners to the research.

1980 *Mathematics Subject Classification*. Primary 20-02, 01A70.

Luzin began this elective course in the theory of functions of a real variable in September 1914. Besides me, his audience included Aleksandrov, Khintchine and others. We became his first students. A year later M. Ya. Souslin joined us.

Luzin's course was very captivating. At this time he was actively involved with investigations in the theory of functions, and he frequently posed questions touching on unsolved problems to his listeners. Moreover, he often turned to the audience for help when a gap appeared in the course of his presentations. This was a valuable pedagogical device which stimulated activity on the part of the listeners. Remarkable were not only his lectures, but also the ensuing discussions he readily entered into with students in the hallway or his office and later even in his apartment.

I recall that he got the idea of organizing a seminar dealing with a comparison of the two integrals of Denjoy and Borel. I began thinking intensely about this problem and solved it shortly thereafter, establishing that the Denjoy integral is more general than the Borel integral. I told Luzin about this, and he said that I had exhausted the topic and he would not organize the seminar. My example of a function not Borel integrable but Denjoy integrable was published in "Matematicheskiĭ Sbornik" as my first research article. Up to this point I have meant the Denjoy integral in the narrow sense. My fellow student Khintchine, who also studied under Luzin, gave an important generalization of the concept of the Denjoy integral, later called the Denjoy-Khintchine integral because Denjoy himself had independently arrived at the same generalization.

It was at this time that the small nucleus of the future Moscow school of the theory of functions was forming around Luzin. He was already close to its older representatives V. V. Stepanov and I. I. Privalov, and in 1914 and 1915 the students Aleksandrov and Khintchine also joined his circle, then later Souslin, V. S. Fedorov, Urysohn, and several others.

In this period the younger representatives of the Moscow school of the theory of functions made their first discoveries and published them in part. Besides the results already cited, I mention my example of a trigonometric series with nonzero coefficients that converges to zero almost everywhere, i.e., an example of an M-set of measure zero, in a later terminology. Luzin regarded this result of mine with great interest. He submitted my paper with this result along with his recommendation to Hadamard in Paris for publication in "Comptes Rendus". Luzin also sent Hadamard papers by others of my comrades for the same journal.

In a reply Hadamard expressed satisfaction that serious research was being done in Moscow, and it was clear that he put a high value on Luzin's role in organizing this work.

Aleksandrov's paper on the cardinality of Borel sets was sent to "Comptes Rendus" at the same time as my article. To derive his main theorem Aleksandrov used what is now called the A-operation, which shortly after was given applications extremely important for the subsequent development of set theory, as well as topology. In my view, Aleksandrov's paper was, in general, the most consequential of all our papers sent at that time to Paris. Somewhat later Souslin used Aleksandrov's operation to effectively construct sets that were not B-sets, and he called them A-sets in honor of Aleksandrov. Luzin checked Souslin's result, appraised it highly, and suggested that a paper be written. This article, published in "Comptes

Rendus" in 1917, contains a number of essential propositions about properties of A-sets. Due to the suggestion of Aleksandrov, A-sets are now often called Souslin sets. Unfortunately, Souslin himself passed away in 1919 and was not able to take part in the subsequent evolution of the theory of A-sets. This theory was later developed in depth by Luzin, the Polish mathematician Sierpiński (who was then living in Moscow), the German mathematician Hausdorff, and many others. Incidentally, Sierpiński returned to Poland after the war and developed a school of mathematics there which to a significant degree took its theme from the Luzin school.

Such were the initial fruits of activity of the first students and followers of Luzin. As is clear, he strongly stimulated our investigations, introducing us to the problems of the theory of functions close to him, encouraging us to work independently, and assisting in the most rapid publication of our first papers. Moreover, he invariably conducted himself with us almost as an equal, distinguishing himself only by his great erudition and by a special kind of mental sharpness, perhaps not very quick, but extremely profound and penetrating. It is true that some of our results turned out to be unexpected for him and refuted his initial conjectures.

It should be said that Luzin himself carried out very active research during this period. It was at this time, in the spring of 1916, that he brilliantly defended his dissertation "The integral and trigonometric series," for which he was awarded the degree of Doctor of Pure Mathematics, bypassing the master's degree.*

Luzin worked in intensive spurts. With this style of work, of course, he would utterly exhaust himself. Therefore, after the results he desired were obtained, there would ensue now and then periods of weariness and scientific passivity.

The years when the Moscow school in the theory of functions was being established and then its subsequent development at the beginning of the 1920's undoubtedly constituted a flourishing period for the research and pedagogical activities of Luzin.

In the 1920's Luzin's school was replenished by new young talent. Lyusternik, N. K. Bari, Lavrent'ev, Schnirelmann, Novikov, L. V. Keldysh, Kolmogorov, and V. I. Glivenko became his students. It was this energetic generation of Luzin's students that made up the famous "Lusitania" about which much has been written.

At the same time there began a period of independent research by many of Luzin's older students and the formation of the daughter schools of research which I mentioned at the beginning of this paper. Problems going directly back to Luzin in the area of the metric and descriptive theory of functions were enriched by profound results. Luzin's school received worldwide recognition.

Since those times the remarkable traditions of the Luzin school have never been interrupted. These traditions are now alive in the activities of many Soviet mathematicians, including numerous specialists in the theory of functions, of which the majority are direct scientific descendents of Luzin.

Translated by H. H. McFADEN

**Editor's note*. The Russian is Степень Магистра (literally "master's degree"). Before the revolution it was roughly equivalent to the Ph.D. in an American university.

Труды Матем. Инст. Стеклов.
164(1983)

Proceedings of the Steklov
Institute of Mathematics
1985, Issue 2

UNIFORM ASYMPTOTIC REPRESENTATIONS OF ORTHOGONAL POLYNOMIALS

UDC 517.512

V. M. BADKOV

ABSTRACT. The author establishes uniform asymptotic representations as $n \to \infty$, on the whole circumference $\Gamma_1 = \{z\colon |z| = 1\}$, for the polynomials $\varphi_n(z)$ that are orthogonal on Γ_1 with respect to a weight function with a finite number of singularities each of which is a singularity of power type. The asymptotic representations involve the Jacobi polynomials. As simple corollaries of these representations similar results are obtained for trigonometric orthogonal polynomials and for polynomials orthogonal on an interval. In particular, previous results of S. N. Bernstein on asymptotic representations of generalized Jacobi polynomials are strengthened and generalized.

Bibliography: 28 titles.

§1. Introduction

This paper establishes representations, asymptotic for $n \to \infty$, and uniform on the whole circumference $\Gamma_1 = \{z\colon |z| = 1\}$, for the polynomials $\{\varphi_n(z)\}_0^\infty$ that are orthogonal on Γ_1 with respect to a weight function with a finite number of singular points, each of which is a singularity of power type. These representations involve the Jacobi polynomials. We obtain, as simple corollaries, similar results for orthogonal trigonometric polynomials and for polynomials that are orthogonal on an interval. In this way, in particular, we can sharpen and generalize Bernstein's results on asymptotic representations of the generalized Jacobi polynomials. The main results of the paper were reported at the International Conference on the Constructive Theory of Functions (Varna, 1981).

1. Here we introduce some notation and discuss the connections between the three kinds of systems of orthogonal polynomials that we shall consider.

Throughout the paper, $\mathbf{C}$ is the complex plane; $\mathbf{R} = (-\infty, \infty)$; $\mathbf{Z} = \{0, \pm 1, \pm 2, \ldots\}$; $\mathbf{Z}_+ = \{0, 1, 2, \ldots\}$; $\mathbf{N} = \{1, 2, 3, \ldots\}$; $L^r[a, b]$ is the space of Lebesgue measurable complex-valued functions with finite norm $\|f\|_{L^r[a,b]}$, where, for $1 \leqslant r < \infty$,

$$\|f\|_{L^r[a,b]} = \left\{\int_a^b |f(t)|^r\, dt\right\}^{1/r}, \qquad \|f\|_{L^\infty[a,b]} = \operatorname*{ess\,sup}_{a \leqslant t \leqslant b} |f(t)|;$$

1980 *Mathematics Subject Classification*. Primary 42C05, 33A65.

0081-5438/85 $1.00 + $.25 per page

L^r is the space of measurable 2π-periodic functions with finite norm $\|f\|_r = (2\pi)^{-1/r}\|f\|_{L^r[0,2\pi]}$; and $C_{2\pi}$ is the space of continuous 2π-periodic functions with norm $\|f\|_{C_{2\pi}} = \|f\|_\infty$.

Let $\alpha(t)\colon [-1,1] \to \mathbf{R}$ and $\sigma(\tau)\colon [0,2\pi] \to \mathbf{R}$ be bounded nondecreasing functions with infinitely many values. These functions uniquely determine the following three systems of orthogonal polynomials [1]–[3]:

1) the system of algebraic polynomials

$$p_{\alpha,n}(t) = k_{\alpha,n}t^n + \cdots, \qquad \deg p_{\alpha,n} = n, \qquad k_{\alpha,n} > 0\ (n \in \mathbf{Z}_+)$$

that are orthogonal on $[-1,+1]$ in the sense that

$$\int_{-1}^{1} p_{\alpha,m}(t)p_{\alpha,n}(t)\,d\alpha(t) = \delta_{m,n} \qquad (m,n \in \mathbf{Z}_+);$$

2) the system of algebraic polynomials

$$\varphi_{\sigma,n}(z) = \kappa_{\sigma,n}z^n + \cdots, \qquad \deg \varphi_{\sigma,n} = n, \qquad \kappa_{\sigma,n} > 0\ (n \in \mathbf{Z}_+)$$

that are orthogonal on Γ_1 with respect to $d\sigma(\tau)$:

$$\frac{1}{2\pi}\int_0^{2\pi} \varphi_{\sigma,m}(e^{i\tau})\overline{\varphi_{\sigma,n}(e^{i\tau})}\,d\sigma(\tau) = \delta_{m,n} \qquad (m,n \in \mathbf{Z}_+);$$

3) the system of trigonometric polynomials $\{T_{\sigma,n}(\tau)\}_{n=0}^\infty$ that are obtained by orthogonalizing the sequence 1, $\cos\tau$, $\sin\tau$, $\cos 2\tau$, $\sin 2\tau,\ldots$ by the Gram-Schmidt method with respect to $d\sigma(\tau)$ on $[0,2\pi]$. In this system $T_{\sigma,0} = \text{const} > 0$; $\forall\, n \in \mathbf{N}$, $T_{\sigma,2n-1}$, and $T_{\sigma,2n}$ are trigonometric polynomials of order n with positive coefficients for $\cos n\tau$ and $\sin n\tau$, respectively, and in addition

$$\frac{1}{2\pi}\int_0^{2\pi} T_{\sigma,m}(\tau)T_{\sigma,n}(\tau)\,d\sigma(\tau) = \delta_{m,n} \qquad (m,n \in \mathbf{Z}_+). \tag{1.1}$$

There are close connections among these systems. For example, if

$$\alpha(-1) = 0, \qquad \sigma(\tau) = \alpha(\cos\tau)\operatorname{sgn}(\tau - \pi),$$

a theorem of Szegö [1], [2] states that

$$p_{\alpha,n}\left(\frac{1}{2}\left(z + \frac{1}{z}\right)\right) = \frac{[\varphi_{\sigma,2n}(z) + \varphi^*_{\sigma,2n}(z)]z^{-n}}{\sqrt{2\pi[1 + \varphi_{\sigma,2n}(0)/\kappa_{\sigma,2n}]}} \qquad (n \in \mathbf{Z}_+, z \in \mathbf{C}), \tag{1.2}$$

where

$$\varphi^*_{\sigma,n}(z) = z^n\overline{\varphi}_{\sigma,n}(1/z) = \overline{\varphi_{\sigma,n}}(0)z^n + \cdots + \kappa_{\sigma,n}. \tag{1.3}$$

It was shown in [4] that each pair of systems $\{\varphi_{\sigma,n}(z)\}_{n=0}^\infty$ and $\{T_{\sigma,n}(\tau)\}_{n=0}^\infty$ satisfies, for all $n \in \mathbf{N}$ and $\tau \in \mathbf{R}$, the equations

$$\begin{aligned} T_{\sigma,2n-1}(\tau) &= c_n \operatorname{Re}\{e^{-i(n-1)\tau}\varphi_{\sigma,2n-1}(e^{i\tau})\}, \\ T_{\sigma,2n}(\tau) &= c_n \operatorname{Im}\{e^{-in\tau}\varphi_{\sigma,2n}(e^{i\tau})\}, \end{aligned} \tag{1.4}$$

where

$$c_n = (2\kappa_{\sigma,2n})^{1/2}[\kappa_{\sigma,2n} - \operatorname{Re}\varphi_{\sigma,2n}(0)]^{-1/2}. \tag{1.5}$$

Szegö [3] had already observed that in addition to the system $\{T_{\sigma,n}(\tau)\}_{n=0}^\infty$ there is an infinite set of systems $\{t_{\sigma,n}(\tau)\}_{n=0}^\infty$ of orthogonal trigonometric polynomials that satisfy orthogonality conditions of type (1.1) and have the properties that $t_{\sigma,0} = \text{const}$

and $t_{\sigma,2n-1}$ and $t_{\sigma,2n}$ are polynomials of order n for all $n \in \mathbf{N}$. Szegö also observed that the polynomials of order n in any such system are linear combinations of the polynomials of the same order in any other such system; and that one such system is the system $A_0, A_1(\tau), B_1(\tau), A_2(\tau), B_2(\tau), \ldots,$ where

$$A_n(\tau) = \sqrt{2}\left\{1 + |\varphi_{\sigma,2n}(0)|/\kappa_{\sigma,2n}\right\}^{-1/2} \operatorname{Re}\left\{e^{-i(n\tau+\gamma_{2n})}\varphi_{\sigma,2n}(e^{i\tau})\right\} \qquad (n \in \mathbf{Z}_+),$$

$$B_n(\tau) = \sqrt{2}\left\{1 + |\varphi_{\sigma,2n}(0)|/\kappa_{\sigma,2n}\right\}^{-1/2} \operatorname{Im}\left\{e^{-i(n\tau+\gamma_{2n})}\varphi_{\sigma,2n}(e^{i\tau})\right\} \qquad (n \in \mathbf{N}),$$

and the parameter $\gamma_{2n} \in \mathbf{R}$ is determined by the condition that $e^{-i\gamma_{2n}}\varphi_{\sigma,2n} \in \mathbf{R}$. This system is not, in general, the same as $\{T_{\sigma,n}(\tau)\}_{n=0}^{\infty}$, and the formulas defining it are less convenient to use than (1.4)–(1.5), since they involve the parameters γ_{2n}, about whose behavior almost nothing is known.

Because of (1.2)–(1.5), it is convenient to study the systems 1)–3) in the framework of a single theory in which the principal role is played by the polynomials that are orthogonal on the unit circumference.

We shall be concerned only with absolutely continuous functions α and σ. Instead of $\alpha'(t)$, $p_{\alpha,n}(t)$, $\sigma'(\tau)$, $\varphi_{\sigma,n}(z)$, and $T_{\sigma,n}(\tau)$ we shall write $p(t)$, $p_n(t)$, $\varphi(\tau)$, $\varphi_n(z)$, and $\Phi_n(\tau)$. The functions $p(t)$ and $\varphi(\tau)$ are determined only up to sets of measure zero, and $\varphi(\tau)$ is assumed to be 2π-periodic. The system $\{p_n(t)\}_0^{\infty}$ is orthonormal on $[-1,1]$ with weight function p, and $\{\varphi_n(z)\}_0^{\infty}$ and $\{\Phi_n(\tau)\}_0^{\infty}$ are respectively orthonormal on Γ_1 and $[0, 2\pi]$ with weight function φ.

2. We begin with some remarks on the history of the problems that we shall consider. In applications of the theory of orthogonal polynomials it is often necessary to know the asymptotic behavior ofo the polynomials as $n \to \infty$. It is particularly useful to have uniform asymptotic representations of orthogonal polynomials on the entire set on which they are orthogonal. The problem of obtaining such representations is especially difficult when the weight function has singular points (for the systems under consideration singular points are points at which $p(t)\sqrt{1-t^2}$ and $\varphi(\tau)$ are zero or infinite). This problem was first solved for the Jacobi polynomials $\{P_n^{\alpha,\beta}(t)\}_{n=0}^{\infty}$, which are orthonormal on $[-1,1]$ with weight function

$$p(t) = (1-t)^{\alpha}(1+t)^{\beta} \qquad (\alpha, \beta > -1) \tag{1.6}$$

and the normalization $P_n^{\alpha,\beta}(1) = \binom{n+\alpha}{n}$. This was done by Hilb for $\alpha = \beta = 0$, and then by Szegö and Rau in the general case (see [1], pp. 195, 197). The representation has the form

$$\left(\sin\frac{\tau}{2}\right)^{\alpha}\left(\cos\frac{\tau}{2}\right)^{\beta} P_n^{(\alpha,\beta)}(\cos\tau)$$
$$= \left(N_n^{\alpha,\beta}\right)^{-\alpha}\frac{\Gamma(n+\alpha+1)}{\Gamma(n+1)}\left(\frac{\tau}{\sin\tau}\right)^{1/2} J_{\alpha}\left(N_n^{\alpha,\beta}\tau\right) + O\left(r_{1,n}^{(\alpha)}(\tau)\right), \tag{1.7}$$

where

$$r_{1,n}^{(\alpha)}(\tau) = \begin{cases} \tau^{1/2}n^{-3/2}, & cn^{-1} \leqslant \tau \leqslant \pi - \varepsilon, \\ \tau^{\alpha+2}n^{\alpha}, & 0 < \tau \leqslant cn^{-1}, \end{cases} \tag{1.8}$$

$\varepsilon \in (0, \pi)$ and $c > 0$ are fixed, $N_n^{\alpha,\beta} = n + (\alpha + \beta + 1)/2$, and $J_\alpha(z)$ is the Bessel function of the first kind. For $\tau \in [\varepsilon, \pi]$ the corresponding representation is easily obtained from (1.7), since (see [1], p. 59)

$$P_n^{(\alpha,\beta)}(-t) = (-1)^n P_n^{(\alpha,\beta)}(t). \tag{1.9}$$

The representation (1.7) is obtained by the method of Liouville and Steklov, which requires the knowledge of a simple differential equation for $p_n(t)$ which can be transformed into a Sturm-Liouville equation. Such equations are known only for the classical orthogonal polynomials and some related polynomials (for example, polynomials that are orthogonal on $[-1,1]$ with weight function $|(1 - t^2)^\alpha| + |t|^\gamma$, where $\alpha, \gamma > -1$ [5]), whose asymptotic properties are easily deduced from the relationship without using the theory of differential equations. Because the Liouville-Steklov method is inapplicable in the general case, there has been a considerable tendency to discuss other problems, among them the asymptotic behavior of an orthogonal system of polynomials as obtained from the corresponding properties of another system that is, in some sense, close to it.

Thus Bernstein [6] discussed generalized Jacobi polynomials, i.e. $\{p_n(t)\}_0^\infty$ with

$$p(t) = H(t)(1 - t)^\alpha(1 + t)^\beta \qquad (\alpha, \beta > -1), \tag{1.10}$$

where H is a nonnegative measurable function, bounded away from zero and infinite. Under some supplementary restrictions on H he studied the asymptotic connection between these polynomials and the Jacobi polynomials $\{\hat{P}_n^{\alpha,\beta}(t)\}_{n=0}^\infty$ which are orthonormal on $[-1,1]$ with the weight function (1.6). Assuming that

$$\begin{aligned} &H \in C[-1,1], \qquad H(t) > 0 \qquad (|t| \leqslant 1),\\ &\omega(H'; \delta)_{[-1,1]} = O(\ln^{-2}\delta) \qquad (\delta \to +0), \end{aligned} \tag{1.11}$$

where

$$\omega(f; \delta)_{[a,b]} = \sup\{|f(t) - f(x)|; t, x \in [a,b], |t - x| \leqslant \delta\}$$

is the modulus of continuity of f on $[a, b]$, Bernstein found, for $\alpha, \beta \in (-1, -\frac{1}{2})$, an asymptotic representation, uniform for $\tau \in [0, \pi]$, but without an error estimate, for $p_n(\cos\tau)$ in terms of

$$\hat{P}_n^{\alpha,\beta}(\cos\tau), \qquad \hat{P}_{n-1}^{\alpha+1,\beta+1}(\cos\tau)\sin\tau. \tag{1.12}$$

Also under (1.11), and assuming that $\max(\alpha, \beta) \geqslant -\frac{1}{2}$, Bernstein found a uniform asymptotic representation (without an error estimate) for the product $w(\cos\tau) \cdot p_n(\cos\tau)$, where

$$w(t) = (1 - t)^\rho(1 + t)^{\rho_1}, \quad \rho = \max(0, \alpha/2 + 1/4), \quad \rho_1 = \max(0, \beta/2 + 1/4),$$

in terms of the product $\hat{P}_n^{\alpha,\beta}(\cos\tau)w(\cos\tau)$ and its derivative with respect to τ. From this result one cannot obtain the asymptotic form of $p_n(t)$ at $t = 1$ if $\rho > 0$ or at $t = -1$ if $\rho_1 > 0$.

When $\alpha = \beta = -\frac{1}{2}$ the weight function (1.10) does not have singularities. In this case a uniform asymptotic representation of $p_n(\cos\tau)$ was obtained in [6] under restrictions on H considerably weaker than (1.11):

$$\exists\, \varepsilon > 0 \quad \omega(H; \delta)_{[-1,1]} = O(\ln^{-1-\varepsilon}\delta^{-1}) \qquad (\delta \to +0); \tag{1.13}$$

it was also shown there that the error in this representation is $O(\ln^{-\varepsilon} n)$. The same result was obtained somewhat later by Szegö (see [1], p. 297) by using (1.2) in the form of a trivial corollary of the corresponding result which he established by a simpler method for the polynomials $\varphi_n(z)$. This may be stated as follows: if $\varphi \in C_{2\pi}$, $\varphi(\tau) > 0$ ($\tau \in \mathbf{R}$) (consequently φ has no singular points!) and if, for some $\varepsilon > 0$, the ordinary modulus of continuity $\omega(\varphi; \delta)$ satisfies

$$\omega(\varphi; \delta) = O(\ln^{-1-\varepsilon}\delta^{-1}) \qquad (\delta \to +0), \tag{1.14}$$

then

$$\varphi_n(e^{i\Theta}) = e^{in\Theta}\overline{\pi(\varphi; e^{i\Theta})} + \varepsilon_n(\theta), \tag{1.15}$$

where $\varepsilon_n(\Theta) = O(\ln^{-\varepsilon} n)$ (uniformly in $\Theta \in \mathbf{R}$), and $\pi(\varphi;\ e^{i\Theta})$ are the radial boundary values of

$$\pi(\varphi; z) = \exp\left\{-\frac{1}{4\pi}\int_0^{2\pi}\frac{e^{i\tau}+z}{e^{i\tau}-z}\ln\varphi(\tau)\,d\tau\right\} \qquad (|z|<1) \tag{1.16}$$

(which is now often called the Szegö function).

Geronimus and Golinskiĭ [7] remarked that the proof of this result in [8] remains valid if (1.14) is replaced by the more general hypothesis $\omega(\varphi,; \tau)\tau^{-1} \in L^1[0, \pi]$. In [9], $\varepsilon_n(\Theta)$ is estimated in an even more general case.

Besides the results cited above, we now know many other results on the asymptotic behavior of orthogonal polynomials (for details, see, for example, [10]–[12]) that are not directly connected with our problem of the uniform asymptotic approximation of the polynomials $\varphi_n(e^{i\tau})$, $p_n(t)$, and $\Phi_n(t)$ on the whole set of orthogonality when there are singularities of the weight function.

3. In this paper we obtain uniform asymptotic representations in terms of Jacobi polynomials on the whole interval $[-1,1]$ for the polynomials $\{p_n(t)\}_0^\infty$, with weight function satisfying

$$p(t) = H(t)(1-t)^\alpha(1+t)^\beta|t-x_1|^{\delta_1}\cdots|t-x_M|^{\delta_M}; \tag{1.17}$$

$$H \in C[-1,1], \qquad H(t) > 0 \quad (|t| \leqslant 1), \qquad \omega(H; t)_{[-1,1]}t^{-1} \in L^1[0,1]; \tag{1.18}$$

$$-1 < x_1 < \cdots < x_M < 1; \qquad \alpha, \beta, \delta_1, \ldots, \delta_M > -1. \tag{1.19}$$

As simple corollaries of our results we can obtain all of Bernstein's results quoted above except for the case $\alpha = \beta = -\frac{1}{2}$. For a weight function (1.17) with a singular point inside $(-1,1)$, Bernstein's method is not applicable. This difficulty can be overcome by solving a similar problem for the polynomials $\{\varphi_n(z)\}_0^\infty$ with weight function satisfying

$$\varphi(\tau) = h(\tau)\left|\sin[(\tau-\Theta_1)/2]\right|^{\gamma_1}\cdots\left|\sin[(\tau-\Theta_m)/2]\right|^{\gamma_m}, \tag{1.20}$$

$$h \in C_{2\pi}, h(\tau) > 0 \quad (\tau \in \mathbf{R}), \qquad \omega(h,\tau)\tau^{-1} \in L^1[0,\pi], \tag{1.21}$$

$$-\pi < \Theta_1 < \cdots < \Theta_m \leqslant \pi; \qquad \gamma_1, \ldots, \gamma_m > -1. \tag{1.22}$$

An asymptotic representation for $p_n(\cos\Theta)$ is easily obtained by a known method [1], [2] from the corresponding representation for $\varphi_{2n}(e^{i\Theta})$ and formula (1.2). A similar corollary for $\Phi_n(\Theta)$ can be obtained by using (1.4) and (1.5).

For solving our problem for $\varphi_n(e^{i\Theta})$ we use the two-sided estimate

$$|\varphi_{n-1}(e^{i\Theta})| \asymp \prod_{\nu=1}^{m} \left(|\sin[(\Theta - \Theta_\nu)/2]| + n^{-1}\right)^{-\gamma_\nu/2} \qquad (\Theta \in \mathbf{R}, n \in \mathbf{N}) \tag{1.23}$$

which was established by the author in [4] under conditions (1.20)–(1.22) (with reference to which it should be observed that the implied upper bound for $|\varphi_{n-1}(e^{i\Theta})|$ follows easily from the analogous inequality for $|\Phi_n(\Theta)|$ that we established in [13], and was also obtained by Golinskiĭ [14] by the method of our paper [15], which contains an upper bound for $|p_n(x)|$).

We briefly outline the solution of our problem for $\varphi_n(e^{i\Theta})$ and state some of the results that follow from its solution.

Our construction begins in §2 with the discussion of some properties of the polynomials $\{\psi_n^{\alpha,\beta}(z)\}_{n=0}^{\infty}$ which are orthonormal on Γ_1 with weight function

$$\psi^{\alpha,\beta}(\tau) = (1 - \cos\tau)^{\alpha+1/2}(1 + \cos\tau)^{\beta+1/2}, \qquad (\alpha, \beta > -1) \tag{1.24}$$

and to some extent play the role of the Jacobi polynomials on Γ_1. By Szegö's formula ([1], p. 294) the polynomials $\psi_{2n-1}^{\alpha,\beta}(e^{i\Theta})$ and $\psi_{2n}^{\alpha,\beta}(e^{i\Theta})$ can be expressed very simply in terms of the polynomials (1.12). By analogy with the generalized Jacobi polynomials we are led to consider the polynomials $\{\varphi_n(z)\}_0^{\infty}$ which are orthonormal on Γ_1 with the weight function

$$\varphi(\tau) = \psi^{\cdot\beta}(\tau)h(\tau) \qquad (\alpha, \beta > -1, \tau \in \mathbf{R}). \tag{1.25}$$

In §3 we prove the following theorem which establishes an asymptotic connection between these polynomials and $\{\psi_n^{\alpha,\beta}(z)\}_{n=0}^{\infty}$ on Γ_1.

THEOREM 1. *Let the weight function* φ *have the form* (1.25), *where* h *satisfies* (1.21), *and let*

$$\lambda_n \in \mathbf{R}, \quad n^{-1} \leqslant \lambda_n \leqslant 1 \ (n \in \mathbf{N}), \qquad \lambda_n \downarrow 0 \quad (n \to \infty). \tag{1.26}$$

Then

$$\varphi_n(e^{i\Theta}) = \psi_n^{\alpha,\beta}(e^{i\Theta})\overline{\pi(h; e^{i\Theta})}\left[1 + O(\mu_{h,n}^{\alpha,\beta})\right] \tag{1.27}$$

as $n \to \infty$, *uniformly for* $\Theta \in \mathbf{R}$, *where* $\pi(h; e^{i\Theta})$ *is the radial boundary function of the Szegö function* (*see* (1.16)),

$$\mu_{h,n}^{\alpha,\beta} = \omega\left(h; \frac{1}{n}\right)\ln n + \int_0^{\lambda_n} \frac{\omega(h;\tau)}{\tau}\,d\tau + \left[\delta_n(\varphi) + \delta_n(\psi^{\alpha,\beta})\right]\left[\int_{\lambda_n}^{\pi} \frac{\omega^2(h;\tau)}{\tau^2}\,d\tau\right]^{1/2}, \tag{1.28}$$

$$\delta_n(\varphi) = \inf_{c_0,\dots,c_n \in C} \left\| \left[\pi(\varphi; e^{i\tau}) - c_0 - \cdots - c_n e^{in\tau}\right]\sqrt{\varphi(\tau)} \right\|_2 \tag{1.29}$$

and $\delta_n(\psi^{\alpha,\beta})$ *is defined similarly.*

As $n \to \infty$ the functions $\delta_n(\varphi)$ and $\delta_n(\psi^{\alpha,\beta})$ in (1.28) tend to zero, since the corresponding weight functions belong to the class

$$S = \{\varphi\colon \varphi \in L^1, \varphi(\tau) \geqslant 0 \ (\tau \in \mathbf{R}), \ln\varphi(\tau) \in L^1\} \tag{1.30}$$

and, as we know ([2], §1.2, limiting relation 15)

$$\varphi \in S \Rightarrow \delta_n(\varphi) \to 0 \qquad (n \to \infty). \tag{1.31}$$

Therefore the sequence λ_n can always be chosen (and we shall suppose that this has been done) so that it satisfies not only (1.26) but also

$$\mu_{h,n}^{\alpha,\beta} \to 0 \qquad (n \to \infty). \tag{1.32}$$

Theorem 1 is in a certain sense analogous to Bernstein's result cited above for the generalized Jacobi polynomials with $\alpha, \beta \in (-1, -\frac{1}{2})$. In his proof Bernstein used the Christoffel formula ([1], §2.5). There is no analogous formula for the polynomials $\varphi_n(z)$. Hence Bernstein's method is not available for proving Theorem 1. Our proof is based on a different, also known ([2], proof of Theorem 4.5), but more trivial relation,

$$\begin{aligned}\varphi_n(e^{i\Theta}) &= \psi_n(e^{i\Theta})\frac{1}{2\pi}\int_0^{2\pi} \varphi_n(e^{i\tau})\overline{\psi_n(e^{i\tau})}\psi(\tau)\,d\tau \\ &+ \frac{1}{2\pi}\int_{\Theta-\pi}^{\Theta+\pi} \varphi_n(e^{i\tau})\left[\frac{\psi(\tau)}{\varphi(\tau)} - \frac{\psi(\Theta)}{\varphi(\Theta)}\right] K_{\psi,n-1}(e^{i\Theta}, e^{i\tau})\varphi(\tau)\,d\tau,\end{aligned} \tag{1.33}$$

where $\{\psi_n(z)\}_{n=0}^{\infty}$ are the polynomials that are orthonormal on Γ_1 with weight function ψ, and

$$K_{\psi,n}(z_1, z_2) = \psi_0(z_1)\overline{\psi_0(z_2)} + \cdots + \psi_n(z_1)\overline{\psi_n(z_2)}.$$

By the Christoffel-Darboux formula ([2], formula (1.7))

$$K_{\psi,n-1}(e^{i\Theta}, e^{i\tau}) = (1 - e^{i(\Theta-\tau)})^{-1}\left[\psi_n^*(e^{i\Theta})\overline{\psi_n^*(e^{i\tau})} - \psi_n(e^{i\Theta})\overline{\psi_n(e^{i\tau})}\right]. \tag{1.34}$$

We take ψ to be the weight function $\psi^{\alpha,\beta}$. The main idea of the proof of (1.27) is 1) to use the property

$$\varphi \in S \Rightarrow \left\|\left[\varphi_n^*(e^{i\tau}) - \pi(\varphi; e^{i\tau})\right]\sqrt{\varphi(\tau)}\right\|_2 = O(\delta_n(\varphi)) = o(1)(n \to \infty), \tag{1.35}$$

(see [2], Theorem 5.8 and Chapter II) to replace the functions $\varphi_n(e^{i\tau})$, $\psi_n^*(e^{i\tau})$, and $\psi_n(e^{i\tau})$ that appear in the integrands on the right-hand side of (1.33) by the functions $e^{in\tau}\overline{\pi(\varphi; e^{i\tau})}$, $\pi(\psi; e^{i\tau})$, and $e^{-in\tau}\pi(\psi; e^{i\tau})$ which are close to them in the sense of (1.35); 2) to estimate the error resulting from this replacement by using (1.35) and Cauchy's inequality; and finally 3) to decompose the new integrands into sums of terms whose integrals differ only slightly from integrals that are easily calculated. For the first integral in (1.33) there is no difficulty in carrying out this idea. For the second integral in (1.33) the application of this idea is complicated by the smallness of the denominator in (1.31) near the point Θ, so that the suggested change can be made only for $\tau \notin [\Theta - \lambda_n, \Theta + \lambda_n]$, while for other values of τ the function is bounded above by using (1.23). In addition, when τ is close to singular points of ψ we need to bound $|\psi_n^*(e^{i\Theta})| = |\psi_n(e^{i\Theta})|$ above by using (1.23). We then apply (1.23) again to the resulting majorant, but so as to yield a lower bound for $|\psi_n(e^{i\Theta})|$.

A similar idea, applied not to (1.33) but to the corresponding integral representation for $\pi_N(e^{i\Theta})\varphi_n(e^{i\Theta})$, where π_N is a given polynomial of degree N, is applied in §4 to prove a fundamental lemma which is used under certain restrictions on a weight function (not necessarily with the same singularities) to find the asymptotic connection between two systems of polynomials that are orthogonal on Γ_1 at points of an arc, containing singular points of the weight functions, on which their ratio is strictly positive and sufficiently smooth.

The following theorem will be proved in §5 by using this lemma, Theorem 1, and the obvious fact that for the weight function $\varphi(\tau) = \psi(\tau - \Theta_0)$ $(\Theta_0 \in \mathbf{R})$

$$\varphi_n(z) = e^{in\Theta_0}\psi_n(e^{-i\Theta_0}z) \qquad (n \in \mathbf{Z}_+, z \in \mathbf{C}).$$

THEOREM 2. *Let the weight function* φ *satisfy* (1.20)–(1.22), $l \in (1,\dots,m)$, *and let the given sequence* $\{\lambda_n\}_1^\infty$ *satisfy* (1.26). *Then*

$$\varphi_n(z) = 2^{\gamma_l/4}e^{in\Theta_l}\psi_n^{(\gamma_l+1)/2,-1/2}(e^{-i\Theta_l}z)$$
$$\times\bar{\pi}\left(\varphi(\tau)|\sin[(\tau-\Theta_l)/2]|^{-\gamma_l}; z^{-1}\right)[1+O(\mu_n)]$$

as $n \to \infty$, *for each fixed* $\varepsilon \in (0,1)$, *uniformly for* $z \in Q_{l,\varepsilon}$, *where*

$$Q_{l,\varepsilon} = \left\{z : z \in \mathbf{C},\, |z| \geqslant 1;\, \forall\, \nu \in \{1,\dots,m\}\setminus\{l\}\, |z - e^{i\Theta_\nu}| \geqslant \varepsilon\right\},$$

and $\bar{\pi}(g;\, z)$ *is the function, analytic in* $|z| < 1$, *whose Maclaurin coefficients are the complex conjugates of those of* $\pi(g;\, z)$:

$$\mu_n = \omega\left(h;\, \frac{1}{n}\right)\ln n$$
$$+ \int_0^{\lambda_n} \frac{\omega(h,\tau)}{\tau}d\tau + \left[\frac{1}{\sqrt{n}} + \omega\left(h;\, \frac{1}{n}\right)\right]\left[1 + \int_{\lambda_n}^{\pi} \frac{\omega^2(h;\tau)}{\tau^2}d\tau\right]^{1/2}. \qquad (1.37)$$

It is evident that $\mu_n \to 0$ as $n \to \infty$ if the sequence $\{\lambda_n\}_1^\infty$ satisfying (1.26) is suitably chosen.

In §6 the right-hand side of (1.36) is transformed by using the asymptotic formulas for the Jacobi polynomials. Then, as corollaries, we obtain representations for the polynomials $\Phi_n(\Theta)$ and $p_n(\cos\Theta)$ from the corresponding asymptotic representations of $\varphi_n(e^{i\Theta})$. In addition, in §6 we establish that the results of this paper are, in a certain sense, best possible.

§2. Polynomials orthogonal on the unit circle, related to the Jacobi polynomials

Here we list a number of properties of the polynomials $\psi_n^{\alpha,\beta}(z) = \kappa_n^{\alpha,\beta}z^n + \cdots$ $(n \in \mathbf{Z}_+)$ that are orthonormal on Γ_1 with the weight function (1.24), on account of their fundamental role in our further work.

1) *The parameters*

$$a_n^{\alpha,\beta} = -\overline{\psi_{n+1}^{\alpha,\beta}(0)}\,/\kappa_{n+1}^{\alpha,\beta} \qquad (2.1)$$

are real and can be found from the formula [16]

$$a_n^{\alpha,\beta} = -\left[\alpha + 1/2 + (-1)^{n+1}(\beta + 1/2)\right]/(n + \alpha + \beta + 2) \qquad (n \in \mathbf{Z}_+). \qquad (2.2)$$

We present a proof of this formula, since it is given incorrectly in [16], with $(-1)^n$ instead of $(-1)^{n+1}$. The following proposition, which follows from a more general result of Geronimus ([17], §31), is used in [16] to calculate $a_n^{\alpha,\beta}$: if $\varphi(\tau) = p(\cos\tau)|\sin\tau|$, where p and φ are the weight functions of the systems $\{p_n(t)\}_0^\infty$ and $\{\varphi_n(z)\}_0^\infty$, then the parameters $a_n^{(\varphi)} = -\overline{\varphi_{n-1}(0)}/\kappa_{n+1}^{(\varphi)}$ of the second system are given by the formulas

$$a_{2n}^{(\varphi)} = (V_n - U_n)/(V_n + U_n), \qquad a_{2n+1}^{(\varphi)} = 1 - U_{n+1} - V_{n+1}, \tag{2.3}$$

where $n \in \mathbf{Z}_+$, $U_n = \tilde{p}_{n+1}(1)/\tilde{p}_n(1)$, $V_n = -\tilde{p}_{n+1}(-1)/\tilde{p}_n(1)$, and $\tilde{p}_n(t) = p_n(t)/k_n^{(p)} = t^n + \cdots$.

In particular, we can take p and φ to be the weight functions (1.6) and (1.24) and note that the Jacobi polynomials $\{\tilde{P}_n^{(\alpha,\beta)}(t)\}_{n=0}^\infty$ with leading coefficient 1 satisfy ([18], pp. 271, 283)

$$\tilde{P}^{(\alpha,\beta)}(1) = \frac{2^n\Gamma(n+\alpha+\beta+1)(n+\alpha+1)}{\Gamma(\alpha+1)\Gamma(2n+\alpha+\beta+1)}, \quad \tilde{P}_n^{(\alpha,\beta)}(-1) = (-1)^n\tilde{P}_n^{(\beta,\alpha)}(1).$$

Then we obtain (2.2) from (2.3).

2) *The leading coefficient $\kappa_n^{\alpha,\beta}$ of $\psi_n^{\alpha,\beta}$ is given by the formulas* ($n \in \mathbf{Z}_+$)

$$\kappa_{2n}^{(\alpha,\beta)} = \frac{2^{(\alpha+\beta+1)/2}\Gamma(n+1+(\alpha+\beta)/2)\Gamma(n+1+(\alpha+\beta+1)/2)}{\sqrt{\Gamma(n+1)\Gamma(n+\alpha+1)\Gamma(n+\beta+1)\Gamma(n+\alpha+\beta+2)}}, \tag{2.4}$$

$$\begin{aligned}\kappa_{2n+1}^{\alpha,\beta} &= \kappa_{2n}^{d\alpha,\beta}\left(1+\left|a_{2n}^{\alpha,\beta}\right|^2\right)^{-1/2} \\ &= \frac{2^{(\alpha+\beta+1)/2}\Gamma(n+2+(\alpha+\beta)/2)\Gamma(n+1+(\alpha+\beta+1)/2)}{\sqrt{\Gamma(n+1)\Gamma(n+\alpha+2)\Gamma(n+\beta+2)\Gamma(n+\alpha+\beta+2)}}.\end{aligned} \tag{2.5}$$

For the proof, use the formulas ([2], §8.1)

$$\left(\kappa_0^{(\alpha,\beta)}\right)^{-2} = (2\pi)^{-1}\int_0^{2\pi}\psi^{\alpha,\beta}(\tau)\,d\tau, \tag{2.6}$$

$$\left(\kappa_n^{(\alpha,\beta)}\right)^{-2} = \left(\kappa_0^{\alpha,\beta}\right)^{-2}\prod_{k=0}^{n-1}\left\{1-\left|a_k^{\alpha,\beta}\right|^2\right\} \qquad (n \in \mathbf{N}), \tag{2.7}$$

calculate the right-hand side of (2.6), substitute the result into (2.7), and then use (2.2).

3) *The coefficient $\kappa_n^{\alpha,\beta}$ satisfies*

$$\kappa_n^{\alpha,\beta} = 2^{(\alpha+\beta+1)/2}\{1+O(n^{-1})\} \qquad (n \to \infty). \tag{2.8}$$

The proof follows from (2.4), (2.5) and the formula ([19], §1.18, formula (4))

$$\Gamma(t+a)/\Gamma(t) = t^a\left[1+O(t^{-1})\right] \qquad (t \to \infty). \tag{2.9}$$

4) *Setting $x = (z+z^{-1})/2$, $z \in \mathbf{C}$, $z \neq 0$, we have*

$$\begin{aligned}\hat{P}_n^{\alpha,\beta}(x) &= \sqrt{\frac{2n+\alpha+\beta+1}{4\pi(n+\alpha+\beta+1)}}\left\{z^{-n}\psi_{2n}^{\alpha,\beta}(z) + z^n\psi_{2n}^{\alpha,\beta}(z^{-1})\right\} \\ &= \sqrt{\frac{2n+\alpha+\beta+1}{4\pi n}}\left\{z^{-n+1}\psi_{2n-1}^{\alpha,\beta}(z) + z^{n-1}\psi_{2n-1}^{\alpha,\beta}(z^{-1})\right\} \qquad (n \in \mathbf{N})\end{aligned} \tag{2.10}$$

(*the first equation is also valid for* $n = 0$), *and*

$$\hat{P}_{n-1}^{\alpha+1,\beta+1}(x) = \sqrt{\frac{2n+\alpha+\beta+1}{\pi n}} \cdot \frac{z^{-n}\psi_{2n}^{\alpha,\beta}(z) + z^n\psi_{2n}^{\alpha,\beta}(z^{-1})}{z - z^{-1}}$$
$$= \sqrt{\frac{2n+\alpha+\beta+1}{\pi(n+\alpha+\beta+1)}} \cdot \frac{z^{-n+1}\psi_{2n-1}^{\alpha,\beta}(z) - z^{n-1}\overline{\psi}_{2n-1}^{\alpha,\beta}(z^{-1})}{z - z^{-1}}$$
$$(n \in \mathbf{N}). \qquad (2.11)$$

Formulas (2.10) and (2.11) follow from the general formulas of Szegö ([1], p. 294), on substituting the values of a_{2n-1} calculated in accordance with (2.2).

5) *The values* $\psi_n^{\alpha,\beta}(1)$ *are given by the formulas*

$$\psi_{2n}^{\alpha,\beta}(1) = \frac{1}{\Gamma(\alpha+1)}\sqrt{\frac{\pi}{2^{\alpha+\beta+1}} \cdot \frac{\Gamma(n+\alpha+\beta+2)\Gamma(n+\alpha+1)}{\Gamma(n+1)\Gamma(n+\beta+1)}} \qquad (n \in \mathbf{Z}_+), \qquad (2.12)$$

$$\psi_{2n-1}^{\alpha,\beta}(1) = \frac{1}{\Gamma(\alpha+1)}\sqrt{\frac{\pi}{2^{\alpha+\beta+1}} \cdot \frac{\Gamma(n+\alpha+\beta+1)\Gamma(n+\alpha+1)}{\Gamma(n)\Gamma(n+\beta+1)}} \qquad (n \in \mathbf{N}). \qquad (2.13)$$

For the proof it is enough to apply (2.10) for $z = 1$ and the equation ([18], pp. 273, 283)

$$\hat{P}_n^{\alpha,\beta}(1) = \sqrt{\frac{2n+\alpha+\beta+1}{2^{\alpha+\beta+1}} \cdot \frac{\Gamma(n+1)\Gamma(n+\alpha+\beta+1)}{\Gamma(n+\alpha+1)\Gamma(n+\beta+1)}}\binom{n+\alpha}{n}. \qquad (2.14)$$

6) *For* $x = (z + z^{-1})/2$, $z \in \mathbf{C}$, $z \neq 0$, *and* $n \in \mathbf{N}$

$$\psi_{2n}^{\alpha,\beta}(z) = z^n\left\{\sqrt{\frac{\pi(n+\alpha+\beta+1)}{2n+\alpha+\beta+1}}\,\hat{P}_n^{\alpha,\beta}(x) + \frac{z-z^{-1}}{2}\sqrt{\frac{\pi n}{2n+\alpha+\beta+1}}\,\hat{P}_{n-1}^{\alpha+1,\beta+1}(x)\right\}, \qquad (2.15)$$

$$\psi_{2n-1}^{\alpha,\beta}(z) = z^{n-1}\left\{\sqrt{\frac{\pi n}{2n+\alpha+\beta+1}}\,\hat{P}_n^{\alpha,\beta}(x) + \frac{z-z^{-1}}{2}\sqrt{\frac{\pi(n+\alpha+\beta+1)}{2n+\alpha+\beta+1}}\,\hat{P}_{n-1}^{\alpha+1,\beta+1}(x)\right\}. \qquad (2.16)$$

When $z = e^{i\tau}$ *these take the form*

$$\psi_{2n}^{\alpha,\beta}(e^{i\tau}) = e^{in\tau}\left\{\sqrt{\frac{\pi(n+\alpha+\beta+1)}{2n+\alpha+\beta+1}}\,\hat{P}_n^{\alpha,\beta}(\cos\tau) + i\sqrt{\frac{\pi n}{2n+\alpha+\beta+1}}\,\hat{P}_{n-1}^{\alpha+1,\beta+1}(\cos\tau)\sin\tau\right\}, \qquad (2.17)$$

$$\psi_{2n-1}^{d\alpha,\beta}(e^{i\tau}) = e^{i(n-1)\tau}\left\{\sqrt{\frac{\pi n}{2n+\alpha+\beta+1}}\,\hat{P}_n^{\alpha,\beta}(\cos\tau) + i\sqrt{\frac{\pi(n+\alpha+\beta+1)}{2n+\alpha+\beta+1}}\,\hat{P}_{n-1}^{\alpha+1,\beta+1}(\cos\tau)\sin\tau\right\}. \qquad (2.18)$$

These formulas follow from (2.10) and (2.11).

7) *There is an expansion*

$$\psi_n^{\alpha,-1/2}(z) = d_1(\alpha, n)\left[c_{n,0}^{(\alpha)} + c_{n,1}^{(\alpha)}z + \cdots + c_{n,n}^{(\alpha)}z^n\right], \tag{2.19}$$

where $z \in \mathbf{C}$, $n \in \mathbf{Z}_+$ *and*

$$d_1(\alpha, n) = 2^{\alpha/2+1/4}\{\Gamma(\alpha + 1/2)\}^{-1}\{\Gamma(n+1)/\Gamma(n + 2\alpha + 2)\}^{1/2},$$

$$c_{n,\nu}^{(\alpha)} = \Gamma(\nu + \alpha + 3/2)\Gamma(n - \nu + \alpha + 3/2)/\{\Gamma(\nu + 1)\Gamma(n + 1 - \nu)\}.$$

It is sufficient to give the proof for $z = e^{i\tau}$. In this case we can apply (2.17) and (2.18) to the formulas

$$P_n^{(\alpha,-1/2)}(\cos\tau) = \frac{\Gamma(n + \alpha + 1)\Gamma(2n + 1)}{\Gamma(2n + \alpha + 1)\Gamma(n + 1)} P_{2n}^{(\alpha,\alpha)}\left(\cos\frac{\tau}{2}\right), \tag{2.20}$$

$$P_{2n}^{(\alpha,\alpha)}(t) = \frac{\Gamma(2\alpha + 1)\Gamma(2n + \alpha + 1)}{\Gamma(\alpha + 1)\Gamma(2n + 2\alpha + 1)} P_n^{(\alpha+1/2)}(t), \tag{2.21}$$

$$P_{2n}^{(\alpha+1/2)}\left(\cos\frac{\tau}{2}\right) = \alpha_n^2 + 2\sum_{k=1}^{n} \alpha_{n+k}\alpha_{n-k}\cos k\tau, \qquad \alpha_k = \binom{k + \alpha - 1/2}{k}, \tag{2.22}$$

$$\Gamma(z)\Gamma(z + 1/2) = 2^{1/2-2z}(2\pi)^{1/2}\Gamma(2z) \tag{2.23}$$

(see [1], pp. 59, 82, 95, 14). It follows from (2.20)–(2.23) that

$$\hat{P}_n^{\alpha,-1/2}(\cos\tau) = \Gamma\left(\alpha + \frac{1}{2}\right)\left\{\frac{(2n + \alpha + 1/2)\Gamma(2n + 1)}{\pi 2^{1/2-\alpha}\Gamma(2n + 2\alpha + 1)}\right\}^{1/2} \times\left\{\alpha_n^2 + 2\sum_{k=1}^{n}\alpha_{n+k}\alpha_{n-k}\cos k\tau\right\}. \tag{2.24}$$

Differentiating (2.24) with respect to τ and using the formula ([18], p. 282)

$$\hat{P}_{n-1}^{\alpha+1,\beta+1}(t) = [n(n + \alpha + \beta + 1)]^{-1/2}\, d\hat{P}_n^{\alpha,\beta}(t)/dt \tag{2.25}$$

we find that

$$\sqrt{n\left(n + \alpha + \frac{1}{2}\right)}\, \hat{P}_{n-1}^{\alpha+1,\beta+1}(\cos\tau)\sin\tau = 2\Gamma\left(\alpha + \frac{1}{2}\right)\left\{\frac{(2n + \alpha + 1/2)\Gamma(2n + 1)}{\pi 2^{1/2-\alpha}\Gamma(2n + 2\alpha + 1)}\right\}^{1/2}\sum_{k=1}^{n} k\alpha_{n+k}\alpha_{n-k}\sin k\tau. \tag{2.26}$$

Substituting (2.24) and (2.26) into (2.17) and (2.18), after some manipulation we obtain (2.19) for $z \in \Gamma_1$ and $n \in \mathbf{N}$. The validity of (2.19) for $n = 0$ can be verified directly.

Similarly, by applying (2.17), (2.18), (2.21)–(2.23), (2.25), and the formula (see [1], p. 95)

$$P_{2n-1}^{(\alpha+1/2)}(\cos\tau) = 2\sum_{k=1}^{n}\alpha_{n-k}\alpha_{n+k-1}\cos(2k - 1)\tau,$$

we can show that

8)

$$\psi_{2n}^{\alpha,\alpha}(z)=\frac{2^{\alpha+1/2}}{\Gamma(\alpha+1/2)}\left\{\frac{\Gamma(n+1)}{\Gamma(n+2\alpha+2)}\right\}^{1/2}$$
$$\times\sum_{\nu=0}^{n}\frac{\Gamma(\nu+\alpha+3/2)\Gamma(n-\nu+\alpha+1/2)}{\Gamma(\nu+1)\Gamma(n+1-\nu)}z^{2\nu}. \tag{2.27}$$

Multiplying by z on the left-hand side of (2.27), we obtain a corresponding expansion for $\psi_{2n+1}^{\alpha,\alpha}(z)$, since

9)

$$\psi_{2n+1}^{(\alpha,\alpha)}(z)=z\psi_{2n}^{(\alpha,\alpha)}(z). \tag{2.28}$$

The identity (2.28) follows from the formula

$$\kappa_{\sigma,n+1}\varphi_{\sigma,n+1}(z)=\kappa_{\sigma,n}z\varphi_{\sigma,n}(z)+\varphi_{\sigma,n+1}(0)\varphi_{\sigma,n}^{*}(z)$$

(see [2], formula (1.2)), since $\psi_{2n+1}^{\alpha,\alpha}(0)=0$ by (2.1) and (2.2), and consequently (see [2], formula (1.5)) $\kappa_{2n+1}^{\alpha,\alpha}=\kappa_{2n}^{\alpha,\alpha}$.

10) *We have an analog of* (1.9), *namely*

$$\psi_{n}^{\alpha,\beta}(-z)=(-1)^{n}\psi_{n}^{\beta,\alpha}(z). \tag{2.29}$$

This can be verified directly by using the orthogonality relation.

11) *If* $\alpha,\beta>-1$, $n\in\mathbf{N}$, *and* $f_n(\tau)=e^{-i(n-1)\tau}\psi_{2n-1}^{\alpha,\beta}(e^{i\tau})$ *or* $e^{-in\tau}\psi_{2n}^{\alpha,}(e^{i\tau})$, *then we have, as* $n\to\infty$, *uniformly for* $\tau\in[-\pi+\varepsilon,\pi-\varepsilon]$ $(0<\varepsilon<\pi)$

$$f_n(\tau)=2^{-(\alpha+\beta+2)/2}\pi k(|\tau|)\left[J_\alpha\left(N_n^{\alpha,\beta}|\tau|\right)+iJ_{\alpha+1}\left(N_n^{\alpha,\beta}|\tau|\right)\operatorname{sgn}\tau\right]$$
$$\times\left(N_n^{\alpha,\beta}|\tau|\right)^{1/2}\left[1+O(n^{-1})\right] \tag{2.30}$$

and, uniformly for $\tau\in[-\pi,-\varepsilon]\cup[\varepsilon,\pi]$,

$$f_n(\tau)=(-1)^n2^{-(\alpha+\beta+2)/2}\pi k(\pi-|\tau|)$$
$$\times\left[J_\beta\left(N_n^{\alpha,\beta}(\pi-|\tau|)\right)+iJ_{\beta+1}\left(N_n^{\alpha,\beta}(\pi-|\tau|)\right)\operatorname{sgn}\tau\right]$$
$$\times\left[N_n^{\alpha,\beta}(\pi-|\tau|)\right]^{1/2}\left[1+O(n^{-1})\right], \tag{2.31}$$

where $N_n^{\alpha,\beta}=n+(\alpha+\beta+1)/2$, *and*

$$k(\tau)=\pi^{-1/2}[\sin(\tau/2)]^{-\alpha-1/2}[\cos(\tau/2)]^{-\beta-1/2}. \tag{2.32}$$

We can take $\varepsilon=0$ *when* $\beta=-\frac{1}{2}$ *in* (2.30) *and when* $\alpha=-\frac{1}{2}$ *in* (2.31).

PROOF. Because of the evident identity $f_n(-\tau)=\overline{f_n(\tau)}$ it is enough to consider the case $\tau\in[0,\pi]$. In addition, (2.29) easily reduces the case $\tau\in[\varepsilon,\pi]$ to the case $\tau\in[0,\pi-\varepsilon]$. It remains only to prove (2.30) for $[0,\pi-\varepsilon]$. For these values of τ we use (2.17), (2.18) and (1.7). It is clear from the proof of (1.7) in [1] that when $\beta=-\frac{1}{2}$ it remains valid for $\varepsilon=0$. Since (see [1], p. 68)

$$\hat{P}_n^{\alpha,\beta}(t)=\left\{h_n^{(\alpha,\beta)}\right\}^{-1/2}P_n^{(\alpha,\beta)}(t), \tag{2.33}$$

$$h_n^{(\alpha,\beta)}=\frac{2^{\alpha+\beta+1}\Gamma(n+\alpha+1)\Gamma(n+\beta+1)}{(2n+\alpha+\beta+1)\Gamma(n+\alpha+\beta+1)\Gamma(n+1)},$$

and $h_n^{\alpha,\beta}\asymp n^{-1}$ by (2.9), it follows from (1.7) that as $n\to\infty$

$$\hat{P}_n^{\alpha,\beta}(\cos\tau)=(\pi/2)^{1/2}k(\tau)\left(N_n^{\alpha,\beta}\tau\right)^{1/2}g_n^{\alpha,\beta}\left[J_\alpha\left(N_n^{\alpha,\beta}\tau\right)+O\left(r_{1,n}^{(\alpha)}(\tau)\right)\right], \tag{2.34}$$

uniformly for $\tau \in [0, \pi - \varepsilon]$, where $r_{1,n}^{(\alpha)}(\tau)$ is defined by (1.8), and

$$g_n^{\alpha,\beta} = \left(N_n^{\alpha,\beta}\right)^{-\alpha-1/2}\left\{h_n^{\alpha,\beta}\right\}^{-1/2}\Gamma(n+\alpha+1)/\Gamma(n+1) \asymp 1.$$

If we replace α by $\alpha+1$, β by $\beta+1$, and n by $n-1$ in (2.34), and use the equations

$$N_{n-1}^{\alpha+1,\beta+1} = N_n^{\alpha,\beta}, \qquad r_{1,n-1}^{(\alpha+1)}(\tau) = O\left(r_{1,n}^{(\alpha)}(\tau)\right) \qquad (n \to \infty),$$

we obtain, uniformly for $\tau \in [0, \pi - \varepsilon]$, the asymptotic formula

$$\begin{aligned}\hat{P}_{n-1}^{\alpha+1,\beta+1}&(\cos\tau)\sin\tau\\ &= (\pi/2)k(\tau)\left(N_n^{\alpha,\beta}\tau\right)^{1/2} 2g_{n-1}^{\alpha+1,\beta+1}\left[J_{\alpha+1}\left(N_n^{\alpha,\beta}(\tau)\right) + O\left(r_{1,n}^{(\alpha)}(\tau)\right)\right].\end{aligned} \tag{2.35}$$

It follows from (2.17), (2.34) and (2.35), and the easily verified equation

$$2g_{n-1}^{\alpha+1,\beta+1} = g_n^{\alpha,\beta}\left(N_n^{\alpha,\beta}\right)^{-1}\sqrt{n(n+\alpha+\beta+1)}$$

that, uniformly for $\tau \in [0, \pi - \varepsilon]$,

$$\begin{aligned}f_n(\tau) &= (\pi/2)k(\tau)\left(N_n^{\alpha,\beta}\tau\right)^{1/2} g_n^{\alpha,\beta}\left[(n+\alpha+\beta+1)/N_n^{\alpha,\beta}\right]^{1/2}\\ &\quad\times\left[J_\alpha\left(N_n^{\alpha,\beta}\tau\right) + in\left(N_n^{\alpha,\beta}\right)^{-1}J_{\alpha+1}\left(N_n^{\alpha,\beta}\tau\right) + O\left(r_{1,n}^{(\alpha)}(\tau)\right)\right]\end{aligned} \tag{2.36}$$

where $f_n(\tau) = e^{-in\tau}\psi_{2n}^{\alpha,\beta}(e^{i\tau})$. By (2.9),

$$g_n^{\alpha,\beta}\left[(n+\alpha+\beta+1)/N_n^{\alpha,\beta}\right]^{1/2} = \left[1 + O(n^{-1})\right]2^{-(\alpha+\beta)/2}. \tag{2.37}$$

From the relations (see [1], pp. 15–16)

$$J_\nu(t) \asymp t^\nu \qquad (t \to +0), \tag{2.38}$$

$$J_\nu(t) = \left(\frac{2}{\pi t}\right)^{1/2}\cos\left(t - \frac{\nu\pi}{2} - \frac{\pi}{4}\right) + O(t^{-3/2}) \qquad (t \to +\infty) \tag{2.39}$$

it follows that $J_{\alpha+1}(t)$ is bounded on $[0, \infty)$. Consequently

$$n\left(N_n^{\alpha,\beta}\right)^{-1}J_{\alpha+1}\left(N_n^{\alpha,\beta}\tau\right) = J_{\alpha+1}\left(N_n^{\alpha,\beta}\tau\right) + O(n^{-1}) \qquad (n \to \infty). \tag{2.40}$$

uniformly for $\tau \in [0, \pi - \varepsilon]$. Since $J_\alpha(t)$ and $J_{\alpha+1}(t)$ have no positive zeros in common, we have, for a given $c > 0$,

$$\left|J_\alpha(t) + iJ_{\alpha+1}(t)\right| \asymp \begin{cases} t^\alpha, & 0 \leqslant t \leqslant c, \\ t^{-1/2}, & t \geqslant c. \end{cases} \tag{2.41}$$

It follows from (2.36), (2.37), (2.40), and (2.41) that (2.30) is valid for $\tau \in [0, \pi - \varepsilon]$ and $f_n(\tau) = e^{-in\tau}\psi_{2n}^{\alpha,\beta}(e^{i\tau})$. The discussion for $f_n(\tau) = e^{-i(n-1)\tau}\psi_{2n-1}^{\alpha,\beta}$ is completely analogous.

§3. Proof of Theorem 1

1. Because of the trivial identity

$$\varphi_n(e^{i\Theta}) = \psi_n^{\alpha,\beta}(e^{i\Theta})I_1 + I_2, \tag{3.1}$$

where

$$I_1 = \frac{1}{2\pi}\int_0^{2\pi}\varphi_n(e^{i\tau})\,\overline{\psi_n^{\alpha,\beta}(e^{i\tau})}\,\psi^{\alpha,\beta}(\tau)\,d\tau,$$

$$I_2 = \frac{1}{2\pi}\int_0^{2}\varphi_n(e^{i\tau})\sum_{k=0}^{n-1}\psi_k^{\alpha,\beta}(e^{i\Theta})\,\overline{\psi_k^{\alpha,\beta}(e^{i\tau})}\,\psi^{\alpha,\beta}(\tau)\,d\tau,$$

formula (1.27) will follow from

$$I_1 = \pi(h; 0) + O(\delta_n(\varphi) + \delta_n(\psi^{\alpha,\beta})), \tag{3.2}$$

$$I_2 = \psi_n^{\alpha,\beta}(e^{i\Theta})\left[\overline{\pi(h; e^{i\Theta})} - \pi(h; 0) + O(\mu_{h,n}^{\alpha,\beta})\right]. \tag{3.3}$$

2. To prove (3.2) we estimate the difference $I_1 - \pi(h; 0)$. By Cauchy's integral formula

$$\pi(h; 0)\frac{1}{2\pi i}\int_{|\zeta|=1} \frac{\pi(h; \zeta)}{\zeta} d\zeta = \frac{1}{2\pi}\int_0^{2\pi} \pi(h; e^{i\tau})\, d\tau.$$

Therefore (since $\pi(h; 0) \in \mathbf{R}$)

$$|I_1 - \pi(h; 0)| \leqslant \frac{1}{2\pi}\int_0^{2\pi} \left|\varphi_n(e^{i\tau})\,\overline{\psi_n^{\alpha,\beta}(e^{i\tau})}\,\psi^{\alpha,\beta}(\tau) - \overline{\pi(h; e^{i\tau})}\right| d\tau. \tag{3.4}$$

When estimating the right-hand side of (3.4), and later, we will use the following familiar facts. If $w \in S$ (see (1.30)) then for almost all $\tau \in [0, 2\pi]$ there exist radial boundary values ([2], formula (2.6))

$$\pi(w; e^{i\tau}) = \lim_{r\to 1-0} \pi(w; re^{i\tau}) = \{w(\tau)\}^{-1/2} e^{i\gamma(w;\tau)}, \tag{3.5}$$

$$\gamma(\omega; \tau) = \arg \pi(w; e^{i\tau}) = \frac{1}{4\pi}\int_0^{2\pi} \cot\frac{t-\tau}{2} \ln w(t)\, dt \tag{3.6}$$

(the integral is a Cauchy principal value).

As we know, it follows from (3.5) that

$$w \in S \Rightarrow |\pi(w; e^{i\tau})| = [w(\tau)]^{-1/2} \quad \text{a.e. on } [0, 2\pi]. \tag{3.7}$$

If the weight functions u, v, and $w = uv \in S$ then by (3.5) and (1.16)

$$\pi(uv; e^{i\tau}) = \pi(u; e^{i\tau})\pi(v; e^{i\tau}) \quad \text{a.e. on } [0, 2\pi]. \tag{3.8}$$

If $w \in S$ then the expression

$$r_n(w; \tau) = \left[\pi(w; e^{i\tau}) - e^{in\tau}\overline{w_n(e^{i\tau})}\right]\sqrt{w(\tau)} \tag{3.9}$$

is defined a.e. on $[0, 2\pi]$, and by (1.35)

$$\|r_n(w; \tau)\|_2 \leqslant C_1(w)\delta_n(w) \qquad (n \in \mathbf{Z}_+). \tag{3.10}$$

It follows from (3.7)–(3.9) that, a.e. on $[0, 2\pi]$,

$$\begin{aligned} &\left|\varphi_n(e^{i\tau})\,\overline{\psi_n^{\alpha,\beta}(e^{i\tau})}\,\psi^{\alpha,\beta}(\tau) - \overline{\pi(h; e^{i\tau})}\right| \\ &\quad = \left|\varphi_n(e^{i\tau})\,\overline{\psi_n^{\alpha,\beta}(e^{i\tau})}\,\psi^{\alpha,\beta}(\tau) - \overline{\pi(\varphi; e^{i\tau})}\,\pi(\psi^{\alpha,\beta}; e^{i\tau})\psi^{\alpha,\beta}(\tau)\right| \\ &\quad \leqslant \left\{\left|r_n(\varphi; \tau)\psi_n^{\alpha,\beta}(e^{i\tau})\sqrt{\psi^{\alpha,\beta}(\tau)}\right| + \left|r_n(\psi^{\alpha,\beta}; \tau)\right|\right\}\{h(\tau)\}^{-1/2}. \end{aligned} \tag{3.11}$$

By (3.4), (3.11), Cauchy's inequality, and (3.10), we have

$$|I_1 - \pi(h; 0)| \leqslant C_2(\varphi)\|1/h\|_\infty\left[\delta_n(\varphi) + \delta_n(\psi^{\alpha,\beta})\right],$$

and (3.2) follows.

3. To begin the proof of (3.3), we set

$$I_3(T) = \frac{1}{2\pi}\int_T \varphi_n(e^{i\tau}) \sum_{k=0}^{n-1} \psi_k^{\alpha,\beta}(e^{i\Theta})\,\overline{\psi_k^{\alpha,\beta}(e^{i\tau})}\left[\frac{1}{h(\tau)} - \frac{1}{h(\Theta)}\right]\varphi(\tau)\, d\tau, \tag{3.12}$$

where T is an arbitrary measurable subset of $E = [\Theta - \pi, \Theta + \pi]$. Because φ_n is orthogonal with weight function φ to all polynomials of lower degree,

$$I_2 = I_3(E). \tag{3.13}$$

We decompose E into $E_1 = [\Theta - \lambda_n, \Theta + \lambda_n]$ and $E_2 = E \setminus E_1$, where λ_n satisfies (1.26), and show that

$$I_3(E_1) \leqslant C_3(\varphi) \int_0^{\lambda_n} \omega(h; \tau) \tau^{-1} \, d\tau \left|\psi_n^{\alpha,\beta}(e^{i\Theta})\right|. \tag{3.14}$$

We shall need the following lemma.

LEMMA 1. *Let the weight function φ satisfy* (1.20)–(1.22). *Then there is a constant $C_4(\varphi)$ such that*

$$\left(|\sin[(\tau - \Theta)/2]| + n^{-1}\right)\left|K_{\varphi,n-1}(e^{i\Theta}, e^{i\tau})\right| \leqslant C_4(\varphi)\left|\varphi_n(e^{i\Theta})\varphi_n(e^{i\tau})\right| \tag{3.15}$$

for all Θ, $\varepsilon \in \mathbf{R}$, and $n \in \mathbf{N}$, where $K_{\varphi,n-1}(e^{i\Theta}, e^{i\tau})$ is defined by (1.34).

PROOF. Inequality (3.17) is a corollary of the inequalities

$$\left|\sin[(\tau - \Theta)/2] \, K_{\varphi,n-1}(e^{i\Theta}, e^{i\tau})\right| \leqslant \left|\varphi_n(e^{i\Theta})\varphi_n(e^{i\tau})\right| \tag{3.16}$$

and

$$n^{-1}\left|K_{\varphi,n-1}(e^{i\Theta}, e^{i\tau})\right| \leqslant C_5(\varphi)\left|\varphi_n(e^{i\Theta}) - \varphi_n(e^{i\tau})\right|. \tag{3.17}$$

Inequality (3.16) follows from (1.34). To prove (3.17) we apply both sides of (1.23). Let

$$a_{\nu,n}(\tau) = \left(|\sin[(\tau - \Theta_\nu)/2]| + n^{-1}\right)^{-\gamma_\nu/2} \qquad (\nu = 1, \ldots, m),$$

and $\Delta_\nu = [\tau_\nu, \tau_{\nu+1}]$ $(\nu = 1, \ldots, m)$, where the fixed number τ_ν satisfies

$$-\pi < \tau_1 < \Theta_1 < \tau_2 < \cdots < \tau_m < \Theta_m, \qquad \tau_{m+1} = \tau_1 + 2\pi. \tag{3.18}$$

If $\mu, \nu \in \{1, \ldots, m\}$, $\Theta \in \Delta_\mu$, and $\tau \in \Delta_\nu$, then by (1.34)

$$\frac{1}{n}\left|K_{\varphi,n-1}(e^{i\Theta}, e^{i\tau})\right| \leqslant C_y(\varphi)\frac{1}{n}\sum_{k=1}^{n} a_{\mu,k}(\Theta) a_{\nu,k}(\tau) \equiv \sigma_{\mu,\nu}. \tag{3.19}$$

To prove (3.17) it is enough, by (1.34) and (3.19), to show that

$$\sigma_{\mu,\nu} \leqslant C_7(\varphi) a_{\mu,n}(\Theta) a_{\nu,n}(\tau). \tag{3.20}$$

For the proof of (3.20) we consider four cases:

1) When γ_μ and $\gamma_\nu \geqslant 0$, (3.20) follows because $a_{\mu,k}(\Theta)$ and $a_{\nu,k}(\Theta)$ increase with k.

2) If $\gamma_\mu \geqslant 0$ and $-1 < \gamma_\nu < 0$ then by (3.19)

$$\begin{aligned} \sigma_{\mu,\nu} &\leqslant C_8(\varphi) a_{\mu,n}(\Theta) \frac{1}{n} \sum_{k=1}^{n} \left(\left|\sin \frac{\tau - \Theta_\nu}{2}\right|^{-\gamma_\nu/2} + k^{\gamma_\nu/2} \right) \\ &\leqslant C_9(\varphi) a_{\mu,n}(\Theta) a_{\nu,n}(\tau). \end{aligned}$$

3) The case $-1 < \gamma_\mu < 0$, $\gamma_\nu \geqslant 0$ is similar to the preceding case.

4) When γ_μ and $\gamma_\nu \in (-1, 0)$ we deduce from (3.19) that

$$\begin{aligned} \sigma_{\mu,\nu} \leqslant C_{10}(\varphi)\frac{1}{n}\sum_{k=1}^{n} \Bigg\{ &\left|\sin\frac{\Theta - \Theta_\mu}{2}\right|^{-\gamma_\mu/2} \left|\sin\frac{\tau + \Theta_\nu}{2}\right|^{-\gamma_\nu/2} + k^{(\gamma_\mu + \gamma_\nu)/2} \\ &+ k^{\gamma_\mu/2}\left|\sin\frac{\tau - \Theta_\nu}{2}\right|^{-\gamma_\nu/2} + k^{\gamma_\nu/2}\left|\sin\frac{\tau - \Theta_\mu}{2}\right|^{-\gamma_\mu/2} \Bigg\}, \end{aligned}$$

which is equivalent to (3.20).

This completes the proof of the lemma.

Now by using (3.8) and Lemma 1 we find that

$$|I_3(E_1)| \leqslant C_{11}(\varphi) I_4 |\psi_n^{\alpha,\beta}(e^{i\Theta})|, \tag{3.21}$$

where

$$I_4 = \int_{E_1} |\varphi_n(e^{i\tau})\psi_n^{\alpha,\beta}(e^{i\tau})| \omega(h; |\tau - \Theta|)(|\tau - \Theta| + n^{-1})^{-1} \varphi(\tau)\, d\tau. \tag{3.22}$$

If $e_k = [k\pi - n^{-1}, k\pi + n^{-1}]$ ($k \in \mathbf{Z}$) and $e = \bigcup_{-\infty}^{\infty} e_k$, then by (1.23)

$$|\varphi_{n-1}(e^{i\tau})| \leqslant C_{12}(\varphi) \begin{cases} n^{\alpha+1/2}, & \tau \in e_{2k}, \\ n^{\beta+1/2}, & \tau \in e_{2k+1}, \\ \{\varphi(\tau)\}^{-1/2}, & \tau \in \mathbf{R} \setminus e \end{cases} \tag{3.23}$$

(in particular, when $h \equiv 1$ we may write φ instead of $\psi^{\alpha,\beta}$). By (3.22), (3.23), and the inequality

$$\omega(h; \tau_2)\tau_2^{-1} \leqslant 2\omega(h; \tau_1)\tau_1^{-1} \qquad (0 < \tau_1 \leqslant \tau_2), \tag{3.24}$$

we have

$$\begin{aligned} |I_4| &\leqslant C_{13}(\varphi)\Big\{ \omega\Big(h; \frac{1}{n}\Big) n \int_0^{1/n} (n\tau)^{2\alpha+1}\, d\tau \\ &\qquad + \omega\Big(h; \frac{1}{n}\Big) n \int_\pi^{\pi+1/n} [n(\tau - \pi)]^{2\beta+1}\, d\tau + \int_0^{\lambda_n} \frac{\omega(h; \tau)}{\tau}\, d\tau \Big\} \\ &\leqslant C_{14}(\varphi)\Big\{ \omega\Big(h; \frac{1}{n}\Big) + \int_0^{\lambda_n} \frac{\omega(h; \tau)}{\tau}\, d\tau \Big\} \leqslant C_{15}(\varphi) \int_0^{\lambda_n} \frac{\omega(h; \tau)}{\tau}\, d\tau. \end{aligned}$$

Hence (3.14) follows in view of (3.21).

4. Now we establish the inequality

$$|I_3(E_2) - I_5| \leqslant C_{16}(\varphi)\Big\{ \delta_n(\varphi) \Big[\int_{\lambda_n}^{\pi} \frac{\omega^2(h; \tau)}{\tau^2}\, d\tau \Big]^{1/2} + \omega\Big(h; \frac{1}{n}\Big) \Big\} |\psi_n^{\alpha,\beta}(e^{i\Theta})|, \tag{3.25}$$

where

$$I_5 = \frac{1}{2\pi} \int_{E_2} \frac{e^{in\tau}}{\pi(\varphi; e^{i\tau})} \sum_{k=0}^{n-1} \psi_k^{\alpha,\beta}(e^{i\Theta}) \overline{\psi_k^{\alpha,\beta}(e^{i\tau})} \Big[\frac{1}{h(\tau)} - \frac{1}{h(\Theta)} \Big]\, d\tau. \tag{3.26}$$

By (3.9), (3.7), (3.12) and Lemma 1, we have

$$|I_3(E_2) - I_5| \leqslant C_{17}(\varphi) I_6(E_2) |\psi_n^{\alpha,\beta}(e^{i\Theta})|, \tag{3.27}$$

where

$$I_6(T) = \int_T \Big| r_n(\varphi; \tau)\psi_n^{\alpha,\beta}(e^{i\tau}) \sqrt{\psi^{\alpha,\beta}(\tau)} \Big| \omega(h; |\tau - \Theta|)(|\tau - \Theta| + n^{-1})^{-1}\, d\tau.$$

Applying (3.23) and (3.24), we find that

$$|I_6(E_2 \cap e)|$$
$$\leqslant C_{18}(\varphi)\int_{E_2\cap e}\frac{\omega(h;|\tau-\Theta|)}{|\tau-\Theta|+n^{-1}}\left\{|\psi_n^{\alpha,\beta}(e^{i\tau})|\sqrt{\varphi(\tau)}+|\varphi_n(e^{i\tau})\psi_n^{\alpha,\beta}(e^{i\tau})|\varphi(\tau)\right\}d\tau$$
$$\leqslant C_{19}(\varphi)n\omega\left(h;\frac{1}{n}\right)\left\{\int_0^{1/n}\left[(n\tau)^{\alpha+1/2}+(n\tau)^{2\alpha+1}\right]d\tau\right.$$
$$\left.+\int_\pi^{\pi+1/n}\left[n^{\beta+1/2}(\tau-\pi)^{\beta+1/2}+n^{2\beta+1}(\tau+\pi)^{2\beta+1}\right]d\tau\right\}$$
$$\leqslant C_{20}(\varphi)\omega\left(h;\frac{1}{n}\right). \tag{3.28}$$

By (3.10), (3.23) and Cauchy's inequality,

$$|I_6(E_2\setminus e)|\leqslant C_{21}(\varphi)\delta_n(\varphi)\left\{\int_{E_2}\frac{\omega^2(h;|\tau-\Theta|)}{|\tau-\Theta|^2}d\tau\right\}^{1/2}$$
$$\leqslant C_{22}(\varphi)\delta_n(\varphi)\left\{\int_{\lambda_n}^{\pi}\frac{\omega^2(h;\tau)}{\tau^2}d\tau\right\}^{1/2}. \tag{3.29}$$

Then (3.25) follows from (1.26) and (3.27)–(3.29).

5. Transforming the right-hand side of (3.26) by using (1.34) and the formula

$$\varphi_n^*(e^{i\tau})=e^{in\tau}\overline{\varphi_n(e^{i\tau})} \tag{3.30}$$

(see (1.3)) we find that

$$I_5=e^{in\Theta}\overline{\psi_n^{\alpha,\beta}(e^{i\Theta})}\,I_7-\psi_n^{\alpha,\beta}(e^{i\Theta})I_8, \tag{3.31}$$

where

$$I_7=\frac{1}{2\pi}\int_{E_2}\frac{\psi_n^{\alpha,\beta}(e^{i\tau})}{\pi(\varphi;e^{i\tau})[1-e^{i(\Theta-\tau)}]}\left[\frac{1}{h(\tau)}-\frac{1}{h(\Theta)}\right]d\tau, \tag{3.32}$$

$$I_8=\frac{1}{2\pi}\int_{E_2}\frac{e^{in\tau}\overline{\psi_h^{\alpha,\beta}(e^{i\tau})}}{\pi(\varphi;e^{i\tau})[1-e^{i(\theta-\tau)}]}\left[\frac{1}{h(\tau)}-\frac{1}{h(\Theta)}\right]d\tau. \tag{3.33}$$

Introducing the notation

$$I_9=\frac{1}{2\pi}\int_{E_2}\frac{e^{in\tau}\overline{\pi(\psi^{\alpha,\beta};e^{i\tau})}}{\pi(\varphi;e^{i\tau})[1-e^{i(\Theta-\tau)}]}\left[\frac{1}{h(\tau)}-\frac{1}{h(\Theta)}\right]d\tau, \tag{3.34}$$

$$I_{10}(T)=\frac{1}{2\pi}\int_T\left[\frac{1}{h(\tau)}-\frac{1}{h(\Theta)}\right]\frac{d\tau}{\pi(h;e^{i\tau})[1-e^{i(\Theta-\tau)}]}, \tag{3.35}$$

we have, by (3.7)–(3.10) and (3.32)–(3.35)

$$|I_7-I_9|+|I_8-I_{10}(E_2)|\leqslant C_{23}(\varphi)\int_{E_2}|r_n(\psi^{\alpha,\beta};\tau)|\frac{\omega(h;|\tau-\Theta|)}{|\tau-\Theta|}d\tau$$
$$\leqslant C_{24}(\varphi)\delta_n(\psi^{\alpha,\beta})\left[\int_{\lambda_n}^{\pi}\omega^2(\tau)\tau^{-2}\,d\tau\right]^{1/2}. \tag{3.36}$$

Then by (1.21), (3.7), and (3.35),

$$|I_{10}(E_1)|\leqslant C_{25}(h)\int_0^{\lambda_n}\omega(h;\tau)\tau^{-1}\,d\tau. \tag{3.37}$$

It follows from (3.13), (3.14), (3.25), (3.31), (3.36), and (3.37) that

$$\left|I_2 - e^{in\Theta}\overline{\psi^{\alpha,\beta}(e^{i\Theta})}\, I_9 + \psi_n^{\alpha,\beta}(e^{i\Theta}) I_{10}(E)\right|$$

$$\leqslant |I_3(E_1)| + |I_3(E_2) + I_5| + |\psi_n^{\alpha,\beta}(e^{i\Theta})|\{|I_7 - I_9| + |I_8 - I_{10}(E_2)| + |I_{10}(E_1)|\}$$

$$\leqslant C_{26}(\varphi)\left\{\int_0^{\lambda_n} \frac{\omega(h;\tau)}{\tau}\,d\tau + [\delta_n(\varphi)\right.$$

$$\left. + \delta_n(\psi^{\alpha,\beta})]\left[\int_{\lambda_n}^{\pi} \frac{\omega^2(h;\tau)}{\tau^2}\,d\tau\right]^{1/2} |\psi_n^{d\alpha,\beta}(e^{i\Theta})|\right\}. \tag{3.38}$$

6. Let us evaluate the integral $I_{10}(E)$. We transform the integrand in (3.35) by taking $z = e^{i\Theta}$ and $\zeta = e^{i\tau}$:

$$[\pi(h; e^{i\tau})]^{-1}\{[h(\tau)]^{-1} - [h(\Theta)]^{-1}\}[1 - e^{i(\Theta-\tau)}]^{-1}$$

$$= [\overline{\pi(h;\zeta)} - \overline{\pi(h;z)}]\left(1 - \frac{\bar\zeta}{\bar\zeta - \bar z}\right) - \frac{1}{h(\Theta)}\left[\frac{1}{\pi(h;\zeta)} - \frac{1}{\pi(h;z)}\right]\frac{\zeta}{\zeta - z}.$$

From this and (3.35) it follows that

$$I_{10}(E) = \bar I_{11} - \bar I_{12} - \{h(\Theta)\}^{-1} I_{13}, \tag{3.39}$$

where

$$\begin{aligned} I_{11} &= \frac{1}{2\pi i}\int_{|\zeta|=1} \frac{\pi(h;\zeta) + \pi(h;z)}{\zeta}\,d\zeta = \pi(h;0) - \pi(h;z), \\ I_{12} &= \frac{1}{2\pi i}\int_{|\zeta|=1} \frac{\pi(h;\zeta) - \pi(h;z)}{\zeta - z}\,d\zeta, \\ I_{13} &= \frac{1}{2\pi i}\int_{|\zeta|=1} \left[\frac{1}{\pi(h;\zeta)} - \frac{1}{\pi(h;z)}\right]\frac{d\zeta}{\zeta - z}. \end{aligned} \tag{3.40}$$

It is known ([2], Notes on §4.5, formula (28)) that if $\delta \geqslant 1$, $|\zeta_0| = 1$, and $f \in H^\delta$, then

$$\frac{1}{2\pi i}\int_{|\zeta|=1} \frac{f(\zeta) - f(\zeta_0)}{\zeta - \zeta_0}\,d\zeta = 0. \tag{3.41}$$

Therefore $I_{12} = I_{13} = 0$, which with (3.39) yields

$$I_{10}(E) = \bar I_{11} = \pi(h;0) - \overline{\pi(h;e^{i\Theta})}. \tag{3.42}$$

7. Finally we establish the inequality

$$|I_9| \leqslant C_{27}(\varphi)\left\{\omega(h;n^{-1})\ln n + \int_0^{1/n} \omega(h;\tau)\tau^{-1}\,d\tau\right\}. \tag{3.43}$$

By (3.7) and (3.34)

$$I_9 = (2\pi)^{-1} I_{14}(E_2), \tag{3.44}$$

where

$$I_{14}(T) = \int_T e^{i[n\tau + \gamma(\tau)]} F(\tau)\,d\tau, \qquad F = f_1 f_2 f_3,$$

$$f_1(\tau) = [1 - e^{i(\Theta-\tau)}]^{-1}, \quad f_2(\tau) = \frac{1}{h(\tau)} - \frac{1}{h(\Theta)}, \quad f_3(\tau) = \frac{1}{\pi(h; e^{i\tau})},$$

$$\gamma(\tau) = \arg\frac{\overline{\pi(\psi^{\alpha,\beta}; e^{i\tau})}}{\pi(\psi^{\alpha,\beta}; e^{i\tau})} = \arg\frac{(1 - e^{i\tau})^{\alpha+1/2}(1 + e^{i\tau})^{\beta+1/2}}{(1 - e^{-i\tau})^{\alpha+1/2}(1 + e^{-i\tau})^{\beta+1/2}}. \tag{3.45}$$

In general the function $\gamma(\tau)$ is undefined at $\tau = k\pi$ $(k \in \mathbf{Z})$. It is easily verified that

$$\gamma(\tau) = \begin{cases} \arg e^{i(\alpha+1/2)(\tau-\pi)+i(\beta+1/2)\tau}, & 0 < \tau < \pi, \\ \arg e^{i(\alpha+1/2)(\tau-\pi)+i(\beta+1/2)(\tau-2\pi)}, & \pi < \tau < 2\pi. \end{cases}$$

Hence it follows that the set E_2 falls into a finite number (say l) of intervals $[a_k, b_k]$ $(k = 1,\dots,l)$ in whose interiors $\gamma(\tau)$ is linear:

$$\gamma(\tau) = c_k\tau + d_k \qquad (a_k < \tau < b_k;\ k = 1,\dots,l).$$

Let us estimate the integrals $I_{14}([a_k, b_k])$. We set

$$I_{15} = \int_{a_k}^{b_k} e^{i[n\tau+\gamma(\tau)]}F(\tau + \pi n^{-1})\,d\tau.$$

After the change of variable $\tau = u + \pi n^{-1}$ we find that

$$I_{15} = -e^{ic_k\pi/n}I_{14}([a_k - \pi n^{-1}, b_k - \pi n^{-1}]),$$

and consequently

$$|I_{14}([a_k, b_k]) + I_{15}| \leqslant C_{28}(\varphi)[\omega(h; n^{-1}) + n^{-1}]. \tag{3.46}$$

Then evidently

$$|I_{14}([a_k, b_k]) - I_{15}| \leqslant C_{29}(h)(I_{16} + I_{17} + I_{18}), \tag{3.47}$$

where

$$I_{16} = \int_{a_k}^{b_k} \left|f_2\left(\tau + \frac{\pi}{n}\right)f_3\left(\tau + \frac{\pi}{n}\right)\Delta_{\pi/n}f_1(\tau)\right|d\tau, \qquad \Delta_\lambda f(\tau) = f(\tau + \lambda) - f(\tau),$$

$$I_{17} = \int_{a_k}^{b_k} \left|f_1(\tau)f_3\left(\tau + \frac{\pi}{n}\right)\Delta_{\pi/n}f_2(\tau)\right|d\tau, \qquad I_{18} = \int_{a_k}^{b_k} |f_1(\tau)f(\tau)\Delta_{\pi/n}f_3(\tau)|\,d\tau.$$

Using (3.24) and (3.45), we find that

$$I_{16} \leqslant C_{30}(h)\frac{1}{n}\int_{E_2} \frac{\omega(h; |\tau + \pi n^{-1} - \Theta|)\,d\tau}{|(\tau + \pi n^{-1} - \Theta)(\tau - \Theta)|} \leqslant C_{31}(h)\,\omega\left(h; \frac{1}{n}\right)\ln\frac{1}{\lambda_n}, \tag{3.48}$$

$$I_{17} \leqslant C_{32}(h)\,\omega\left(h; \frac{1}{n}\right)\int_{E_2} \frac{d\tau}{|\tau - \Theta|} \leqslant C_{33}(h)\,\omega\left(h; \frac{1}{n}\right)\ln\frac{1}{\lambda_n}, \tag{3.49}$$

$$I_{18} \leqslant C_{34}(h)\,\omega\left(f_3; \frac{1}{n}\right)\int_{E_2} \frac{\omega(h; |\tau - \Theta|)}{|\tau - \Theta|}\,d\tau \leqslant C_{35}(h)\,\omega\left(f_3; \frac{1}{n}\right). \tag{3.50}$$

The inequality

$$|I_{14}(E_2)| \leqslant C_{36}(\varphi)\{\omega(h; n^{-1})\ln n + \omega(f_3; n^{-1})\}. \tag{3.51}$$

follows from (3.46)–(3.50) and (1.26).

Now we prove by a standard argument that

$$\omega(f_3; n^{-1}) \leqslant C_{37}(h)\left\{\int_0^{1/n} \omega(h; \tau)\tau^{-1}\,d\tau + n^{-1}\int_{1/n}^{\pi} \omega(h; \tau)\tau^{-2}\,d\tau\right\}. \tag{3.52}$$

In fact, since

$$\left|\frac{e^{z_2} - e^{z_1}}{z_2 - z_1}\right| = \left|\sum_{k=1}^{\infty} \frac{1}{k!} \sum_{\nu=0}^{k-1} z_1^\nu z_2^{k-1-\nu}\right| \leqslant e^{\max(|z_1|, |z_2|)}$$

for arbitrary z_1 and $z_2 \in \mathbf{C}$, we have, for arbitrary τ_1 and $\tau_2 \in \mathbf{R}$,

$$|f_3(\tau_2) - f_3(\tau_1)| \leqslant C_{38}(h)|\ln f_3(\tau_2) - \ln f_3(\tau_1)|$$
$$\leqslant C_{39}(h)\left\{|\ln h(\tau_2) - \ln h(\tau_1)| + \left|\widetilde{\ln} h(\tau_2) - \widetilde{\ln} h(\tau_1)\right|\right\}.$$

Therefore

$$\omega(f_3; \delta) \leqslant C_{40}(h)\left[\omega(\ln h; \delta) + \omega(\widetilde{\ln} h; \delta)\right]. \tag{3.53}$$

By (1.21)

$$\omega(\ln h; \delta) \leqslant C_{41}(h)\omega(h; \delta). \tag{3.54}$$

It follows from (3.54) and a theorem of Zygmund ([20], p. 121) that

$$\omega(\widetilde{\ln}\, h; \delta) \leqslant C_{42}(h)\left[\int_0^\delta \omega(h; \tau)\tau^{-1}\,d\tau + \delta\int_\delta^\pi \omega(h; \tau)\tau^{-2}\,d\tau\right]. \tag{3.55}$$

Then (3.52) follows from (3.53)–(3.55). Finally, we deduce (3.43) from (3.44), (3.51), (3.52) and (3.24).

8. A corollary of (3.38), (3.42), (3.43), and (1.28) is the inequality

$$\left|I_2 - \psi_n^{\alpha,\beta}(e^{i\Theta})\left[\overline{\pi(h; e^{i\Theta})} - \pi(h; 0)\right]\right| \leqslant C_{43}(\varphi)\mu_{h,n}^{\alpha,\beta}|\psi_n^{\alpha,\beta}(e^{i\Theta})|,$$

which is equivalent to (3.3). Then (1.27) follows from (3.1)–(3.3). Therefore Theorem 1 is established.

9. To conclude this section we give two corollaries of Theorem 1.

COROLLARY 1. *Under the hypotheses of Theorem* 1,

$$\varphi_n^*(z) = z^n\psi_n^{\alpha,\beta}(z^{-1})\pi(h; z)\left[1 + O\left(\mu_{h,n}^{\alpha,\beta}\right)\right] \tag{3.56}$$

as $n \to \infty$, *uniformly for* z *in the disk* $|z| \leqslant 1$.

In fact, (3.56) is equivalent to

$$\varphi_n^*(z)\left[z^n\psi_n^{\alpha,\beta}(z^{-1})\pi(h; z)\right]^{-1} - 1 = O\left(\mu_{h,n}^{\alpha,\beta}\right), \tag{3.57}$$

whose left-hand side is analytic in $|z| < 1$ since the zeros of $\psi_n^{\alpha,\beta}(z)$ are in the interior of this disk ([1], p. 292). By the maximum principle it is enough to verify (3.57) for $z \in \Gamma_1$, and for these values of z it follows from (1.27) and (3.30).

By (1.3), Corollary 1 implies the following corollary.

COROLLARY 2. *Under the hypotheses of Theorem* 1

$$\varphi_n(z) = \psi_n^{\alpha,\beta}(z)\bar{\pi}(h; z^{-1})\left\{1 + O\left(\mu_{h,n}^{\alpha,\beta}\right)\right\} \tag{3.58}$$

as $n \to \infty$, *uniformly in* z *for* $|z| \geqslant 1$. *Here* $\bar{\pi}(g; z)$ *has the same meaning as in Theorem* 2.

§4. Fundamental lemma

1. In this section we establish the fundamental lemma of this paper.

LEMMA 2. *Let the weight functions* u *and* $v \in S$ *(see* (1.30)*) and let* $\{u_n(z)\}_0^\infty$ *and* $\{v_n(z)\}_0^\infty$ *be the corresponding polynomials orthonormal on* Γ_1. *Suppose that for certain integers* N_1 *and* $N \in \mathbf{Z}_+$, $N_1 \leqslant N$, *there are polynomials* $Q_N(z)$ *and* $Q_{N_1}(z)$ *of respective degrees* N *and* N_1, *where* Q_N *is divisible by* Q_{N_1}, *and*

$$\lambda(\tau) \equiv Q_N(e^{i\tau})u(\tau)\left[\overline{Q_{N_1}(e^{i\tau})}\,v(\tau)\right]^{-1} \in \operatorname{Lip} 1, \tag{4.1}$$

$$v(\tau)[u(\tau)]^{-1}\left|Q_{N_1}(e^{i\tau})\right|^2 \leqslant C_1, \quad u(\tau)[v(\tau)]^{-1}\left|Q_N(e^{i\tau})\right|^2 \leqslant C_2 \qquad (\tau \in \mathbf{R}). \tag{4.2}$$

Then

$$Q_N(e^{i\Theta})v_n(e^{i\Theta})[u_n(e^{i\Theta})]^{-1} = Q_N(e^{i\Theta})\overline{\pi(v; e^{i\Theta})}\left[\overline{\pi(u; e^{i\Theta})}\right]^{-1} + O\big(\delta_n(v) + \delta_{n-N_1}(u) + n^{-1}\ln n + \omega(f_1; n^{-1})\big) \tag{4.3}$$

as $n \to \infty$, *uniformly for* $\Theta \in \mathbf{R}$, *where* $\delta_n(v)$ *and* $\delta_{n-N_1}(u)$ *are defined by* (1.29) *and*

$$f_1(\tau) = \overline{\pi(u; e^{i\tau})}\left[\pi(v; e^{i\tau})\right]^{-1}\overline{Q_{N_1}(e^{i\tau})}. \tag{4.4}$$

PROOF. Expanding $Q_N v_n$ in terms of $\{u_n\}_{n=0}^{n+N}$, we have

$$\begin{aligned} Q_N(e^{i\Theta})v_n(e^{i\Theta}) &= \frac{1}{2\pi}\int_0^{2\pi} Q_N(e^{i\tau})v_n(e^{i\tau})K_{u,n+N}(e^{i\Theta}, e^{i\tau})u(\tau)\,d\tau \\ &= I_1 + I_2, \end{aligned} \tag{4.5}$$

where

$$\begin{aligned} I_1 &= \frac{1}{2\pi}\int_0^{2\pi} v_n(e^{i\tau})v(\tau)K_{u,n+N}(e^{i\Theta}, e^{i\tau})\,\overline{Q_{N_1}(e^{i\tau})}\,d\tau\lambda(\Theta), \\ I_2 &= \frac{1}{2\pi}\int_0^{2\pi} v_n(e^{i\tau})v(\tau)K_{u,n+N}(e^{i\theta}, e^{i\tau})\,\overline{Q_{N_1}(e^{i\tau})}\,[\lambda(\tau) - \lambda(\Theta)]\,d\tau. \end{aligned} \tag{4.6}$$

The polynomials v_n are orthonormal, with weight function v, to all polynomials of lower degree. Therefore when $n > N_1$

$$I_1 = \frac{1}{2\pi}\int_0^{2\pi} v_n(e^{i\tau})v(\tau)\sum_{k=n-N_1}^{n+N} u_k(e^{i\Theta})\,\overline{u_k(e^{i\tau})}\;\overline{Q_{N_1}(e^{i\tau})}\,d\tau\cdot\lambda(\Theta).$$

By (3.30) we find from this that

$$I_1 = \sum_{k=n-N_1}^{n+N}\frac{1}{2\pi}\int_0^{2\pi} e^{i(n-k)\tau}\,\overline{v_n^*(e^{i\tau})}\,v(\tau)u_k^*(e^{i\tau})\,\overline{Q_{N_1}(e^{i\tau})}\,d\tau\cdot\lambda(\Theta)u_k(e^{i\Theta}). \tag{4.7}$$

Setting

$$\begin{aligned} I_3 &= \sum_{k=n-N_1}^{n+N}\frac{1}{2\pi}\int_0^{2\pi} e^{i(n-k)\tau}\frac{1}{\pi(v; e^{i\tau})}u_k^*(e^{i\tau})\,\overline{Q_{N_1}(e^{i\tau})}\,d\tau\cdot\lambda(\Theta)u_k(e^{i\Theta}), \\ I_4 &= \sum_{k=n-N_1}^{n+N}\frac{1}{2\pi}\int_0^{2\pi} e^{i(n-k)\tau}\frac{\pi(u; e^{i\tau})}{\pi(v; e^{i\tau})}\,\overline{Q_{N_1}(e^{i\tau})}\,d\tau\cdot\lambda(\Theta)u_k(e^{i\Theta}), \end{aligned} \tag{4.8}$$

we have, by (4.7), (3.7), (3.9), (3.10), (4.1) and (4.2),

$$\begin{aligned} |I_1 - I_4| &\leqslant |I_1 - I_3| + |I_3 - I_4| \\ &\leqslant C_3\delta_n(v)\sum_{k=n-N_1}^{n+N}\left\|\sqrt{v(\tau)}\,u_k(e^{i\tau})Q_{N_1}(e^{i\tau})\right\|_2|\lambda(\Theta)u_k(e^{i\Theta})| \\ &\quad + C_4\sum_{k=n-N_1}^{n+N}\delta_k(u)\left\|\sqrt{\frac{v(\tau)}{u(\tau)}}\,Q_{N_1}(e^{i\tau})\right\|_2|\lambda(\Theta)u_k(e^{i\Theta})| \\ &\leqslant C_5\left[\delta_n(v) + \delta_{n-N_1}(u)\right]\left\{|u_{n-N_1}(e^{i\Theta})| + \cdots + |u_{n+N}(e^{i\Theta})|\right\}. \end{aligned} \tag{4.9}$$

2. Now we establish the inequality

$$\left|u_{n-N_1}(e^{i\Theta})\right| + \cdots + \left|u_{n+N}(e^{i\Theta})\right| \leqslant C_6\left|u_n(e^{i\Theta})\right| \qquad (n \geqslant N_1, \Theta \in \mathbf{R}). \tag{4.10}$$

By the equation (see [1], p. 293)

$$\kappa_n^{(u)} z u_n(z) = \kappa_{n+1}^{(u)} u_{n+1}(z) - u_{n+1}(0) u_{n+1}^*(z), \tag{4.11}$$

where $\kappa_n(u)$ is the coefficient of z^n in $u_n(z)$, we have

$$\frac{e^{im\theta}u_n(e^{i\theta})}{u_{n+m}(e^{i\theta})} = \frac{\kappa_{n+m}^{(u)}}{\kappa_n^{(u)}} \prod_{k=1}^{m}\left[1 + \overline{a_{n+k+1}^{(u)}}\frac{u_{n+k}^*(e^{i\Theta})}{u_{n+k}(e^{i\Theta})}\right], \tag{4.12}$$

where $a_n^{(u)} = -\overline{u_{n+1}(0)}/\kappa_{n+1}^{(u)}$. Since $u \in S$, we have $\kappa_n^{(u)} \uparrow \kappa^{(u)} < \infty$ as $n \to \infty$ (see [2], §1.2). Therefore

$$0 \leqslant \frac{\kappa_{n+m}^{(u)}}{\kappa_n^{(u)}} - 1 \leqslant \left[\frac{\kappa_{n+m}^{(u)}}{\kappa_n^{(u)}}\right]^2 - 1 = \frac{\left[\kappa_{n+m}^{(u)}\right]^2 - \left[\kappa_n^{(u)}\right]^2}{\left[\kappa_n^{(u)}\right]^2}$$

$$= \left[\kappa_n^{(u)}\right]^{-2}\left\{\left|u_{n+1}(0)\right|^2 + \cdots + \left|u_{n+m}(0)\right|^2\right\} \leqslant C_7\delta_n^2(u) \tag{4.13}$$

(see [2], formula (2.18)). Since $|a_n^{(u)}| \leqslant C_8\delta_n(u)$ (see [2], formula (8.15′)), it follows from (4.12) and (4.13) by using (3.30) that

$$e^{im\Theta}u_n(e^{i\Theta})\left[u_{n+m}(e^{i\Theta})\right]^{-1} = 1 + O(\delta_n(u)) \qquad (n \to \infty), \tag{4.14}$$

uniformly for $\Theta \in \mathbf{R}$. If we replace n by $n - m$, we have

$$u_{n-m}(e^{i\Theta}) = e^{-im\Theta}u_n(e^{i\Theta})[1 + O(\delta_{n-m}(u))] \qquad (n \to \infty). \tag{4.15}$$

In addition, it follows from (4.14) that

$$u_{n+m}(e^{i\Theta}) = e^{im\Theta}u_n(e^{i\Theta})[1 + O(\delta_n(u))] \qquad (n \to \infty). \tag{4.16}$$

We can easily deduce (4.10) from (4.15) and (4.16).

3. By (4.9) and (4.10) we have

$$|I_1 - I_4| \leqslant C_9(u; N_1, N)\left[\delta_n(v) + \delta_{n-N_1}(u)\right]\left|u_n(e^{i\Theta})\right| \tag{4.17}$$

for all $n \geqslant N_1$ and $\Theta \in \mathbf{R}$. Let us show that

$$I_4 = u_{n+N+1}(e^{i\Theta})\left[I_5 + O\left(\delta_{n-N_1}(u)\right)\right], \tag{4.18}$$

uniformly as $n \to \infty$ for $\Theta \in \mathbf{R}$, where

$$I_5 = \sum_{k=0}^{N+N_1} \frac{1}{2\pi i}\int_{|\zeta|=1} \frac{\pi(u;\zeta)}{\pi(v;\zeta)} Q_{N_1}^*(\zeta)\zeta^{-k-1}\,d\zeta\, e^{i(k-1)\Theta}\frac{Q_N^*(e^{i\Theta})u(\Theta)}{Q_{N_1}^*(e^{i\Theta})v(\Theta)}. \tag{4.19}$$

By (4.8), (3.30), (4.15), (4.16) and (4.2),

$$I_4 = \sum_{k=0}^{N+N_1} \frac{1}{2\pi i}\int_{|\zeta|=1} \frac{\zeta^{-k-1}\pi(u;\zeta)Q_{N_1}^*(\zeta)}{\pi(v;\zeta)}\,d\zeta\, u_{n+k-N_1}(e^{i\Theta})\lambda(\Theta)$$

$$= \sum_{k=0}^{N+N_1} \frac{1}{2\pi i}\int_{|\xi|=1} \frac{\pi(u;\zeta)Q_{N_1}^*(\zeta)\,d\zeta}{\pi(v;\zeta)\zeta^{k+1}}\,\frac{u_{n+N+1}(e^{i\theta})\lambda(\Theta)}{e^{i(N+N_1+1-k)\Theta}}\left[1 + O\left(\delta_{n-N_1}(u)\right)\right].$$

from this we can deduce (4.18) by using (4.1) and (4.19).

4. Now we show that

$$\begin{aligned} I_2 = & -I_5 u_{n+N+1}(e^{i\Theta}) \\ & + \overline{\pi(v;e^{i\Theta})}\left[\overline{\pi(u;e^{i\Theta})}\right]^{-1} Q_N(e^{i\Theta}) u_n(e^{i\Theta})[1+O(\delta_n(u))] \\ & + u_n(e^{i\Theta}) O\left(n^{-1}\ln n + \omega(f_1;n^{-1}) + \delta_n(v)\right) \end{aligned} \tag{4.20}$$

as $n \to \infty$, uniformly for $\Theta \in \mathbf{R}$. Setting

$$I_6 = \frac{1}{2\pi}\int_0^{2\pi} \frac{e^{in\tau}}{\pi(v;e^{i\tau})} K_{u,n+N}(e^{i\Theta}, e^{i\tau}) \overline{Q_{N_1}(e^{i\tau})}\,[\lambda(\tau)-\lambda(\Theta)]\,d\tau, \tag{4.21}$$

we have

$$|I_2 - I_6| \leqslant C_{10}|u_{n+N+1}(e^{i\Theta})|\delta_n(v) \tag{4.22}$$

by (4.6), (3.7), (3.10), (1.34), (4.1) and (4.2). It follows from (4.21) and (1.34) that

$$I_6 = u^*_{n+N+1}(e^{i\Theta}) I_7 - u_{n+N+1}(e^{i\Theta}) I_8, \tag{4.23}$$

where

$$I_7 = \frac{1}{2\pi}\int_0^{2\pi} \frac{e^{in\tau}}{\pi(v;e^{i\tau})} \overline{u^*_{n+N+1}(e^{i\tau})}\,\overline{Q_{N_1}(e^{i\tau})}\,\frac{\lambda(\tau)-\lambda(\Theta)}{1-e^{i(\Theta-\tau)}}\,d\tau,$$

$$I_8 = \frac{1}{2\pi}\int_0^{2\pi} \frac{e^{in\tau}}{\pi(v;e^{i\tau})} \overline{u_{n+N+1}(e^{i\tau})}\,\overline{Q_{N_1}(e^{i\tau})}\,\frac{\lambda(\tau)-\lambda(\Theta)}{1-e^{i(\Theta-\tau)}}\,d\tau.$$

Setting

$$I_9 = \frac{1}{2\pi}\int_0^{2\pi} e^{in\tau}\frac{\overline{\pi(u;e^{i\tau})}}{\pi(v;e^{i\tau})}\,\overline{Q_{N_1}(e^{i\tau})}\,\frac{\lambda(\tau)-\lambda(\Theta)}{1-e^{i(\Theta-\tau)}}\,d\tau, \tag{4.24}$$

$$I_{10} = \frac{1}{2\pi}\int_0^{2\pi} e^{-i(N+1)\tau}\frac{\pi(u;e^{i\tau})}{\pi(v;e^{i\tau})}\,\overline{Q_{N_1}(e^{i\tau})}\,\frac{\lambda(\tau)-\lambda(\Theta)}{1-e^{i(\Theta-\tau)}}\,d\tau, \tag{4.25}$$

we have, by (3.9) and (3.10),

$$|I_7 - I_9| + |I_8 - I_{10}| = O(\delta_{n+N+1}(u)).$$

Hence it follows from (4.22) and (4.23) that

$$\begin{aligned} I_2 = & \, u^*_{n+N+1}(e^{i\Theta}) I_9 - u_{n+N+1}(e^{i\Theta}) I_{10} \\ & + O(\delta_n(v) + \delta_{n+N+1}(u))|u_{n+N+1}(e^{i\Theta})|. \end{aligned} \tag{4.26}$$

uniformly for $\Theta \in \mathbf{R}$.

5. Let us show that

$$I_{10} = I_5 - e^{-i\Theta}\pi(u;e^{i\Theta})[\pi(v;e^{i\Theta})]^{-1}\,\overline{Q^*_N(e^{i\Theta})}\,u(\Theta)[v(\Theta)]^{-1}. \tag{4.27}$$

We can verify, by using (4.1) and (3.7), that

$$e^{iN\tau}\frac{\pi(u;e^{i\tau})}{\pi(v;e^{i\tau})}\,\overline{Q_{N_1}(e^{i\tau})}\,[\lambda(\tau)-\lambda(\Theta)] = \sum_{k=1}^{3} F_k(e^{i\tau}, e^{i\Theta}),$$

where

$$F_1(e^{i\tau}, e^{i\Theta}) = \overline{Q_N^*(e^{i\tau})}\ \overline{\pi(v; e^{i\tau})}\left[\overline{\pi(u; e^{i\tau})}\right]^{-1} - \overline{Q_N^*(e^{i\Theta})}\ \overline{\pi(v; e^{i\theta})}\left[\overline{\pi(u; e^{i\Theta})}\right]^{-1},$$

$$F_2(e^{i\tau}, e^{i\Theta}) = \frac{\overline{Q_N^*(e^{i\Theta})}}{Q_{N_1}^*(e^{i\Theta})}\frac{u(\Theta)}{v(\Theta)}\left[Q_{N_1}^*(e^{i\Theta})\frac{\pi(u; e^{i\Theta})}{\pi(v; e^{i\Theta})} - Q_{N_1}^*(e^{i\tau})\frac{\pi(u; e^{i\tau})}{\pi(v; e^{i\tau})}\right],$$

$$F_3(e^{i\tau}, e^{i\Theta}) = \frac{\overline{Q_N^*(e^{i\Theta})}}{Q_{N_1}^*(e^{i\Theta})}\frac{u(\Theta)}{v(\Theta)}Q_{N_1}^*(e^{i\tau})\frac{\pi(u; e^{i\tau})}{\pi(v; e^{i\tau})}[1 - e^{i(N+N_1)(\tau-\Theta)}].$$

Hence we find by using (4.25) that

$$I_{10} = \bar{I}_{11} - I_{12} + I_{13}, \tag{4.28}$$

where

$$I_{11} = \frac{1}{2\pi}\int_0^{2\pi}\frac{e^{i\tau}\overline{F_1(e^{i\tau}, e^{i\Theta})}\,d\tau}{1 - e^{i(\tau-\Theta)}}, \qquad I_{12} = \frac{1}{2\pi}\int_0^{2\pi}\frac{e^{-i\tau}F_2(e^{i\tau}, e^{i\Theta})}{1 - e^{i(\Theta-\tau)}}d\tau,$$

$$I_{13} = \frac{1}{2\pi}\int_0^{2\pi}\frac{e^{-i\tau}F_3(e^{i\tau}, e^{i\Theta})}{1 - e^{-i(\tau-\Theta)}}d\tau. \tag{4.29}$$

By (4.2) and (3.41)

$$I_{11} = \frac{1}{2\pi i}\int_{|\zeta|=1}\overline{F_1(\zeta; e^{i\Theta})}\ \frac{d\zeta}{e^{i\Theta} - \zeta}e^{i\Theta} = 0. \tag{4.30}$$

Since

$$\left[(\zeta - e^{i\Theta})e^{i\Theta}\right]^{-1} = -(\zeta e^{i\Theta})^{-1} + \left[e^{i\Theta}(\zeta - e^{i\Theta})\right]^{-1}$$

we have

$$I_{12} = -e^{-i\Theta}\frac{1}{2\pi i}\int_{|\zeta|=1}F_2(\zeta; e^{i\Theta})\,\frac{d\zeta}{\zeta} + e^{-i\Theta}\frac{1}{2\pi i}\int_{|\zeta|=1}\frac{F_2(\zeta; e^{i\Theta})}{\zeta - e^{i\Theta}}d\zeta.$$

In view of (4.2) and (3.41) the second integral vanishes. Therefore

$$I_{12} = -e^{-i\Theta}\,\overline{Q_N^*(e^{i\Theta})}\,u(\Theta)[v(\Theta)]^{-1}\pi(u; e^{i\Theta})[\pi(v; e^{i\Theta})]^{-1} + e^{-i\Theta}\frac{\overline{Q_N^*(e^{i\Theta})}\,u(\Theta)}{Q_{N_1}(e^{i\Theta})v(\Theta)}\frac{1}{2\pi i}\int_{|\zeta|=1}Q_{N_1}^*(\zeta)\frac{\pi(u; \zeta)\,d\zeta}{\pi(v; \zeta)\zeta}. \tag{4.31}$$

Finally, it is easy to transform (4.29) into the form

$$I_{13} = \frac{\overline{Q_N^*(e^{i\Theta})}\,u(\Theta)}{Q_{N_1}^*(e^{i\Theta})v(\Theta)}\sum_{k=1}^{N+N_1}\frac{1}{2\pi i}\int_{|\zeta|=1}\frac{Q_{N_1}^*(\zeta)\pi(u; \zeta)}{\zeta^{k+1}\pi(v; \zeta)}d\zeta e^{i(k-1)\Theta}. \tag{4.32}$$

Then (4.27) follows from (4.28) and (4.30)–(4.32). By using (3.30) and (3.7) we can put (4.27) into the form

$$I_{10} = I_5 - \overline{\pi(v; e^{i\Theta})}\left[\overline{\pi(u; e^{i\Theta})}\right]^{-1}e^{-i(N+1)\Theta}Q_N(e^{i\Theta}). \tag{4.33}$$

6. Let us find an upper bound for $|I_9|$. We write (4.24) in the form

$$I_9 = \frac{1}{2\pi}\int_0^{2\pi} e^{in\tau}F(\tau)\,d\tau, \qquad F(\tau) = f_1(\tau)f_2(\tau)f_3(\tau), \tag{4.34}$$

where $f_1(\tau)$ is defined by (4.4), $f_2(\tau) = \lambda(\tau) - \lambda(\Theta)$ and $f_3(\tau) = [1 - e^{i(\Theta-\tau)}]^{-1}$. Evidently

$$4\pi I_9 = \int_{\Theta-\pi}^{\Theta+\pi} e^{in\tau}F(\tau)\,d\tau - \int_{\Theta-\pi}^{\Theta+\pi} e^{in\tau}F\left(\tau + \frac{\pi}{n}\right)d\tau.$$

Therefore

$$4\pi|I_9| \leqslant I_{14} + I_{15}, \tag{4.35}$$

where

$$I_{14} = \int_{e_{1,n}}\left[|F(\tau)| + \left|F\left(\tau+\frac{\pi}{n}\right)\right|\right]d\tau, \qquad I_{15} = \int_{e_{2,n}}\left|F(\tau) - F\left(\tau + \frac{\pi}{n}\right)\right|d\tau,$$

$$e_{1,n} = \{\tau\colon |\tau - \Theta| \leqslant 2\pi n^{-1}\}, \qquad e_{2,n} = [\Theta - \pi, \Theta + \pi]\setminus e_{1,n}.$$

Since $\lambda(\tau) \in \mathrm{Lip}\,1$, we have

$$|f_2(\tau)f_3(\tau)| \leqslant C_{11} \qquad (\tau, \Theta \in \mathbf{R}). \tag{4.36}$$

By (4.4) and (4.2),

$$|f_1(\tau)| = \{v(\tau)/u(\tau)\}^{1/2}|Q_{N_1}(e^{i\tau})| \leqslant \sqrt{C_1} \qquad (\tau \in \mathbf{R}). \tag{4.37}$$

Consequently

$$I_{14} \leqslant C_{12}n^{-1} \qquad (\Theta \in \mathbf{R}, n \in \mathbf{N}). \tag{4.38}$$

To estimate I_{15} we apply the identity

$$a_1a_2a_3 - b_1b_2b_3 = (a_1 - b_1)a_2a_3 + b_1(a_2 - b_2)a_3 + b_1b_2(a_3 - b_3).$$

We will then have

$$I_{15} \leqslant I_{16} + I_{17} + I_{18}, \tag{4.39}$$

where

$$I_{16} = \int_{e_{2,n}} |f_2(\tau)f_3(\tau)\Delta_{\pi/n}f_1(\tau)|\,d\tau, \tag{4.40}$$

$$I_{17} = \int_{e_{2,n}} |f_1(\tau)f_3(\tau)\Delta_{\pi/n}f_2(\tau)|\,d\tau, \tag{4.41}$$

$$I_{18} = \int_{e_{2,n}} \left|f_1\left(\tau+\frac{\pi}{n}\right)f_2\left(\tau+\frac{\pi}{n}\right)\Delta_{\pi/n}f_3(\tau)\right|d\tau. \tag{4.42}$$

By (4.36) and (4.40)

$$I_{16} \leqslant C_{13}\omega(f_1; n^{-1}) \qquad (\Theta \in \mathbf{R}, n \in \mathbf{N}). \tag{4.43}$$

It follows from (4.37), (4.41) and (4.42) that

$$I_{17} \leqslant C_{14}n^{-1}\int_{e_{2,n}} \frac{d\tau}{|\tau - \Theta|} \leqslant C_{15}n^{-1}\ln n \qquad (\Theta \in \mathbf{R}, n \in \mathbf{N}), \tag{4.44}$$

$$\begin{aligned} I_{18} &\leqslant C_{16}\int_{e_{2,n}} \left|\left(\tau + \frac{\pi}{n} - \Theta\right)\frac{e^{i(\Theta-\tau)} - e^{i(\Theta-\tau-\pi/n)}}{[1 - e^{i(\Theta-\tau)}][1 - e^{i(\Theta-\tau-\pi/n)}]}\right|d\tau \\ &\leqslant C_{17}\int_{e_{2,n}} \frac{d\tau}{|\tau - \Theta|} \leqslant C_{18}n^{-1}\ln n \qquad (\Theta \in \mathbf{R}, n \in \mathbf{N}). \end{aligned} \tag{4.45}$$

From (4.35), (4.38), (4.39) and (4.43)–(4.45) we deduce that

$$|I_9| \leqslant C_{19}\left[n^{-1}\ln n + \omega\left(f_1; n^{-1}\right)\right] \qquad (\Theta \in \mathbf{R}, n \in \mathbf{N}). \tag{4.36}$$

Then we obtain (4.20) by using (4.26), (4.33), (4.46), (4.15) and (4.16).

Finally, we deduce (4.3) from (4.5), (4.17), (4.18) and (4.20). This completes the proof of Lemma 2.

§5. Proof of Theorem 2

1. By Theorem 1, if we set

$$w(\tau) = |\sin(\tau/2)|^{\gamma_l} h_1(\tau), h_1(\tau) = h(\tau + \Theta_l)$$

we have the asymptotic representation

$$w_n(e^{i\Theta}) = \psi_n(e^{i\Theta})\overline{\pi(h_1; e^{i\Theta})}\left[1 + O\left(\mu_{h,n}^{(\gamma_l - j1)/2, -1/2}\right)\right], \tag{5.1}$$

uniformly for $\Theta \in \mathbf{R}$, where $n \to \infty$ and $\psi_n(z) = 2^{\gamma_l/4}\psi_n^{(\gamma_l - 1)/2, -1/2}(z)$. We set

$$u(\tau) = w(\tau - \Theta_l) = \left|\sin[(\tau - \Theta_l)/2]\right|^{\gamma_l} h(\tau). \tag{5.2}$$

Then, as is easily verified, $u_n(z) = e^{in\Theta_l} w_n(ze^{i-\Theta_l})$. Hence it follows from (5.2) that

$$u_n(e^{i\Theta}) = e^{in\Theta_l}\psi_n(e^{i(\Theta - \Theta_l)})\overline{\pi(h; e^{i\Theta})}\left[1 + O\left(\mu_{h,n}^{(\gamma_l - 1)/2, -1/2}\right)\right] \tag{5.3}$$

as $n \to \infty$, uniformly for $\Theta \in \mathbf{R}$.

Now we apply Lemma 2, taking $\nu(\tau) = \varphi(\tau)$, $N = 2N_1 = 2(B_1 + \cdots + B_m)$, $B_\nu \in \mathbf{Z}_+$ $(\nu = 1, \ldots, m)$, $B_l = 0$ and

$$Q_N(z) = Q_{N_1}^2(z) = (1 - ze^{-i\Theta_l})^{2B_1} \cdots (1 - ze^{-i\Theta_m})^{2B_m}.$$

If all the B_ν with $\nu \neq l$ are sufficiently large, then the hypotheses of Lemma 2 are satisfied and consequently we have the asymptotic representation (4.3). Therefore

$$Q_N(e^{i\Theta})\varphi_n(e^{i\Theta}) = u(e^{i\Theta})\left[Q_N(e^{i\Theta})\overline{\pi(\varphi/u; e^{i\Theta})} + O(\mu_{1,n})\right]$$

as $n \to \infty$, uniformly for $\Theta \in \mathbf{R}$, where

$$\mu_{1,n} = \delta_n(\varphi) + \delta_{n-N_1}(u) + n^{-1}\ln n + \omega\left(f_1; n^{-1}\right). \tag{5.4}$$

It follows from this representation, by (5.3), that as $n \to \infty$

$$\begin{aligned} &Q_N(e^{i\Theta})\varphi_n(e^{i\Theta})\left[e^{in\Theta_l}\psi_n(e^{i(\Theta - \Theta_l)})\right]^{-1} \\ &\quad = Q_N(e^{i\Theta})\pi\left(\varphi(\tau)\left|\sin[(\tau - \Theta_l)/2]\right|^{-\gamma_l}; e^{i\Theta}\right) + O(\mu_{2,n}), \end{aligned} \tag{5.5}$$

where the remainder term is uniform for $\Theta \in \mathbf{R}$, and

$$\mu_{2,n} = \mu_{1,n} + \mu_{h,n}^{(\gamma_l - 1)/2, -1/2}. \tag{5.6}$$

Taking complex conjugates in (5.5), we find by (3.30) that

$$Q_N^*(z)\left\{\frac{\varphi_n^*(z)}{\psi_n^*(ze^{-i\Theta_l})} - \pi\left(\frac{\varphi(\tau)}{\left|\sin[(\tau - \Theta_l)/2]\right|^{\gamma_l}}; z\right)\right\} = O(\mu_{2,n}) \tag{5.7}$$

as $n \to \infty$, uniformly for $z \in \Gamma_1$. The left-hand side is continuous in the disk $|z| \leqslant 1$ and analytic in its interior. Therefore by the maximum principle (5.7) holds uniformly for z in the disk $|z| \leqslant 1$.

2. Let us show that $\mu_{2,n} = O(\mu_n)$. We need the following simple lemma (which is implicit in [21]).

LEMMA 3. *Let* $\{\xi_n(z)\}_0^\infty$, $\{\eta_n(z)\}_0^\infty$, *and* $\{g_n(z)\}_0^\infty$ *be systems of polynomials that are orthonormal on* Γ_1 *with respective weight functions* ξ, η, *and* $g = \xi\eta$; *let* $\xi, \eta \in S$ (*see* (1.30)) *and let there be a constant* $C_1(\xi)$ *such that*

$$\left|\xi_n(e^{i\Theta})\sqrt{\xi(\Theta)}\right| \leqslant C_1(\xi) \qquad (\Theta \in \mathbf{R}, n \in \mathbf{Z}_+). \tag{5.8}$$

Then there is a constant $C_2(\xi)$ *such that*

$$\delta_n(g) \leqslant \delta_{[n/2]}(\xi) + C_2(\xi)\delta_{[n/2]}(\eta) \qquad (n \in \mathbf{Z}_+). \tag{5.9}$$

PROOF. We introduce the notation

$$U_n^{(\varphi)}(\tau) = \kappa_n^{(\varphi)}[\kappa^{(\varphi)}]^{-1}\varphi_n^*(e^{i\tau}), \qquad \kappa^{(\varphi)} = \lim_{n\to\infty} \kappa_n^{(\varphi)}.$$

Since (see [2], formula (2.8))

$$\delta_n(\varphi) = \left\|\left[\pi(\varphi; e^{i\tau}) - U_n^{(\varphi)}(\tau)\right]\varphi(\tau)\right\|_2,$$

by (1.29) we have

$$\begin{aligned}\delta_n(g) &\leqslant \left\|\pi(g; e^{i\tau}) - U_{[n/2]}^{(\xi)}(\tau)U_{[n/2]}^{(\eta)}(\tau)\sqrt{g(\tau)}\right\|_2\\ &\leqslant \left\|\left[\pi(g; e^{i\tau}) - U_{[n/2]}^{(\xi)}(\tau)\pi(\eta; e^{i\tau})\right]\sqrt{g(\tau)}\right\|_2\\ &\quad+ \left\|U_{[n/2]}^{(\xi)}(\tau)\left[\pi(\eta; e^{i\tau}) - U_{[n/2]}^{(\eta)}(\tau)\right]\sqrt{g(\tau)}\right\|_2.\end{aligned} \tag{5.10}$$

By (3.7) and (3.8) the first of the last two terms is $\delta_{[n/2]}(\xi)$, and by (5.8) the second is majorized by

$$\kappa_n^{(\xi)}[\kappa^{(\xi)}]^{-1}C_1(\xi)\delta_{[n/2]}(\eta) \leqslant C_2(\xi)\delta_{[n/2]}(\eta).$$

Hence (5.9) follows from (5.10).

This completes the proof of the lemma.

3. Let us take $g = \varphi$ and $\xi = h$; then, as we know, (5.8) is satisfied (see [8] or [9]) and with $\eta = \varphi/h$ we have $\delta_n(\varphi) = O(n^{-1/2})$ (see [21]). Therefore by Lemma 3

$$\delta_n(\varphi) = O\left(n^{-1/2} + \delta_{[n/2]}(h)\right) \qquad (n \to \infty). \tag{5.11}$$

In addition, if $\omega_2(h, \delta)$ is the mean-square modulus of continuity of h, it is known ([2], formula (3.9) and Table I) that

$$\delta_n(h) \leqslant C_3\omega_2(h; n^{-1}) \leqslant C_4\omega(h; n^{-1}),$$

whence we find, by (5.11), that under (1.20)–(1.22)

$$\delta_n(\varphi) = O\left(n^{-1/2} + \omega(h; n^{-1})\right) \qquad (n \to \infty).$$

In particular,

$$\mu_{h,n}^{(\gamma_l-1)/2,-1/2} = O\left(n^{-1/2} + \omega(h; n^{-1})\right) \qquad (n \to \infty),$$

and the same estimate holds for the weight function (5.2). Therefore by (5.4) and (5.6),

$$\mu_{2,n} = O\left(n^{-1/2} + \omega(h; n^{-1}) + n^{-1}\ln n + \omega\left(f_1 n^{-1}\right)\right) \tag{5.12}$$

as $n \to \infty$, where according to (4.4), (5.2) and (3.8),

$$f_1(\tau) = f_4(\tau)\,\overline{\pi(h; e^{i\tau})}\,\left[\pi(h; e^{i\tau})\right]^{-1}, \tag{5.13}$$
$$f_4(\tau) = \overline{\pi(u/h; e^{i\tau})}\;\overline{Q_{N_1}(e^{i\tau})}\,\left[\pi(\varphi/h; e^{i\tau})\right]^{-1}.$$

If all the B_ν with $\nu \neq l$ are sufficiently large, then

$$f_4(\tau) \in \operatorname{Lip} 1. \tag{5.14}$$

In the proof of Theorem 1 we established the inequality (3.52) for the modulus of continuity of $f_3(\tau) = [\pi(h;\, e^{i\tau})]^{-1}$ (see (3.45)). Since $\pi(h;\, e^{i\tau}) \in C_{2\pi}$ and is not zero, an estimate similar to (3.52) holds for the modulus of continuity of $\overline{\pi(h;\, e^{i\tau})}$. It follows, by (5.13) and (5.14), that

$$\omega(f_1; n^{-1}) \leqslant C_5\left\{\int_0^{1/n} \omega(h;\tau)\tau^{-1}\,d\tau + n^{-1}\int_{n^{-1}}^{\pi} \omega(h;\tau)\tau^{-2}\,d\tau\right\}$$
$$\leqslant C_6\left\{\int_0^{1/n} \omega(h;\tau)\tau^{-1}\,d\tau + \omega(h; n^{-1})\ln n\right\}. \tag{5.15}$$

From (5.12), (5.15), and (1.37) we find that $\mu_{2,n} = O(\mu_n)$.

4. Hence it follows from (5.7) that

$$\begin{aligned}&\varphi_n^*(z)\left[2^{\gamma_l/4}e^{-in\Theta_l}z^n\psi_n^{(\gamma_l-1)/2,-1/2}(z^{-1}e^{i\Theta_l})\right]^{-1}\\ &\quad= \pi\left(\varphi(\tau)\left|\sin[(\tau-\Theta_l)/2]\right|^{-\gamma_l}; z\right) + O\left(\mu_n\left|Q_N^*(z)\right|^{-1}\right)\end{aligned} \tag{5.16}$$

as $n \to \infty$, uniformly for z in the disk $|z| \leqslant 1$. Replacing z by $\bar{z}^{-1}$ in (5.16) and taking complex conjugates in the result, we find that

$$\varphi_n(z)\left[e^{in\Theta_l}2^{\gamma_l/4}\psi_n^{(\gamma_l-1)/2,-1/2}(ze^{i\Theta_l})\right]^{-1} = \bar{\pi}\left(\varphi(\tau)\left|\sin[(\tau-\Theta_l)/2]\right|^{-\gamma_l}; z^{-1}\right)$$
$$+\,O\left(\mu_n|z|^N\left|Q_N(z)\right|^{-1}\right),$$

uniformly in z for $|z| \geqslant 1$. Hence the conclusion of Theorem 2 follows because of the uniform boundedness of $z^N[Q_N(z)]^{-1}$ for $z \in Q_l$.

5. We remark that when $m = 1$ the set $Q_{l,\varepsilon}$ in Theorem 2 coincides with $\{z\colon |z| \geqslant 1\}$ and that Suetin [22] has recently obtained, for the weight function (1.20) with $m = 1$ and a smoother factor h, a formula similar to (1.36) but only for

$$z \in \left\{\zeta\colon |\zeta| \geqslant 1,\ \left|\zeta - e^{i\Theta_1}\right| \geqslant n^{-1}\right\}.$$

No asymptotic formula for $\varphi_n(e^{i\Theta_1})$ was obtained in [22].

§6. Some corollaries of Theorems 1 and 2

In this section we establish some results, which follow from Theorems 1 and 2, on the asymptotic behavior of the polynomials $\varphi_n(e^{i\Theta})$, $\Phi_n(\Theta)$ and $p_n(\cos\Theta)$; and we also prove that, in a certain sense, the results obtained in this paper cannot be improved.

1. When $z = e^{i\Theta}$ we can easily deduce the following corollary of Theorem 2.

COROLLARY 3. *Let the hypotheses of Theorem* 2 *be satisfied and let the intervals* $\Delta_\nu = [\tau_\nu, \tau_{\nu+1}]$ $(\nu = 1,\ldots,m)$ *satisfy* (3.18). *Then*

$$\varphi_n(e^{i\Theta}) = 2^{\gamma_l/4} e^{in\Theta_l} \overline{\pi\left(\varphi(\tau)|\sin[(\tau - \Theta_l)/2]|^{-\gamma_l}; e^{i\Theta}\right)} \times \psi^{(\gamma_l - 1)/2, -1/2}\left(e^{i(\Theta - \Theta_l)}\right)[1 + O(\mu_n)] \tag{6.1}$$

as $n \to \infty$, *uniformly for* $\Theta \in \Delta_l$, *where* μ_n *is defined by* (1.37).

If we use (2.30) to express the right-hand side of (6.1) in terms of Bessel functions we can easily obtain the following corollary from Corollary 3.

COROLLARY 4. *Let the hypotheses of Corollary* 3 *be satisfied, and let*

$$g_{1,n}(\Theta) = e^{-i(n-1)\Theta}\varphi_{2n-1}(e^{i\Theta}), \qquad g_{2,n}(\Theta) = e^{-in\Theta}\varphi_{2n}(e^{i\Theta}). \tag{6.2}$$

Then

$$g_{k,n}(\Theta) = (\pi/2)^{1/2} e^{in\Theta_l} F_{(\gamma_l - 1)/2}((n + \gamma_l/4)(\Theta - \Theta_l))|\sin[(\Theta - \Theta_l)/2]|^{-\gamma_l/2} \times \overline{\pi\left(\varphi(\tau)|\sin[(\tau - \Theta_l)/2]|^{-\gamma_l}; e^{i\Theta}\right)}\,[1 + O(\mu_{k,n})], \tag{6.3}$$

as $n \to \infty$, *uniformly for* $\Theta \in \Delta_l$, *where* $\mu_{1,n} = \mu_{2n-1}$, $\mu_{2,n} = \mu_{2n}$, μ_n *is defined by* (1.37); $k = 1, 2$; *and*

$$F_\alpha(\tau) = \left[J_\alpha(|\tau|) + iJ_{\alpha+1}(|\tau|)\operatorname{sgn}\tau\right]\sqrt{|\tau|}\,. \tag{6.4}$$

By using (2.39) we can use Corollary 4 to obtain a somewhat weaker but simpler result.

COROLLARY 5. *Let* $l \in (1,\ldots,m)$, *and let the weight function* φ *satisfy* (1.20)–(1.22) *without becoming either zero or infinity for* $\Theta \in [a, b]\setminus\{\Theta_l\}$. *Then*

$$\varphi_n(e^{i\Theta}) = e^{in\Theta}\overline{\pi(\varphi; e^{i\Theta})}\left\{1 + O\left(\mu_n + n^{-1}|\Theta - \Theta_l|^{-1}\right)\right\}, \tag{6.5}$$

$$\varphi_n^*(e^{i\Theta}) = \pi(\varphi; e^{i\Theta})\left\{1 + O\left(\mu_n + n^{-1}|\Theta - \Theta_l|^{-1}\right)\right\}, \tag{6.6}$$

as $n \to \infty$, *uniformly for* $\Theta \in [a, b]\setminus(\Theta_l - cn^{-1}, \Theta_l + cn^{-1})$ *with a given* c.

PROOF. It follows from (2.39) and (6.4) that

$$F_{(\gamma_l - 1)/2}((n + \gamma_l/4)\tau) = (2/\pi)^{1/2} e^{i[(4n+\gamma_l)\tau - \pi\gamma_l \operatorname{sgn}\tau]/4} + O\left(|n\tau|^{-1}\right), \tag{6.7}$$

for $|\tau| > cn^{-1}$. It is easily verified that if $0 < |\Theta - \Theta_l| < 2\pi$ then

$$\pi\left(|\sin[(\tau - \Theta_l)/2]|^{\gamma_l}; e^{i\Theta}\right) = |\sin[(\Theta - \Theta_l)/2]|^{-\gamma_l/2} e^{-i\gamma_l[\Theta - \Theta_l - \pi\operatorname{sgn}(\Theta - \Theta_l)]/4}. \tag{6.8}$$

Then (6.5) and the equivalent (because of (1.3)) result (6.6) follows from (6.2), (6.3) and (6.8).

2. Consider the problem of how essential it is to have

$$\omega(h; \tau)\tau^{-1} \in L^1[0, \pi] \tag{6.9}$$

in Theorems 1 and 2 and Corollaries 1–5. We shall need the following lemma.

LEMMA 4. *For every* $\Theta_0 \in (0, \pi)$ *and every modulus of continuity* $\omega(\delta)$ *for which*

$$\omega(\tau)\tau^{-1} \notin L^1[0, \pi], \tag{6.10}$$

there is an even positive function $h \in C_{2\pi}$ *such that*

$$\omega(h; \delta) = O(\omega(\delta)) \qquad (\delta \to +0) \tag{6.11}$$

and moreover the function

$$F(\Theta) = \operatorname{sgn} \pi(h; e^{i\Theta}) \equiv |\pi(h; e^{i\Theta})|^{-1} \pi(h; e^{i\Theta}) \tag{6.12}$$

has discontinuities at $\Theta = \pm\Theta_0$.

PROOF. Let $h(\tau) = \exp\{f(\tau)\}$, where $f \in C_{2\pi}$, is even, and equal to zero on $[0, \Theta_0]$ and to $\Omega(\tau - \Theta_0)$ on $[\Theta_0, \pi]$, where

$$\Omega(t) = t^{-1}\int_0^t \omega(u)\, du = \int_0^1 \omega(vt)\, dv. \tag{6.13}$$

It is easy to show (see [14], Lemma 5.1) that $\Omega(t)$ is a modulus of continuity on $[0, \infty)$ and satisfies

$$4^{-1}\omega(t) \leqslant \Omega(t) \leqslant \omega(t). \tag{6.14}$$

Hence (6.11) follows.

It is clear from (6.13) that $\Omega'(t)$ exists and is continuous on $(0, \infty)$. Therefore $f(\Theta)$ is continuously differentiable on the intervals $[-\pi, -\Theta_0]$, $(-\Theta_0, \Theta_0)$, and $(\Theta_0, \pi]$. For Θ on these intervals, $F(\Theta)$ can be expressed by the formula (see [1], pp. 277–278)

$$F(\Theta) = \exp\left\{\frac{i}{4\pi} g(\Theta)\right\}, \qquad g(\Theta) = \int_{-\pi}^{\pi} [f(\tau) - f(\Theta)] \cot\frac{\tau - \Theta}{2}\, d\tau. \tag{6.15}$$

Let $\Theta = \Theta_0 - \delta$ where $\delta \in i_0 = (0, \delta_0]$ and $0 < \delta_0 < 2^{-1}\min(\Theta_0, \pi - \Theta_0)$. Then $f(\Theta) = 0$; $f(\tau) = 0$ on $[-\Theta_0, \Theta_0]$; and on $[-\pi, -\Theta_0]$

$$|f(\tau)| \leqslant \omega(\pi - \Theta_0), \qquad |\cot[(\tau - \Theta)/2]| \leqslant C_1(\Theta_0).$$

From this and (6.14) for $\delta \in i_0$

$$\begin{aligned} g(\Theta) &= \left\{\int_{-\pi}^{-\Theta_0} + \int_{\Theta_0}^{\pi}\right\} f(\tau) \cot\frac{\tau - \Theta}{2}\, d\tau \\ &\geqslant \frac{1}{2}\int_{\Theta_0 + \delta}^{\pi - \delta} \omega(\tau - \Theta_0) \cot\frac{\tau - \Theta}{2}\, d\tau - C_2(\Theta_0) \\ &= \frac{1}{2}\int_{\delta}^{(\pi - \Theta_0)/2} \omega(2u - \delta) \cot u\, du - C_2(\Theta_0) \\ &\geqslant \frac{1}{2}\int_{\delta}^{\delta_0} \omega(u) \cot u\, du - C_2(\Theta_0). \end{aligned}$$

Hence it follows by (6.10) that $g(\Theta) \to +\infty$ as $\delta \to +0$. Since $g(\Theta)$ is continuous on $(-\Theta_0, \Theta_0)$, there is a number $C_3(\Theta_0) > 0$ such that $g(\Theta)$ takes every value on the interval $(C_3(\Theta_0), +\infty)$ as $\delta \to +0$. Therefore, by (6.15), $F(\Theta)$ is discontinuous at Θ_0. Since h is even, it is evident that $\pi(h; e^{-i\Theta}) = \overline{\pi(h; e^{i\Theta})}$ and consequently, by (6.12), $F(-\Theta) = \overline{F(\Theta)}$. Therefore $F(\Theta)$ is also discontinuous at $\Theta = -\Theta_0$.

This completes the proof of Lemma 4.

It follows from Lemma 4 that the conclusion of Corollary 5 fails without the restriction (6.9). More precisely, we have the following lemma.

LEMMA 5. *For every modulus of continuity satisfying* (6.10) *there are an even positive function $h \in C_{2\pi}$ with modulus of continuity satisfying* (6.12) *and an interval* $[c, d] \subset \Delta_l$ *not containing the point Θ_l such that the asymptotic formula* (6.6) *cannot hold uniformly for* $\Theta \in [c, d]$.

In fact, let $\varepsilon \in (0, \Theta_l - \varepsilon_l)$, $c = \tau_l$, $d = \Theta_l - \varepsilon$, and $\Theta_0 = |d|$. Then $c < d$, and $[c, d] \subset \Delta_l$ and does not contain Θ_l. Moreover, $\Theta_0 \in (0, \pi)$. Let $h(\tau)$ be the function defined in Lemma 4. Then if, with this h, the relation (6.6) were satisfied uniformly for $\Theta \in [c, d]$, the function $\pi(\varphi;\, e^{i\Theta})$ would be continuous on $[c, d]$. Hence, because

$$\pi(h;\, e^{i\Theta}) = \pi(\varphi;\, e^{i\Theta})/\pi(\varphi/h;\, e^{i\Theta}), \qquad \pi(\varphi/h;\, e^{i\Theta}) \in C[c, d],$$
$$\pi(\varphi/h;\, e^{i\Theta}) \neq 0 \qquad (\Theta \in [c, d])$$

it would follow that $\pi(h;\, e^{i\Theta})$ would be continuous on $[c, d]$, and so would the function (6.12). But this would contradict Lemma 4, according to which $F(\Theta)$ is discontinuous at the point $\pm\Theta_0$, one of which coincides with d.

Since Theorem 2 ⇒ Corollary 3 ⇒ Corollary 4 ⇒ Corollary 5, the conclusions of Theorem 2 and Corollaries 3 and 4 fail without (6.9) (in the same sense as for Corollary 5). Similar remarks apply to Theorem 1 and its Corollaries 1 and 2.

3. By using (1.4) and (1.5) we can deduce from the asymptotic formulas for $\varphi_n(e^{i\Theta})$ the corresponding results for $\Phi_n(\Theta)$.

The next corollary follows from Corollary 3 by using (1.4) and (1.5).

COROLLARY 6. *Under the hypotheses of Corollary* 3,

$$\begin{aligned}\Phi_{2n-1}(\Theta) = 2^{\gamma_l/4}\sqrt{\pi}\,\{ &X_{n,l}(\Theta)[\cos\eta_{n,l}(\Theta) + O(\mu_{2n-1})] \\ &+ Y_{n,l}(\Theta)[\sin\eta_{n,l}(\Theta) + O(\mu_{2n-1})]\} \\ &\times\sqrt{|\sin[(\Theta-\Theta_l)/2]|^{\gamma_l}/\varphi(\Theta)}\,,\end{aligned} \tag{6.16}$$

$$\begin{aligned}\Phi_{2n}(\Theta) = 2^{\gamma_l/4}\sqrt{\pi}\,\{ &-X_{n,l}(\Theta)[\sin\eta_{n,l}(\Theta) + O(\mu_{2n})] \\ &+ Y_{n,l}(\Theta)[\cos\eta_{n,l}(\Theta) + O(\mu_{2n})]\} \\ &\times\sqrt{|\sin[(\Theta-\Theta_l)/2]|^{\gamma_l}/\varphi(\Theta)}\,,\end{aligned} \tag{6.17}$$

as $n \to \infty$, uniformly for $\Theta \in \Delta_l$, where

$$X_{n,l}(\Theta) = \hat{P}_n^{(\gamma_l-1)/2,-1/2}(\cos(\Theta-\Theta_l)),$$
$$Y_{n,l}(\Theta) = \hat{P}_{n-1}^{(\gamma_l+1)/2,1/2}(\cos(\Theta-\Theta_l))\sin(\Theta-\Theta_l),$$
$$\eta_{n,l}(\Theta) = \arg\pi\Big(\varphi(\tau)\big|\sin[(\tau-\Theta_l)/2]\big|^{-\gamma_l};\, e^{i\Theta}\Big) - n\Theta_l. \tag{6.18}$$

PROOF. It is known ([2], Theorem 8.3) that

$$\varphi_{2n}(0)/\kappa_{2n}^{(\varphi)} = O(\delta_{2n-1}(\varphi)) \qquad (n \to \infty).$$

Therefore in (1.5)

$$c_n = \sqrt{2}\,[1 + O(\delta_{2n-1}(\varphi))] \qquad (n \to \infty). \tag{6.19}$$

It was shown in the proof of Theorem 2 that

$$\delta_n(\varphi) = O(\omega(h; n^{-1}) + n^{-1/2}). \tag{6.20}$$

By (6.20) and (1.37)

$$\delta_{2n-1}(\varphi) = O(\mu_{2n-1}), \qquad \delta_{2n-1}(\varphi) = O(\mu_{2n}). \tag{6.21}$$

Formulas (6.16) and (6.17) follow from (1.4), (1.5), (6.1), (6.19), (6.21), (2.17) and (2.18).

Using (1.4) and (1.5), we deduce the next corollary by a proof similar to that of Corollary 4.

COROLLARY 7. *Under the hypotheses of Corollary* 3,

$$\begin{aligned}\Phi_{2n-1}(\Theta) = \{\pi/\varphi(\Theta)\}^{1/2}\{W_{n,l}(\Theta)[\cos\eta_{n,l}(\Theta) + O(\mu_{2n-1})] \\ + Z_{n,l}(\Theta)[\sin\eta_{n,l}(\Theta) + O(\mu_{2n-1})]\},\end{aligned} \tag{6.22}$$

$$\begin{aligned}\Phi_{2n}(\Theta) = \{\pi/\varphi(\Theta)\}^{1/2}\{-W_{n,l}(\Theta)[\sin\eta_{n,l}(\Theta) + O(\mu_{2n})] \\ + Z_{n,l}(\Theta)[\cos\eta_{n,l}(\Theta) + O(\mu_{2n})]\},\end{aligned} \tag{6.23}$$

as $n \to \infty$, *uniformly for* $\Theta \in \Delta_l$, *where* $\eta_{n,l}(\Theta)$ *and* μ_n *are defined by* (6.18) *and* (1.37),

$$W_{n,l}(\Theta) = \sqrt{(n+\gamma_l/4)|\Theta - \Theta_l|}\, J_{(\gamma_l - 1)/2}((n+\gamma_l/4)|\Theta - \Theta_l|), \tag{6.24}$$

and

$$\begin{aligned}Z_{n,l}(\Theta) = \sqrt{(n+\gamma_l/4)|\Theta - \Theta_l|} \\ \times J_{(\gamma_l + 1)/2}((n+\gamma_l/4)|\Theta - \Theta_l|)\operatorname{sgn}(\Theta - \Theta_l).\end{aligned} \tag{6.25}$$

Finally, we obtain the following corollary by using Corollary 5 and (1.4) and (1.5).

COROLLARY 8. *Under the hypotheses of Corollary* 5,

$$\begin{aligned}\Phi_{2n-1}(\Theta) = \{2/\varphi(\Theta)\}^{1/2} \\ \times\left\{\cos[n\Theta - \gamma(\varphi;\Theta)] + O\left(\mu_{2n} + n^{-1}|\Theta - \Theta_l|^{-1}\right)\right\},\end{aligned} \tag{6.26}$$

$$\begin{aligned}\Phi_{2n}(\Theta) = \{2/\varphi(\Theta)\}^{1/2} \\ \times\left\{\sin[n\Theta - \gamma(\varphi;\Theta)] + O\left(\mu_{2n} + n^{-1}|\Theta - \Theta_l|^{-1}\right)\right\},\end{aligned} \tag{6.27}$$

as $n \to \infty$, *uniformly for* $\Theta \in [a, b]\setminus(\Theta_l - cn^{-1}, \Theta_l + cn^{-1})$, *where* $\gamma(\varphi, \Theta)$ *and* μ_n *are defined by* (3.6) *and* (1.37).

By applying (1.4), (1.5) and (4.16) (for $m = 1$) it is easy to show that the relations

$$\Phi_{2n-1}(\Theta) = \{2/\varphi(\Theta)\}^{1/2}\{\cos[n\Theta - \gamma(\varphi;\Theta)] + o(1)\} \qquad (n \to \infty),$$

$$\Phi_{2n}(\Theta) = \{2/\varphi(\Theta)\}^{1/2}\{\sin[n\Theta - \gamma(\varphi;\Theta)] + o(1)\} \qquad (n \to \infty),$$

uniformly for $\Theta \in [c, d]$, imply (6.6) uniformly for $\Theta \in [c, d]$. Hence it follows from Lemma 5 that the remarks on the essential nature of the restriction (6.9), made with reference to Theorems 1 and 2 and Corollaries 1–5, apply also to Corollary 8.

Since it can be shown that Corollary 6 $\Rightarrow$ Corollary 7 $\Rightarrow$ Corollary 8, a similar comment applies to Corollaries 6 and 7.

To conclude this subsection we note a corollary that follows from Theorem 1 by (2.17), (2.18), (1.4), (1.5), (2.1), (2.2), (6.19) and (6.20).

COROLLARY 9. *Under the hypotheses of Theorem* 1

$$\Phi_{2n-1}(\Theta) = \{\pi/h(\Theta)\}^{1/2}\{\hat{P}_n^{\alpha,\beta}(\cos\Theta)[\cos\gamma(h;\Theta) + \varepsilon_{1,2n+1}(\Theta)] + \hat{P}_{n-1}^{\alpha+1,\beta+1}(\cos\Theta)\sin\Theta[\sin\gamma(h;\Theta) + \varepsilon_{2,2n-1}(\Theta)]\}, \quad (6.28)$$

$$\Phi_{2n}(\Theta) = \{\pi/h(\Theta)\}^{1/2}\{-\hat{P}_n^{\alpha,\beta}(\cos\Theta)[\sin\gamma(h;\Theta) + \varepsilon_{1,2n}(\Theta)] + \hat{P}_{n-1}^{\alpha+1,\beta+1}(\cos\Theta)\sin\Theta[\cos\gamma(h;\Theta) + \varepsilon_{2,2n}(\Theta)]\}, \quad (6.29)$$

where $\varepsilon_{k,n}(\Theta) = O(n^{-1} + \mu_{h,n}^{\alpha,\beta})$ $(n \to \infty;\ k = 1,2)$, *uniformly for* $\Theta \in \mathbf{R}$, *where* $\gamma(h;\Theta)$ *is defined by* (3.6).

It is easy to verify that (6.9) is also essential for this corollary.

4. Finally we consider the problem of uniform asymptotic representations of the polynomials $p_n(\cos\Theta)$ with a weight function satisfying (1.17)–(1.19). Such representations can be deduced rather easily by means of (1.2) from the corresponding results (above) for $\varphi_n(e^{i\Theta})$ (Theorem 1, Corollaries 3–5) since the weight function defined by the formula

$$\varphi(\tau) = p(\cos\tau)|\sin\tau|, \quad (6.30)$$

can be represented in the form (1.20) by using (1.17)–(1.21). However, we can derive the asymptotic representations for $p_n(\cos\Theta)$ more simply by means of the representation for $\Phi_n(\Theta)$ established in the preceding subsection, using the fact that if φ and p are connected by (6.30), and $\{q_n(t)\}_0^\infty$ is the system of polynomials that are orthogonal on $[-1,1]$ with weight function $q(t) = (1-t^2)p(t)$, then the system $\{\Phi_n(\Theta)\}_0^\infty$ has the form $\Phi_0 = \sqrt{\pi}\,p_0$,

$$\Phi_{2n-1}(\Theta) = \sqrt{\pi}\,p_n(\cos\Theta), \quad \Phi_{2n}(\Theta) = \sqrt{\pi}\,q_{n-1}(\cos\Theta)\sin\Theta \qquad (n \in \mathbf{N}). \quad (6.31)$$

Formulas (6.31) can be verified directly. Their existence was first noticed in [3], although they were not given explicitly. See [4] and [13] for applications of (6.31).

The next corollary follows from Corollary 6 and the first formula (6.31).

COROLLARY 10. *Let the weight function p satisfy* (1.17)–(1.19). *Set* $\delta_0 = 2\beta + 1$, $\delta_{M+1} = 2\alpha + 1$, $x_0 = -1$, $x_{M+1} = 1$, *and* $\overline{\Theta}_\nu = \arccos x_\nu$ $(\nu = 0,1,\ldots,M+1)$. *Choose a point* $\bar{\tau}_\nu \in (\overline{\Theta}_{\nu+1}, \overline{\Theta}_\nu)$ *and set* $G_\nu = [\bar{\tau}_\nu, \bar{\tau}_{\nu+1}]$ $(\nu = 0,1,\ldots,M+1)$, $G_0 = [\bar{\tau}_0, \pi]$, *and* $G_{M+1} = [0, \bar{\tau}_M]$. *Then for each* $l = 0,1,\ldots,M+1$

$$p_n(\cos\Theta) = 2^{\delta_l/4}\{\hat{P}_n^{(\delta_l-1)/2,-1/2}(\cos(\Theta - \overline{\Theta}_l))[\cos\bar{\eta}_{n,l}(\Theta) + O(\bar{\mu}_{2n-1})] + \hat{P}_{n-1}^{(\delta_l+1)/2,1/2}(\cos(\Theta-\overline{\Theta}_l))\sin(\Theta - \overline{\Theta}_l)[\sin\bar{\eta}_{n,l}(\Theta) + O(\bar{\mu}_{2n-1})]\} \times\{f_l(\Theta)\}^{-1/2} \quad (6.32)$$

as $n \to \infty$, *uniformly for* $\Theta \in G_l$, *where* $\bar{\mu}_n$ *denotes the number* (1.37) *for*

$$h(\tau) = C_0 H(\cos\tau), \qquad C_0 = 2^{1+\alpha+\beta+\delta_1+\cdots+\delta_M}, \tag{6.33}$$

$$\bar{\eta}_{n,l}(\Theta) = \arg \pi\left(f_l; e^{i\Theta}\right) - n\bar{\Theta}_l,$$

$$f_l(\Theta) = \left|\sin\left[\left(\Theta - \bar{\Theta}_l\right)/2\right]\right|^{-\delta_l} p(\cos\Theta)|\sin\Theta|.$$

The derivations of the following corollaries are similar to those of Corollaries 7–9.

COROLLARY 11. *With the hypotheses and notation of Corollary* 10 *the asymptotic representation*

$$p_n(\cos\Theta) = \Big\{\left[\cos\bar{\eta}_{n,l}(\Theta) + O(\bar{\mu}_{2n-1})\right] J_{(\delta_l-1)/2}\left((n+\delta_l/4)\left|\Theta - \bar{\Theta}_l\right|\right)$$
$$+\left[\sin\bar{\eta}_{n,l}(\Theta) + O(\bar{\mu}_{2n-1})\right] J_{(\delta_l+1)/2}\left((n+\delta_l/4)\left|\Theta - \bar{\Theta}_l\right|\right)$$
$$\times \operatorname{sgn}\left(\Theta - \bar{\Theta}_l\right)\Big\} s_{n,l}(\Theta) \tag{6.34}$$

holds as $n \to \infty$, *uniformly for* $\Theta \in G_l$, *where*

$$s_{n,l}(\Theta) = \left\{(n+\delta_l/4)\left|\Theta - \bar{\Theta}_l\right|/\varphi(\Theta)\right\}^{1/2}$$

and $\varphi(\Theta)$ *is defined by* (6.30).

COROLLARY 12. *Let the weight function* p *satisfy* (1.17)–(1.19), *let* $\bar{\Theta}_\nu$ $(\nu = 0, 1, \ldots, M+1)$ *be the points defined in Corollary* 10; *and let the interval* $[a, b] \subset (0, \pi)$ *contain the point* $\bar{\Theta}_l$ *but not any other points* $\bar{\Theta}_\nu$. *Then*

$$p_n(\cos\Theta) = \left\{2/\left[\pi\varphi(\Theta)\right]\right\}^{1/2}\Big\{\cos\left[n\Theta - \gamma(\varphi;\Theta)\right]$$
$$+O\left(\bar{\mu}_{2n-1} + n^{-1}\left|\Theta - \bar{\Theta}_l\right|^{-1}\right)\Big\}, \tag{6.35}$$

as $n \to \infty$, *uniformly for* $\Theta \in [a, b] \setminus (\bar{\Theta}_l - cn^{-1}, \bar{\Theta}_l + cn^{-1})$, *where* $\varphi(\Theta)$ *and* $\gamma(\varphi, \Theta)$ *are defined by* (6.30) *and* (3.6) *and* $\bar{\mu}_n$ *has the same meaning as in Corollaries* 10 *and* 11.

COROLLARY 13. *Let the weight function p satisfy* (1.10) *and* (1.18). *Then*

$$p_n(\cos\Theta) = \{h(\Theta)\}^{-1/2}\Big\{\hat{P}_n^{\alpha,\beta}(\cos\Theta)\left[\cos\bar{\eta}_n(\Theta) + \bar{\varepsilon}_{1,n}(\Theta)\right]$$
$$+\hat{P}_{n-1}^{\alpha+1,\beta+1}(\cos\Theta)\sin\Theta\left[\sin\bar{\eta}_n(\Theta) + \bar{\varepsilon}_{2,n}(\Theta)\right]\Big\}, \tag{6.36}$$

where $\bar{\eta}_n(\Theta) = \arg \pi(h; e^{i\Theta})$, $\bar{\varepsilon}_{k,n}(\Theta) = O(\mu_{h,2n-1}^{\alpha,\beta})$ $(n \to \infty; k = 1, 2)$, *uniformly for* $\Theta \in [0, \pi]$, *and* h *and* $\mu_{h,n}^{\alpha,\beta}$ *are defined by* (6.33) *and* (1.28).

Corollary 13 contains all of Bernstein's results quoted in §1 on the asymptotic behavior of generalized Jacobi polynomials on their interval of orthogonality, except in the case $\alpha = \beta = -1/2$.

Corollary 11 has points of contact with the results of [23] and [24], where, for the case of the weight function

$$p(t) = (1-t)^\alpha(1+t)^\beta|t - x_0|^\gamma \qquad (-1 < x_0 < 1;\ \alpha, \beta > -1, \gamma > 0) \tag{6.37}$$

the following formula was announced for $p_n(t)$ (no complete proof was published):

$$p_n(t) \cong c_{1,n}c_2(x_0)\{p(t)\sqrt{1-t^2}\}^{-1/2} J_\nu(b_n g(t))\sqrt{g(t)} \qquad (n \to \infty), \tag{6.38}$$

where $t, x_0 \in [a, b] \subset (-1, 1)$ and

$$\nu = \begin{cases} (\gamma+1)/2 + o(1) & (n \to \infty), p_n(x_0) = 0, \\ (\gamma-1)/2 + o(1) & (n \to \infty), p_n(x_0) \neq 0, \end{cases}$$

$$c_{1,n} = O(n^{1/2}), \qquad b_n = \{n(n+\alpha+\beta+\gamma+1)\}^{1/2},$$

$$c_2(x) = (1-x)^{(\alpha+1)/2}(1+x)^{(\beta+1)/2}, \qquad g(t) = |\arcsin t - \arcsin x_0|.$$

It was pointed out in [25] that this result had never been given a rigorous proof. In [10] it was remarked that [23] and [24] contain several inaccuracies. Here we show that (6.38) is erroneous.

In fact, let $p(t)$ and $\varphi(\tau)$ be defined by (6.37) and (6.30); let $\Theta_0 = \arccos x_0$ and $\Theta \in (0, \Theta_0)$. Then, as is easily verified,

$$\pi(\varphi; e^{i\Theta}) = \{\varphi(\Theta)\}^{-1/2} e^{-i\bar{\gamma}(\Theta)},$$

$$\bar{\gamma}(\Theta) = (\alpha+\beta+\gamma+1)\Theta/2 - (\alpha/2 + 1/4)\pi.$$

By a theorem of Szegö ([1], Theorem 12.1.6)

$$p_n(\cos\Theta) = (2/\pi)^{1/2}\{\varphi(\Theta)\}^{-1/2}\cos[n\Theta + \bar{\gamma}(\Theta)] + o(1) \qquad (n \to \infty). \tag{6.39}$$

On the other hand, if we replace the Bessel function in (6.38) by its asymptotic form according to (2.39) (the possibility of this replacement is justified by the uniform dependence of the remainder term in (2.39) on $\nu \in [\nu_1, \nu_2]$ for given ν_1 and ν_2, which follows from the remarks in [26], §7.3, on the behavior of the remainder terms of the asymptotic expansions of $J_{\pm\nu}(t)$), we find that for $\Theta \in (0, \Theta_0)$, according to (6.38), the polynomial $p_n(\cos\Theta)$ must coincide with one of the expressions

$$\frac{\sqrt{2}\, c_{1,n}c_2(x_0)}{\sqrt{\pi b_n \varphi(\Theta)}} \cos\left[\left(n + \frac{\alpha+\beta+\gamma+1}{2}\right)(\Theta_0 - \Theta) - \frac{\gamma+1}{4}\pi - \frac{\pi}{4}\right] + o(1). \tag{6.40}$$

It follows from (6.39) that $c_{1,n} \asymp \sqrt{n}$ in (6.40).

Let Θ_0/π be irrational and let $\alpha - (\gamma+1)/2 \notin \mathbf{Z}$. Then by the first assumption the sequence

$$u_n = \cos[n\Theta_0/2 + \bar{\gamma}(\Theta_0/2)] \qquad (n \in \mathbf{N})$$

is everywhere dense on $[-1, 1]$, so that there is a subsequence $n_k \uparrow \infty$ for which $u_{n_k} \to 0$ and consequently, by (6.39), $p_{n_k}(\cos\Theta_0/2) = o(1)$ $(k \to \infty)$. But on the other hand, by the condition $\alpha - (\gamma+1)/2 \notin \mathbf{Z}$, neither of the expressions (6.40), with n replaced by n_k and Θ by $\Theta_0/2$, would approach zero as $k \to \infty$.

In conclusion we remark that an analog of (6.38) for a weight function with one-sided singular points was announced in [27] and [28]. However, the method of proof suggested there was based on an incorrect differential equation for $p_n(t)$.

BIBLIOGRAPHY

1. Gabor Szegö, *Orthogonal polynomials*, 3rd ed., Amer. Math. Soc., Providence, R. I., 1967.

2. Ya. L. Geronimus, *Polynomials, orthogonal on a circumference and on an interval. Estiames, asymptotic formulas, orthogonal series*, Fizmatgiz, Moscow, 1958; English transl., *Polynomials orthogonal on a circle and interval*, Pergamon Press, 1960; English transl., *Orthogonal polynomials: Estimates, asymptotic formulas, and series of polynomials orthogonal on the unit circle and on an interval*, Consultants Bureau, New York, 1961.

3. Gabor Szegö, *On bi-orthogonal systems of trigonometric polynomials*, Magyar Tud. Akad. Mat. Kutató Int. Közl. (1963), 255–273 (1964).

4. V. M. Badkov, *Approximations of functions in the uniform metric by Fourier sums of orthogonal polynomials*, Trudy Mat. Inst. Steklov. **145** (1980), 20–62; English transl. in Proc. Steklov Inst. Math. **1981**, no. 1 (145).

5. K. V. Lashchenov,* a) *On a class of orthogonal polynomials*, Leningrad. Gos. Ped. Inst. Uchen. Zap. **89** (1953), 167–189. (Russian)

b) *On interpolation with the roots of orthogonal polynomials of weight* $(1-x^2)^p|x|^q$, Leningrad. Gos. Ped. Inst. Uchen. Zap. **89** (1953), 191–206. (Russian)

6. Serge Bernstein, *Sur les polynomes orthogonaux relatifs à un segment fini*. I, II, J. Math. Pures Appl. (9) **9** (1930), 127–177; (9) **10** (1931), 219–286.

7. Ya. L. Geronimum and B. L. Golinskiĭ, *Asymptotic formulas for orthogonal polynomials*, Teor. Funktsiĭ, Funktsional. Anal. i Prilozhen. Vyp. 1 (1965), 141–163. (Russian)

8. Ulf Grenander and Gabor Szegö, *Toeplitz forms and their applications*, Univ. of California Press, Los Angeles, Calif., 1958.

9. B. L. Golinskiĭ, *On the rapidity of convergence of a sequence of orthogonal polynomials to the limit function*, Ukrain. Mat. Zh. **19** (1967), no. 4, 11–28; English transl. in Ukrainian Math. J. **19** (1967).

10. P. K. Suetin, *V. A. Steklov's problem in the theory of orthogonal polynomials*, Itogi Nauki: Mat. Anal., vol. 15, VINITI, Moscow, 1977, pp. 5–82; English transl. in J. Soviet Math. **12** (1979), no. 6.

11. V. M. Badkov, *The asymptotic behavior of orthogonal polynomials*, Mat. Sb. **109(151)** (1979), 46–59; English transl. in Math. USSR Sb. **37** (1980).

12. B. L. Golinskiĭ, *Asymptotic representation of orthogonal polynomials*, Uspekhi Mat. Nauk **35** (1980), no. 2 (212), 145–196; English transl. in Russian Math. Surveys **35** (1980).

13. V. M. Badkov, *Approximation properties of Fourier series with respect to orthogonal polynomials*, Uspekhi Mat. Nauk **33** (1978), no. 4 (202), 51–106; English transl. in Russian Math. Surveys **33** (1978).

14. B. L. Golinskiĭ, *Polynomials orthogonal on the unit circle with generalized Jacobi weight functions*, Izv. Akad. Nauk Armyan. SSR Ser. Mat. **13** (1978), 87–99. (Russian)

15. V. M. Badkov, *Convergence in the mean and almost everywhere of Fourier series in polynomials orthogonal on an interval*, Mat. Sb. **95 (137)** (1974), 229–262; English transl. in Math. USSR Sb. **24** (1974).

16. B. L. Golinskiĭ, *On V. A. Steklov's problem in the theory of orthogonal polynomials*, Mat. Zametki **15** (1974), 21–32; English transl. in Math. Notes **15** (1974).

17. Ya. L. Geronimus, *Polynomials orthogonal on a circle and their applications*, Khar'kov. Gos. Univ. Uchen. Zap. **24** = Zap. Nauchno-Issled. Inst. Mat. Mekh. i Khar'kov. Mat. Obshch. (4) **19** (1948), 35–120; English transl. in Amer. Math. Soc. Transl. (1) **3** (1962).

18. P. K. Suetin, *Classical orthogonal polynomials*, 2nd ed., "Nauka", Moscow, 1979. (Russian)

19. A. Erdélyi et al., *Higher transcendental functions*. Vol. 1, McGraw-Hill, 1953.

20. A. Zygmund, *Trigonometric series*, 2nd rev. ed., Vol. 1, Cambridge Univ. Press, 1959.

21. I. A. Ibraginov and V. N. Solev, *Asymptotic behavior of the prediction error for a stationary sequence with spectral density of a special form*, Teor. Veroyatnost. i Primenen. **13** (1968), 746–751; English transl. in Theor. Probab. Appl. **13** (1968).

22. P. K. Suetin, *Asymptotic properties of polynomials orthogonal on a circle with algebraic singularities of the weight function*, All-Union Sympos. Theory of Approximation of Functions in the Complex Domain, Abstracts of Reports, Bashkir. Gos. Univ., Ufa, 1980, pp. 132–133. (Russian)

23. V. P. Konoplev, *On polynomials orthogonal with respect to weight functions that vanish or become infinijte at isolated points of the interval of orthogonality*, Dokl. Akad. Nauk SSSR **141** (1961), 781–784; English ransl. in Soviet Math. Dokl. **2** (1961).

24. ______, *On polynomials orthogonal with respect to weight functions that vanish or become infinite at isolated points of the interval of orthogonality*, Saratov. Gos. Ped. Inst. Uchen. Zap. Mat. **1962**, 70–75. (Russian)

**Editor's note*. The Russian gives the title of a) and the page span of b).

25. V. M. Badkov, *Boundedness in the mean of orthonormal polynomials*, Mat. Zametki **13** (1973), 759–770; English transl. in Math. Notes **13** (1973).

26. G. N. Watson, *A treatise on the theory of Bessel functions*, 2nd ed., Cambridge Univ. Press, Cambridge, and Macmillan, New York, 1944.

27. V. P. Konoplev, *On the asymptotic behavior of orthogonal polynomials at one-sided singular points of the weight functions (algebraic singularities)*, Dokl. Akad. Nauk SSSR **160** (1965), 997–1000; English transl. in Soviet Math. Dokl. **6** (1965).

28. ______, *On the asymptotic behavior of orthonormal polynomials at one-sided singular points of the weight functions (algebraic singularities)*, Volgh. Mat. Sb. Vyp. 4 (1966), 93–102. (Rusisan)

Translated by H. P. BOAS

Труды Матем. Инст. Стеклов.
164(1983)

Proceedings of the Steklov
Institute of Mathematics
1985, Issue 2

GIBBS CONSTANTS FOR THE PARTIAL SUMS OF FOURIER-WALSH SERIES AND THEIR $(C, 1)$ MEANS

UDC 517.52

L. A. BALASHOV AND V. A. SKVORTSOV

ABSTRACT. The Gibbs phenomenon is studied for series with respect to the Walsh system.

It is shown that the value of the Gibbs constant for the partial sums of Fourier-Walsh series depends on the distribution of the points of discontinuity of the function and, for binary-irrational points, varies between $4/3$ and $3/2$. The value $3/2$ occurs almost everywhere, but the lower bound is attained for certain points (for example, for $x = 1/3$).

It is proved that the Gibbs phenomenon for the arithmetic means of the partial sums of Fourier-Walsh series occurs for almost all points of $[0, 1]$. Here the upper bound of the Gibbs constant equals $1 + (12 + 2\sqrt{34})^{-1}$ and is attained for almost all binary-irrational points. However, there are binary-irrational points (for example, $x = 1/3$) for which the Gibbs phenomenon does not occur for the $(C, 1)$ means of Fourier-Walsh series.

Bibliography: 6 titles.

In this paper we study the Gibbs constants for Fourier series with respect to the Walsh system, both for ordinary convergence and for $(C, 1)$ summability. Since, as we shall see in what follows, the size of the constants depends on the point at which the Gibbs phenomenon is considered, we shall use the term "Gibbs functions."

We recall the corresponding definitions.

Let x_0 be an isolated discontinuity of $f(x)$, a function of bounded variation in a neighborhood of x_0, let $|f(x_0 + 0) - f(x_0 - 0)| = 2d$, and let $\{S_n(x)\}$ be the sequence of partial sums of the Fourier series of $f(x)$ or, more generally, any sequence of functions converging to $f(x)$ at every point of a deleted neighborhood of x_0. Then the value of the Gibbs function for $\{S_n(x)\}$ at x_0 is defined by

$$l(x_0) = \overline{\lim_{\substack{x \to x_0 \\ n \to \infty}}} \frac{1}{d}\left|S_n(x) - \frac{f(x_0 + 0) + f(x_0 - 0)}{2}\right|.$$

If $l(x_0) > 1$, we say that $\{S_n(x)\}$ exhibits the Gibbs phenomenon at x_0.

The Gibbs phenomenon has been extensively studied for Fourier series with respect to the trigonometric system (see, for example, [1]).

The existence of the Gibbs phenomenon for Fourier series with respect to Walsh system was established by Zubakin in [2], where it was shown that it occurs at every binary irrational (the absence of the Gibbs phenomenon at binary-rational points is evident).

In the first section of this paper we show that the Gibbs function for the partial

---

1980 *Mathematics Subject Classification*. Primary 42C10.

sums of a Fourier series with respect to the Walsh system is not constant on the set of binary irrationals, and we find the exact upper and lower bounds of this function; it turns out that the upper bound is attained almost everywhere.

More precisely, the fundamental results obtained in §1 are as follows.

THEOREM 1. *The Gibbs function for the partial sums of a Fourier series with respect to the Walsh system satisfy, at every binary irrational* x, *the inequality* $4/3 \leqslant l(x) \leqslant 3/2$, *and* $l(x) = 3/2$ *almost everywhere. Moreover, there are binary irrationals (for example,* $x = 1/3$*) at which* $l(x) = 4/3$.

In §2 we consider Fourier-Walsh series summed by the method of arithmetic means. We show that the Gibbs phenomenon for the $(C, 1)$ means of the partial sums of a Fourier series with respect to the Walsh system fails to occur at some binary irrationals, but does occur at others. Moreover, the set of points where the Gibbs phenomenon occurs has full measure.

These results make up the content of the following theorem.

THEOREM 2. *The Gibbs phenomenon for the* $(C, 1)$ *means of the partial sums of a Fourier series with respect to the Walsh system occurs almost everywhere. The Gibbs function* $l_C(x)$ *for these means satisfies the inequality* $1 \leqslant l_C(x) \leqslant 1 = [12 + 2\sqrt{34}\,]^{-1}$, *and* $l_C(x) = 1[12 + 2\sqrt{34}\,]^{-1}$ *almost everywhere. Moreover, there are binary irrationals (for example,* $x = 1/3$*) at which* $l_C(x) = 1$, *i.e. the Gibbs phenomenon does not occur.*

The Walsh system is usually defined on the group G of sequences $x = \{x_0, x_1, x_2, \ldots\}$ of zeros and ones with the group operation $x \dot{=} y$ defined as coordinatewise addition module 2. The group G can be mapped in a one-to-one way on the "modified" interval [0, 1] whose elements are binary expansions of the form

$$\tilde{x} = \sum_{i=0}^{\infty} x_i 2^{-i-1},$$

where the two different expansions of binary rational points are considered as different points of the interval.

In what follows we shall not distinguish between elements of the group and the corresponding points of the modified interval. Since Haar measure on G coincides with the Lebesgue measure of the corresponding sets of the interval (see [3]), integration over subsets of G will be identified with ordinary integration over subsets of the interval.

The set of points x of G for which the first n coordinates $x_0, x_1, \ldots, x_{n-1}$ are given will be called the interval

$$\Delta_p^n = \left[\frac{p}{2n}, \frac{p+1}{2^n}\right],$$

where $0 \leqslant p \leqslant 2^n - 1$, $n = 0, 1, 2, \ldots,$ and $p = \sum_{i=0}^{n-1} x_i 2^{n-i-1}$.

We denote the set of expansions of binary irrationals by J.

The system of Walsh functions (in Paley's indexing) is defined for each $k \geqslant 0$ by the equation

$$w_k(x) = (-1)^{\sum_{i+0}^{\infty} k_i x_i},$$

where $k = \sum_0^n k_i 2^i$, $k_i = 0$ or 1.

1. Here we study the Gibbs phenomenon for the partial sums of Fourier series with respect to the Walsh system. The partial sums of the Fourier series of $f(x)$ can be represented in the form

$$S_k(x) = \int_0^1 D_k(t) f(x+t)\,dt,$$

where $D_k(x) = \sum_0^{k-1} w_j(x)$ is the Dirichlet kernel for the Walsh system.

We recall that the Walsh function $w_k(x)$ is constant on each interval Δ_p^n for $2^{n-1} \leqslant k < 2^n$, and the kernel $D_k(x)$ is constant on Δ_p^n for $2^{n-1} < k \leqslant 2^n$.

We shall need a well-known inequality for the Dirichlet kernel (see [4]):

$$|D_k(x)| \leqslant 1/x, \qquad x \neq 0,\ k = 1, 2, \ldots, \tag{1}$$

where x is to be interpreted as an ordinary point of [0, 1].

We also observe that

$$\Delta_{2^n}(x) = \begin{cases} 2^n & \text{for } x \in \Delta_0^n, \\ 0 & \text{for } x \notin D_0^n. \end{cases} \tag{2}$$

For $k = 2^n + m$, $1 \leqslant m \leqslant 2^n$,

$$D_k(x) = D_{2^n}(x) + w_{2^n} D_m(x), \tag{3}$$

whence, in particular, it follows easily that

$$D_{2^n+2^{n-1}}(x) = \begin{cases} 2^n + 2^{n-1} & \text{for } x \in \Delta_0^{n+1}, \\ 2^{n-1} & \text{for } x \in \Delta_1^{n+1} \cup \Delta_2^{n+1}, \\ -2^{n-1} & \text{for } x \in \Delta_3^{n+1}, \\ 0 & \text{for } x \notin \Delta_0^{n-1}. \end{cases} \tag{4}$$

As for the trigonometric system, the study of the Gibbs phenomenon in the general case reduces to the consideration of functions with a single discontinuity of the first kind (see [1]). For the Walsh system, it is convenient to consider $s_a(s) = \operatorname{sgn}(x - a)$ as the simplest such function.

Let us study the behavior of the partial sums of the Fourier series of this function.

THEOREM 3. *Let $S_k(x)$ be the partial sums of the Fourier series of $s_a(x)$ with respect to the Walsh system, where $a \in J$. Then there are a sequence of points $\{x_i\}$, $x_i \to a$, and a sequence of positive integers k_i, $k_i \to \infty$, such that*

$$\left|S_{k_i}^{(x_i)}\right| \geqslant 4/3.$$

PROOF. Since $a \in J$, the expansion $a = \sum_0^\infty a_i/2^{i+1}$ contains infinitely many changes in the value of a_i, for example from 0 to 1.

Let $a_{n-1} = 0$, $a_n = 1$. Then for some s, $0 \leqslant s < 2^{n-1}$, we have $a \in \Delta_{4s+1}^{(n+1)} \subset \Delta_{2s}^{(n)}$. There are two possible cases:

$$1)\ \ a \leqslant \sum_{i=0}^{n} \frac{a_i}{2^{i+1}} + \frac{1}{3}\frac{1}{2^{n+1}}, \qquad 2)\ \ a > \sum_{i=0}^{n} \frac{a_i}{2^{i+1}} + \frac{1}{3}\frac{1}{2^{n+1}}.$$

In case 1) we consider the values of the sums $S_{2^n+2^{n-1}}(x)$ at the point $x = a \dotplus 2^{-n} = a + 2^{-n} \in \Delta_{4s+3}^{(n+1)} \subset \Delta_{2s+1}^{(n)}$. We observe that when $t \in \Delta_0^n$ we have $(a \dotplus 2^{-n} \dotplus t) \in \Delta_{2s+1}^n$ and therefore $s_a(a \dotplus 2^{-n} \dotplus t) = 1$. When $t \in \Delta_2^{(n+1)}$, we have $(a \dotplus 2^{-n} \dotplus t) \in \Delta_{4s+1}^{(n+1)}$ and, by 1), we obtain

$$\operatorname{meas} E_1^{(+)} \equiv \operatorname{meas}\{t; t \in \Delta_2^{(n+1)}; s_a(a + 2^{-n} + t) = 1\} \geqslant 2^{-n}/3,$$
$$\operatorname{meas} E_1^{(-)} \equiv \operatorname{meas}\{t; t \in \Delta_2^{(n+1)}; s_a(a + 2^{-n} + t) = -1\} \leqslant 2^{-(n+1)}/3.$$

When $t \in \Delta_3^{(n+1)}$ we have $(a \dotplus 2^{-n} \dotplus 1) \in \Delta_{4s}^{(n+1)}$ and therefore $s_a(a \dotplus 2^{-n} \dotplus t) = -1$. Then, taking account of (4), we obtain

$$\begin{aligned} S_{2^n+2^{n-1}}(a + 2^{-n}) &= \int_0^1 D_{2^n+2^{n-1}}(t) s_a(a + 2^n + t)\, dt \\ &= \int_{\Delta_2^{(n)}} + \int_{\Delta_2^{(n+1)}} + \int_{\Delta_3^{(n+1)}} D_{2^n+2^{n-1}}(t) s_a(a + 2^{-n} + t)\, dt \\ &= \int_{\Delta_0^{(n)}} + \int_{E_1^{(+)}} + \int_{E_1^{(-)}} + \int_{\Delta_3^{(n+1)}} \\ &\geqslant 1 + \frac{2^n}{3 \cdot 2^{n+1}} - \frac{2^{n-1}}{3 \cdot 2^{n+1}} + \frac{2^{n-1}}{2^{n+1}} = \frac{4}{3}. \end{aligned}$$

In case 2) we have

$$a > \sum_{i=0}^{n-2} \frac{a_i}{2^{i+1}} + \frac{1}{3 \cdot 2^n}. \tag{5}$$

There are two possibilities: 2α) $a_{n-2} = 0$, and 2β) $a_{n-2} = 1$. In case 2α), we have $(a \dotplus 2^{-n} \dotplus 2^{-(n-1)} \dotplus t) > a$ for $t \in \Delta_0^{(n-1)}$ and therefore $s_a(a \dotplus 2^{-n} \dotplus 2^{-n+1} \dotplus t) = 1$. For $t \in \Delta_2^{(n)}$ we have $(a \dotplus 2^{-n+1} \dotplus 2^{-n} \dotplus t) \leftarrow \Delta_{2s+1}^{(n)}$ and therefore

$$s_a(a \dotplus 2^{-n} \dotplus 2^{-n+1} \dotplus t) = 1.$$

For $t \in \Delta_3^{(n)}$ we have $(a \dotplus 2^{-n} \dotplus 2^{-n+1} \dotplus t) \in \Delta_{2s}^{(n)}$ and then, taking account of (5), we have

$$\operatorname{meas} E_2^{(+)} = \operatorname{meas}\{t; t \in \Delta_2^{(n)}; s_a(a + 2^{-n+1} + 2^{-n} + t) = 1\} \leqslant 2^{-n}/3,$$
$$\operatorname{meas} E_2^{(-)} = \operatorname{meas}\{t; t \in \Delta_3^{(n)}; s_a(a + 2^{-n+1} + 2^{-n} + t) = -1\} \geqslant 2^{-n+1}/3.$$

Hence, using (4) with n replaced by $(n - 1)$, we obtain

$$\begin{aligned} S_{2^{n-1}+2^{n-2}}(a + 2^{-n+1} + 2^{-n}) &= \int_0^1 D_{2^{n-1}+2^{n-2}}(t) \cdot s_a(a + 2^{-n} + 2^{-n+1} + t)\, dt \\ &= \int_{\Delta_0^{(n-1)}} + \int_{\Delta_2^{(n)}} + \int_{E_2^{(+)}} + \int_{E_2^{(-)}} \\ &\geqslant 1 + \frac{2^{n-2}}{2^n} + \frac{2^{n-2}}{2^n}\frac{2}{3} - \frac{1}{3}\frac{2^{n-2}}{2^n} = \frac{4}{3}. \end{aligned}$$

In case 2β) we have to consider the value of $S_{2^{n-1}=2^{n-2}}(x)$ at the point $a \dotplus 2^{-n+1}$. Here calculations similar to those in case 2α) lead to the inequality

$$S_{2^{n-1}+2^{n-2}}(a + 2^{-n+1}) \leqslant -4/3.$$

To complete the proof of the theorem it now remains only to recall that the situation under consideration, that in which the value of a_i changes from 0 to 1, recurs, in accordance with hypothesis, infinitely often as i increases. This lets us find the required sequence.

This completes the proof of the theorem.

The theorem provides a lower estimate for the Gibbs function.

THEOREM 4. *The partial sums of the Fourier-Walsh series of $s_a(x)$ satisfy the inequality $|S_k(x)| \leqslant 3/2$ for all x and k.*

PROOF. Let $k = 2^n + m$ and $1 \leqslant m \leqslant 2^n$. We observe that $0 \leqslant D_m(t) \leqslant 2^n$ for $t \in \Delta_0^{(n)}$. Hence, by (2) and (3), $D_k(t) \geqslant 0$ for $t \in \Delta_0^{(n)}$, whence

$$\int_{\Delta_0^{(n)}} |D_k(t)|\, dt = \int_{\Delta_0^{(n)}} D_k(t)\, dt = 1. \tag{6}$$

It is also clear from (2) and (3) that

$$\int_{\Delta_i^{(n)}} D_k(t)\, dt = 0 \quad \text{for } 1 \leqslant k \leqslant 2^n - 1. \tag{7}$$

It follows from (1) and from the constancy of $D_m(t)$ on the intervals $\Delta_i^{(n)}$ that $|D_m(t)| \leqslant 2^{n-1}$ for $t \in [0,1] \setminus \Delta_0^{(n)}$. Hence, using (2) and (3), we have

$$|D_k(t)| \leqslant 2^{n-1} \quad \text{for } t \in [0,1] \setminus \Delta_0^{(n)}. \tag{8}$$

Let $(x \dotplus a) \in \Delta_{i_0}^{(n)}$. If t takes all values on $\Delta_i^{(n)}$ for some $i \neq i_0$ then $(x \dotplus t)$ takes values on an interval on which $s_a(x \dotplus t)$ is constant. Therefore by (7)

$$\int_{\Delta_i^{(n)}} D_k(t) s_a(x \dotplus t)\, dt = 0 \quad \text{for } 1 \leqslant i \leqslant 2^n - 1,\ i \neq i_0. \tag{9}$$

Hence, if we use (6), (8) and (9), we obtain, if $i_0 \neq 0$,

$$\left|\int_0^{\Delta} D_k(t) s_a(x \dotplus t)\, dt\right| = \left|\int_{\Delta_0^n}\right| + \left|\int_{\Delta_{i_s}^n}\right| + \sum_{\substack{i=1 \\ i \neq i_0}} \left|\int_{\Delta_i^n}\right|$$

$$\leqslant \int_{\Delta_0^n} |D_k(t)| dt + \int_{|\Delta_{i_o}^n} |D_k(t)| dt \leqslant 1 + \frac{2^{n-1}}{2^n} = \frac{3}{2}.$$

If $i_0 = 0$ the upper bound is only improved, since we do not have to divide the integral over $\Delta_{i_0}^n$ into separate terms.

This completes the proof of the theorem.

If we analyze the proof of the preceding theorem it is easy to see that the estimate for the integral $\int_{\Delta_{i_0}^n}$ can be made precise for certain values of a, for example for $a = 1/3$. Taking account of the change of sign of $s_a(x \dotplus t)$ on this interval, we obtain

$$\left|\int_{\Delta_{i_0}^n}\right| \leqslant 2^{n-1}\left(\frac{1}{2^{n+1}s} + \frac{1}{3}\frac{1}{2^{n+1}}\right) = \frac{1}{3},$$

whence $|S_k(x)| \leqslant 4/3$. This inequality, together with Theorem 3, shows that the following theorem holds.

THEOREM 5. *For the partial sums of the Fourier-Walsh series of $s_a(x)$ for $a = 1/3$ we have*

$$\overline{\lim_{\substack{k \to \infty \\ x \to a}}} |S_k(x)| = 4/3.$$

Of course the conclusion of Theorem 5 is valid not only for $a = 1/3$ but also for an infinite set of points. However, as we shall now see, a different result holds for almost all values of a.

LEMMA 1. *Let the binary expansion of the number a be*

$$a = \sum_{i=0}^{\infty} \frac{a_i}{2^{i+1}}, \qquad a_{n-1} = 0,\, a_n = 1,\, a_{n+1} = \cdots + a_{n+r-1} = 0\ (r \geqslant 1).$$

Then the partial sums of the Fourier-Walsh series of $s_a(x)$ satisfy

$$\left| s_{2^n+2^{n-1}}\left(a + \frac{1}{2^n}\right)\right| \geqslant \frac{3}{2} - \frac{1}{2^r}.$$

PROOF. If $t \in \Delta_0^n$ then $a \dotplus 1/2^n \dotplus t$ and $a \dotplus 1/2^n$ belong to the same Δ_i^n, different from that containing a. Hence since $a \dotplus 1/2^n = a + 1/2^n > a$ we see that $a \dotplus 1/2^n \dotplus t > a$ for all $t \in \Delta_0^n$ and therefore $s_a(a \dotplus 1/2^n \dotplus t) = 1$. If $t \in \Delta_3^{n+1}$ then $a \dotplus 1/2^n \dotplus t < a$ and therefore $s_a(a \dotplus 1/2^n \dotplus t) = -1$. Consider the set

$$E_3^{(-1)} = \left\{ t \in \Delta_2^{n+1},\, s_a\left(a + \frac{1}{2^n} \dotplus t\right) = -1\right\}.$$

It follows easily from the hypotheses of the lemma that

$$a < \sum_{i=0}^{n-1} \frac{a_i}{2^{i+1}} + \frac{1}{2^{n+r}},$$

whence we find that meas $E_3^{(-1)} < 1/2^{n+r}$.

Now we have, using (4),

$$\begin{aligned} S_{2^n \dotplus 2^{n-1}}\left(a \dotplus \frac{1}{2^n}\right) &= \int_0^1 D_{2^n+2^{n-1}}(t) s_a\left(a \dotplus \frac{1}{2^n} \dotplus t\right) dt \\ &= \int_{\Delta_0^n} |D_{2^n+2^{n-1}}(t)|\, dt + \int_{\Delta_2^{n+1}} |D_{2^n+2^{n-1}}(t)|\, dt \\ &\quad + \int_{\Delta_3^{n+1}} |D_{2^n+2^{n-1}}(t)|\, dt - 2\int_{E_3^{(-1)}} |D_{2^n+2^{n-1}}(t)|\, dt \\ &\geqslant \frac{3}{2} - 2 \cdot 2^{n-1} \frac{1}{2^{n+r}} = \frac{3}{2} - \frac{1}{2^r}. \end{aligned}$$

This completes the proof of the lemma.

We shall need the following proposition, which is a corollary of results of Borel ([5], pp. 37–47).

LEMMA 2. *Let $\{\lambda_j\}$ and $\{n_j\}$ be sequences of positive integers with $n_{j+1} \geqslant n_j + \lambda_j$. Let there be given an infinite sequence of finite sets of zeros and ones for which the jth set has λ_j elements. Then the condition $\sum_{j=1}^{\infty} 2^{-\lambda_j} = \infty$ is necessary and sufficient for almost every binary expansion $0.a_0a_1a_2 \cdots$ to have the property that a_{n_j} belongs to the jth set for infinitely many values of j.*

This lemma, together with Lemma 1 and Theorem 4, establishes the following theorem.

THEOREM 6. *The partial sums of the Fourier-Walsh series of* $s_a(x)$ *satisfy*

$$\varlimsup_{\substack{x\to a\\ k\to\infty}} |S_k(x)| = 3/2$$

for almost all a.

Theorem 1, stated in the Introduction, is an immediate consequence of Theorems 3–6.

2. Here we consider the behavior of the arithmetic means $\sigma_k(x)$ of the partial sums of the Fourier-Walsh series of $s_a(x)$.

It is known that

$$\sigma_k(x; f) = \int_0^1 f(x \dot{+} t) K_k(t)\, dt,$$

where $K_k(t) = (1/k)\Sigma_i^k D_i(t)$ is the Fejér-Walsh kernel.

We recall some properties of $K_k(t)$ and prove some auxiliary lemmas. In what follows we simplify the formulas by systematically using the notation

$$F = [0,1] \Bigg\backslash \left(\Delta_0^n \cup \bigcup_{r=0}^{n-1} \Delta_{2^r}^n \right).$$

The Fejér-Walsh kernel with index 2^n has the form (see [6])

$$K_{2^n}(t) = \begin{cases} 2^{n-1} + 1/2 & \text{for } t \in \Delta_0^n; \\ 2^{n-r-2} & \text{for } t \in \Delta_{2^r}^n,\ r = 0, 1, \dots, n-1; \\ 0 & \text{for } t \in F. \end{cases} \tag{10}$$

If $k = 2^n + m$, $1 \leqslant m \leqslant 2^n$, then

$$kK_k(t) = 2^n K_{2^n}(t) + mD_{2^n}(t) + w_{2^n}(t) \sum_{i=1}^{ms} D_i(t). \tag{11}$$

LEMMA 3. *For* $t \in \Delta_1^s$, $s = 1, \dots, n$,

$$\left| \sum_{i=1}^{m} D_i(t) \right| \leqslant 2^n \cdot 2^{s-2} \tag{12}$$

for $1 \leqslant m \leqslant 2^n$.

PROOF. It was shown in [4] that the Dirichlet-Walsh kernel satisfies

$$D_{j\cdot 2^s}(t) = 0, \qquad \left| D_{j\cdot 2^s + 2^{s-1}}(t) \right| = 2^{s-1} \quad \text{for all } j \geqslant 0,\ t \in \Delta_1^s. \tag{13}$$

It follows easily from (13) and the fact that $|w_k(t)| = 1$ for all k and t that

$$\begin{aligned} \left| D_{j\cdot 2^s + i}(t) \right| &= i \quad \text{for } 0 \leqslant 2^{s-1},\ t \in \Delta_1^s; \\ \left| D_{j\cdot 2^s + 2^{s-1} + is}(t) \right| &= 2^{s-1} - i \quad \text{for } 0 \leqslant i \leqslant 2^{s-1},\ t \in \Delta_1^2. \end{aligned} \tag{14}$$

Fix s, $1 \leqslant s \leqslant n$, and notice that $m \leqslant 2^s \cdot 2^{n-s}$. Then if we apply (14) we obtain, for $t \in \Delta_1^s$,

$$\left| \sum_{i=1}^{m} D_i(t) \right| \leqslant \sum_{j=0}^{2^{n-s}-1} \sum_{i=1}^{2^s} \left| D_{j\cdot 2^s + i} j(t) \right| = 2^{n-s} \left(\sum_{i=1}^{2^{s-1}} i + \sum_{i=1}^{2^{s-1}-1} i \right)$$
$$= 2^{n-s} \cdot 2^{s-2} = 2^n \cdot 2^{s-2},$$

i.e. we obtain (12).

LEMMA 4. *The inequality* $K_{2^n+m}(t) \geqslant 0$ *is valid for* $t \in \Delta^n_{2^r}$ $r = 0, 1, \ldots, n-1$, *and* $1 \leqslant m \leqslant 2^n$.

PROOF. If we observe that $\Delta^n_{2^r} \subset \Delta^{n-r}_1$ and that according to (10) $2^n \cdot K_{2^n}(t) = 2^{2n-r-2}$ for $t \in \Delta^n_{2^r-1}$, we obtain

$$\left|\sum_{i=1}^{m} D_i(t)\right| \leqslant 2^n K_{2^n}(t).$$

from Lemma 3 for $s = n - r$. The conclusion of the lemma now follows from (10), (11) and the preceding inequality.

LEMMA 5. *For* $2^n \leqslant k \leqslant 2^{n+1}$

$$\int_{\Delta^n_i} kK_k(t)\,dt = \int_{\Delta^n_i} 2^n K_{2^n}(t)\,dt \tag{15}$$

for $i = 1, \ldots, 2^n - 1$. *In particular*, $\int_{\Delta^n_i} kK_k(t)\,dt = 0$ *for* $\Delta^n_i \subset F$.

PROOF. Applying (11), we obtain

$$\int_{\Delta^n_i} kK_k(t)\,dt = \int_{\Delta^n_i} 2^n K_{2^n}(t)\,dt + m\int_{\Delta^n_i} D_{2^n}(t)\,dt$$
$$+ \int_{\Delta^n_i} w_{2^n}(t) \sum_{i=1}^{m} D_i(t)\,dt = J_1 + J_2 + J_3. \tag{16}$$

Since $i \geqslant 1$, and $D_{2^n}(t) = 0$ for $t \notin \Delta^n_0$, we have $J_2 = 0$. Now we observe that it follows from the definition of the Walsh functions that the function $\sum_1^m D_i(t)$ is constant on Δ^n_i and that $w_{2^n}(t)$ is orthogonal to 1 on Δ^n_i. Hence $J_3 = 0$. Then (15) follows from this and (16).

We denote by $f(\Delta)$ the constant value taken by f on Δ.

LEMMA 6. *For a given* $n \geqslant 2$, *the maximum of* $|K_{2^n+m}(t)|$ *for* $t \in F$ *and* m *on the interval* $1 \leqslant m \leqslant 2^n$ *is attained for* $t \in \Delta^n_3$ *and* m_n *such that* $2^{n-2} < m < 2^{n-1}$. *Here*

$$A_n = \left|K_{2^n+m_n}(\Delta^n_3)\right| = \frac{2^{n-1}}{6+\sqrt{34}} + o(2^n). \tag{17}$$

PROOF. It follows from (10) and (11) that $K_{2^n}(t) = D_{2^n}(t) = 0$ for $t \in F$, and therefore

$$|K_{2^n+m}(t)| = \left|\frac{\sum_1^m D_i(t)}{2^n + m}\right|. \tag{18}$$

We first determine the maximum of this quantity for m in $1 \leqslant m \leqslant 2^n$ and $t \in \Delta^n_3$. Noticing that $\Delta^n_3 \subset \Delta^{n-1}_1$, we apply (14) for $s = n-1$ and find that $|K_{2^n+m}(\Delta^n_3)|$ increases monotonically with m if $1 \leqslant m \leqslant 2^{n-2}$. In addition, it is easy to see from (3) that $D_{2^{n-1}+j}(t) = -D_j(t)$ on Δ^n_3 for $1 \leqslant j \leqslant 2^{n-1}$, i.e. $D_i(\Delta^n_3)$ changes sign on crossing the point $i = 2^{n-1}$. Hence, taking account of (14), we conclude that the sum $|\sum_1^m D_i(\Delta^n_3)|$, and hence also $|K_{2^n+m}(\Delta^n_3)|$, decreases as m increases for $2^{n-1} \leqslant m \leqslant 2^n$.

It follows from what we have said that the maximum of (18) over m is attained for m on the interval $2^{n-2} \leqslant n \leqslant 2^{n-1}$. Using (14) again, we find that (18) equals

$$\frac{2^{2n-4} + 2^{n-2} + 2^{n-1}u - u^2 - u}{2(2^n + 2^{n-2} + u)} \tag{19}$$

on this interval, where $u = m - 2^{n-2}$ and therefore $0 \leqslant n \leqslant 2^{n-2}$.

Studying the function (19), we find that it attains its maximum strictly inside the specified interval, for some u_n, $0 < u_n < 2^{n-2}$, and that

$$|K_{2^n+m_n}(\Delta_3^n)| = \left[\frac{2^{n-1}}{6+\sqrt{34}} + o(2^n)\right] \equiv A_n$$

for $m_n = 2^{n-2} + u_n$.

Now we consider (18) for t in the other intervals Δ_j^n contained in the set F.

As we have already noticed, all $D_i(t)$ have constant sign on Δ_j^n for $1 \leqslant i \leqslant 2^{n-1}$. Hence if we use the monotonicity of $|D_i(t)|$ with respect to t (see [4]), we obtain

$$\left|\sum_{i=1}^m D_i(\Delta_j^n)\right| \leqslant \sum_{i=1}^m |D_i(\Delta_j^n)| \leqslant \sum_{i=1}^m |D_i(\Delta_3^n)| = \left|\sum_{i=1}^m D_i(\Delta_3^n)\right|$$

for $1 \leqslant m \leqslant 2^{n-1}$, from which it follows that for these values of m

$$|K_{2^n+m}(\Delta_j^n)| \leqslant A_n. \tag{20}$$

For $2^{n-1} \leqslant m \leqslant 2^n$ we use Lemma 3. Noticing that

$$|K_{2^n+2^{n-1}}(\Delta_3^n)| = \frac{2^{2n-4}}{2^n + 2^{n-1}},$$

and finding, for $\Delta_j^n \subset F$, the interval Δ_1^s, $s \leqslant n-2$, for which $\Delta_j^n \subset \Delta_1^s$, we obtain

$$|K_{2^n+m}(\Delta_j^n)| = \frac{|\Sigma_1^m D_i(\Delta_j^n)|}{2^n+m} \leqslant \frac{2^{n+s-2}}{2^n+m} \leqslant \frac{2^{2n-4}}{2^n+2^{n-1}} = |K_{2^n+2^{n-1}}(\Delta_3^n)| \leqslant A_n.$$

Hence we have proved that (20) is satisfied for all m, $1 \leqslant m \leqslant 2^n$, and all $\Delta_j^n \subset F$.

This completes the proof of the lemma.

Now we prove the following theorem.

THEOREM 7. *For every $a \in J$ the means $\sigma_k(x)$ of the Fourier-Walsh series of $s_a(x)$ satisfy*

$$|\sigma_k(x)| \leqslant 1 + \frac{1}{2(6+\sqrt{34})} + o(1) \tag{21}$$

for all $x \in [0,1]$, where the term of $o(1)$ tends to 0, uniformly in x, as $k \to \infty$.

PROOF. Let $2^n \leqslant k \leqslant 2^{n+1}$ and $(a \dot{-} x) \in \Delta_i^n$. If $\Delta_{i_0}^n \subset CF$, where CF is the complement of F with respect to $[0,1]$, then $s_a(x \dot{+} t)$ has constant sign for $t \in \Delta_i^n$, where Δ_i^n is any interval contained in F. Hence it follows, taking account of Lemma 5, that for any interval $\Delta_i^n \subset F$ we have

$$\left|\int_{\Delta_i^n} K_k(t) s_a(x \dot{+} t)\, dt\right| = \left|\int_{\Delta_i^n} K_k(t)\, dt\right| = 0.$$

From this and Lemma 4 we find that

$$\begin{aligned}|\sigma_k(x)k| &\leqslant \left|\int_0^1 K_k(t) s_a(x \dot{+} t)\, dt\right| \leqslant \int_{CF} |K_k(t)|\, dt + \left|\int_F K_k(t)\, dt\right| \\ &= \int_0^1 K_k(t)\, dt = 1 \end{aligned} \tag{22}$$

for $\Delta_{i_0}^n \subset CF$. Hence in this case we obtain (21) automatically.

Now let $\Delta^n_{i_0} \subset F$. Then, using Lemmas 4, 5, and 6, we obtain

$$|\sigma_k(x)| \leqslant \int_{CF} |K_k(t)\ dt|\ dt + \sum_{\Delta^n_i \subset F} \left| \int_{\Delta^n_i} K_k(t)\ dt \right|$$
$$+ \int_{\Delta^n_{i_0}} |K_k(t)|\ dt \leqslant 1 + \left[2(6+\sqrt{34})\right]^{-1} + o(1).$$

This completes the proof of the theorem.

THEOREM 8. *Let $a = 1/3 = 0.(01)$. Then the Gibbs phenomenon does not occur for the means $\sigma_k(x)$ of the Fourier-Walsh series of $s_a(x)$.*

PROOF. We may suppose without loss of generality that $x > a$. Let r satisfy $(x \dot{-} a) \in \Delta^r_1 \subset \Delta^{r-1}_0$. Such an r is determined uniquely by x. Let $k = 2^n + m$. If $r \geqslant n$, we have either $(x \dot{-} a) \in \Delta^n_1$ or $(x - a) \in \Delta^n_0$ (for $r > n$). In this case inequality (22) holds for $\sigma_k(x)$.

Now let $r < n$. In this case we consider

$$k\sigma_k(x) = \int_{\Delta^r_0} + \int_{\Delta^r_1} + \sum_{j=1}^{r-1} \int_{\Delta^{r-j}_1} kK_k(t) s_a(x+t)\ dt = J_1 + J_2 + J_3. \tag{23}$$

Using (10) and (11), and the fact that $x \dot{+} t > a$ for $t \in \Delta^r_0$ and therefore $s_a(x \dot{+} t) = 1$, we obtain

$$J_1 = \int_{\Delta^r_0} \left[2^n K_{2^n}(t) + mD_{2^n}(t) + w_{2^n}(t) \sum_{j=1}^{m} D_j(t) \right] dt$$
$$= \frac{2^n+1}{2} + \sum_{j=1}^{n-r} 2^{n-j-1} + m = k - 2^{r-1} + \frac{1}{2}. \tag{24}$$

We now estimate J_2. Let $x \dot{-} a \in \Delta^n_{i_0} \subset \Delta^r_1$. Then the inequality

$$|J_2| \leqslant \sum_{\Delta^n_i \subset \Delta^r_1} \left| \int_{\Delta^n_i} kK_k(t)\ dt \right| = \int_{\Delta^n_{2^{n-r}}} 2^n K_{2^n}(t)\ dt + \int_{\Delta^n_{i_0}} \left| \sum_{j=1}^{m} D_j(t) \right| dt$$
$$\leqslant 2^{r-2} + 2^{r-2} = 2^{r-1} \tag{25}$$

follows from (10), (11), and Lemmas 5 and 3.

Finally, using the distribution of the intervals $x \dot{+} \Delta^{r-j}_1$, which follows from the arithmetic structure of the number $a = 0.(01)$, and also, (10) and (11) and Lemma 5, we obtain

$$J_3 = \sum_{j=1}^{r-1} \int_{\Delta^{r-j}_1} kK_k(t) s_a(x+t)\ dt = \sum_{j=1}^{r-1} (-1)^n 2^{r-j-2}$$
$$= -2^{r-3} \cdot \frac{2}{3}\left(1 - (-2)^{-(r-1)}\right). \tag{26}$$

Substituting (24)–(26) into (23), we find that for sufficiently large r

$$|\sigma_k(x)| \leqslant 1 - \frac{2^{r-3}/3 - 2^{-1}}{k} < 1.$$

Hence the theorem is established.

We now show that there is a point a such that the Gibbs phenomenon does occur for the means $\sigma_k(x)$ of the Fourier-Walsh series of $s_a(x)$. More precisely, we have the following theorem.

THEOREM 9. *For almost all $a \in J$ the Gibbs function for the $(C,1)$ means $\sigma_k(x)$ of the Fourier-Walsh series of $s_a(x)$ is equal to $1 + (12 + 2\sqrt{34})^{-1}$.*

First we establish the following lemma.

LEMMA 7. *Let the binary expansion $a = \sum_0^\infty a_i/2^{i+1}$ satisfy, for some n, p and r, with $2 < p \leqslant n$ and $0 < r$, the equations*

$$a_{n-p} = a_{n-p+1} + \cdots + a_{n-1} = a_{n+1} = \cdots = a_{n+r} = 0, \qquad a_n = 1. \quad (27)$$

Then

$$\sigma_{2^n+m_n}\left(a + \frac{1}{2^n} + \frac{1}{2^{n-1}}, s_a\right) \geqslant 1 + \frac{1-2^{-r}}{12+2\sqrt{34}} - \frac{1}{5\cdot 2^{p-2}s} + o(1),$$

where m_n is the number in Lemma 6.

PROOF OF THE LEMMA. It follows easily from (27) that $(a \dotplus 2^{-n} \dotplus 2^{-n+1} \dotplus t) > a$ for $t \in \Delta_0^{n-p} \setminus \Delta_3^n$, and therefore $s_a J(a \dotplus 2^{-n} \dotplus 2^{-n+1} \dotplus t) = 1$. For $t \in \Delta_7^{n+1}$ we have $(a \dotplus 2^{-n} \dotplus 2^{-n+1} \dotplus t) < a$ and therefore $s_a J(a \dotplus 2^{-n} \dotplus 2^{-n+1} \dotplus t) = -1$. Since $2^{n-2} < m_n < 2l^{n-1}$, it is easy to calculate by using (11) that $K_{2^n+m_n}(t) < 0$ for $t \in \Delta_7^{n-1}$. Notice that

$$a < \sum_{i=0}^{n} \frac{a_i}{2^{i+1}} + \frac{1}{2^{n+r+1}}.$$

Hence the set

$$E_4^{(-)} = \left\{t \in \Delta_6^{n+1};\, s_a(a + 2^{-n} + 2^{-n+1} + t) = -1\right\}$$

satisfies the inequality

$$\text{meas } E_4^{(-)} < 2^{-(n+r+1)}. \quad (28)$$

We have

$$\sigma_{2^n+m_n}(a + 2^{-n} + 2^{-n+1}) = \int_0^1 K_{2^n+M_n}(t) s_a(a + 2^{-n} + 2^{-n+1} + t)\, dt$$

$$= \int_{\Delta_0^{n-p}\setminus\Delta_2^n} + \int_{\Delta_3^n} + \int_{[0,1]\setminus\Delta_0^{n-p}} = J_1 + J_2 + J_3.$$

Since $\int_{\Delta_3^n} K_{2^n+m_n}(t)\, dt = 0$ (10) and (11), we have

$$J_1 = \int_{\Delta_0^{n-p}\setminus\Delta_3^n} K_{2^n+m_n}(t)\, dt = \int_{\Delta_0^{n-p}} = \int_0^1 - \int_{[0,1]\setminus\Delta_0^{n-p}} sK_{2^n+m_n}(t)\, dt.$$

To estimate J_3 we use the fact that $s_a(a \dotplus 2^{-n} \dotplus 2^{-n+1} \dotplus t)$ has constant sign on the intervals $\Delta_i^n < [0,1] \setminus \Delta_0^{n-p}$. Therefore

$$|J_3| \leqslant \left|\int_{[0,1]\setminus\Delta_0^{n-p}} K_{2^n+m_n}(t)\, dt\right|.$$

Taking account of Lemma 5 and (11), we obtain

$$J_1 + J_3 \geqslant \int_0^1 K_{2^n+m_n}(t)\,dt - 2\left|\int_{[0,1]\setminus\Delta_0^{n-p}} K_{2^n+m_n}(t)\,dt\right|$$

$$= 1 - \frac{2^{n+1}}{2^n+m_n}\int_{[0,1]\setminus\Delta_0^{n-p}} K_{2^n}(t)\,dt = 1 - \frac{2^{n+1}}{2^n+m_n}\frac{1}{2^{p+1}} \geqslant 1 - \frac{1}{2^{p-2}\cdot 5}.$$

The following inequality is obtained by using (17) and (28):

$$J_2 = \int_{\Delta_6^{n+1}} + \int_{\Delta_7^{n+1}} = \int_{\Delta_6^{n+1}} K_{2^n+m_n}(t)\,dt + \int_{\Delta_7^{n+1}} |K_{2^n+m_n}(t)|\,dt$$

$$-2\int_{E_4^{(-1)}} |K_{2^n+m_n}(t)|\,dt \geqslant A_n\left(\frac{1}{2^n} - \frac{2}{2^{n+r+1}}\right).$$

Combining inequalities, we obtain

$$J_1 + J_2 + J_3 \geqslant 1 + \frac{A_n}{2^n}\left(1 - \frac{1}{2^r}\right) - \frac{1}{2^{p-2}\cdot 5}.$$

This completes the proof of the lemma.

Applying Lemma 2, we obtain Theorem 9 from Lemma 7 and Theorem 7. The statement of Theorem 2 in the introduction is a corollary of Theorems 7–9.

Bibliography

1. A. Zygmund, *Trigonometric series*, 2nd rev. ed., Vol. 1, Cambridge Univ. Press, 1959.
2. A. M. Zubakin, *The Gibbs phenomenon for multiplicative systems of Walsh or Vilenkin-Dzhofarli type*, Sibirsk. Mat. Zh. **12** (1971), 147–157; English transl. in Siberian Math. J. **12** (1971).
3. G. N. Agaev et al., *Multiplicative systems of functions and harmonic analysis on zero-dimensional groups*, "Èlm", Baker, 1981. (Russian)
4. A. A. Shneider, *On series of Walsh functions with monotonic coefficients*, Izv. Akad. Nauk SSSR Ser. Mat. **12** (1948), 179–192. (Russian)
5. Émile Borel, *Traité du calcul des probabilités et de ses applications*. Tome II, Fasc 1: *Applications à l'arithmétique et à la théorie des fonctions*, Gauthier-Villars, Paris, 1926.
6. Shigeki Yano, *On approximation by Walsh functions*, Proc. Amer. Math. Soc. **2** (1951), 962–967.

Translated by R. P. BOAS

Труды Матем. Инст. Стеклов.
164(1983)

Proceedings of the Steklov
Institute of Mathematics
1985, Issue 2

CONSTRUCTION OF POLYNOMIAL BASES IN FINITE-DIMENSIONAL SPACES OF FUNCTIONS ANALYTIC IN THE DISK

UDC 517.5

S. V. BOCHKAREV

ABSTRACT. It is established that spaces of nth-degree polynomials defined in the unit disk of the complex plane have bases with bounded basis constant. A method is developed for constructing polynomials orthogonal on the unit circle and forming in the space A_n a basis whose basis constant does not depend on the dimension of A_n. Moreover, it is proved that in H_n^1 the polynomial system thus constructed forms an unconditional basis with uniformly bounded unconditional basis constant. Also considered is the orthogonal system trigonometically conjugate to the Franklin system, and inequalities of the same type as established earlier by the author for the Franklin system are proved for it.

Bibliography: 12 titles.

The purpose of this article is to construct bases in the finite-dimensional spaces A_n and H_n^1 of functions analytic in the disk, formed by the functions z^k, $0 \leqslant k \leqslant n-1$, and equipped with the A- and H^1-norms. The existence of a basis in the disk algebra A was established by the author in [1]. The Franklin system was used in constructing this basis. Then Maurey [2] obtained a nonconstructive proof that H^1 has an unconditional basis. Carleson [3] gave a construction of an unconditional basis in BMO whose dual basis is an unconditional basis in H^1, and Wojtaszczyk [4] applied Carleson's proof to the Franklin system and established that it is an unconditional basis in H^1.

However, since the norms of the projections onto the spaces A_n and H_n^1 grow like $\log n$, the question of the existence of bases with uniformly bounded basis constants in these spaces was very problematical. We remark that this question was posed in 1981 by Pełczyński at a conference on Banach spaces in Oberwolfach. In the present article we construct an orthogonal system of polynomials that forms a basis with bounded basis constant in A_n and H_n^1. Our results were announced in [5].

In §1 the Gram-Schmidt orthogonalization method is used to construct a discrete orthonormal system on the class of residues modulo n, and properties of it are established.

In §2 the discrete orthonormal system is employed to construct a system of polynomials orthonormal on the unit circle. Here estimates are obtained for the

1980 *Mathematics Subject Classification*. Primary 46B15, 46E15; Secondary 42A50.

0081-5438/85 $1.00 + $.25 per page

polynomial system thus constructed. It turns out that in this case the orthogonalization essentially improves the structural properties of the system.

In §3 we establish the existence of a basis with bounded basis constant in A_n. It is proved that the system of orthogonal polynomials constructed in §2 forms in A_n a basis whose basis constant does not depend on the dimension of A_n.

In §4 it is established that the system of orthogonal polynomials we have obtained forms an unconditional basis in H_n^1 with uniformly bounded unconditional basis constant.

In §5 we study the system trigonometrically conjugate to the Franklin system. Here estimates are obtained for the conjugate Franklin functions and then used to establish for series in the conjugate Franklin system inequalities of the same type as obtained earlier by the author for series in the Franklin system (see [1] and [9]).

An independent enumeration of formulas is employed in each section. If reference is made to a formula in another section, then the number of that section is placed first.

§1. Definition of a discrete orthogonal system and properties of it

Let $n = 2^s + q$, $0 \leqslant q < 2^s$. We define the numbers $R_k^{(m)}$, with $k = 0,\dots,m$ for $m = s$, and $k = 0,\dots,2^m$ for $m < s$, as follows. Let

$$R_0^{(0)} = n, \quad R_0^{(1)} = 0, \quad R_1^{(1)} = n - \left[\frac{n}{2}\right], \quad R_2^{(1)} = \left[\frac{n}{2}\right],$$

where $[x]$ denotes the integer part of a number x.

Suppose that the numbers $R_k^{(p)}$ are defined for $p \leqslant m$ and $m < s - 1$. Then

$$R_{2k-1}^{(m+1)} = R_k^{(m)} - \left[R_k^{(m)}/2\right], \tag{1}$$

$$R_{2k}^{(m+1)} = \left[R_k^{(m)}/2\right]. \tag{2}$$

Moreover, for $m = s$ let $R_k^{(s)} = 1$ for all $k = 1,\dots,n$.

For any m the numbers $R_k^{(m)}$ form a partition of the interval of positive integers 0, $1,\dots,n$. We prove that the $R_k^{(m)}$ satisfy for $m < s$ the inequalities

$$\left[n/2^m\right] \leqslant R_k^{(m)} \leqslant \left[n/2^m\right] + 1, \qquad k < 1,\dots,2^m. \tag{3}$$

Indeed, the inequalities (3) hold for $R_1^{(1)}$ and $R_2^{(1)}$. Assume that they hold for $m - 1$. Then, by (1) and (2), the greatest of the numbers $R_k^{(m)}$ does not exceed the number $[n/2^{m-1}] + 1 - [([n/2^{m-1}] + 1)/2]$, while the smallest is equal to $[n/2^m]$. If $[n/2^{m-1}]$ is even, then $[([n/2^{m-1}] + 1)/2] = [n/2^m]$, and (3) is valid.

But if $[n/2^{m-1}]$ is odd, then $[([n/2^{m-1}] + 1)/2] = [n/2^m] + 1$, and (3) is again valid.

We shall define n functions $g_k^{(m)}(l)$ on the class of residues modulo n. Let $p_k^{(m)} = \sum_{i=0}^k R_i^{(m)}$. We set $g_0^{(0)}(l) \equiv 1$.

Let $n = 2^s + q$, $0 \leqslant q < 2^s$, $m < s$ and $k = 1,\dots,2^{m-1}$. Then

$$g_k^{(m)}(l) = \begin{cases} 0 & \text{for } l \in \left[0, p_{2k-2}^{(m)}\right],\ l \in \left[p_{2k}^{(m)}, n\right], \\ 1 & \text{for } l = p_{2k-1}^{(m)}, \\ \text{linear on } \left[p_{2k-2}^{(m)}, p_{2k-1}^{(m)}\right] \text{ and } \left[p_{2k-1}^{(m)}, n\right]. \end{cases} \tag{4}$$

For $m = s$ and $k = 1,\dots,n$ let $g_k^{(s)}(l) \equiv 0$ if $k = p_i^{(s-1)}$ for some $i = 0,\dots,2^{s-1}$. Otherwise, let

$$g_k^{(s)}(l) = \begin{cases} 1 & \text{for } l = k, \\ 0 & \text{for the remaining } l. \end{cases} \tag{5}$$

Moreover, we introduce a system of functions $h_k^{(m)}(l)$, $k = 1,\dots,2^{m-1}$, by setting

$$h_k^{(m)}(l) = \begin{cases} 0 & \text{if } l \in \left[0, p_{k-1}^{(m)}\right], l \in \left[p_{k+1}^{(m)}, n\right], \\ 1 & \text{if } l = p_k^{(m)}, \\ \text{linear on } \left[p_{k-1}^{(m)}, p_k^{(m)}\right], \left[p_k^{(m)}, p_{k+1}^{(m)}\right] \end{cases} \tag{6}$$

for $m < s$, and

$$h_k^{(s)}(l) = \begin{cases} 1 & \text{if } l = k, \\ 0 & \text{if } l \neq k \end{cases} \tag{7}$$

for $m = s$, where $k = 1,\dots,n-1$.

We apply the Gram-Schmidt orthogonalization process to the system of functions $g_k^{(m)}(l)$ (see (4) and (5)). Denote the system of functions obtained by $v_k^{(m)}(l)$. For these functions we prove

THEOREM 1. *The functions $v_k^{(m)}(l)$ satisfy the inequality*

$$\left|v_k^{(m)}(l)\right| \leqslant B\left(\frac{2^m}{n}\right)^{1/2} 2^{-\min(|p_{2k-1}^{(m)} - l|, n - |p_{2k-1}^{(m)} - l|)/([n/2^{m-1}]+1)} \quad \textit{for } m < s, \tag{8}$$

$$\left|v_k^{(s)}(l)\right| \leqslant B \cdot 2^{-\min(|k-l|, n-|k-l|)/4}, \tag{8'}$$

where B is a positive constant.

PROOF. Since the last group of functions $v_k^{(s)}(l)$ occupies a special position, we consider first the case when $m < s$. The function $v_k^{(m)}(l)$ must be orthogonal to the functions $g_r^{(q)}(l)$ for $q < m$ and $r = 1,\dots,2^{q-1}$, as well as to the functions $g_r^{(m)}(l)$ for $r < k$. Consequently, $v_k^{(m)}(l)$ is orthogonal to the functions $h_r^{(m)}(l)$ for $r \leqslant 2k - 3$ and to the functions $h_r^{(m-1)}(l)$ for $r \geqslant k$, because these functions are linear combinations of the indicated functions $g_r^{(q)}(l)$ and the functions $g_r^{(m)}(l)$, $r < k$.

We adjoin to these functions the function

$$h_{k-1}^{(m-1)}(l) - h_{k-1}^{(m-1)}\left(p_{2k-3}^{(m)}\right) h_{2k-3}^{(m)}(l) \tag{9}$$

and the function $h_0^{(m)}(l)$ defined as follows: for $k > 1$

$$h_0^{(m)}(l) = \begin{cases} 1 & \text{if } l = 0, l = n, \\ 0 & \text{for } l \in \left[p_1^{(m)}, p_{2^{m-1}-1}^{(m-1)}\right], \\ \text{linear on } \left[0, p_1^{(m)}\right] \text{ and } \left[p_{2^{m-1}-1}^{(m-1)}, n\right], \end{cases} \tag{10}$$

while for $k = 1$

$$h_0^{(m)}(l) = \begin{cases} 1 & \text{if } l = 0, l = n, \\ 1 & \text{for } l \in \left[p_1^{(m-1)}, p_{2^{m-1}-1}^{(m-1)}\right], \\ \text{linear on } \left[0, p_1^{(m-1)}\right], \left[p_{2^{m-1}-1}^{(m-1)}, n\right]. \end{cases} \tag{11}$$

The resulting system of functions forms a basis in the subspace spanned by the functions $g_r^{(q)}(l)$, $q < m$, $r = 1, \dots, 2^{q-1}$, and $g_r^{(m)}(l)$, $r < k$. Therefore, the $v_k^{(m)}(l)$ must satisfy the relations

$$\left(v_k^{(m)}, h_r^{(m)}\right) = 0 \quad \text{for } 1 \leqslant r \leqslant 2k-3, \tag{12}$$

$$\left(v_k^{(m)}, h_r^{(m-1)}\right) = 0 \quad \text{for } k \leqslant r \leqslant 2^{m-1}-1, \tag{13}$$

$$\left(v_k^{(m)}, h_{k-1}^{(m-1)}\right) - h_{k-1}^{(m-1)}\left(p_{2k-3}^{(m)}\right)\left(v_k^{(m)}, h_{2k-3}^{(m)}\right) = 0, \tag{14}$$

$$\left(v_k^{(m)}, h_0^{(m)}\right) = 0. \tag{15}$$

The function $v_k^{(m)}(l)$ is linear on the intervals $[p_{r-1}^{(m)}, p_r^{(m)}]$ for $1 \leqslant r \leqslant 2k$ and on $[p_r^{(m-1)}, p_{r+1}^{(m-1)}]$ for $k \leqslant r < 2^{m-1}$. Therefore, it suffices to know its values at the points $p_r^{(m)}$, $0 \leqslant r \leqslant 2k$, and $p_r^{(m-1)}$, $k < r < 2^{m-1}$. We denote these values by

$$x_0, x_1, \dots, x_{2k-1}, y_{2^{m-1}-1}, y_{2^{m-1}-k-1}, \dots, y_1. \tag{16}$$

Then the relations (12)–(15) give the following system of equations for the inequalities (16). The relations (12) imply for $1 \leqslant r \leqslant 2k-3$ the equations

$$\alpha_r^{(m)} x_{r-1} + \beta_r^{(m)} x_r + \gamma_r^{(m)} x_{r+1} = 0, \tag{17}$$

where

$$\alpha_r^{(m)} = \tfrac{1}{6}\left(\left(R_r^{(m)}\right)^2 - 1\right), \tag{18}$$

$$\beta_r^{(m)} = \frac{\left(R_r^{(m)}\right)^2}{3} + \frac{R_r^{(m)} R_{r+1}^{(m)}}{3} + \frac{R_r^{(m)}}{6R_{r+1}^{(m)}} + \frac{1}{6}, \tag{19}$$

$$\gamma_r^{(m)} = \frac{R_r^{(m)}}{6R_{r+1}^{(m)}}\left(\left(R_{r+1}^{(m)}\right)^2 - 1\right). \tag{20}$$

It follows from (18)–(20) that for all $m = 1, \dots, s-1$ and $r = 1, \dots, 2^m - 1$

$$\left(\beta_r^{(m)} - \alpha_r^{(m)}\right)/\gamma_r^{(m)} \geqslant 2, \tag{21}$$

$$\left(\beta_r^{(m)} - \gamma_r^{(m)}\right)/\alpha_r^{(m)} \geqslant 2. \tag{22}$$

The relations (13) give the equations (see (18)–(20))

$$\alpha_{2^{m-1}-r}^{(m-1)} y_{r+1} + \beta_{2^{m-1}-r}^{(m-1)} y_r + \gamma_{2^{m-1}-r}^{(m-1)} y_{r-1} = 0, \tag{23}$$

where $2 \leqslant r \leqslant 2^{m-1} - k - 1$.

The coefficients in (23) also satisfy (21) and (22). Besides these two systems of equations there are two transition equations and two boundary equations. The first transition equation is (see (14))

$$a_k^{(m)} x_{2k-3} + b_k^{(m)} x_{2k-2} + c_k^{(m)} x_{2k-1} + d_k^{(m)} y_{2^{m-1}-k} = 0, \tag{24}$$

where

$$a_k^{(m)} = \tfrac{1}{6}\left(\left(R_{2k-2}^{(m)}\right)^2 - 1\right), \tag{25}$$

$$\begin{aligned} b_k^{(m)} = \left(6\left(R_{2k-1}^{(m)} + R_{2k}^{(m)}\right)\right)^{-1}\Big[&3R_{2k-2}^{(m)} R_{2k-1}^{(m)} R_{2k}^{(m)} + 2\left(R_{2k-2}^{(m)}\right)^2 R_{2k-1}^{(m)} \\ &+ 2\left(R_{2k-2}^{(m)}\right)^2 R_{2k}^{(m)} + R_{2k-2}^{(m)} R_{2k-1}^{(m)} + 7R_{2k-2}^{(m)} + R_{2k-1}^{(m)} + R_{2k}^{(m)}\Big], \end{aligned} \tag{26}$$

$$c_k^{(m)} = \frac{R_{2k-2}^{(m)}}{6\left(R_{2k-1}^{(m)} + R_{2k}^{(m)}\right)}\left[3R_{2k-1}^{(m)} R_{2k}^{(m)} + \left(R_{2k-1}^{(m)}\right)^2 + 2\left(R_{2k}^{(m)}\right)^2\right], \tag{27}$$

$$d_k^{(m)} = \frac{R_{2k-2}^{(m)}}{6\left(R_{2k-1}^{(m)} + R_{2k}^{(m)}\right)}\left(\left(R_{2k}^{(m)}\right)^2 - 1\right). \tag{28}$$

The second transition equation is (see (14))

$$e_k^{(m)}x_{2k-2} + i_k^{(m)}x_{2k-1} + j_k^{(m)}y_{2^m-k} + f_k^{(m)}y_{2^m-k-1} = 0, \tag{29}$$

where

$$e_k^{(m)} = \tfrac{1}{6}\left(\left(R_{2k-1}^{(m)}\right)^2 - 1\right), \tag{30}$$

$$i_k^{(m)} = \tfrac{1}{6}\left(2\left(R_{2k-1}^{(m)}\right)^2 + \left(R_{2k}^{(m)}\right)^2 + 3R_{2k-1}^{(m)}R_{2k}^{(m)}\right), \tag{31}$$

$$\begin{aligned} j_k^{(m)} = {} & \tfrac{1}{3}\left(R_{2k}^{(m)}\right)^2 k + \tfrac{1}{3}R_{2k-1}^{(m)}R_{k+1}^{(m-1)} + \tfrac{1}{3}R_{2k}^{(m)}R_{k+1}^{(m-1)} \\ & + \tfrac{1}{2}R_{2k-1}^{(m)}R_{2k}^{(m)} + \left(6R_{k+1}^{(m-1)}\right)^{-1}\left(R_{2k-1}^{(m)} + R_{2k}^{(m)}\right) + \tfrac{1}{6}, \end{aligned} \tag{32}$$

$$f_k^{(m)} = \frac{1}{6}\left(R_{2k-1}^{(m)}R_{k+1}^{(m-1)} + R_{2k}^{(m)}R_{k+1}^{(m-1)} - \frac{R_{2k-1}^{(m)} + R_{2k}^{(m)}}{R_{k+1}^{(m-1)}}\right). \tag{33}$$

Eliminating the unknown x_{2k-1} from (24) and (29), we obtain the equation (see (25)–(28) and (30)–(33))

$$a_k^{(m)}x_{2k-3} + B_k^{(m)}x_{2k-2} + C_k^{(m)}y_{2^m-k} + D_k^{(m)}y_{2^m-k-1} = 0, \tag{34}$$

where

$$B_k^{(m)} = b_k^{(m)} - \frac{c_k^{(m)}}{i_k^{(m)}}e_k^{(m)}, \quad C_k^{(m)} = d_k^{(m)} - \frac{c_k^{(m)}}{i_k^{(m)}}j_k^{(m)}, \quad D_k^{(m)} = -\frac{c_k^{(m)}}{i_k^{(m)}}f_k^{(m)}.$$

The relation (15) gives the first boundary condition (see (10) and (11))

$$\alpha_0^{(m)}y_1 + \beta_0^{(m)}x_0 + \gamma_0^{(m)}x_1 = 0, \tag{35}$$

where

$$\alpha_0^{(m)} = \tfrac{1}{6}\left(\left(R_{2^{m-1}}^{(m-1)}\right)^2 - 1\right), \tag{36}$$

$$\beta_0^{(m)} = \begin{cases} \tfrac{1}{3}\left(R_{2^{m-1}}^{(m-1)}\right)^2 + \tfrac{1}{3}R_1^{(m)}R_{2^{m-1}}^{(m-1)} + R_{2^{m-1}}^{(m-1)}/6R_1^{(m)} + \tfrac{1}{6} & \text{for } k > 1, \\ \tfrac{1}{3}\left(R_{2^{m-1}}^{(m-1)}\right)^2 + \tfrac{1}{3}R_1^{(m-1)}R_{2^{m-1}}^{(m-1)} + R_{2^{m-1}}^{(m-1)}/6R_1^{(m-1)} + \tfrac{1}{6} & \text{for } k = 1, \end{cases} \tag{37}$$

$$\gamma_0^{(m)} = \begin{cases} R_{2^{m-1}}^{(m-1)}\left(\left(R_1^{(m)}\right)^2 - 1\right)/6R_1^{(m)} & \text{for } k > 1, \\ R_{2^{m-1}}^{(m-1)}\left(\left(R_1^{(m-1)}\right)^2 - 1\right)/6R_1^{(m-1)} & \text{for } k = 1. \end{cases} \tag{38}$$

The second boundary condition is the following equation (see (13)):

$$\alpha_{2^{m-1}-1}^{(m-1)}y_2 + \beta_{2^{m-1}-1}^{(m-1)}y_1 + \gamma_{2^{m-1}-1}^{(m-1)}x_0 = 0, \tag{39}$$

whose coefficients are determined by (18)–(20). Note that the coefficients in the boundary conditions (35) and (39) also satisfy (21) and (22).

It is convenient for what follows to rename the unknowns, setting $x_k' = (-1)^k x_k$ and $y_k' = (-1)^{k+1}y_k$. Moreover, it can be assumed that $x_0 \geqslant 0$. We prove that at least one of the sequences $x_0', x_1', \dots, x_{2k-1}'$ and $y_1', y_2', \dots, y_{2^{m-1}-k}'$ satisfies the inequalities

$$x_{k+1}' \geqslant 2x_k', \qquad k = 0, \dots, 2k-3, \tag{40}$$

$$|y_{k+1}'| \geqslant 2|y_k'|, \qquad k = 1, \dots, 2^{m-1} - k - 1. \tag{41}$$

Indeed, the boundary equations (35) and (39) take the form

$$\beta_0^{(m)}x_0' - \gamma_0^{(m)}x_1' + \alpha_0^{(m)}y_1' = 0, \tag{42}$$

$$\gamma_{2^{m-1}-1}^{(m-1)}x_0' - \alpha_{2^{m-1}-1}^{(m-1)}y_2' + \beta_{2^{m-1}-1}^{(m-1)}y_1' = 0. \tag{43}$$

If $y_1' \geqslant 0$, then (see (42) and (21))

$$\beta_0^{(m)}x_0' - \gamma_0^{(m)}x_1' \leqslant 0, \quad \text{i.e., } x_1' \geqslant \beta_0^{(m)}x_0'/\gamma_0^{(m)} \geqslant 2x_0'. \tag{44}$$

Assume that $x_r \geqslant x_{r-1}$. Then we get from (17) and (21) that

$$\alpha_r^{(m)}x_r' - \beta_r^{(m)}x_r' + \gamma_r^{(m)}x_{x+1}' \geqslant 0,$$

i.e.,

$$x_{r+1}' \geqslant \frac{\beta_r^{(m)} - \alpha_r^{(m)}}{\gamma_r^{(m)}}x_r' \geqslant 2x_r'.$$

Thus, (21), (44) and (17) give us inequality (40).

Assume now that $y_1 < 0$. We consider the two cases $|y_1| \leqslant x_0$ and $|y_1| > x_0$. In the first case, we get from (42) and (21) that

$$\beta_0^{(m)}x_0' - \gamma_0^{(m)}x_1' - \alpha_0^{(m)}x_0' \leqslant 0, \quad \text{i.e., } x_1' \geqslant \frac{\beta_0^{(m)} - \alpha_0^{(m)}}{\gamma_0^{(m)}}x_0' \geqslant 2x_0',$$

and this implies (40) by the preceding. In the second case (43) and (22) imply that

$$-\gamma_{2^{m-1}-1}^{(m-1)}y_1' - \alpha_{2^{m-1}-1}^{(m-1)}y_2' + \beta_{2^{m-1}-1}^{(m-1)}y_1' \geqslant 0, \tag{45}$$

i.e.,

$$-y_2' \geqslant -\frac{\beta_{2^{m-1}-1}^{(m-1)} - \gamma_{2^{m-1}-1}^{(m-1)}}{\alpha_{2^{m-1}-1}^{(m-1)}}y_1' \geqslant -2y_1'.$$

Suppose that $0 \geqslant y_{r-1} \geqslant y_r$ for some $2 \leqslant r \leqslant 2^{m-1} - k - 1$. Then it follows from (23) and (22) that

$$\alpha_{2^{m-1}-r}^{(m-1)}y_{r+1}' - \beta_{2^{m-1}-r}^{(m-1)}y_r' + \gamma_{2^{m-1}-r}^{(m-1)}y_r' \leqslant 0,$$

i.e.,

$$y_{r+1}' \leqslant \frac{\beta_{2^{m-1}-r}^{(m-1)} - \gamma_{2^{m-1}-r}^{(m-1)}}{\alpha_{2^{m-1}-r}^{(m-1)}}y_r' \quad \text{or} \quad |y_{r+1}'| \geqslant 2|y_r'|.$$

Hence, the inequalities (41) follow from (23), (22) and (45). Thus, either the inequalities (40) or the inequalities (41) always hold.

Assume, for example, that (40) is valid. Consider the homogeneous system of equations for the unknowns $y_1', \dots, y_{2^m-1}'$ (see (23)). Denote by $\{Q_k\}$ and $\{q_k\}$ the two particular solutions of this system that satisfy the following boundary conditions:

$$Q_1 = Q_2 = 1, \tag{46}$$

$$q_{2^m-k} = q_{2^m-k-1}, \qquad q_1 = 1. \tag{47}$$

It follows from (21) and (22) that $\{Q_r\}$ and $\{q_r\}$ satisfy the inequalities (see (46) and (47))

$$Q_r \geqslant 2Q_{r-1}, \qquad 3 \leqslant r \leqslant 2^m - k, \tag{48}$$

$$q_r \leqslant \tfrac{1}{2}q_{r-1}, \qquad 2 \leqslant r \leqslant 2^m - k - 1. \tag{49}$$

Since $\{Q_r\}$ and $\{q_r\}$ are linearly independent, the general solution of the homogeneous equation

$$\alpha^{(m)}_{2^{m-1}-r}y'_{r+1} - \beta^{(m)}_{2^{m-1}-r}y'_r + \gamma^{(m)}_{2^{m-1}-r}y'_{r-1} = 0$$

is

$$y'_r = A_1Q_r + A_2q_r, \tag{50}$$

where A_1 and A_2 are constants. These constants are determined from the boundary equation (39) and the transition equation (34).

Assume that the constants A_1 and A_2 satisfy the relation $|A_2| \leqslant 2|A_1|$. Then (see (46), (47) and (50))

$$|y_r| \leqslant 3|A_1|Q_r \leqslant 3|A_1|Q_{2^{m-1}-k} \cdot 2^{k+r-2^{m-1}} \leqslant 6|y_{2^{m-1}-k}| \cdot 2^{k+r-2^{m-1}}. \tag{51}$$

On the other hand, it follows from (40) that

$$x_r \leqslant x_{2k-2} \cdot 2^{r+2-2k} \quad \text{for } 0 \leqslant r \leqslant 2k-2. \tag{52}$$

Therefore, the functions $v_k^{(m)}(l)$, which are normalized in l_n^2, satisfy (8).

But if $2|A_1| < |A_2|$, then either $2|A_1|Q_r < |A_2|q_r$ for all $1 \leqslant r \leqslant 2^{m-1} - k$, or there exists an index p such that $2|A_1|Q_p \geqslant |A_2|q$ but $2|A_1|Q_{p-1} < |A_2|q_{p-1}$. In the first case we have (see (42) and (50))

$$|y_r| \leqslant 2|A_2|q_r \leqslant 2|A_2|q_1 \cdot 2^{1-r} \leqslant 4|y_1| \cdot 2^{1-r} \leqslant Bx_1 \cdot 2^{-r}. \tag{53}$$

In the second case (53) holds for $r < p$ and (51) for $r > p$; consequently, (8) again holds. Thus, Theorem 1 is proved for the functions $v_k^{(m)}(l)$ when $m < s$.

We now get an estimate for the functions in the last group, i.e., for the functions $v_k^{(s)}(l)$. These functions are not identically zero if the functions $g_k^{(s)}(l)$ are nonzero (see (5)). Therefore, we assume that $k \neq p_i^{(s-1)}$, $i = 0,\dots,2^{s-1}$. Consider first the case when $k \geqslant 4$. Let $k \in [p_{m-1}^{(s-1)}, p_m^{(s-1)}]$ for some $2 \leqslant m \leqslant 2^{s-1}$. Then $v_k^{(s)}(l)$ is orthogonal to the functions $h_r^{(s)}(l)$ for $r < p_{m-1}^{(s-1)}$, and, consequently (see (7)), $v_k^{(s)}(l) = 0$ for $1 \leqslant l \leqslant p_{m-1}^{(s-1)}$. Moreover, $v_k^{(s)}(l)$ is orthogonal to the functions $h_r^{(s-1)}(l)$ for $r \geqslant m$. The function $v_k^{(s)}(l)$ has nodes at the points $p_r^{(s-1)}$ for $m \leqslant r \leqslant 2^{s-1}$, as well as at the points $q \in [p_{m-1}^{(s-1)}, p_m^{(s-1)}]$, $q \leqslant k$. The number of such points q does not exceed $R_m^{(s-1)}$; hence (see (3)) there can be two, three, or four such points q. For definiteness consider the case when there are three such points q, and $R_m^{(s-1)} = 3$ and $R_{m+1}^{(s-1)} = 3$. The remaining cases can be handled similarly. The values of $v_k^{(s)}(l)$ at the nodal points are denoted by

$$x_{2^{s-1}-m+3}, x_{2^{s-1}-m+2}, \dots, x_1, x_0. \tag{54}$$

Since $v_k^{(s)}(l)$ is orthogonal to $h_{k-1}^{(s)}(l)$, it follows that $x_{2^{s-1}-m+2} = 0$, and the first boundary condition is

$$3x_{2^{s-1}-m+3} + x_{2^{s-1}-m+1} = 0. \tag{55}$$

The second boundary condition is

$$x_1 + 5x_0 = 0 \quad \text{if } R_{2^{s-1}}^{(s-1)} = 2, \tag{56}$$

$$2x_1 + 7x_0 = 0 \quad \text{if } R_{2^{s-1}}^{(s-1)} = 3. \tag{57}$$

The orthogonality condition $v_k^{(s)}(l) \perp h_m^{(s-1)}(l)$ gives the equation

$$3x_{2^{s-1}-m+1} + 7x_{2^{s-1}-m} + 2x_{2^{s-1}-m-1} = 0. \tag{58}$$

And the orthogonality conditions $(v_k^{(s)}, h_r^{(s-1)}) = 0$ for $m < r < 2^{s-1}$ give a system of equations of the form (23) for the unknowns $x_{2^{s-1}-m}, \ldots, x_1, x_0$. Since the coefficients in these equations satisfy (22), the second boundary condition (see (56) and (57)) implies that $|x_r| \geqslant 2^r |x_0|$, which yields inequality (8) (see (55) and (58)). Finally, it remains to consider the first three functions $v_k^{(s)}(l)$, $k = 1, 2, 3$.

We confine ourselves to the function $v_1^{(s)}(l)$ and take $R_1^{(s-1)} = 3$, $R_2^{(s-1)} = 2$ and $R_{2^{s-1}}^{(s-1)} = 2$ for definiteness. The situation is analogous in the remaining cases. In this case there are $2^{s-1} + 1$ unknowns $x_{2^{s-1}}, \ldots, x_1, x_0$ and they satisfy the equations

$$10x_{2^{s-1}} + 2x_{2^{s-1}-1} + 3x_1 + 15x_0 = 0, \tag{59}$$

$$8x_{2^{s-1}} + 19x_{2^{s-1}-1} + 3x_{2^{s-1}-2} = 0. \tag{60}$$

Moreover, a system of $2^{s-1} - 2$ equations of the form (23) must hold for the unknowns $x_{2^{s-1}-1}, \ldots, x_1, x_0$. Note that the coefficients in this system (including (60)) satisfy (21) and (22).

We introduce the two particular solutions $\{Q_r\}$ and $\{q_r\}$ of the indicated system that satisfy the boundary conditions

$$Q_{2^{s-1}} = Q_{2^{s-1}-1} = 1, \tag{61}$$

$$q_1 = q_0 = 1. \tag{62}$$

The general solution can be represented in the form $x_r = B_1 Q_r + B_2 q_r$. The constants B_1 and B_2 are determined from (59) and the normalization condition. By (21) and (22) for the solutions $\{Q_r\}$ and $\{q_r\}$, we have $|Q_{r-1}| \geqslant 2|Q_r|$, $1 \leqslant r < 2^{s-1}$, and $|q_r| \geqslant 2|q_{r-1}|$, $1 < r \leqslant 2^{s-1}$. Since either $|B_1 Q_r| \geqslant |B_2 q_r|$ for all $0 \leqslant r \leqslant 2^{s-1}$, or there is an index m such that $|B_1 Q_m| \geqslant |B_2 q_m|$ but $|B_1 Q_{m+1}| < |B_2 q_{m+1}|$, the inequality (8′) is a consequence of the indicated estimates for $\{Q_r\}$ and $\{q_r\}$.

Theorem 1 is proved.

REMARK 1. In addition to the estimate of the functions $v_k^{(m)}(l)$ contained in Theorem 1, we shall need the boundedness of the Legesgue functions for this system. This fact can be established by starting from variational considerations and using the method presented by Ciesielski [6] for the Franklin system.

§2. Definition and properties of the orthogonal system of polynomials

Denote by $D_n(z)$ the complex Dirichlet kernel of order $n - 1$, i.e.,

$$D_n(z) = \sum_{k=0}^{n-1} z^k. \tag{1}$$

The kernel $D_n(z)$ has $n - 1$ zeros, located at the points $e^{i\alpha_k}$, $\alpha_k = 2\pi k/n$, $k = 1, \ldots, n-1$. The functions $D_n(re^{i(x-\alpha_k)})$ form in A_n and H_n^1 a basis orthogonal on the unit circle, but not having the property that the basis constants are bounded.

Using the discrete orthogonal system $v_k^{(m)}(l)$ constructed in §1, we define a new orthogonal system of polynomials $V_k^{(m)}(z)$ in the following way. Let

$$V_k^{(m)}(re^{ix}) = \frac{1}{\sqrt{2\pi n}} \sum_{l=0}^{n-1} v_k^{(m)}(l) D_n(re^{i(x-\alpha_l)}). \tag{2}$$

Since the system $v_k^{(m)}(l)$ is orthogonal, the system of polynomials $V_k^{(m)}(z)$ forms an orthonormal system on the unit circle. The boundary values $V_k^{(m)}(e^{ix})$ and $D_n(e^{i(x-\alpha_l)})$ of the polynomials will be denoted by $V_k^{(m)}(x)$ and $D_n(x - \alpha_l)$. Denote by $\rho(x, y)$ the distance between x and y on the unit circle. Moreover, let $\beta_k^{(m)} = 2\pi p_{2k-1}^{(m)}/n$ for $m < s$, and let $\beta_k^{(s)} = \alpha_k$.

THEOREM 2. *The polynomials* $V_k^{(m)}(x)$ *satisfy the inequality*

$$\left|V_k^{(m)}(x)\right| \leqslant B \min\left(2^{m/2}, 2^{-3/2m}\left[\rho\left(x, \beta_k^{(m)}\right)\right]^{-2}\right). \tag{3}$$

PROOF. We first prove (3) for the functions $V_k^{(m)}(x)$ with $m < s$. The last group of functions $V_k^{(s)}(x)$ will be considered separately. Since the $v_k^{(m)}(l)$ have nodes at the points $p_r^{(m)}$, $0 \leqslant r \leqslant 2k$, and at $p^{(m-1)}$, $k < r < 2^{m-1}$, the polynomial $V_k^{(m)}(x)$ can be represented in the form

$$\sqrt{2\pi n} \cdot V_k^{(m)}(x) = \sum_{l=0}^{2k-1} v_k^{(m)}\left(p_l^{(m)}\right) \sum_{q=0}^{n-1} h_l^{(m)}(q) D_n(x - \alpha_q)$$

$$+ v_k^{(m)}\left(p_{2k}^{(m)}\right) \sum_{q=0}^{n-1} \left(h_k^{(m-1)}(q) - h_k^{(m-1)}\left(p_{2k-1}^{(m)}\right) h_{2k-1}^{(m)}(q)\right) D_n(x - \alpha_q)$$

$$+ \sum_{l=k+1}^{2^{m-1}-1} v_k^{(m)}\left(p_l^{(m-1)}\right) \sum_{q=0}^{n-1} h_l^{(m-1)}(q) D_n(x - \alpha_q), \tag{4}$$

where

$$h_0^{(m)}(q) = \begin{cases} 1 & \text{for } q = 0,\ q = n, \\ 0 & \text{for } p_1^{(m)} \leqslant q \leqslant p_{2^{m-1}-1}^{(m-1)}, \\ \text{linear on } \left[0, p_1^{(m)}\right], \left[p_{2^{m-1}-1}^{(m-1)} n\right], \end{cases} \tag{5}$$

and the functions $h_k^{(m)}(q)$ are defined by (1.6) for $k \geqslant 1$.

Assume that $k \geqslant 2^{m-2}$. We introduce the notation

$$x_l = \begin{cases} p_{l+k-2^{m-2}}^{(m)} & \text{for } 0 \leqslant l \leqslant 2^{m-2} + k, \\ p_{l-2^{m-2}}^{(m-1)} & \text{for } 2^{m+2} + k < l \leqslant 3 \cdot 2^{m-2}, \\ p_{l-3\cdot 2^{m-2}}^{(m)} & \text{for } 3 \cdot 2^{m-2} < l < 2^{m-1} + k, \end{cases} \tag{6}$$

$$A_l^{(m)} = \begin{cases} \sum_{q=0}^{n-1} h_{l+k-2^{m-2}}^{(m)}(q), & \text{for } 0 \leqslant l < 2^{m-2} + k, \\ \sum_{q=0}^{n-1}\left(h_k^{(m-1)}(q) - h_k^{(m-1)}\left(p_{2k-1}^{(m)}\right) h_{2k-1}^{(m)}(q)\right) & \text{for } l = 2^{m-2} + k, \\ \sum_{q=0}^{n-1} h_{l-2^{m-2}}^{(m-1)}(q) & \text{for } 2^{m-2} + k < l \leqslant 3 \cdot 2^{m-2} \\ \sum_{q=0}^{n-1} h_{l-3\cdot 2^{m-2}}^{(m)}(q) & \text{for } 3 \cdot 2^{m-2} < l < 2^{m-1} + k. \end{cases} \tag{7}$$

Moreover, denote by $H_l^{(m)}(x)$, $l = 0, \ldots, 2^{m-1} + k - 1$, the function

$$\sqrt{2\pi n} \cdot H_l^{(m)}(x) = \begin{cases} \left(A_{2^{m-2}+k}\right)^{-1} \sum_{q=0}^{n-1}\left(h_k^{(m-1)}(q) - h_k^{(m-1)}\left(p_{2k-1}^{(m)}\right) h_{2k-1}^{(m)}(q)\right) \\ \qquad \times D_n(x - \alpha_q) \quad \text{for } l = 2^{m-2} + k, \\ \left(A_l^{(m)}\right)^{-1} \sum_{q=0}^{n-1} h_s^{(m)}(q) D_n(x - \alpha_q) \quad \text{for the remaining } l, \end{cases} \tag{8}$$

where the index s is determined from l just as in (7).

It follows from (4)–(8) that

$$V_k^{(m)}(x) = \sum_{l=0}^{2^{m-1}+k-1} v_k^{(m)}(x_l) A_l^{(m)} H_l^{(m)}(x). \tag{9}$$

Applying the Abel transformation, we now get that (see (9))

$$\begin{aligned} V_k^{(m)}(x) = &\sum_{l=0}^{2^{m-1}+k-2} \left(H_l^{(m)}(x) - H_{l+1}^{(m)}(x)\right) \sum_{q=0}^{l} v_k^{(m)}(x_q) A_q^{(m)} \\ &+ \left(\sum_{l=0}^{2^{m-1}+k-1} v_k^{(m)}(x_l) A_l^{(m)} \right) H_{2^{m-1}+k-1}^{(m)}(x). \end{aligned} \tag{10}$$

Let us prove that

$$\sum_{l=0}^{2^{m-1}+k-1} v_k^{(m)}(x_l) A_l^{(m)} = 0. \tag{11}$$

Indeed, if $l \in [x_q, x_{z+1}]$, then

$$v_k^{(m)}(l) = \frac{x_{q+1} - l}{x_{q+1} - x_q} v_k^{(m)}(x_q) + \frac{l - x_q}{x_{q+1} - x_q} v_k^{(m)}(x_{q+1})$$

and, consequently, the coefficient with which $v_k^{(m)}(x_q)$ appears in the sum $\sum_{l=0}^{n-1} v_k^{(m)}(l)$ is equal to $A_q^{(m)}$. Thus,

$$\sum_{l=0}^{2^{m-1}+k-1} v_k^{(m)}(x_l) A_l^{(m)} = \sum_{l=0}^{n-1} v_k^{(m)}(l) = 0,$$

because $v_k^{(m)}(l)$ is orthogonal to a constant.

Next we must estimate the quantity $\sum_{q=0}^{l} v_k^{(m)}(x_q) A_q^{(m)}$. The function $f_l^{(m)}(q)$ is defined as follows on the class of residues modulo n.

For $0 \leqslant l \leqslant 2^{m-2} + k - 2$ let (see (6))

$$f_l^{(m)}(q) = \begin{cases} 1 & \text{if } x_1 \leqslant q \leqslant x_l, \\ 0 & \text{if } x_{l+2} \leqslant q \leqslant n,\ 0 \leqslant q \leqslant x_0, \\ \text{linear on } [x_0, x_1] \text{ and } [x_l, x_{l+2}]. \end{cases} \tag{12}$$

If $l = 2^{m-2} + k - 1$, then let

$$f_l^{(m)}(q) = \begin{cases} 1 & \text{for } x_1 \leqslant q \leqslant x_{l-1}, \\ 0 & \text{for } x_{l+1} \leqslant q \leqslant n,\ 0 \leqslant q \leqslant x_0, \\ \text{linear on } [x_{l-1}, x_{l+1}],\ [x_0, x_1]. \end{cases} \tag{13}$$

If $2^{m-2} + k \leqslant l \leqslant 3 \cdot 2^{m-2}$, then

$$f_l^{(m)}(q) = \begin{cases} 1 & \text{for } x_1 \leqslant q \leqslant x_l, \\ 0 & \text{for } x_{l+1} \leqslant q \leqslant n,\ 0 \leqslant q \leqslant x_0, \\ \text{linear on } [x_0, x_1],\ [x_l, x_{l+1}]. \end{cases} \tag{14}$$

And if $3 \cdot 2^{m-2} < l < 2^{m-1} + k$, then

$$f_l^{(m)}(q) = \begin{cases} 1 & \text{for } x_1 \leqslant q \leqslant n,\ 0 \leqslant q \leqslant x_{l-1}, \\ 0 & \text{for } x_l \leqslant q \leqslant x_0, \\ \text{linear on } [x_{l-1}, x_l],\ [x_0, x_1]. \end{cases} \tag{15}$$

The function $f_l^{(m)}(q)$ defined by (12)–(15) is orthogonal to $v_k^{(m)}(q)$ for all $l = 0, \dots, 2^{m-1} + k - 1$. Suppose, for example, that $2^{m-2} + k < l \leqslant 3 \cdot 2^{m-2}$ (see (14)). In this case we have (see (6), (7) and (11))

$$\begin{aligned} &\sum_{q=0}^{l} v_k^{(m)}(x_q) A_q^{(m)} \\ &= \sum_{q=x_0}^{x_l} v_k^{(m)}(q) + v_k^{(m)}(x_0) \sum_{q=0}^{x_0-1} h_{k-2^{m-2}}^{(m)}(q) + v_k^{(m)}(x_l) \sum_{q=x_l+1}^{n} h_{l-2^{m-2}}^{(m-1)}(q) \\ &= \left(v_k^{(m)}, f_l^{(m)}\right) + \sum_{q=x_0}^{n} v_k^{(m)}(q) h_{k-2^{m-2}}^{(m)}(q) - \sum_{q=x_l+1}^{n} v_k^{(m)}(q) h_{l-2^{m-2}}^{(m-1)}(q) \\ &\quad + v_k^{(m)}(x_l) \sum_{q=x_l+1}^{n} h_{l-2^{m-2}}^{(m-1)}(q). \end{aligned} \tag{16}$$

To estimate the sums appearing in (16) we use Theorem 1. By (1.8), the first sum satisfies the estimate (see (1.3), (1.6) and (6))

$$\left| \sum_{q=x_0}^{n} v_k^{(m)}(q) h_{k-2^{m-2}}^{(m)}(q) \right| \leqslant B \left(\frac{n}{2^m} \right)^{1/2} \cdot 2^{-2^{m-3}}. \tag{17}$$

The same estimate holds for the third term.

For the second sum we obtain

$$\left| \sum_{q=x_l+1}^{n} v_k^{(m)}(q) h_{l-2^{m-2}}^{(m-1)}(q) \right| \leqslant B \left(\frac{n}{2^m} \right)^{1/2} \cdot 2^{-2/3|l-k-2^{m-2}|}, \tag{18}$$

and an analogous inequality is valid for the fourth sum.

Thus (see (16)–(18)),

$$\left| \sum_{q=0}^{l} v_k^{(m)}(x_q) A_q^{(m)} \right| \leqslant B \left(\frac{n}{2^m} \right)^{1/2} \left(2^{-2/3|l-k-2^{m-2}|} + 2^{-2^{m-3}} \right). \tag{19}$$

Let us now estimate $|H_l^{(m)}(x) - H_{l+1}^{(m)}(x)|$. First of all we prove that for all $x \in [0, 2\pi]$ and $0 \leqslant l \leqslant 2^{m-1} + k - 1$

$$\left| H_l^{(m)}(x) \right| \leqslant B \cdot 2^m \cdot n^{-1/2}. \tag{20}$$

Indeed,

$$\frac{1}{n} \sum_{q=0}^{n-1} h_l^{(m)}(q) D_n(x - \alpha_q) = \sum_{k=0}^{n-1} c_k^{(n)} e^{ikx},$$

where the $c_k^{(n)}$ are the Fourier-Lagrange coefficients of $h_l^{(m)}(tn/2\pi)$, i.e.,

$$c_k^{(n)} = \int_0^{2\pi} h_l^{(m)}\left(\frac{tn}{2\pi} \right) e^{-ikt}\, d\mu_n(t),$$

while $\mu_n(t)$ is a measure concentrated at the points $\{\alpha_q\}$. The Fourier-Lagrange coefficients satisfy the inequality (see [7], Chapter 8, formula (4.10))

$$\sum_{k=0}^{n-1} \left|c_k^{(n)}\right| \leqslant \sum_{l=-\infty}^{\infty} |\gamma_l|,$$

where $\{\gamma_l\}$ are the Fourier coefficients of $h_l^{(m)}(tn/2\pi)$. But the Fourier coefficients of $h_l^{(m)}(tn/2\pi)$ have the estimate

$$|\gamma_l| \leqslant B \min\left\{\text{meas supp}\, h_l^{(m)},\ \left(\text{meas supp}\, h_l^{(m)}\right)^{-1} l^{-2}\right\}.$$

The inequality (20) follows from this.

Next, (6)–(8) give us that

$$\begin{aligned} H_l^{(m)}(x) - H_{l+1}^{(m)}(x) &= \sqrt{\frac{n}{2\pi}} \int_0^{2\pi} \left(\frac{1}{A_l^{(m)}} h_{l-2^{m-2}}^{(m-1)}\left(\frac{tn}{2\pi}\right) - \frac{1}{A_{l+1}^{(m)}} h_{l+1-2^{m-2}}^{(m-1)}\left(\frac{tn}{2\pi}\right)\right) D_n(x-t)\, d\mu_n(t) \\ &= \sqrt{\frac{n}{2\pi}} \int_0^{2\pi} \left(\frac{1}{A_l^{(m)}} h_{l-2^{m-2}}^{(m-1)}\left(\frac{tn}{2\pi}\right) - \frac{1}{A_{l+1}^{(m)}} h_{l+1-2^{m-2}}^{(m-1)}\left(\frac{tn}{2\pi}\right)\right) \\ &\quad \times \left(D_n(x-t) - D_n\left(x - \frac{2\pi x_l}{n}\right)\right) d\mu_n(t). \end{aligned} \tag{21}$$

Since (see (6))

$$\begin{aligned} \left|D_n(x-\alpha_k) - D_n\left(x - \frac{2\pi x_l}{n}\right)\right| &\leqslant \left|e^{inx} - 1\right| \cdot \left| \frac{1}{e^{i(x-\alpha_k)} - 1} - \frac{1}{e^{i(x-2\pi x_l/n)} - 1}\right| \\ &\leqslant \frac{2\pi^2 \rho(\alpha_k, 2\pi x_l/n)}{\rho(0, x-\alpha_k)\rho(0, x - 2\pi x_l/n)} \end{aligned}$$

for any α_k $(k = 0, 1, \dots, n-1)$, (21) yields the estimate

$$\left|H_l^{(m)}(x) - H_{l+1}^{(m)}(x)\right| \leqslant B n^{-1/2} \cdot 2^{-m} \left[\rho\left(x, \frac{2\pi x_l}{n}\right)\right]^{-2} \tag{22}$$

for $\rho(x, 2\pi x_l/n) \geqslant 2^{-m+2}$.

An inequality of type (19) and inequality (22) are obtained in the same way for the remaining values of l (see (12)–(15)). Combining these inequalities and using (20), (10) and (11), we have that for $x \in [x_q, x_{q+1}]$

$$\begin{aligned} V_k^{(m)}(x) &\leqslant B \Bigg(2^{-3/2m} \sum_{l=0,\, l\neq q,\, q+1}^{2^{m-1}+k-1} 2^{-1/3 \min(|l-k-2^{m-2}|, n-|l-k-2^{m-2}|)} \\ &\quad \times \left[\rho\left(x, \frac{2\pi x_l}{n}\right)\right]^{-2} + 2^{3/2m} \cdot 2^{-2^{m-3}} + 2^{m/2} \cdot 2^{-1/3 \min(|q-k-2^{m-2}|, n-|q-k-2^{m-2}|)} \Bigg) \\ &\leqslant B \min\left(2^{m/2}, 2^{-3/2m} \left[\rho\left(x, \frac{2\pi x_{k+2^{m-2}}}{n}\right)\right]^{-2}\right). \end{aligned} \tag{23}$$

This (see (6), (23)) proves inequality (3) for $m < s$ and $k \geqslant 2^{m-2}$. The proof of (3) is analogous for $k < 2^{m-2}$. Here the initial point of the summation (the point x_0) should be chosen in such a way as to make the first and third sums in (16) small and obtain an estimate of the form (19).

Accordingly, it remains to estimate the functions $V_k^{(s)}(x)$. Assume that $0 \leqslant k \leqslant [n/4]$. Then, writing $V_k^{(s)}(x)$ in the form

$$V_k^{(s)}(x) = \frac{1}{\sqrt{2\pi n}} \sum_{l=[n/2]}^{n+[n/2]-1} v_k^{(s)}(l) D_n(x - \alpha_l)$$

and using the Abel transformation, we get

$$V_k^{(s)}(x) = \frac{1}{\sqrt{2\pi n}} \sum_{l=[n/2]}^{n+[n/2]-2} \sum_{q=[n/2]}^{l} v_k^{(s)}(q)(D_n(x - \alpha_l)$$
$$- D_n(x - \alpha_{l+1})) + \frac{1}{\sqrt{2\pi n}} \sum_{q=[n/2]}^{n+[n/2]-1} v_k^{(s)}(q) D_n(x - \alpha_{n+[n/2]-1}). \tag{24}$$

Since $v_k^{(s)}(q)$ is orthogonal to a constant, the second term in (24) is equal to zero. To estimate the first sum, we introduce the functions $f_l^{(s)}(q)$. Choose numbers m and r such that $p_m^{(s-1)} < [n/2] \leqslant p_{m+1}^{(s-1)}$ and $p_r^{(s-1)} < l \leqslant p_{r+1}^{(s-1)}$. For $[n/2] \leqslant l \leqslant n$ let

$$f_l^{(s)}(q) = \begin{cases} 1 & \text{if } p_{m+1}^{(s-1)} \leqslant q \leqslant p_r^{(s-1)}, \\ 0 & \text{if } 0 \leqslant q \leqslant p_m^{(s-1)},\ p_{r+1}^{(s-1)} \leqslant q \leqslant n, \\ \text{linear on } \left[p_m^{(s-1)}, p_{m+1}^{(s-1)}\right], \left[p_r^{(s-1)}, p_{r+1}^{(s-1)}\right]. \end{cases} \tag{25}$$

And for $0 < l < [n/2]$ let

$$f_l^{(s)}(q) = \begin{cases} 1 & \text{if } p_{m+1}^{(s-1)} < q \leqslant n \text{ or } 0 \leqslant q \leqslant p_r^{(s-1)}, \\ 0 & \text{for } p_{r+1}^{(s-1)} \leqslant q \leqslant p_m^{(s-1)}, \\ \text{linear on } \left[p_r^{(s-1)}, p_{r+1}^{(s-1)}\right], \left[p_m^{(s-1)}, p_{m+1}^{(s-1)}\right]. \end{cases} \tag{26}$$

Now if, for example, $[n/2] < l \leqslant n$, then (see (1.6) and (25))

$$\sum_{q=[n/2]}^{l} v_k^{(s)}(q) = \left(v_k^{(s)}, f_l^{(s)}\right) - \sum_{q=0}^{[n/2]-1} h_{m+1}^{(s-1)}(q) v_k^{(s)}(q)$$
$$+ \sum_{q=[n/2]}^{p_{m+1}^{(s-1)}} h_m^{(s-1)}(q) v_k^{(s)}(q) + \sum_{q=0}^{l} h_{r+1}^{(s-1)}(q) v_k^{(s)}(q)$$
$$- \sum_{q=l+1}^{n} h_r^{(s-1)}(q) v_k^{(s)}(q). \tag{27}$$

From this, considering that $v_k^{(s)}(l)$ is orthogonal to all the functions $f_r^{(s)}(q)$, $r = 0, \ldots, n-1$, and using Theorem 1, we get (see (1.8) and (27))

$$\left| \sum_{q=[n/2]}^{l} v_k^{(s)}(q) \right| \leqslant B\left(2^{-n/16} + 2^{-1/4 \min(|l-k|,\, n-|l-k|)}\right). \tag{28}$$

Inequality (28) can be proved similarly for $0 < l < [n/2]$ by using the functions $f_l^{(s)}(q)$ in the form (26).

On the other hand, the Dirichlet kernels satisfy the inequality

$$|D_n(x - \alpha_l) - D_n(x - \alpha_{l+1})| \leqslant B \min\left(n, n^{-1}[\rho(x, \alpha_l)]^{-2}\right). \tag{29}$$

Combining (24) and (27)–(29), we obtain inequality (3). Finally, if $[n/4] < k < n$, then, since $v_k^{(s)}(l) = 0$ for $l \leqslant k - 4$, the lower limit of the summation in (24) can be taken to be the origin of coordinates, and the functions $f_l^{(s)}(q)$ can be defined in the corresponding way. The rest of the proof stays unchanged. Thus, Theorem 2 is proved.

REMARK 2. Theorem 2 shows that in this case orthogonalization not only preserves the structural properties of the functions, but even improves them essentially. For example, after orthogonalization the functions of the last group have order of decrease $1/x^2$ instead of order of decrease $1/x$, which holds for the Dirichlet kernels.

§3. Uniform boundedness of the basis constants in the spaces A_n

In this section it is established that the spaces A_n have a basis with bounded basis constant. Namely, it is proved that the system $\{V_k^{(m)}(z)\}$ of polynomials introduced in §2 forms a basis in A_n for which the basis constant does not depend on the dimension of A_n. Along with double enumeration of the functions $\{V_k^{(m)}(x)\}$ and $\{v_k^{(m)}(l)\}$, we shall also use single enumeration $\{V_k(x)\}_{k=0}^{n-1}$ and $\{v_k(l)\}_{k=0}^{n-1}$ in what follows.

THEOREM 3. *Let $\Phi(z) \in A_n$. Then*

$$\left\|S_p(\Phi)\right\|_A \leqslant B\|\Phi\|_A \tag{1}$$

for any $0 \leqslant p \leqslant n-1$, where the $S_p(\Phi, z)$ are the partial sums of the Fourier series of $\Phi(z)$ in the system $\{V_k(z)\}$, and B is an absolute constant.

PROOF. In the formula

$$S_p(\Phi, x) = \sum_{k=0}^{p} (\Phi, \overline{V}_k) V_k(x) = \int_0^{2\pi} \Phi(y) \sum_{k=0}^{p} V_k(x)\overline{V}_k(y)\, dy \tag{2}$$

for the partial sums we pass to integration with respect to a discrete measure. This can be done, because (see (2.2))

$$\begin{aligned}(\Phi, \overline{V}_k) &= \frac{1}{n^{3/2}\sqrt{2\pi}} \int_0^{2\pi} \left(\sum_{l=0}^{n-1} \Phi(\alpha_l) D_n(y-\alpha_l) \sum_{q=0}^{n-1} v_k(q) \overline{D}_n(y-\alpha_q) \right) dy \\ &= \sqrt{\frac{2\pi}{n}} \sum_{l=0}^{n-1} \Phi(\alpha_l) v_k(l) \\ &= \frac{2\pi}{n} \sum_{l=0}^{n-1} \Phi(\alpha_l) \overline{V}_k(\alpha_l) = 2\pi \int_0^{2\pi} \Phi(y) \overline{V}_k(y)\, d\mu_n(y) \end{aligned} \tag{3}$$

and, consequently (see (2) and (3)),

$$\begin{aligned} S_p(\Phi, x) &= 2\pi \int_0^{2\pi} \Phi(y) \sum_{k=0}^{p} V_k(x)\overline{V}_k(y)\, d\mu_n(y) \\ &= 2\pi \int_0^{2\pi}\int_0^{2\pi} \Phi(y) D_n(x-t) \sum_{k=0}^{p} V_k(t)\overline{V}_k(y)\, d\mu_n(y)\, d\mu_n(t). \end{aligned} \tag{4}$$

For determining the domain of integration let $|x|$ denote the shortest distance between the point e^{ix} and the point $(1,0)$ along the unit circle. We denote by $F_p(x, y)$ and $f_p(l, q)$ the kernels

$$F_p(x, y) = \sum_{k=0}^{p} V_k(x)\overline{V}_k(y), \tag{5}$$

$$f_p(l, q) = \sum_{k=0}^{p} v_k(l)v_k(q) \tag{6}$$

and represent $S_p(\Phi, x)$ in the form

$$\begin{aligned}\frac{1}{2\pi}S_p(\Phi, x) &= \int_0^{2\pi}\Phi(y)\int_{|t-x|\leqslant 2\pi/p} F_p(t, y)D_n(x-t)\,d\mu_n(t)\,d\mu_n(y)\\ &+\int_0^{2\pi}\Phi(y)\int_{|t-x|>2\pi/p} F_p(t, y)D_n(x-t)\,d\mu_n(t)\,d\mu_n(y)\\ &+\frac{1}{2\pi}\int_{|t-y|>2\pi/p}\Phi(y)D_n(x-y)\,d\mu_n(y)\\ &-\frac{1}{2\pi}\int_{|t-y|>2\pi/p}\Phi(y)D_n(x-y)\,d\mu_n(y).\end{aligned} \tag{7}$$

Breaking up the inside integral of the first term in (7) into two parts, we get

$$\begin{aligned}&\int_{|t-x|\leqslant 2\pi/p} F_p(t, y)D_n(x-t)\,d\mu_n(t)\\ &= \int_{|t-x|\leqslant 2\pi/n} F_p(t, y)D_n(x-t)\,d\mu_n(t)\\ &\qquad+\int_{2\pi/n<|t-x|\leqslant 2\pi/p} F_p(t, y)D_n(x-t)\,d\mu_n(t)\\ &= \int_{|t-x|\leqslant 2\pi/n} F_p(t, y)D_n(x-t)\,d\mu_n(t)\\ &\qquad+(e^{inx}-1)\int_{2\pi/n<|t-x|\leqslant 2\pi/p}\frac{F_p(t, y)}{e^{i(x-t)}-1}\,d\mu_n(t).\end{aligned} \tag{8}$$

Let x_0 be the point in $\{\alpha_l\}$ closest to x. Then for $p \geqslant 2$ the second integral in (8) can be estimated as follows:

$$\begin{aligned}&\left|\int_{2\pi/n<|t-x|\leqslant 2\pi/p}\frac{F_p(t, y)}{e^{i(x-t)}-1}\,d\mu_n(t)\right|\\ &\leqslant \left(\frac{1}{n}\int_{\pi/n<|t-x|}\frac{|F_p(t, y)|}{(x-t)^2}\,d\mu_n(t) + \left|\int_{\pi/n<|t-x_0|\leqslant 3\pi/p}\frac{F_p(t, y)}{e^{i(x_0-t)}-1}\,d\mu_n(t)\right|\right)\\ &\leqslant B\left(\frac{1}{n}\int_{\pi/n<|t-x|}\frac{|F_p(t, y)|}{(x-t)^2}\,d\mu_n(t)\right.\\ &\qquad+\int_{\pi/n<t<3\pi/p}\left|\frac{F_p(x_0+t, y)-F_p(x_0-t, y)}{t}\right|d\mu_n(t)\\ &\qquad\left.+\left|\int_{\pi/n<|t|\leqslant 3\pi/p}F_p(x_0+t, y)\left(\frac{1}{e^{it}-1}-\frac{1}{it}\right)d\mu_n(t)\right|\right).\end{aligned} \tag{9}$$

Since (see (2.2), (5) and (6)) $F_p(\alpha_l, \alpha_q) = (n/2\pi) f_p(l, q)$, while the functions $f_p(l, q)$ are polygonal, we have

$$\int_{\pi/n < t < 3\pi/p} \left| \frac{F_p(x_0 + t, y) - F_p(x_0 - t, y)}{t} \right| d\mu_n(t) \leqslant Bp \int_{-10\pi/p}^{10\pi/p} \left| F_p(x_0 + t, y) \right| d\mu_n(t) \tag{10}$$

Combining (8)–(10) and considering that the Lebesgue functions for the system $\{v_k(l)\}$ are bounded, we get

$$\begin{aligned} &\left| \int_0^{2\pi} \Phi(y) \int_{|t-x| \leqslant 2\pi/p} F_p(t, y) D_n(x - t)\, d\mu_n(t)\, d\mu_n(y) \right| \\ &\leqslant B \int_0^{2\pi} |\Phi(y)| \int_{|t-x| \leqslant 2\pi/p} |F_p(t, y)|\, d\mu_n(t)\, d\mu_n(y) \\ &+ n \int_{|t-x| \leqslant 2\pi/n} d\mu_n(t) \int_{k0}^{2\pi} |\Phi(y) F_p(t, y)|\, d\mu_n(y) \\ &+ \frac{1}{n} \int_{\pi/n < |t-x|} \frac{d\mu_n(t)}{(x-t)^2} \int_0^{2\pi} |\Phi(y) F_p(t, y)|\, d\mu_n(y) \\ &+ p \int_{-10\pi/p}^{10\pi/p} d\mu_n(t) \int_0^{2\pi} |\Phi(y) F_p(x_0 + t, y)|\, d\mu_n(y) \leqslant B \|\Phi\|_A. \end{aligned} \tag{11}$$

Let us now estimate the next two terms in (7). We have (see (2.2)) that for $y = \alpha_q$

$$\int_{|x-t| > 2\pi/p} D_n(x - t) F_p(y, t)\, d\mu_n(t) = \frac{1}{2\pi} \sum_{|x - \alpha_l| > 2\pi/p} D_n(x - \alpha_l) \sum_{k=0}^{p} v_k(l) v_k(q). \tag{12}$$

The expression

$$\sum_{|x - \alpha_l| > 2\pi/p} D_n(x - \alpha_l) \sum_{k=0}^{p} v_k(l) v_k(q)$$

is the partial sum of order p for the function

$$D_n^*(x - \alpha_q) = \begin{cases} e^{inx-1} / e^{i(x-\alpha_q)-1} & \text{for } |x - \alpha_q| > 2\pi/p, \\ 0 & \text{for } |x - \alpha_q| \leqslant 2\pi/p. \end{cases}$$

The boundedness of the Lebesgue functions of the discrete system $\{v_k(l)\}$ gives us that the partial sum of order p in the discrete system $\{v_k(l)\}$ approximates the function $D_n^*(x - \alpha_q)$ just like the polynomial of best approximation with respect to the same system $\{v_k(l)\}$. Therefore (see (12))

$$\left| \int_{2\pi} \Phi(y) \int_{|x-t| \geqslant 2\pi/p} D_n(x - t) F_p(t, y)\, d\mu_n(t)\, d\mu_n(y) - \frac{1}{2\pi} \int_{|x-y| \geqslant 2\pi/p} \Phi(y) D_n(x - y)\, d\mu_n(y) \right| \leqslant B \|\Phi\|_A. \tag{13}$$

We now take the Jackson interpolation polynomial (see [7], Chapter 8, §6) of order p for the functions $\Phi(x+\pi/n)$ and represent it in the form

$$\begin{aligned}
&\int_0^{2\pi} \Phi\left(x+\frac{\pi}{n}\right) K_p(t-x)\, d\mu_n(x) \\
&\quad = \int_0^{2\pi} \Phi(y) \int_0^{2\pi} K_p(t-x) D_n\left(x+\frac{\pi}{n}-y\right) d\mu_n(x)\, d\mu_n(y) \\
&\quad = -2\int_{|t-y|\leqslant 1/p} \Phi(y) \int_0^{2\pi} K_p(t-x) \frac{1}{e^{i(x+\pi/n-y)}-1}\, d\mu_n(x)\, d\mu_n(y) \\
&\quad\quad - \int_{|t-y|>1/p} \Phi(y)\, d\mu_n(y) \\
&\quad\quad \times \int_0^{2\pi} K_p(t-x)\left(\frac{1}{e^{i(x+\pi/n-y)-1}} - \frac{1}{e^{i(t+\pi/n-y)-1}}\right) d\mu_n(x) \\
&\quad\quad -2\int_{|t-y|>1/p} \Phi(y) \frac{1}{e^{i(t+\pi/n-y)-1}}\, d\mu_n(y). \qquad (14)
\end{aligned}$$

Using estimates for the Fejér kernel and its derivative, we get

$$\begin{aligned}
&\left|\int_0^{2\pi} K_p(t-x) \frac{1}{e^{i(x+\pi/n-y)-1}}\, d\mu_n(x)\right| \\
&\quad \leqslant B\left(\int_{|x|\leqslant 1/p} \left|\frac{K_p(t-x-y)-K_p(t+x-2\pi/n-y)}{x+\pi/n}\right| d\mu_n(x)\right. \\
&\quad\quad \left. + \int_{|x-y|>1/p} \left|\frac{K_p(t-x)}{e^{i(x+\pi/n-y)-1}}\right| d\mu_n(x)\right) \leqslant B\left(p+\frac{1}{p}\int_{1/p} \frac{dx}{x^3}\right) \leqslant Bp, \qquad (15)
\end{aligned}$$

and, consequently, the first term in (14) does not exceed $B\|\Phi\|_A$.

Let $|t-x| \leqslant 1/2p$. Then

$$\begin{aligned}
&\left|\int_{|t-y|>1/p} \Phi(y)\, d\mu_n(y)\right. \\
&\quad\quad \left. \times \int_{|t-x|\leqslant 1/2p} K_p(t-x)\left(\frac{1}{e^{i(x+\pi/n-y)-1}} - \frac{1}{e^{i(t+\pi/n-y)-1}}\right) d\mu_n(x)\right| \\
&\quad \leqslant \frac{B}{p}\int_{|t-x|\leqslant 1/2p} K_p(t-x)\, d\mu_n(x) \int_{|t-y|>1/2p} |\Phi(y)| \frac{1}{(t-y)^2}\, d\mu_n(y) \\
&\quad \leqslant B\|\Phi\|_A. \qquad (16)
\end{aligned}$$

The estimates for the Fejér kernel and its derivative can be used as in the proof of (15) to obtain the estimate

$$\int_{|x-y|<1/p} K_p(t-x)\left(\frac{1}{e^{i(x+\pi/n-y)-1}} - \frac{1}{e^{i(t+\pi/n-y)-1}}\right) d\mu_n(x) \leqslant \frac{B}{p|t-y|^2} \qquad (17)$$

for $|t-y| > 1/p$.

Accordingly, it remains to estimate the second term in (14) when the domain of integration satisfies the inequalities $|x-y| > 1/2p$, $|t-y| > 1/2p$ and $|t-x| > 1/2p$. This domain breaks up into the union of the domains where the quantities $|x-y|$, $|t-y|$ and $|t-x|$ are ordered in some way. For example, consider the domain

$$G_t = \{x, y; |x-t| < |t-y| < |x-y|\}.$$

In this case the second term in (14) does not exceed

$$B\|\Phi\|_A \frac{1}{p} \iint_{G_t} \frac{d\mu_n(x)\,d\mu_n(y)}{|x-t|\,|t-y|\,|x-y|} \leqslant B\|\Phi\|_A \frac{1}{p} \int_{1/2p}^{\pi} \frac{dy}{y^2} \int_{1/2p}^{y} \frac{dx}{x} \leqslant B\|\Phi\|_A. \tag{18}$$

The integrals over the other domains of this type can be estimated similarly. It is thereby established (see (14)–(18)) that the last term in (7) also admits an estimate of the type (1).

Theorem 3 is proved.

§4. Existence of an unconditional basis in H_n^1 with bounded basis constant

In this section we establish that the orthogonal system of polynomials $V_k(z)$ introduced in §2 forms in H_n a basis such that the unconditional basis constant does not depend on the dimension of H_n^1.

THEOREM 4. *Let $\Phi(z) \in H_n^1$. Then for any choice of values $\varepsilon_k = \pm 1$*

$$\left\| \sum_{k=0}^{n-1} \varepsilon_k (\Phi, \overline{V}_k) V_k(z) \right\|_{H^1} \leqslant B\|\Phi\|_{H^1}, \tag{1}$$

where the $(\Phi, \overline{V}_k)$ are the Fourier coefficients of $\Phi(z)$ in the system $V_k(z)$.

To prove Theorem 4 we introduce the following equivalent normalization in H_n^1.

THEOREM 5. *The norm of functions $\Phi(z) \in H_n^1$ is equivalent to the norm*

$$\frac{1}{n} \sum_{k=0}^{n-1} |\Phi(\alpha_k)| + \frac{1}{n^2} \left\| \left\{ \sum_{k=0}^{n-1} \Phi(\alpha_k) \frac{1}{e^{i(\alpha_p - \alpha_k + \pi/n)} - 1} \right\} \right\|_1. \tag{2}$$

PROOF. We establish that

$$\|\Phi\|_1 \leqslant B\left(\frac{1}{n} \sum_{k=0}^{n-1} |\Phi(\alpha_k)| + \frac{1}{n_2} \left\| \left\{ \sum_{k=0}^{n-1} \Phi(\alpha_k) \frac{1}{e^{i(\alpha_p - \alpha_k + \pi/n)} - 1} \right\} \right\|_1 \right). \tag{3}$$

For $x \in [\alpha_p, \alpha_{p+1}]$ let $\Phi(x)$ be represented in the form

$$\begin{aligned}
\Phi(x) = {} & \frac{1}{n}(e^{inx} - 1)\Phi(\alpha_p)\left(\frac{1}{e^{i(x-\alpha_p)} - 1} - \frac{1}{e^{i\pi/n} - 1} \right) \\
& + \frac{1}{n}(e^{inx} - 1)\Phi(\alpha_{p+1})\left(\frac{1}{e^{i(x-\alpha_{p+1})} - 1} - \frac{1}{e^{-i\pi/n} - 1} \right) \\
& + \frac{1}{n}(e^{inx} - 1) \sum_{\substack{k=0 \\ k \neq p, p+1}}^{n-1} \Phi(\alpha_k)\left(\frac{1}{e^{i(x-\alpha_k)} - 1} - \frac{1}{e^{i(\alpha_p - \alpha_k + \pi/n)} - 1} \right) \\
& + \frac{1}{n}(e^{inx} - 1) \sum_{k=0}^{n-1} \Phi(\alpha_k) \frac{1}{e^{i(\alpha_p - \alpha_k + \pi/n)} - 1}.
\end{aligned}$$

This implies that for $x \in [\alpha_p, \alpha_{p+1}]$

$$|\Phi(x)| \leqslant B\left(|\Phi(\alpha_p)| + \sum_{\substack{k=0\\k\neq p}}^{n-1} \frac{|\Phi(\alpha_k)|}{(p-k)^2} + \frac{1}{n}\left|\sum_{k=0}^{n-1} \Phi(\alpha_k)\frac{1}{e^{i(\alpha_p-\alpha_k+\pi/n)}-1}\right|\right). \quad (4)$$

Using (4), we get

$$\begin{aligned}\|\Phi\|_1 &\leqslant \frac{1}{n}\sum_{p=0}^{n-1} \max_{x\in[\alpha_p,\alpha_{p+1}]} |\Phi(x)| \\ &\leqslant B\left(\frac{1}{n}\sum_{p=0}^{n-1}|\Phi(\alpha_p)| + \frac{1}{n}\sum_{p=0}^{n-1}\sum_{\substack{k=0\\k\neq p}}^{n-1}\frac{|\Phi(\alpha_k)|}{(p-k)^2}\right. \\ &\qquad \left. + \frac{1}{n^2}\sum_{p=0}^{n-1}\left|\sum_{k=0}^{n-1}\Phi(\alpha)\frac{1}{e^{i(\alpha_p-\alpha_k+\pi/n)}-1}\right|\right) \\ &\leqslant B\left(\frac{1}{n}\sum_{p=0}^{n-1}|\Phi(\alpha_p)| + \frac{1}{n^2}\sum_{p=0}^{n-1}\left|\sum_{k=0}^{n-1}\Phi(\alpha_k)\frac{1}{e^{i(\alpha_p-\alpha_k+\pi/n)}-1}\right|\right).\end{aligned} \quad (5)$$

This (see (5)) proves (3). The reverse inequality

$$\frac{1}{n}\sum_{k=0}^{n-1}|\Phi(\alpha_k)| + \frac{1}{n^2}\left\|\left\{\sum_{k=0}^{n-1}\Phi(\alpha_k)\frac{1}{e^{i(\alpha_p-\alpha_k+\pi/n)}-1}\right\}\right\|_1 \leqslant B\|\Phi\|_1 \quad (6)$$

is obtained by applying the inequality (see [7], Chapter 8, formula (7.11))

$$\left\{\int_0^{2\pi} |P(e^{it})|^p \, d\mu_n(t)\right\}^{1/p} \leqslant B\left(\int_0^{2\pi} |P(e^{it})|^p \, dt\right)^{1/p}, \qquad 1 \leqslant p \leqslant \infty,$$

where $P(z) = c_1 + \cdots + c_n z^{n-1}$, first to $\Phi(z)$, and then to $\Phi(re^{i(t+\pi/n)})$, and by taking into account that

$$\begin{aligned}\left|\Phi\left(\alpha_p + \frac{\pi}{n}\right)\right| &= \frac{1}{n}\left|(e^{in(\alpha_p+\pi/n)}-1)\sum_{k=0}^{n-1}\Phi(\alpha_k)\frac{1}{e^{i(\alpha_p-\alpha_k-\pi/n)}-1}\right| \\ &= \frac{2}{n}\left|\sum_{k=0}^{n-1}\Phi(\alpha_k)\frac{1}{e^{i(\alpha_p-\alpha_k+\pi/n)}-1}\right|.\end{aligned}$$

Thus, Theorem 5 is proved.

Moreover, in proving Theorem 4 we shall use an estimate for the conjugate system

$$\tilde{v}_k^{(m)}(l) = \frac{1}{n}\sum_{q=0}^{n-1}\frac{v_k^{(m)}(q)}{\tan 1/2(\alpha_l - \alpha_q + \pi/n)}.$$

THEOREM 6. *The conjugate system $\tilde{v}_k^{(m)}(l)$ satisfies the inequality*

$$|v_k^{(m)}(l)| \leqslant B \cdot 2^{-3/2m} n^{-1/2}\left[\rho(\beta_k^{(m)}, \alpha)\right]^{-2}. \quad (7)$$

PROOF. This theorem follows from Theorem 2. Indeed, according to (2.2),

$$\begin{aligned}V_k^{(m)}\left(\alpha_l + \frac{\pi}{n}\right) &= \frac{e^{in(\alpha_l+\pi/n)}-1}{\sqrt{2\pi n}}\sum_{q=0}^{n-1} v_k^{(m)}(q)\frac{1}{e^{i(\alpha_l-\alpha_q+\pi/n)}-1} \\ &= \frac{1}{\sqrt{2\pi n}}\sum_{q=0}^{n-1} v_k^{(m)}(q)\left(1 + i\frac{1}{\tan 1/2(\alpha_l-\alpha_q+\pi/n)}\right).\end{aligned}$$

Taking into account that $v_k^{(m)}(q)$ is orthogonal to a constant, we get that

$$\tilde{v}_k^{(m)}(l) = -i\sqrt{\frac{2\pi}{n}} \cdot V_k^{(m)}\left(\alpha_l + \frac{\pi}{n}\right),$$

and (7) follows from this in view of (2.3).

The theorem is proved.

Consider now the discrete space H^1. Coifman and Weiss [8] established that the main properties of H^1—the atomistic characterization, and the duality between H^1 and BMO—remain valid for a broad class of spaces with a measure of homogeneous type in which the distance is regular with respect to the measure. In particular, the group of nth roots of unity with the discrete measure $\mu_n(t)$ is such a space of homogeneous type. Denote by h_n^1 and BMO_n the spaces corresponding to H^1 and BMO, defined on the set of nth roots of unity.

We establish that the discrete orthogonal system introduced in §1 forms an unconditional basis in BMO_n. Moreover, following Carleson and Wojtaszczyk, we prove that the norm in BMO_n is equivalent to a certain unconditional norm. This norm was introduced and used by Carleson [3] in his construction of an unconditional basis in BMO, which is the closure in the BMO norm of the space of continuous functions. Then Wojtaszczyk [4] applied Carleson's proof to the Franklin system and established that the Franklin system is an unconditional basis in BMO and H^1.

An essential role in the proof that the norms are equivalent will be played by the estimates in Theorems 1 and 6 for the discrete system $v_k(l)$ and the conjugate system $\tilde{v}_k(l)$. We introduce some notation used in Theorem 7.

The class of residues modulo n corresponds in a natural way to the nth roots of unity. Denote by $\delta_k^{(m)}$ the intervals forming a partition of the circle and corresponding to the partition of the class of residues that is formed by the numbers $R_k^{(m)}$. The discrete orthonormal system corresponding on the circle to the system $v_k(l)$ and normalized according to the measure $\mu_n(x)$ is denoted by $v_k(x)$. With each function $v_k(x) = v_q^{(m)}(x)$ we associate the interval $\delta_k = \delta_{2q-1}^{(m)}$ for $m < s$, and $\delta_k = \delta_q^{(s)}$ for $v_k(x) = v_q^{(s)}(x)$. Moreover, let Δ denote a nonempty interval on the set of roots of unity.

THEOREM 7. *Let f be a real function defined on the set of roots of unity. Then the norms*

$$\max_{\Delta}\left(\frac{1}{|\Delta|}\int_{\Delta}\left|f(x) - \frac{1}{|\Delta|}\int_{\Delta} f(t)\, d\mu_n(t)\right|^2 d\mu_n(x)\right)^{1/2} \tag{8}$$

and

$$\max_{\Delta}\left(\frac{1}{|\Delta|}\sum_{\delta_k \subseteq \Delta} a_k^2(f)\right)^{1/2}, \tag{9}$$

where $a_k(f) = \int_0^{2\pi} f(x) v_k(x)\, d\mu_n(x)$, are equivalent.

PROOF. The proof of Theorem 7 amounts to reducing one norm to the other and estimating the additional terms which thereby arise. We start by proving that

$$\max_{\Delta}\left(\frac{1}{|\Delta|}\sum_{\delta_k \subseteq \Delta} a_k^2(f)\right)^{-1/2} \leqslant B\|f\|_{\mathrm{BMO}_n}. \tag{10}$$

First of all note that, by the duality between h_n^1 and BMO_n, Theorems 1 and 6 give us

$$|a_k(f)| = \left|\int_0^{2\pi} f(x) v_k(x)\, d\mu_n(x)\right| \leqslant \|f\|_{\mathrm{BMO}_n} \|v_k\|_{h_n^1} \leqslant \frac{B}{\sqrt{k}} \|f\|_{\mathrm{BMO}_n}. \tag{11}$$

Since it follows from (1.3), (1.8), and (11) that

$$\begin{aligned}\left(\int_{C\Delta} \left|\sum_{\delta_k \subseteq \Delta} a_k(f) v_k(x)\right|^2 d\mu_n(x)\right)^{1/2} &\leqslant \sum_{\delta_k \subseteq \Delta} |a_k(f)| \left(\int_{C\Delta} v_k^2(x)\, d\mu_n(x)\right)^{1/2} \\ &\leqslant B\|f\|_{\mathrm{BMO}_n} \sum_{2^{-m} \leqslant |\Delta|} 2^{-m/2} \leqslant B\|f\|_{\mathrm{BMO}_n} |\Delta|^{1/2},\end{aligned} \tag{12}$$

to prove (11) it suffices to prove that

$$\frac{1}{|\Delta|} \int_\Delta \left|\sum_{\delta_k \subseteq \Delta} a_k(f) v_k(x)\right|^2 d\mu_n(x) \leqslant B\|f\|_{\mathrm{BMO}_n}^2. \tag{13}$$

Let α_l and α_q be any two nth roots of unity in Δ. Then, using (1.3), (1.8), and (11) and taking into account the fact that $v_k(x)$ is a polygonal function, we get

$$\begin{aligned}&\left|\sum_{|\delta_k| > |\Delta|} a_k(f) v_k(\alpha_q) - \sum_{|\delta_k| > |\Delta|} a_k(f) v_k(\alpha_l)\right| \\ &\qquad \leqslant B\|f\|_{\mathrm{BMO}_n} \sum_{|\delta_k| > |\Delta|} \frac{1}{\sqrt{k}} (|v_k(\alpha_q) - v_k(\alpha_i)|) \\ &\qquad \leqslant B\|f\|_{\mathrm{BMO}_n} |\Delta| \sum_{2^{-m} > |\Delta|} 2^m \leqslant B\|f\|_{\mathrm{BMO}_n}.\end{aligned} \tag{14}$$

Moreover,

$$\begin{aligned}&\left|\int_\Delta \sum_{|\delta_k| \leqslant |\Delta|} a_k(f) v_k(x)\, d\mu_n(x)\right| \\ &\leqslant B\|f\|_{\mathrm{BMO}_n} \left(\sum_{\substack{|\delta_k| \leqslant |\Delta| \\ \delta_k \subseteq \Delta}} \frac{1}{\sqrt{k}} \left|\int_{C\Delta} v_k(x)\, d\mu_n(x)\right| + \sum_{\substack{|\delta_k| \leqslant |\Delta| \\ \delta_k \subseteq \Delta}} \left|\frac{1}{\sqrt{k}} \int_\Delta v_k(x)\, d\mu_n(x)\right|\right) \\ &\leqslant B\|f\|_{\mathrm{BMO}_n} \sum_{2^{-m} \leqslant |\Delta|} 2^{-m} \leqslant B\|f\|_{\mathrm{BMO}_n} |\Delta|.\end{aligned} \tag{15}$$

By (14) and (15) we get

$$\left|\frac{1}{|\Delta|} \int_\Delta f(x)\, d\mu_n(x) - \sum_{|\delta_k| > |\Delta|} a_k(f) v_k(\alpha_l)\right| \leqslant B\|f\|_{\mathrm{BMO}_n} \tag{16}$$

for $\alpha_l \in \Delta$. But (14) and (16) give us (see (8)) that

$$\int_\Delta \left|\sum_{|\delta_k| \leqslant |\Delta|} a_k(f) v_k(x)\right|^2 d\mu_n(x) \leqslant B|\Delta|\, \|f\|_{\mathrm{BMO}_n}. \tag{17}$$

Since

$$\left(\int_\Delta \left|\sum_{\substack{|\delta_k|\leqslant|\Delta|\\ \delta_k\not\subset\Delta}} a_k(f)v_k(x)\right|^2 d\mu_n(x)\right)^{1/2} \leqslant \sum_{\substack{|\delta_k|\leqslant|\Delta|\\ \delta_k\not\subset\Delta}} |a_k|\left(\int_\Delta |v_k(x)|^2\, d\mu_n(x)\right)^{1/2}$$
$$\leqslant B\|f\|_{\mathrm{BMO}_n} \sum_{2^{-m}\leqslant|\Delta|} 2^{-m/2} \leqslant B\|f\|_{\mathrm{BMO}_n}|\Delta|^{1/2}, \tag{18}$$

inequality (13) follows from (17) and (18).

Inequality (10) is thereby proved (see (12)).

Let $\|f\|_v$ denote the norm defined by (9). To prove the reverse inequality note that inequality (11) and, hence, also (14) and (18) remain in force if the norm $\|f\|_{\mathrm{BMO}_n}$ in them is replaced by the norm $\|f\|_v$. Therefore, for any $\alpha_l \in \Delta$

$$\int_\Delta \left|f(x) - \sum_{|\delta_k|>|\Delta|} a_k(f)v_k(\alpha_k)\right|^2 d\mu_n(x) \leqslant B\|f\|_v^2|\Delta|. \tag{19}$$

Thus (see (9) and (19)), Theorem 7 is proved.

We now proceed to the proof of the main theorem of this section.

PROOF OF THEOREM 4. Let $\Phi(z) \in H_n^1$. Then

$$\Phi(x) = \frac{1}{n}\sum_{k=0}^{n-1} \Phi(\alpha_k)D_n(x-\alpha_k). \tag{20}$$

Let (see (2.2))

$$\Phi(x) = \frac{1}{n}\sum_{k=0}^{n-1} \varepsilon_k(\Phi, \overline{V}_k)V_k(x). \tag{21}$$

It follows from (2.2) and (20) that

$$\left(\Phi, \overline{V}_k\right) = \int_0^{2\pi} \Phi(x)\overline{V}_k(x)\,dx$$
$$= \frac{1}{\sqrt{2\pi}\cdot n^{3/2}}\sum_{l=0}^{n-1}\Phi(\alpha_l)v_k(l) = \sqrt{2\pi}\int_0^{2\pi}\Phi(x)v_k(x)\,d\mu_n(x). \tag{22}$$

Therefore (see (21) and (22)),

$$\Phi^{(\varepsilon)}(\alpha_l) = \sum_{k=0}^{n-1}\varepsilon_k(\Phi, v_k)v_k(\alpha_l), \tag{23}$$

and, consequently (see (21)),

$$\Phi^{(\varepsilon)}(x) = \frac{1}{n}\sum_{l=0}^{n-1}\Phi^{(\varepsilon)}(\alpha_l)D_n(x-\alpha_l) = \frac{1}{n}\sum_{l=0}^{n-1}D_n(x-\alpha_l)\sum_{k=0}^{n-1}\varepsilon_k(\Phi, v_k)v_k(\alpha_l). \tag{24}$$

By Theorem 5,

$$\|\Phi^{(\varepsilon)}(z)\|_{H^1} \leqslant B\left(\frac{1}{n}\sum_{l=0}^{n-1}|\Phi^{(\varepsilon)}(\alpha_l)| + \frac{1}{n^2}\sum_{l=0}^{n-1}\left|\sum_{k=0}^{n-1}\Phi^{(\varepsilon)}(\alpha_k)\frac{1}{e^{i(\alpha_l-\alpha_k+\pi/n)}-1}\right|\right). \tag{25}$$

Since

$$\frac{\Phi^{(\varepsilon)}(\alpha_k)}{e^{i(\alpha_l-\alpha_k+\pi/n)-1}} = \frac{\left(\operatorname{Re}\Phi^{(\varepsilon)}(\alpha_k)+i\operatorname{Im}\Phi^{(\varepsilon)}(\alpha_k)\right)\left(e^{-i(\alpha_l-\alpha_k+\pi/n)}-1\right)}{2(1-\cos(\alpha_l-\alpha_k+\pi/n))}$$

$$= -\frac{1}{2}\left(\operatorname{Re}\Phi^{(\varepsilon)}(\alpha_k) - \frac{\operatorname{Im}\Phi^{(\varepsilon)}(\alpha_k)}{\tan 1/2(\alpha_l-\alpha_k+\pi/n)}\right.$$

$$\left. + i\operatorname{Im}\Phi^{(\varepsilon)}(\alpha_k) + i\frac{\operatorname{Re}\Phi^{(\varepsilon)}(\alpha_k)}{\tan 1/2(\alpha_l-\alpha_k+\pi/n)}\right), \tag{26}$$

it follows (see (25) and (26)) that

$$\left\|\Phi^{(\varepsilon)}(z)\right\|_{H^1} \leqslant B\left(\left\|\operatorname{Re}\Phi^{(\varepsilon)}(\alpha_l)\right\|_{h_n^1} + \left\|\operatorname{Im}\Phi^{(\varepsilon)}(\alpha_l)\right\|_{h_n^1}\right). \tag{27}$$

We estimate $\|\operatorname{Re}\Phi^{(\varepsilon)}(\alpha_l)\|_{h_n^1}$. By the duality between h_n^1 and BMO_n, there exists a function $G(x)$ defined on the set of nth roots of unity such that $\|G(x)\|_{\mathrm{BMO}_n} = 1$ and

$$\left\|\operatorname{Re}\Phi^{(\varepsilon)}(\alpha_l)\right\|_{h_n^1} = \int_0^{2\pi} \operatorname{Re}\Phi^{(\varepsilon)}(x)G(x)\,d\mu_n(x)$$

$$= \int_0^{2\pi} \operatorname{Re}\Phi(x)G^{(\varepsilon)}(x)\,d\mu_n(x) \leqslant B\|\operatorname{Re}\Phi\|_{h_n^1}\left\|G^{(\varepsilon)}(x)\right\|_{\mathrm{BMO}_n}, \tag{28}$$

where

$$G^{(\varepsilon)}(x) = \sum_{k=0}^{n-1} \varepsilon_k(G, v_k)v_k(x).$$

Applying Theorem 5 and 7, we find that (see (28))

$$\left\|\operatorname{Re}\Phi^{(\varepsilon)}(\alpha_l)\right\|_{h_n^1} \leqslant B\|\Phi(z)\|_{H^1}\left\|G^{(\varepsilon)}(x)\right\|_v \leqslant B\|\Phi(z)\|_{H^1}\|G(x)\|_v$$

$$\leqslant B\|\Phi(z)\|_{H^1}\|G(x)\|_{\mathrm{BMO}_n} \leqslant B\|\Phi(z)\|_{H^1}. \tag{29}$$

The norm $\|\operatorname{Im}\Phi^{(\varepsilon)}(\alpha_l)\|_{h_n^1}$ can be estimated similarly. Consequently (see (1), (29) and (27)), Theorem 4 is proved.

§5. The conjugate Franklin system

In this section inequalities of the same type as obtained earlier by the author for the Franklin system (see [1] and [9]) are established for the system $\{\tilde{f}_k(x)\}$ trigonometrically conjugate to the Franklin system. We remark that in considering the system trigonometrically conjugate to the Franklin system it is natural to consider the periodic variant of the Franklin system obtained when the Schauder function $\varphi_1(x) = x$ is excluded from the system under orthogonalization.

First of all we establish that the functions $\{\tilde{f}_k(x)\}$ satisfy the following inequality.

THEOREM 8. *The system $\{\tilde{f}_k(x)\}$ of functions trigonometrically conjugate to the Franklin system satisfies the inequalities*

$$|\tilde{f}_{2^m+k}(x)| \leqslant B\min\left[\sqrt{2^m}, \frac{1}{2^{3/2m}(\rho(x, \pi k/2^{m-1}))^2}\right] \tag{1}$$

for all $m = 0, 1, \dots$ and $k = 1, \dots, 2^m$.

PROOF. The functions $f_{2^m+k}(x)$ have nodes at the points $\pi l/2^m$ for $0 \leqslant l \leqslant 2k$ and $\pi p/2^{m-1}$ for $k < p \leqslant 2^m$. Assume that $2^{m-2} \leqslant k < 3 \cdot 2^{m-2}$, and let

$$F_{2^m+k}(x) = \int_0^x f_{2^m+k}(t)\,dt. \tag{2}$$

If $0 \leqslant k < 2^{m-2}$ or if $3 \cdot 2^{m-2} \leqslant k < 2^m$, then the lower limit of the integration in (2) should be taken to be the point π.

Let $x \in [\pi l/2^{m-1}, \pi(l+1)/2^{m-1}]$. The function $f_{2^m+k}(x)$ is orthogonal to the function $g_{2^m+k}(x)$ defined as follows:

$$g_{2^m+k}(x) = \begin{cases} 1 & \text{for } \pi/2^{m-1} \leqslant x \leqslant \pi l/2^{m-1}, \\ 0 & \text{for } \pi(l+1)/2^{m-1} \leqslant x \leqslant 2\pi, \\ \text{linear on } [0, \pi/2^{m-1}],\ [\pi l/2^{m-1}, \pi(l+1)/2^{m-1}]. \end{cases} \tag{3}$$

Therefore (see (2) and (3)),

$$|F_{2^m+k}(x)| \leqslant \int_0^{\pi/2^{m-1}} |f_{2^m+k}(t)|dt + \int_{\pi l/2^{m-1}}^{\pi(l+1)/2^{m-1}} |f_{2^m+k}(t)|dt. \tag{4}$$

Let us represent $\tilde{f}_{2^m+k}(x)$ in the form

$$\begin{aligned} -\pi\tilde{f}_{2^m+k}(x) &= \int_{x-1/2^m}^{x+1/2^m} \frac{f_{2^m+k}(t)}{2\tan\frac{1}{2}(t-x)}dt + \int_0^{x-1/2^m} \frac{f_{2^m+k}(t)}{2\tan\frac{1}{2}(t-x)}dt \\ &\quad + \int_{x+1/2^m}^{2\pi} \frac{f_{2^m+k}(t)}{2\tan\frac{1}{2}(t-x)}dt \\ &= -\frac{1}{2}\cot\frac{1}{2^{m+1}}\left[F_{2^m+k}\left(x-\frac{1}{2^m}\right) + F_{2^m+k}\left(x+\frac{1}{2^m}\right)\right] \\ &\quad + \int_{x-1/2^m}^{x+1/2^m} \frac{f_{2^m+k}(t)}{2\tan\frac{1}{2}(t-x)}dt \\ &\quad + \frac{1}{4}\int_0^{x-1/2^m} \frac{F_{2^m+k}(t)}{\sin^2\frac{1}{2}(t-x)}dt + \frac{1}{4}\int_{x+1/2^m}^{2\pi} \frac{F_{2^m+k}(t)}{\sin^2\frac{1}{2}(t-x)}dt. \end{aligned} \tag{5}$$

Using the fact that the function $f_{2^m+k}(t)$ is polygonal, we get

$$\left|\int_{x-1/2^m}^{x+1/2^m} \frac{f_{2^m+k}(t)}{2\tan\frac{1}{2}(t-x)}dt\right| \leqslant B\max|f_{2^m+k}(x+t)|, \qquad |t| \leqslant 2^{3-m}. \tag{6}$$

The Franklin system satisfies an inequality of Ciesielski [10], which in the periodic case has the form

$$\left|f_{2^m+k}\left(\frac{\pi n}{2^m}\right)\right| \leqslant B \cdot 2^{m/2} r^{\min(|n-2k|, 2^{m+1}-|n-2k|)}, \qquad 0 < r < 1. \tag{7}$$

By (4) and (7),

$$\left|F_{2^m+k}\left(x-\frac{1}{2^m}\right)+F_{2^m+k}\left(x+\frac{1}{2^m}\right)\right|\cot\frac{1}{2^{m+1}}$$
$$\leqslant B\cdot 2^{m/2}\left(r^{\min(|l-k|,2^m-|l-k|)}+r^{2^{m-2}}\right). \tag{8}$$

It remains to estimate the last two terms in (5). By (4) and (7),

$$\left|\int_0^{x-1/2^m}\frac{F_{2^m+k}(t)}{\sin^2\frac{1}{2}(t-x)}dt\right|+\left|\int_{x+1/2^m}^{2\pi}\frac{F_{2^m+k}(t)}{\sin^2\frac{1}{2}(t-x)}dt\right|$$
$$\leqslant B\cdot 2^{-3/2m}\left[\sum_{p=1}^{l-1}\frac{1}{\sin^2\frac{1}{2}\left(\frac{\pi p}{2^{m-1}}-x\right)}\left(r^{\min(|p-k|,2^m-|p-k|)}+r^{2m-2}\right)\right.$$
$$\left.+\sum_{p=l+2}^{2^m}\frac{1}{\sin^2\frac{1}{2}\left(\frac{\pi p}{2^{m-1}}-x\right)}\left(r^{\min(|p-k|,2^m-|p-k|)}+r^{2^{m-2}}\right)\right]$$
$$\leqslant B\min\left[2^{m/2},2^{-3/2m}\left(\rho\left(\frac{\pi k}{2^{m-1}},x\right)\right)^{-2}\right]. \tag{9}$$

Thus (see (5), (6), (8) and (9)), Theorem 8 is proved.

THEOREM 9. *Let $g(x)\in L(0,2\pi)$. Then for any $y>0$*

$$\operatorname{meas}\left\{\left(\sum_{k=1}^{\infty}a_k^2\tilde{f}_k^2(x)\right)^{1/2}>y\right\}\leqslant\frac{B}{y}\int_0^{2\pi}|g(x)|dx, \tag{10}$$

where $a_k=\int_0^{2\pi}g(x)f_k(x)\,dx$.

PROOF. The proof is based on the estimate of the conjugate Franklin system contained in Theorem 8. A modified version of a well-known decomposition lemma is used in the proof (see [11]).

LEMMA. *Suppose that $g(x)\in L(0,2\pi)$, $g(x)\geqslant 0$, and $y>2\int_0^{2\pi}g(x)\,dx$. Then there exist functions $g_1(x)$ and $g_2(x)$ and an open set $Q\subseteq[0,2\pi]$ such that*

$$Q=\cup\Delta_i, \tag{11}$$

where the $\Delta_i=[\alpha_i,\beta_i]$ are dyadic intervals such that $(\alpha_i,\beta_i)\cap(\alpha_j,\beta_j)=\varnothing$ for $i\neq j$;

$$|\Delta_i|\leqslant\rho(\Delta_i,P)\leqslant 4|\Delta_i|, \tag{12}$$

where $\rho(\Delta_i,P)$ is the distance from Δ_i to the set $P=[0,2\pi]\setminus Q$;

$$g(x)=g_1(x)+g_2(x); \tag{13}$$

$$|g_1(x)|\leqslant By \tag{14}$$

for almost all $x\in[0,2\pi]$;

$$g_2(x)=0\quad\text{for } x\in P; \tag{15}$$

$$\operatorname{meas}Q\leqslant\frac{2}{y}\int_0^2 g(x)\,dx; \tag{16}$$

$$\int_{\Delta_i} g_2(x)\,dx = 0 \quad \textit{and} \quad \int_{\alpha_i}^{\beta_i} dx \int_{\alpha_i}^{x} g_2(t)\,dt = 0 \tag{17}$$

for any Δ_i; *and*

$$|G_2(x)| \leqslant By|\Delta_i| \quad \textit{for } x \in \Delta_i, \tag{18}$$

where $G_2(x) = \int_0^x g_2(t)\,dt$.

The proof of this lemma without the relation (12) is contained in the author's papers [1] and [9]. To obtain (12), take a Whitney decomposition (see [12], Chapter II, §1) of the set Q, where $Q = \{H(x, g) > y\}$, $H(x, g)$ being the Hardy-Littlewood maximal function. The rest of the proof is just as in [1] and [9].

We proceed to the proof of (10). It can be assumed that $g(x) \geqslant 0$ and $y > 2\int_0^{2\pi} g(x)\,dx$. Since $g_1(x)$ satisfies (see (13)–(16))

$$\int_0^{2\pi} g_1^2(x)\,dx = \int_P g^2(x)\,dx + \int_Q g_1^2(x)\,dx$$
$$\leqslant y\int_0^{2\pi} g(x)\,dx + By^2 \operatorname{meas} Q \leqslant By\int_0^{2\pi} g(x)\,dx, \tag{19}$$

it suffices to prove that

$$\sum_{k=1}^{\infty} b_k^2 \int_P \tilde{f}_k^2(x)\,dx \leqslant By\int_0^{2\pi} g(x)\,dx, \tag{20}$$

where $b_k = \int_0^{2\pi} g_2(x) f_k(x)\,dx$.

First of all, note that, in view of (17),

$$\int_{\Delta_i} g_2(x) f_k(x)\,dx = 0 \tag{21}$$

if $|\Delta_i| \leqslant \pi/2^m$. Therefore (see (11), (15), (20) and (21)),

$$b_k = \int_{Q_m} g_2(x) f_k(x)\,dx, \tag{22}$$

where $Q_m = \bigcup_{|\Delta_i| > \pi/2^m} \Delta_i$.

Let $R_l^{(m)} = [(l-1)\pi/2^m, l\pi/2^m]$. Using (18) and (7), we get that for $\pi k/2^{m-1} \in \Delta_{i(k)}$ and $\Delta_{i(k)} \subseteq Q_m$ (see (22))

$$|b_{2^m+k}| = \left| \int_{Q_m} G_2(x) f'_{2^m+k}(x)\,dx \right|$$
$$\leqslant \sum_{l=1}^{2^{m+1}} \left| \int_{R_l^{(m)} \cap Q_m} G_2(x) f'_{2^m+k}(x)\,dx \right| \leqslant By \cdot 2^{m/2} |\Delta_{i(k)}|. \tag{23}$$

By (1), we have for $\pi k/2^{m-1} \in \Delta_{i(k)}$ that (see (12))

$$\int_P \tilde{f}_{2^m+k}^2(x)\,dx \leqslant B\frac{1}{2^{3m}|\Delta_{i(k)}|^3}. \tag{24}$$

Consequently (see (23), (24)),

$$\sum_{m=0}^{\infty} \sum_{\pi k/2^{m-1} \in Q_m} b_{2^m+k}^2 \int_P \tilde{f}_{2^m+k}^2(x)\,dx \leqslant By^2 \sum_{m=0}^{\infty} \frac{1}{2^{2m}} \sum_{\pi k/2^{m-1} \in Q_m} \frac{1}{|\Delta_{i(k)}|}$$
$$\leqslant By^2 \sum_{m=0}^{\infty} \frac{1}{2^m} \sum_{\Delta_i \subseteq Q_m} 1 \leqslant By^2 \operatorname{meas} Q \leqslant By\int_0^{2\pi} g(x)\,dx. \tag{25}$$

Now let $\{\delta_i^{(m)}\}$ be the open intervals adjoining Q_m. Then

$$\sum_{m=0}^{\infty} \sum_{\delta_i^{(m)}} \sum_{\pi k/2^{m-1} \in \delta_i^{(m)}} b_{2^m+k}^2$$

$$\leqslant \sum_{m=0}^{\infty} \sum_{\delta_i^{(m)}} \sum_{\pi k/2^{m-1} \in \delta_i^{(m)}} \left(\sum_{l=1}^{2^{m+1}} \left| \int_{R_l^{(m)} \cap Q_m} G_2(x) f'_{2^m+k}(x)\, dx \right| \right)^2$$

$$\leqslant By^2 \sum_{m=0}^{\infty} 2^m \sum_{\Delta_i \subseteq Q_m} |\Delta_i|^2 \leqslant By^2 \operatorname{meas} Q \leqslant By \int_0^{2\pi} g(x)\, dx. \qquad (26)$$

Thus (see (19), (20), (25) and (26)), Theorem 9 is proved.

The proof of the following result is similar.

THEOREM 10. *Suppose that $g(x) \in L(0, 2\pi)$ and $y > 0$. Then for any choice of signs $\varepsilon_k = \pm 1$*

$$\operatorname{meas}\left\{ \left| \sum_{k=1}^{\infty} \varepsilon_k a_k \tilde{f}_k(x) \right| > y \right\} \leqslant \frac{B}{y} \int_0^{2\pi} |g(x)| dx,$$

where $a_k = \int_0^{2\pi} g(x) f_k(x)\, dx$.

BIBLIOGRAPHY

[1] S. V. Bochkarev, *Existence of a basis in the space of functions analytic in the disk, and some properties of the Franklin system*, Mat. Sb. **95(137)** (1974), 3–18; English transl. in Math. USSR Sb. **24** (1974).

[2] B. Maurey, *Isomorphisms entre espaces H_1*, Acta Math. **145** (1980), 79–120.

[3] Lennart Carleson, *An explicit unconditional basis in H_1*, Bull. Sci. Math. (2) **104** (1980), 405–416.

[4] P. Wojtaszczyk, *The Franklin system is an unconditional basis in H_1*, Ark. Mat. **20** (1982), 293–300.*

[5] S. V. Bochkarev, *Construction of bases in finite-dimensional spaces of functions analytic in the disk*, Dokl. Akad. Nauk SSSR **264** (1982), 1295–1297; English transl. in Soviet Math. Dokl. **25** (1982).

[6] Z. Ciesielski, *Properties of the orthonormal Franklin system*, Studia Math. **23** (1963), 141–157.

[7] A. Zygmund, *Trigonometric series*, 2nd rev. ed., Vol. 2, Cambridge Univ. Press, 1959.

[8] Ronald R. Coifman and Guido Weiss, *Extensions of Hardy spaces and their use in analysis*, Bull. Amer. Math. Soc. **83** (1977), 569–645.

[9] S. V. Bochkarev, *A method of averaging in the theory of orthogonal series and some problems in the theory of bases*, Trudy Mat. Inst. Steklov. **146** (1978); English transl., Proc. Steklov Inst. Math. **1980**, no. 3(146).

[10] Z. Ciesielski, *Properties of the orthonormal Franklin system*. II, Studia Math. **27** (1966), 289–323.

[11] A. Zygmund, *On a theorem of Marcinkiewicz concerning interpolation of operations*, J. Math. Pures Appl. (9) **35** (1956), 223–248.

[12] Elias M. Stein, *Singular integrals and differentiability properties of functions*, Princeton Univ. Press, Princeton, N. J., 1970.

Translated by H. H. McFADEN

**Editor's note*. The Russian cites a preprint of this paper.

Труды Матем. Инст. Стеклов.
164(1983)

Proceedings of the Steklov
Institute of Mathematics
1985, Issue 2

MULTIPLY INTERPOLATING SPLINES OF DEGREE $2k + 1$ AND DEFICIENCY k

UDC 517:518.12

N. L. ZMATRAKOV AND YU. N. SUBBOTIN

ABSTRACT. This paper investigates the approximation properties of polynomial splines of degree $2k + 1$ and deficiency k that interpolate a $(k - 1)$-times differentiable function together with its derivatives up to order $(k - 1)$ at the knots of the net Δ_n: $a = x_0^{(n)} < x_1^{(n)} < \cdots < x_{N_n}^{(n)} = b$. Results that are definitive, in a certain sense, are obtained for the convergence in L_p $(1 \leqslant p \leqslant \infty)$ of the splines and their derivatives under restrictions on the relative lengths of the successive intervals of Δ_n.

Bibliography: 17 titles.

We investigate the interpolation properties of splines of degree $2k + 1$ and deficiency k that interpolate a $(k - 1)$-times continuously differentiable function together with its derivatives up to order $k - 1$ at the knots of the net

$$\Delta_n\colon a = x_0^{(n)} < x_1^{(n)} < \cdots < x_{N_n}^{(n)} = b. \tag{1.1}$$

We consider only periodic functions and periodic splines, and Δ_n is to be continued to the whole line, i.e. $x_i^{(n)} = x_{i+N_n}^{(n)} + b - a$ for all i. Our results can be extended to other types of boundary conditions (see, for example, [1]–[3]).

Our main concern in this paper is the approximation properties of the splines under local restrictions on the sequence of nets Δ_n:

$$\sup_n \sup_{|i-j|=1} h_i^{(n)}/h_j^{(n)} \leqslant \rho < \infty, \qquad h_i^{(n)} = x_i^{(n)} - x_{i-1}^{(n)}, \tag{1.2}$$

although we shall incidentally also consider the global restriction

$$\sup_n \sup_{i,j} h_i^{(n)}/h_j^{(n)} \leqslant M < \infty. \tag{1.3}$$

The case $k = 1$ has been thoroughly investigated (see [1]–[11] and the references there).

Sharma and Meir [4], [5] studied the uniform convergence of cubic splines and their third derivatives under (1.2). The first author [10], [11] studied the convergence of the derivatives in the L_p metric under (1.2).

The case $k = 2$ was first considered in [4] and [12].

We introduce some notation. We denote by C the space of $(b - a)$-periodic functions that are continuous on the real axis, with the norm

$$\|f\| = \|f\|_\infty = \sup_{-\infty < x < \infty} |f(x)|,$$

1980 *Mathematics Subject Classification*. Primary 41A05, 41A15.

and by L_p $(1 \leqslant p < \infty)$ the space of $(b-a)$-periodic functions that are measurable on the real axis, with the finite norm

$$\|f\|_p = \left(\int_a^b |f(x)|^p\,dx\right)^{1/p}.$$

For simplicity of notation, we write $C = L_\infty$.

In addition, we denote by C^l the space of $(b-a)$-periodic functions that are l times continuously differentiable on the real axis ($C^0 = C$), and by W_p^l $(1 \leqslant p \leqslant \infty;$ $l = 1, 2, \ldots;$ $W_\infty^l = C^l)$ the space of $(b-a)$-periodic functions with $(l-1)$th derivative absolutely continuous on the real axis and $f^{(l)} \in L_p$. As usual, if $f \in L_p$ $(1 \leqslant p \leqslant \infty)$ we define its modulus of continuity by

$$\omega(f, \delta)_p = \sup_{|h| \leqslant \delta} \|f(\cdot + h) - f(\cdot)\|_p.$$

For $p = \infty$ we sometimes write $\omega(f, \delta)_\infty = \omega(f, \delta)$.

DEFINITION. We call S_n a *multiply interpolating spline* (m.i.s.) *of degree* $2k+1$ *and deficiency* k *for the function* $f \in C^{k-1}$ $(k = 1, 2, \ldots)$ if

a) $S_n^{(j)}(x_i^{(n)}) = f^{(j)}(x_i^{(n)}), j = 0, 1, \ldots, k-1$, i any integer,

b) $S_n \in C^{k+1}$, and

c) S_n is an algebraic polynomial of degree at most $2k+1$ on each interval $[x_{i-1}^{(n)}, x_i^{(n)}]$.

To simplify the terminology, in this paper multiply interpolating splines of degree $2k+1$ of deficiency k will be referred to simply as splines and the indices n on $x_i^{(n)}$ and $h_i^{(n)}$ will be suppressed whenever this will cause no confusion. In addition, we set $\delta_n = \max_i h_i^{(n)}$. In general, we denote by C_i $(i = 1, 2, \ldots)$ various absolute positive constants.

The results on the uniform convergence of m.i.s. and their derivatives were obtained by Subbotin and announced in [13]. The corresponding results in the L_p metrics $(1 \leqslant p < \infty)$ were established by Zmatrakov.

We note that this paper makes extensive use of the methods developed in [8]–[11], since the corresponding matrices have tridiagonal structure.

We investigate the convergence of the lth $(l = 0, 1, \ldots, 2k+1)$ derivatives of m.i.s. in the L_p metrics $(1 \leqslant p \leqslant \infty)$ for the classes C^{k-1} and W_p^l $(l = k, k+1, \ldots, 2k+1)$. It is shown, in particular, that only in the cases $l = k$, $p = \infty$; $l = k+2$, $p = 1$; and $l = k+1$, $1 \leqslant p \leqslant \infty$ does the lth derivative of the m.i.s. always converge in the L_p metric if merely $\delta_n \to 0$ $(n \to \infty)$, whereas in the remaining cases the convergence depends on the distribution of the knots. It is also shown that under (1.3) the m.i.s. and their derivatives converge for the corresponding classes of functions.

There is also interest in the problem of determining definitive conditions of the type (1.2) that guarantee the convergence of the polynomial splines of degree $2k+1$ and deficiency s $(1 \leqslant s \leqslant k)$ that interpolate an $s-1$ times continuously differentiable function together with its derivatives up to order $s-1$ at the knots of Δ_n. For $s = 1$ a definitive condition on ρ in (1.2) that guarantees the uniform convergence of the interpolatory splines for every continuous function was obtained by Friedland and Micchelli [14], [15], and under (1.3) by de Boor [16]. Splines of maximal deficiency $s = k + 1$ are Hermite splines, whose convergence in the L_p metric was established by Birkhoff, Schultz and Varga [17]. We shall use Hermite splines for intermediate approximations.

§2. Equations for splines

To simplify the formulas we shall understand that terms containing $\nu! = 1 \cdot 2 \cdot \cdots \cdot \nu$ with negative ν are suppressed in sums; $0! = 1$, $C_n^m = n!/m!(n-m)!$ and $C_0^0 = 1$. Let a function $f \in C^{k-1}$ be given. We introduce the notation $t = (x - x_{i-1})/h_i$, $h_i = x_i - x_{i-1}$, $f_j^{(m)} = f^{(m)}(x_j)$, and, for $l = 0, 1, \ldots, 2k+1$,

$$r_i^{(l)}(x) = (k+1) \sum_{m=0}^{k-1} \sum_{\mu=0}^{k-m} h_i^{m-l}$$

$$\times \left\{ f_{i-1}^{(m)} \left[\sum_{j=0}^{l} (-1)^j C_l^j C_{m+\mu}^m \frac{(k+\mu)!}{(k+1-j)!(m+\mu-l+j)!} (1-t)^{k+1-j} t^{m+\mu-l+h} \right] \right.$$

$$+ (-1)^m f_i^{(m)} \left[\sum_{j=0}^{l} (-1)^{l-j} C_l^j C_{m+\mu}^m \frac{(k+\mu)!}{(k+1-j)!(m+\mu-l+j)!} \right.$$

$$\left. \left. \times t^{k+1-j} (1-t)^{m+\mu-l+j} \right] \right\}, \qquad (2.1)$$

$$\Phi_l^1(x) = \sum_{j=0}^{l} (-1)^j C_l^j \frac{(k+1)!}{(k+1-j)!(k-l+j)!} t^{k-l+j} (1-t)^{k+1-j},$$

$$\Phi_l^2(x) = (-1)^l \Phi_l^1(x_{i-1} + x_i - x).$$

If $f \in C^k$, we can construct the corresponding Hermite spline $H(x)$ (see, for example, [3], p. 81) of order $2k+1$, whose lth ($l = 0, 1, \ldots, 2k+1$) derivative has the following form on $[x_{i-1}, x_i]$:

$$H_i^{(l)}(x) = h_i^{k-l} \left[f_{i-1}^{(k)} \Phi_l^1(x) + (-1)^k f_i^{(k)} \Phi_l^2(x) \right] + r_i^{(l)}(x), \qquad (2.2)$$

and if $f \in C^{k-1}$ we obtain, for the lth ($l = 0, 1, \ldots, 2k+1$) derivative of the multiply interpolating spline S_n, on $[x_{i-1}, x_i]$,

$$S_n^{(l)}(x) = h_i^{k-l} \left[m_{i-1} \Phi_l^1(x) + (-1)^k m_i \Phi_l^2(x) \right] + r_i^{(l)}(x). \qquad (2.3)$$

where $m_j = S_n^{(k)}(x_j)$ for all integral j.

If $f_j^{(m)} = 0$ ($j = i-1, i$; $m = 1, \ldots, k$) then

$$H_i^{(k+1)}(x_i) = r_i^{(k+1)}(x_i) = (-1)^k (k+1)! C_{2k+1}^k h_i^{-k-1} (f_i - f_{i-1}),$$

$$H_i^{(k+1)}(x_{i-1}) = r_i^{(k+1)}(x_{i-1}) = (k+1)! C_{2k+1}^k h_i^{-k-1} (f_i - f_{i-1})$$

and therefore in this case

$$\left[r_{i+1}^{(k+1)}(x_i) - r_i^{(k+1)}(x_i) \right]/(k+1) = \left[H_{i+1}^{(k+1)}(x_i) - H_i^{(k+1)}(x_i) \right]/(k+1)$$

$$= k! C_{2k+1}^k \left[\frac{f_{i+1} - f_i}{h_{i+1}^{k+1}} + (-1)^{k+1} \frac{f_i - f_{i-1}}{h_i^{k+1}} \right]. \qquad (2.4)$$

Let

$$\begin{aligned} v_i &= \left[r_{i+1}^{(k+1)}(x_i) - r_i^{(k+1)}(x_i)\right]/(k+1), \\ u_i &= \left[H_{i+1}^{(k+1)}(x_i) - H_i^{(k+1)}(x_i)\right]/(k+1). \end{aligned} \tag{2.5}$$

We have the following representation for the $(k+1)$th one-sided derivatives of S_n:

$$S_n^{(k+1)}(x_i - 0) = \frac{k+1}{h_i}\left[(-1)^{k+1} m_{i-1} + (k+1) m_i\right] + r_i^{(k+1)}(x_i),$$

$$S_n^{(k+1)}(x_i + 0) = \frac{k+1}{h_{i+1}}\left[-(k+1) m_i + (-1)^k m_{i+1}\right] + r_{i+1}^{(k+1)}(x_i).$$

From the condition $S_n \in C^{k+1}$ we obtain the system of equations

$$\begin{aligned} (-1)^{k+1}\frac{h_{i+1}}{h_i + h_{i+1}} m_{i-1} + (k+1) m_i + (-1)^{k+1}\frac{h_i}{h_i + h_{i+1}} \\ = \frac{h_i h_{i+1}}{h_i + h_{i+1}} v_i \qquad (i = 1, 2, \ldots, N). \end{aligned} \tag{2.6}$$

If $f \in C^k$ and $l_i = m_i - f_i^{(k)}$, then in terms of l_i the system (2.6) takes the form

$$\begin{aligned} (-1)^{k+1}\frac{h_{i+1}}{h_i + h_{i+1}} l_{i-1} + (k+1) l_i k + (-1)^{k+1}\frac{h_i}{h_i + h_{i+1}} l_{i+1} \\ = \frac{h_i h_{i+1}}{h_i + h_{i+1}} u_i \qquad (i = 1, 2, \ldots, N). \end{aligned} \tag{2.7}$$

For $f \in W_1^k$ we set

$$\bar{\nu}_i = \bar{\nu}_i(f) = h_i^{-1}\int_{x_{i-1}}^{x_i} f^{(k)}(x)\,dx, \qquad \bar{\nu}_i = \bar{\nu}_{i+N} \text{ for all } i. \tag{2.8}$$

Just as in (2.3) we define the spline $\overline{H}(x) = \overline{H}(f, x)$ whose lth ($l = 1, 2, \ldots, 2k+1$) derivative on $[x_{i-1}, x_i]$ has the form

$$\overline{H}_i^{(l)}(x) = h_i^{k-l}\left[\bar{\nu}_{i-1}\Phi_l^1(x) + \bar{\nu}_i\Phi_l^2(x)\right] + r_i^{(l)}(x). \tag{2.9}$$

Let $\bar{l} = m_i - \bar{\nu}_i$; then we have, by reasoning similar to that for (2.7),

$$\begin{aligned} (-1)^{k+1}\frac{h^{i+1}}{h_i + h_{i+1}}\bar{l}_{i-1} + (k+1)\bar{l}_i + (-1)^{k+1}\frac{h_i}{h_i + h_{i+1}}\bar{l}_{i+1} \\ = \frac{h_i h_{i+1}}{(h_i + h_{i+1})}\left[\overline{H}_{i+1}^{(k+1)}(x_i) - \overline{H}_i^{(k+1)}(x_i)\right] = \frac{h_i h_{i+1}}{h_i + h_{i+1}}\bar{u}_i. \end{aligned} \tag{2.10}$$

For $f \in C^l$ $(k+1 \leqslant l \leqslant 2k+1)$ we introduce the notation

$$R = 2^{l-2k-1}\sum_{j=l-k}^{k+1}(-1)^j C_l^j \frac{(k+1)!}{(k+1-j)!(k-l+j)!},$$

$$\begin{aligned} &y_i = (x_{i-1} + x_i)/2, \quad a_i = S_n^{(l)}(y_i), \quad gb_i = H_i^{(l)}(y_i), \\ &d_i = a_i - b_i, \quad l_i = m_i - f_i^{(k)}, \quad m_i = S_n^{(k)}(x_i). \end{aligned} \tag{2.11}$$

It is easy to show that the constant $R = R(k, l) \neq 0$ for all integers $k \geqslant 1$ and $l \geqslant 0$. From (2.2) and (2.3) we obtain

$$\begin{aligned} a_i &= Rh_i^{(k-l)}\left(m_{i-1} + (-1)^{k+l}m_i\right) + r_i^{(l)}(y_i), \\ b_i &= Rh_i^{k-l}\left(f_{i-1}^{(k)} + (-1)^{k+l}f_i^{(k)}\right) + r_i^{(l)}(y_i), \\ d_i &= a_i - b_i = Rh_i^{k-l}\left(l_{i-1} + (-1)^{k+l}l_i\right). \end{aligned} \tag{2.12}$$

Using (2.7), we have

$$h_{i-1}^{l-k}d_{i-1} = R\left(l_{i-2} + (-1)^{k+l}l_{i-1}\right),$$
$$h_i^{l-k}d_i = R\left(l_{i-1} + (-1)^{k+l}l_i\right),$$
$$(-1)^{k+1}h_i l_{i-2} + (k+1)(h_{i-1} + h_i)l_{i-1} + (-1)^{k+1}h_{i-1}l_i = h_{i-1}h_i u_{i-1}.$$

We multiply these equations respectively by $(-1)^k h_i$, $(-1)^l h_{i-1}$, and $-R$, add them, and divide the result by $h_{i-1} + h_i$. As a result we obtain

$$\begin{aligned} &(-1)^k \frac{h_i h_{i-1}^{l-k}}{h_{i-1} + h_i} d_{i-1} + (-1)^l \frac{h_{i-1} h_i^{l-k}}{h_{i-1} + h_i} d_i \\ &= R\left[k + (-1)^l + 1\right] l_{i-1}\text{-}Ru_{i-1}h_{i-1}h_i/(h_{i-1} + h_i). \end{aligned} \tag{2.13}$$

If we replace i by $i + 1$ in (2.3), multiply by $(-1)^{k+1}$, add the result to the original equation, and use (2.12), we deduce that

$$\begin{aligned} &(-1)^{k+1}\frac{h_{i-1}}{h_{i-1} + h_i}\left(\frac{h_{i-1}}{h_i}\right)^{l-k-1} d_{i-1} + (-1)^{k+1}\frac{h_{i+1}}{h_i + h_{i+1}}\left(\frac{h_{i+1}}{h_i}\right)^{l-k-1} d_{i+1} \\ &\qquad + \left[k + 1 + (-1)^l + (-1)^{l+1}\left(\frac{h_{i-1}}{h_{i-1} + h_i} + \frac{h_{i+1}}{h_i + h_{i+1}}\right)\right] d_i \\ &= Rh_i^{k-l}\left(\frac{h_{i-1}h_i}{h_{i-1} + h_i}u_{i-1} + (-1)^{k+l}\frac{h_i h_{i+1}}{h_i + h_{i+1}}u_i\right). \end{aligned} \tag{2.14}$$

§3. Auxiliary propositions

For a function $f \in W_1^l$ $(l = 1, 2, \ldots)$ we have

$$f(x) = P_{l,i}(x) + \varphi_{l,i}(x), \tag{3.1}$$

where

$$P_{l,i}(x) = f_i + f_i'(x - x_i) + \cdots + f_i^{(l)}\frac{(x - x_i)^{l-1}}{(l-1)!} + \nu_i\frac{(x - x_i)^l}{l!},$$

$$\varphi_{l,i}(x) = \frac{1}{(l-1)!}\int_{x_i}^x (x - y)^{l-1}\left[f^{(l)}(y) - \nu_i\right] dy,$$

$$\begin{aligned} &f_i^{(j)} = f^{(j)}(x_i) \qquad (j = 0, 1, \ldots, l-1), \\ &\nu_i = f_i^{(l)}, \quad \text{if } f \in C^l, \\ &\nu_i = \delta_n^{-1}\int_{x_i^1}^{x_i^2} f^{(l)}(y)\, dy \quad \text{if } f \in W_1^l, \end{aligned} \tag{3.2}$$

and the points z_i^1 and z_i^2 are chosen to satisfy $z_i^2 - z_i^1 - \delta_n$ and $[x_{i-1}, x_i] \subseteq [z_i^1, z_i^2]$.

For $l = 0$ we set

$$P_{0,i}(x) \equiv f_i, \qquad \varphi_{0,i}(x) = f(x) - f_i.$$

In addition, for $0 \leqslant m \leqslant l - 1$ we have

$$\begin{gathered} \varphi_{l,i}^{(m)}(x) = \frac{1}{(l-1-m)!}\int_{x_i}^{x} (x-y)^{l-1-m}\left[f^{(l)}(y) - \nu_i\right] dy, \\ \varphi_{l,i}^{(l)}(x) = f^{(l)}(x) - \nu_i, \qquad \varphi_{l,i}^{(j)}(x_i) = 0 \qquad (j = 0, 1, \dots, l-1). \end{gathered} \tag{3.3}$$

On $[x_{i-1}, x_i]$, we have

$$\left|\varphi_{l,i}^{(m)}(x)\right| \leqslant h_i^{l-1-m}\int_{x_{i-1}}^{x_i} \left|f^{(l)}(y) - \nu_i\right| dy \qquad (m = 0, 1, \dots, l-1), \tag{3.4}$$

$$\left|\varphi_{l,i}^{(l)}(x)\right| = \left|f^{(l)}(x) - \nu_i\right|.$$

For $f \in C^l$, if we take $\nu_i = f_i^{(l)}$ in (3.1), we have

$$\left|\varphi_{l,i}^{(m)}(x)\right| \leqslant h_i^{l-m}\omega\left(f^{(l)}, h_i\right) \qquad (m = 0, 1, \dots, l) \tag{3.5}$$

for $x \in [x_{i-1}, x_i]$.

We consider $r_i(x) = r_i(f, x)$ as a linear operator from $C^{k-1}[x_{i-1}, x_i]$ to itself. Then, using (3.1) for $l = k - 1$ and $\nu_i = f_i^{(k-1)}$, we deduce that

$$r_i(f, x) = r_i(P_{k-1,i}, x) + r_i(\varphi_{k-1,i} x) = P_{k-1,i}(x) + r_i(\varphi_{k-1,i}, x),$$

and consequently

$$r_i^{(k+1)}(f, x) = r_i^{(k+1)}(\varphi_{k-1,i}, x). \tag{3.6}$$

Using (3.5) for $l = k - 1$, we obtain

$$\begin{gathered} \left|r_{i+1}^{(k+1)}(x_i)\right| \leqslant C_1\omega\left(f^{(k-1)}, h_{i+1}\right)/h_{i+1}^2, \\ \left|r_i^{(k+1)}(x_i)\right| \leqslant C_1\omega\left(f^{(k-1)}, h_i\right)/h_i. \end{gathered} \tag{3.7}$$

We introduce the following notation for given $l \geqslant k$, $1 \leqslant p \leqslant \infty$:

$$\begin{gathered} \mu_{i,l}^1 = h_{i-1}^{-1+1/p}\int_{x_{i-1}}^{x_{i-1}} \left|f^{(l)}(y) - \nu_i\right| dy, \\ \mu_{i,l}^2 = h_i^{-1+1/p}\int_{x_{i-1}}^{x_i} \left|f^{(l)}(y) - \nu_i\right| dy, \\ \mu_{i,l}^3 = h_{i+1}^{-1+1/p}\int_{x_i}^{x_{i+1}} \left|f^{(l)}(y) - \nu_i\right| dy. \end{gathered} \tag{3.8}$$

If $f \in W_p^l$, then, as in [10], it is easy to establish the estimates

$$\left(\sum_{i=1}^{N} \left|\mu_{i,l}^j\right|^p\right)^{1/p} \leqslant C_1\omega\left(f^{(l)}, \delta_n\right)_p \qquad (j = 1, 2, 3), \tag{3.9}$$

for $1 \leqslant p < \infty$; and for $p = \infty$ $(1/p = 0, f \in C^l)$ it is evident that

$$\max_{1 \leqslant i \leqslant N} \left|\mu_{i,l}^j\right| \leqslant C_1\omega\left(f^{(l)}, \delta_n\right) \qquad (j = 1, 2, 3). \tag{3.10}$$

For $f \in W_1^k$ it follows from (3.1), with $l = k$, and (2.9) that

$$\overline{H}_i(f, x) = \overline{H}_i(P_{k,i}, x) + \overline{H}_i(\varphi_{k,i}, x) = P_{k,i}(x) + \overline{H}_i(\varphi_{k,i}, x)$$

and also

$$\overline{H}_{i+1}(f, x) = P_{k,i+1}(x) + \overline{H}_{i+1}(\varphi_{k,i+1}, x).$$

Hence we obtain

$$\overline{H}_i^{(k+1)}(f,x)=\overline{H}_i^{(k+1)}(\varphi_{k,i},x),\qquad x\in[x_{i-1},x_i],$$
$$\overline{H}_{i+1}^{(k+1)}(f,x)=\overline{H}_{i+1}^{(k+1)}(\varphi_{k,i+1},x),\qquad x\in[x_i,x_{i+1}].$$

By using (3.4), we deduce that

$$\left|\overline{H}_i^{(k+1)}(x_i)\right|\leqslant C_1\left(h_i^{-1}h_{i-1}^{-1/p}\mu_{i,k}^1+h_i^{-1-1/p}\mu_{i,k}^2\right),$$
$$\left|\overline{H}_{i+1}^{(k+1)}(x_i)\right|\leqslant C_1\left(h_{i+1}^{-1}h_i^{-1/p}\mu_{i+1,k}^1+h_{i+1}^{-1-1/p}\mu_{i+1,k}^2\right)$$

and consequently, by (1.4) and (2.10),

$$|\bar{u}_i|h_{i+1}h_i^{1+1/p}/(h_i+h_{i+1})\leqslant C_1\rho^{1/p}\left(\mu_{i,k}^1+\mu_{i,k}^2+\mu_{i+1,k}^1+\mu_{i+1,k}^2\right).\quad(3.11)$$

Let $f\in W_1^l$ $(l=k+1,k+2,\dots,2k+1)$ or $f\in C^k$. Using the representations (2.2) and (3.1), we obtain

$$H_{i-1}(f,x)=P_{l,i-1}+H_{i-1}(\varphi_{l,i-1},x),\qquad x\in[x_{i-2},x_{i-1}],$$
$$H_i(f,x)=P_{l,i-1}(x)+H_i(\varphi_{l,i-1},x),\qquad x\in[x_{i-1},x_i].$$

Hence we deduce that

$$(k+1)u_{i-1}=H_i^{(k+1)}(\varphi_{l,i-1},x_{i-1})-H_{i-1}^{(k+1)}(\varphi_{l,i-1},x_{i-1}).$$

By (3.4) we have

$$\left|H_{i-1}^{(k+1)}(\varphi_{l,i-1},x_{i-1})\right|\leqslant C_1h_{i-1}^{l-k-2}\int_{x_{i-2}}^{x_{i-1}}\left|f^{(l)}(y)-\nu_{i-1}\right|dy,$$
$$\left|H_i^{(k+1)}(\varphi_{l,i-1},x_{i-1})\right|\leqslant C_1h_i^{l-k-2}\int_{x_{i-1}}^{x_i}\left|f^{(l)}(y)-\nu_{i-1}\right|dy$$

and consequently

$$|u_{i-1}|\leqslant C_1h_{i-1}^{l-k-1-1/p}\mu_{i-1,l}^2+C_1h_i^{l-k-1-1/p}\mu_{i-1,l}^3.$$

Similarly we obtain

$$|u_i|\leqslant C_1h_i^{l-k-1-1/p}\mu_{i,l}^2+C_1h_{i+1}^{l-k-1-1/p}\mu_{i,l}^3.$$

For $l=k$, $p=\infty$ $(f\in C^k)$ we deduce, using (3.10), that

$$|u_i|h_ih_{i+1}/(h_i+h_{i+1})\leqslant C_1\left(\mu_{i,k}^2+\mu_{i,k}^3\right)\leqslant C_2\omega\left(f^{(k)},\delta_n\right).\quad(3.12)$$

For $l=k+1$ and $1\leqslant p\leqslant\infty$, we obtain

$$\begin{aligned}&h_i^{1/p}\left[|u_{i-1}|h_{i-1}/(h_{i-1}+h_i)+|u_i|h_{i+1}/(h_i+h_{i+1})\right]\\&\quad\leqslant C_1\left(\mu_{i-1,l+1}^2+\mu_{i-1,k+1}^3+\mu_{i,k+1}^2+\mu_{i,k+1}^3\right).\end{aligned}\quad(3.13)$$

For $l\geqslant k+2$ and $1\leqslant p\leqslant\infty$, we find by (1.2) that

$$\begin{aligned}&h_i^{k+1-l+1/p}\left[|u_{i-1}|h_{i-1}/(h_{i-1}+h_i)+|u_i|h_{i+1}/(h_i+h_{i+1})\right]\\&\quad\leqslant C_1\rho^{l-k-1-1/p}\left(\mu_{i-1,l}^2+\mu_{i-1,l}^3+\mu_{i,l}^2+\mu_{i,l}^3\right).\end{aligned}\quad(3.14)$$

It is easy to show that the following inequalities hold on $[x_{i-1},x_i]$ for $f\in C^{k-1}$:

$$\left|f^{(k-1)}(x)-r_i^{(k-1)}(x)\right|\leqslant C_1\omega\left(f^{(k-1)},h_i\right),$$
$$\begin{aligned}\left|f^{(k-1)}(x)-S_n^{(k-1)}(x)\right|&\leqslant\left|f^{(k-1)}(x)-r_i^{(k-1)}(x)\right|+\left|r_i^{(k-1)}(x)-S_n^{(k-1)}(x)\right|\\&\leqslant C_1\omega\left(f^{(k-1)},h_i\right)+C_2h_i\max\{|m_{i-1}|,|m_i|\}.\end{aligned}\quad(3.15)$$

Let $\overline{H}(x) = \overline{H}(f, x)$ be defined by (2.9). By (3.1),

$$\overline{H}_i^{(k)}(f, x) = \nu_i + \overline{H}_i^{(k)}(\varphi_{k,i}, x)$$

on $[x_{i-1}, x_i]$.

Let $f \in W_p^k$. Let $\tilde{H}^{(k)}(x)$ denote the $(b-a)$-periodic function that coincides with $\overline{H}_i^k(\varphi_{k,i}, x)$ on $[x_{i-1}, x_i)$, and let $\tilde{f}^{(k)}(x)$ denote the $(b-a)$-periodic piecewise constant function that is equal to ν_i on $[x_{i-1}, x_i)$. Then we have

$$\|f^{(k)} - \overline{H}^{(k)}\|_p \leqslant \|f^{(k)} - \tilde{f}^{(k)}\|_p + \|\tilde{H}^{(k)}\|_p \qquad (1 \leqslant p < \infty).$$

Using (3.4), and taking account of (1.2), (3.8) and (3.9), we obtain

$$\left|\overline{H}_i^{(k)}(\varphi_{k,i}, x)\right| \leqslant C_1\left(h_{i-1}^{1/p}\mu_{i,k}^1 + h_i^{1/p}\mu_{i,k}^2\right),$$

$$\|\tilde{H}^{(k)}\|_p = \left(\sum_{i=1}^N \int_{x_{i-1}}^{x_i} \left|\overline{H}_i^{(k)}(\varphi_{k,i}, x)\right|^p dx\right)^{1/p}$$

$$\leqslant C_1\left(\sum_{i=1}^N \left|h_{i-1}^{-1/p}h_i^{1/p}\mu_{i,k}^1 + \mu_{i,k}^2\right|^p\right)^{1/p} \leqslant C_2(\rho)\,\omega\left(f^{(k)}, \delta_n\right)_p$$

and similarly

$$\|f^{(k)} - \tilde{f}^{(k)}\|_p \leqslant C_3\omega\left(f^{(k)}, \delta_n\right)_p.$$

Hence we easily deduce that

$$\|f^{(k)} - S_n^{(k)}\|_p \leqslant \|f^{(k)} - \overline{H}^{(k)}\|_p + \|\overline{H}^{(k)} + S_n^{(k)}\|_p$$

$$\leqslant C_1(\rho)\,\omega\left(f^{(k)}, \delta_n\right)_p + C_2\left(\sum_{i=1}^N h_i\left(|\bar{l}_{i-1}| + |\bar{l}_i|\right)^p\right)^{1/p},$$

where $\bar{l}_i = m_i - \bar{\nu}_i$. By periodicity, we have, for $1 \leqslant p < \infty$,

$$\|f^{(k)} - S_n^{(k)}\|_p \leqslant C_1(\rho)\,\omega\left(f^{(k)}, \delta_n\right)_p + C_2(\rho)\left(\sum_{i=1}^N h_i|\bar{l}_i|^p\right)^{1/p}. \qquad (3.16)$$

For $p = \infty$ ($f \in C^k$) we have

$$\|f^{(k)} - S_n^{(k)}\| \leqslant C_1\omega\left(f^{(k)}, \delta_n\right) + C_2 \max_{1 \leqslant i \leqslant N} |l_i|, \qquad (3.17)$$

where $l_i - m_i - f_i^{(k)}$.

Let $f \in W_p^l$ ($k = k+1, k+2, \dots, 2k+1$) and let H be the Hermite spline with the knots of Δ_n; then from (2.2) and (2.3), with $1 \leqslant p < \infty$, we have

$$\|f^{(l)} - S_n^{(l)}\|_p \leqslant \|f^{(l)} - H^{(l)}\|_p + \|H^{(l)} - S_n^{(l)}\|_p \leqslant C_1\omega\left(f^{(l)}, \delta_n\right)_p$$

$$+C_2\left(\sum_{i=1}^N h_i^{(k-l)p}\int_{x_{i-1}}^{x_i} \left|\left(f_{i-1}^{(k)} - m_{i-1}\right)\Phi_l^1(x) + \left(f_i^{(k)} - m_i\right)\Phi_l^2(x)\right|^p dx\right)^{1/p}$$

$$\leqslant C_1\omega\left(f^{(l)}, \delta_n\right)_p + C_2\left(\sum_{i=1}^N h_i^{(k-l)p}\int_{x_{i-1}}^{x_i} \left(|l_{i-1}| + |l_i|\right)^p dx\right)^{1/p}$$

$$\leqslant C_1\omega\left(f^{(l)}, \delta_n\right)_p + C_3\left(\sum_{i=1}^N h_i^{(k-l)p+1}\left(|l_{i-1}| + |l_i|\right)^p\right)^{1/2}. \qquad (3.18)$$

If $f \in C^l$, we have the following analog of (3.18):

$$\|f^{(l)} - S_n^{(l)}\| \leqslant C_1\omega\left(f^{(l)}, \delta_n\right) + C_2 \max_{1 \leqslant i \leqslant N} h_i^{k-l}\left(|l_{i-1}| + |l_i|\right). \tag{3.19}$$

From (2.13) we obtain

$$|l_{i-1}| \leqslant |R^{-1}|\left(\frac{h_i h^{l-k} Di - 1}{h_{i-1} + h_i}|d_{i-1}| + \frac{h_{i-1}h_i^{l-k}}{h_{i-1} + h_i}|d_i|\right) + \frac{h_{i-1}h_i}{h_{i-1} + h_i}|u_{i-1}|,$$

$$|l_i| \leqslant |R^{-1}|\left(\frac{h_{i+1}h_i^{l-k}}{h_i + h_{i+1}}|d_i| + \frac{h_i h_{i+1}^{l-k}}{h_i + h_{i+1}}|d_{i+1}|\right) + \frac{h_i h_{i+1}}{h_i + h_{i+1}}|u_i|.$$

For $l = k + 1$, $1 \leqslant p < \infty$, if we take account of (3.13) and (3.9), we find from (3.18) that

$$\|f^{(k+1)} - S_n^{(k+1)}\|_p \leqslant C_1\omega\left(f^{(k+1)}, \delta_n\right)_p + C_2\left(\sum_{i=1}^N h_i|d_i|^p\right)^{1/p}, \tag{3.20}$$

and for $p = \infty$ we deduce from (3.19), by using (3.13) and (3.10), that

$$\|f^{(k+1)} - S_n^{(k+1)}\| \leqslant C_1\omega\left(f^{(k+1)}, \delta_n\right) + C_2 \max_{1 \leqslant i \leqslant N} |d_i|. \tag{3.21}$$

Under (1.2) and the inequality $l \geqslant k + 2$, we have

$$h_l^{k-l+1/p}\left(|l_{i-1}| + |l_i|\right) \leqslant C_1\rho^{l-k-1-1/p}\left(h_{i-1}^{1/p}|d_{i-1}| + h_i^{1/p}|d_i| + h_{i+1}^{1/p}|d_{i+1}|\right)$$
$$+ h_i^{k+1-l+1/p}\left(|u_{i-1}| + |u_i|\right) \qquad (1 \leqslant p < \infty),$$

$$h_i^{k-l}\left(|l_{i-1}| + |l_i|\right) \leqslant C_1\rho^{l-k-1}\left(|d_{i-1}| + |d_i| + |d_{i+1}|\right)$$
$$+ h_i^{k+1-l}\left(|u_{i-1}| + |u_i|\right) \qquad (p = \infty).$$

From (3.20), by using the periodicity of the function and the spline, and (3.14) and (3.9), we deduce the estimates

$$\|f^{(l)} - S_n^{(l)}\|_p \leqslant C_1\rho^{l-k-1-1/p}\left[\omega\left(f^{(l)}, \delta_n\right)_p + \left(\sum_{i=1}^N h_i|d_i|^p\right)^{1/p}\right] \quad (1 \leqslant p < \infty), \tag{3.22}$$

$$\|f^{(l)} - S_n^{(l)}\| \leqslant C_1\rho^{l-k-1}\left[\omega\left(f^{(l)}, \delta_n\right) + \max_{1 \leqslant i \leqslant N} |d_i|\right]. \tag{3.23}$$

§4. Estimates for the elements of inverse matrices

Let there be given the $N \times N$ matrix

$$\mathscr{A} = \left\|\begin{matrix} \gamma_1 & \beta_1 & 0 & 0 & \cdot & 0 & 0 & \alpha_1 \\ \alpha_2 & \gamma_2 & \beta_2 & 0 & \cdot & 0 & 0 & 0 \\ 0 & \alpha_3 & \gamma_3 & \beta_3 & \cdot & 0 & 0 & 0 \\ \cdot & \cdot & \cdot & \cdot & \cdot & \cdot & \cdot & \cdot \\ 0 & 0 & 0 & 0 & \cdot & \alpha_{N-1} & \gamma_{N-1} & \beta_{N-1} \\ \beta_N & 0 & 0 & 0 & \cdot & 0 & \alpha_N & \gamma_N \end{matrix}\right\|. \tag{4.1}$$

We denote by A the determinant of $\mathscr{A}$, and by A_{ij} the determinant of the matix obtained by deleting the ith row and jth column of $\mathscr{A}$. We also introduce the notation

$$P_{ij} = \begin{vmatrix} \gamma_i & \beta_i & 0 & 0 & \cdot & 0 & 0 & 0 \\ \alpha_{i+1} & \gamma_{i+1} & \beta_{i+1} & 0 & \cdot & 0 & 0 & 0 \\ 0 & \alpha_{i+2} & \gamma_{i+2} & \beta_{i+2} & \cdot & 0 & 0 & 0 \\ \cdot & \cdot & \cdot & \cdot & \cdot & \cdot & \cdot & \cdot \\ 0 & 0 & 0 & 0 & \cdot & \alpha_{j-1} & \gamma_{j-1} & \beta_{j-1} \\ 0 & 0 & 0 & 0 & \cdot & 0 & \alpha_j & \gamma_j \end{vmatrix} \tag{4.2}$$

$$(1 \leqslant i \leqslant j \leqslant N), \quad P_{1,0} = P_{N+1,N} = 1, \quad P_{2,0} = P_{N+1,N-1} = 0,$$

$$\mathscr{K}(i+1, j) = \begin{cases} 1 & \text{if } i = j, \\ \alpha_{i+1}\alpha_{i+2} \cdots \alpha_j & \text{if } i < j, \end{cases}$$

$$\mathscr{L}(i, j-1) = \begin{cases} 1 & \text{if } i = j, \\ \beta_i\beta_{i+1} \cdots \beta_{j-1} & \text{if } i < j. \end{cases}$$

Let $\mathscr{A}^{-1} = \{a_{ij}\}_{i,j=1}^N$; then

$$a_{ij} = (-1)^{i+j} A_{ji}/A. \tag{4.3}$$

Developing the determinants A_{ij} and A in the corresponding ways, we obtain

$$A_{ij} = (P_{1,i-1}P_{j+1,N} - \alpha_1\beta_N P_{2,i-1}P_{j+1,N-1})\mathscr{K}(i+1, j) \qquad (1 < i \leqslant j < N),$$

$$A_{iN} = \mathscr{K}(i+1, N)P_{1,i-1} + (-1)^N \beta_N \mathscr{L}(1, i-1)P_{i+1,N-1} \qquad (1 \leqslant i \leqslant N),$$

$$A_{1,j} = \mathscr{K}(2, j)P_{j+1,N} + (-1)^N \beta_N \mathscr{L}(j, N-1)P_{2,j-1} \qquad (1 \leqslant j \leqslant N),$$

$$A_{i,j} = \mathscr{L}(j, i-1)(P_{1,j-1}P_{i+1,N} - \alpha_1\beta_N P_{2,j-1}P_{i+1,N-1}) \qquad (1 < j < i < N),$$

$$A_{i,1} = \mathscr{L}(1, i-1)P_{i+1,N} + (-1)^N \alpha_1 \mathscr{K}(i+1, N)P_{2,i-1} \qquad (1 \leqslant i \leqslant N),$$

$$A_{N,j} = \mathscr{L}(j, N-1)P_{1,j-1} + (-1)^N \alpha_1 \mathscr{K}(2, j)P_{j+1,N-1} \qquad (1 \leqslant j \leqslant N),$$

$$A = P_{1,N} - \alpha_1\beta_N P_{2,N-1} + (-1)^{N+1}\mathscr{K}(1, N).$$

From now on, we shall consider only matrices of the form (4.1) with dominant main diagonal, i.e.

$$\min_{1 \leqslant i \leqslant N} (|\gamma_i| - |\alpha_i| - |\beta_i|) > 0 \tag{4.4}$$

and of the following forms:

a)

$$\alpha_i = (-1)^{k+1} h_{i+1}/(h_i + h_{i+1}), \quad \beta_i = (-1)^{k+1}(1 - |\alpha_i|), \quad \gamma_i = k+1; \tag{4.5}$$

b)

$$\alpha_i = (-1)^{k+1} h_{i-1}/(h_{i-1} + h_i), \qquad \beta_i = (-1)^{k+1} h_{i+1}/(h_i + h_{i+1}),$$

$$\gamma_i = k + 1 + (-1)^l + (-1)^{l+k}(\alpha_i + \beta_i). \tag{4.6}$$

Since $\gamma_i > 0$ ($k = 1, 2, \ldots$) and $\mathscr{A}$ and P_{ij} satisfy (4.4), we have

$$A > 0, P_{ij} > 0 \qquad (j \geqslant i + 1). \tag{4.7}$$

We derive the necessary estimates for case a). We have

$$\begin{aligned} P_{ij} &= \gamma_i P_{i+1,j} - \alpha_{i+1}\beta_i P_{i+2,j} \\ &= \left(k+|\alpha_i|\right)P_{i+1,j} + |\beta_i|\left(P_{i+1,j} - |\alpha_{i+1}|P_{i+2,j}\right), \\ P_{ij} &= \gamma_j P_{i,j-1} - \alpha_j\beta_{j-1}P_{i,j-2} \\ &= \left(k+|\beta_j|\right)P_{i,j-1} + |\alpha_j|\left(P_{i,j-1} - |\beta_{j-1}|P_{i,j-2}\right). \end{aligned} \tag{4.8}$$

Hence we obtain by induction, using (1.2),

$$\begin{aligned} P_{ij} &\geqslant \left(k+|\alpha_i|\right)P_{il+1,j} \geqslant kP_{i+1,j}, \\ P_{ij} &\geqslant \left(k+|\beta_j|\right)P_{i,j-1} \geqslant kP_{i,j-1}, \\ \left[k+1/(\rho+1)\right]^{j-i+1} &\leqslant P_{ij} \leqslant (k+1)P_{i,j-1}, \\ P_{ij} &= P_{i,\nu}P_{\nu+1,j} - \beta_\nu\alpha_{\nu+1}P_{i,\nu-1}P_{\nu+2,j} \\ &\geqslant \left[1-\rho^2/(1+\rho)^2\right]P_{i\nu}P_{\nu+1,j} \qquad (i \leqslant \nu \leqslant j). \end{aligned} \tag{4.9}$$

We consider case b) under condition (1.2). In the same way as for (4.8), we deduce that

$$P_{ij} = \begin{cases} \left(k+|\alpha_i|\right)P_{i+1,j} + |\beta_i|\left(P_{i+1,j} - |\alpha_{i+1}|P_{i+2,j}\right), & \text{if } l \text{ is odd}, \\ \left(k+|\beta_{i-1}|\right)P_{i+1,j} + |\alpha_{i+1}|\left(P_{i+1,j} - |\beta_i|P_{i+2,j}\right), & \text{if } l \text{ is even}, \end{cases}$$

$$P_{ij} = \begin{cases} \left(k+|\beta_j|\right)P_{i,j-1} + |\alpha_j|\left(P_{i,j-1} - |\beta_{j-1}|P_{i,j-2}\right), & \text{if } l \text{ is odd}, \\ \left(k+|\alpha_{j+1}|\right)P_{i,j-1} + |\beta_{j-1}|\left(P_{i,j-1} - |\alpha_j|P_{i,j-2}\right), & \text{if } l \text{ is even}, \end{cases}$$

and, as in (4.9),

$$\begin{aligned} P_{ij} &\geqslant P_{i+1,j}\min\left\{k+|\alpha_i|, k+|\beta_{i-1}|\right\}, \\ P_{ij} &\geqslant P_{i,j-1}\min\left\{k+|\beta_j|, k+|\alpha_{j+1}|\right\}, \\ \left[k+1/(1+\rho)\right]^{j-i+1} &\leqslant P_{ij} \leqslant (k+2)P_{i,j-1}, \\ P_{ij} &\geqslant \tfrac{3}{4}P_{i\nu}P_{\nu+1,j} \qquad (i \leqslant \nu \leqslant j). \end{aligned} \tag{4.10}$$

It follows from (4.9) and (4.10) that, when (1.2) is satisfied, for sufficiently large N,

$$A \geqslant \left[1-\rho^2/(1+\rho)^2\right]P_{1,N} - 1 \geqslant P_{1,N}\left[1-\rho^2/(1+\rho)^2\right]/2. \tag{4.11}$$

It follows from (4.9)–(4.11), in either case a) or b), that for sufficiently large N

$$\begin{aligned} |a_{ji}| &\leqslant C_1|\mathscr{K}(i+1,j)|/P_{i+1,j} \qquad (1 < i \leqslant j < N), \\ |a_{ji}| &\leqslant C_1|\mathscr{L}(j,i-1)|/P_{j,i-1} \qquad (1 < j < i < N), \\ |a_{ji}| &\leqslant C_1\left(|\mathscr{K}(i+1,N)|/P_{i+1,N} + |\mathscr{L}(1,i-1)|/P_{1,i-1}\right) \\ &\qquad (1 \leqslant i \leqslant N; j = 1, N), \end{aligned} \tag{4.12}$$

$$|a_{ij}| \leqslant C_1\left(|\mathscr{K}(2,j)|/P_{2,j} + |\mathscr{L}(j,N-1)|/P_{j,N-1}\right) \qquad (1 < j < N; i = 1, N).$$

Let there be given the matrices $\mathscr{A}^{-1} = \{a_{ij}\}_{i,j=1}^N$, $\mathscr{B}^{-1} = \{b_{ij}\}_{i,j=1}^N$, and a diagonal matrix $\mathscr{D} = \operatorname{diag}(\xi_1,\dots,\xi_N)$ such that $\mathscr{B} = \mathscr{D}\mathscr{A}\mathscr{D}^{-1}$. In addition, as above, A is the determinant of $\mathscr{A}$, B is the determinant of $\mathscr{B}$, and A_{ij} is the determinant of the matrix obtained from $\mathscr{A}$ by deleting the ith row and the jth column. Since $\mathscr{B}^{-1} = \mathscr{D}\mathscr{A}^{-1}\mathscr{D}^{-1}$ and $B = A$, we have

$$b_{ij} = \xi_i/\xi_j a_{ij} = (-1)^{i+j} A_{ji}/A, \qquad |b_{ij}| = |a_{ij}|\,|\xi_i|/|\xi_j|. \tag{4.13}$$

Let l_p^N be the space of vectors $x = (x_1,\dots,x_N)$ with norm

$$\|x\| = \|x\|_\infty = \max_{1\leqslant i\leqslant N} |x_i|, \quad \text{if } p = \infty;$$

$$\|x\|_p = \left(\sum_{i=1}^N |x_i|^p\right)^{1/p} \qquad (1 \leqslant p < \infty).$$

The norm of the matrix $\mathscr{B} = \{b_{ij}\}_{i,j=1}^N$, as an operator from l_p^N to l_p^N, is defined in the usual way:

$$\|\mathscr{B}\|_p = \sup_{\|x\|_p=1} \|\mathscr{B}x\|_p.$$

It is known that

$$\|\mathscr{B}\| = \|\mathscr{B}\|_\infty = \max_{1\leqslant i\leqslant N} \sum_{j=1}^N |b_{ij}|, \qquad \|\mathscr{B}\|_1 = \max_{1\leqslant j\leqslant N} \sum_{i=1}^N |b_{ij}|,$$

and by the Riesz-Thorin theorem we have

$$\|\mathscr{B}\|_p \leqslant \|\mathscr{B}\|_\infty^{1-1/p} \|\mathscr{B}\|_1^{1/p}. \tag{4.14}$$

In addition, if $\mathscr{B}$ has a dominant diagonal, i.e.

$$\theta_i = |b_{ii}| - \sum_{\substack{j=1\\ j\neq i}} |b_{ij}| > 0 \qquad (i = 1,2,\dots,N), \tag{4.15}$$

then (see, for example, [2])

$$\|\mathscr{B}^{-1}\| = \|\mathscr{B}^{-1}\|_\infty \leqslant \max_{1\leqslant i\leqslant N} \theta_i^{-1}. \tag{4.16}$$

If $\mathscr{B}$ satisfies the condition

$$\mathscr{J}_i = |b_{ii}| - \sum_{\substack{j=1\\ j\neq i}}^N |b_{ji}| > 0 \qquad (i = 1,2,\dots,N), \tag{4.17}$$

then (see, for example, [2])

$$\|\mathscr{B}^{-1}\|_1 \leqslant \max_{1\leqslant i\leqslant N} \mathscr{J}_i^{-1}. \tag{4.18}$$

§5. Convergence of derivatives of order l ($l = 0,1,\dots,k$)

Let S_n be the multiply interpolating spline of degree $2k+1$ and deficiency k that interpolates $f \in C^{k-1}$ together with its derivatives up to order $k-1$ at the knots of Δ_n (see (1.1)). We denote by $\rho_{k,p} > 1$ the root of the equation

$$(k+1)(x+1)x^{1/p} = 1 + x^{1+2/p} \qquad (x \geqslant 1,\ 1 \leqslant p < \infty) \tag{5.1}$$

(there is a unique root for $x \geqslant 1$). It is evident that

$$\rho_{k,1} = \left(k + 2 + \sqrt{k(k+4)}\right)/2 \quad \text{and} \quad (k+1)^p < \rho_{k,p} < (k+2)^p.$$

Set

$$\varepsilon_\rho = \left[(k+1)(\rho+1) + \sqrt{(k+1)^2(\rho+1)^2 - 4\rho}\right]/2;$$

then it is easy to verify that $\rho^{1+1/p} = \varepsilon_\rho$ for $\rho = \rho_{k,p}$, whereas $\rho^{1+1/p} < \varepsilon_\rho$ and $\rho^{1+1/p} > \varepsilon_\rho$ for $1 \leqslant \rho \leqslant \rho_{k,p}$ and for $\rho > \rho_{k,p}$, respectively.

THEOREM 1. *For every function $f \in C^{k-1}$ and for every sequence of nets Δ_n satisfying* (1.3) *we have the estimate*

$$\left\|f^{(k-1)} - S_n^{(k-1)}\right\| \leqslant C(M)\,\omega\left(f^{(k-1)}, \delta_n\right), \tag{5.2}$$

and for every sequence of nets Δ_n satisfying (1.2) *we have*

$$\left\|f^{(k-1)} - S_n^{(k-1)}\right\| \leqslant C(\rho)\,\omega\left(f^{(k-1)}, \delta_n\right) \tag{5.3}$$

for $\rho < \rho_{k,1}$.

PROOF. First we deduce (5.2). By (3.15),

$$\left\|f^{(k-1)} - S_n^{(k-1)}\right\| \leqslant C_1\omega\left(f^{(k-1)}, \delta_n\right) + C_2\delta_n\|m\|, \tag{5.4}$$

where $m = (m_1, \ldots, m_N)^{\mathrm{T}}$. Taking account of (2.6), we have the following system of equations for determining the m_i:

$$\begin{aligned} \alpha_i m_{i-1} + (k+1)m_i + \beta_i m_{i+1} &= v_i h_i h_{i+1}/(h_i + h_{i+1}) \\ &= q_i \qquad (i = 1, 2, \ldots, N), \\ m_0 = m_N, \qquad m_{N+1} &= m_1, \end{aligned} \tag{5.5}$$

where $\alpha_i = (-1)^{k+1}h_{i+1}/(h_i + h_{i+1})$ and $\beta_i = (-1)^{k+1}(1 - |\alpha_i|)$.

We write the system (5.5) in the matrix form $\mathscr{A}m = q$, where $\mathscr{A}$ is the matrix (4.1) with $\gamma_i = k + 1$. Then, as above, we obtain

$$\|\mathscr{A}^{-1}\| \leqslant k^{-1}, \quad \|q\| \leqslant C_3\delta_n\omega\left(f^{(k-1)}, \delta_n\right)/h^2, \quad \|m\| \leqslant k^{-1}\|q\|,$$

where $h = \min_{1 \leqslant i \leqslant N} h_i$. Substituting these estimates into (5.4), we obtain (5.2).

Now we establish (5.3). Multiply both sides of (2.6) by h_i and make the change of variables $g_i = h_i m_i$. Then we obtain the following system of equations for determining g_i:

$$\mathscr{B}g = q, \qquad \mathscr{B} = \mathscr{D}\mathscr{A}\mathscr{D}^{-1}, \tag{5.6}$$

where $q_i = v_i h_i^2 h_{i+1}/(h_i + h_{i+1})$, $q = (q_1, \ldots, q_N)^{\mathrm{T}}$, $g = (g_1, \ldots, g_N)^{\mathrm{T}}$, $\mathscr{D}$ is the diagonal matrix $\operatorname{diag}(h_1, \ldots, h_N)$ and $\mathscr{A}$ is the matrix (4.1) with

$$\alpha_i = (-1)^{k+1}h_{i+1}/(h_i + h_{i+1}),$$

$\beta_i = (-1)^{k+1}h_i/(h_i + h_{i+1})$, and $\gamma_i = k + 1$.

Let us estimate $\|\mathscr{B}^{-1}\|$. Let $\mathscr{B}^{-1} = \{b_{ij}\}_{i,j=1}^N$ and $\mathscr{A}^{-1} = \{a_{ij}\}_{i,j=1}^N$, and let B be the determinant of $\mathscr{B}$ and A the determinant of $\mathscr{A}$. Since $B = A$ and $\mathscr{B}^{-1} = \mathscr{D}\mathscr{A}^{-1}\mathscr{D}^{-1}$, if we use (4.3) we obtain

$$b_{ij} = \left(h_i/h_j\right)a_{ij} = (-1)^{i+j}\left(h_i/h_j\right)A_{ji}/A. \tag{5.7}$$

In the present case we use (4.12). By the periodicity of Δ_n and (1.2), we have

$$h_i/h_j \leqslant \rho^{\min\{|i-j|,\, N-|i-j|\}} \qquad (i,\, j = 1, 2, \dots, N) \tag{5.8}$$

and therefore

$$|b_{ji}| \leqslant C_1\left\{\rho^{|i-j|}\frac{|\mathscr{K}(2,\, j)|}{P_{2,j}} + \rho^{N-|i-j|}\frac{|\mathscr{L}(j, N-1)|}{P_{j,N-1}}\right\} \qquad (i,\, j = 1, 2, \dots, N). \tag{5.9}$$

We introduce the notation

$$\Phi(\alpha, i, j) = \frac{|\mathscr{K}(i,\, j)|}{P_{ij}}, \qquad \Psi(\beta, i, j) = \frac{|\mathscr{L}(i,\, j)|}{P_{ij}}, \tag{5.10}$$

where $\alpha = (\alpha_i, \alpha_{i+1}, \dots, \alpha_j)$ and $\beta = (\beta_i, \beta_{i+1}, \dots, \beta_j)$.

Let us find the maxima of Φ and Ψ, respectively, in the domains

$$\begin{aligned} \mathscr{E}_\alpha &= \left\{1/(1+\rho) \leqslant |\alpha_l| \leqslant \rho/(1+\rho),\, i \leqslant l \leqslant j\right\}, \\ \mathscr{E}_\beta &= \left\{\omega/(1+\rho) \leqslant |\beta_l| \leqslant \rho/(1+\rho),\, i \leqslant l \leqslant j\right\}. \end{aligned} \tag{5.11}$$

Since Φ and Ψ are continuous and positive, we shall find the minima of Φ^{-1} and Ψ^{-1} in these domains. Since $\beta_l = (-1)^{k+1}(1 - |\alpha_l|)$, we have

$$\Phi^{-1}(\alpha, i, j) = \begin{vmatrix} \dfrac{k+1}{|\alpha_i|} & \left(\dfrac{(-1)^{k+1}}{|\alpha_i|} + (-1)^k\right) & 0 & \cdot & 0 & 0 & 0 \\ (-1)^{k+1} & \dfrac{k+1}{|\alpha_{i+1}|} & \left(\dfrac{(-1)^{k+1}}{|\alpha_{i+1}|} + (-1)^k\right) & \cdot & 0 & 0 & 0 \\ 0 & (-1)^{k+1} & \dfrac{k+1}{|\alpha_{i+2}|} & \cdot & 0 & 0 & 0 \\ \cdots & \cdots & \cdots & \cdots & \cdots & \cdots & \cdots \\ 0 & 0 & 0 & \cdot & 0 & (-1)^{k+1} & \dfrac{k+1}{|\alpha_j|} \end{vmatrix}$$

It is easy to verify, by an argument similar to that used in [10], that

$$\frac{\partial}{\partial|\alpha_l|}\Phi^{-1}(\alpha, i, j) < 0 \qquad (i \leqslant l \leqslant j)$$

and consequently

$$\begin{aligned} \Phi^{-1}(\alpha, i, j) &\geqslant \Phi^{-1}(\alpha, i, j)\big|_{\alpha_i = \alpha_{i+1} = \cdots = \alpha_j = \rho/(1+p)} \\ &= \frac{\rho^{i-j-1}}{\sqrt{(k+1)^2(\rho+1)^2 - 4\rho}}\left[\varepsilon_\rho^{j-i+2} - (\rho/\varepsilon_\rho)^{j-i+2}\right]. \end{aligned}$$

Since we have $2\rho < \varepsilon_\rho$ for $1 \leqslant \rho < \rho_{k,1}$, we obtain

$$\Phi^{-1}(\alpha, i, j) \geqslant C_1(\varepsilon_\rho/\rho)^{j-i} \tag{5.12}$$

in $\mathscr{E}_\alpha$. Similarly, we obtain

$$\Psi^{-1}(\beta, i, j) \geqslant C_1(\varepsilon_\rho/\rho)^{j-i} \tag{5.13}$$

in $\mathscr{E}_\beta$.

From (5.12) and (5.13) we deduce, taking account of (5.9), that

$$|b_{ji}| \leqslant C(\rho)\left[\left(\rho^2/\varepsilon_\rho\right)^{|j-i|} + \left(\rho^2/\varepsilon_\rho\right)^{N-|j-i|}\right]. \tag{5.14}$$

By using (4.13) we obtain

$$\begin{aligned}\|\mathscr{B}^{-1}\| &\leqslant C(\rho) \max_{1\leqslant j\leqslant N} \sum_{i=1}^{N} \left[\left(\rho^2/\varepsilon_\rho\right)^{|j-i|} + \left(\rho^2/\varepsilon_\rho\right)^{N-|j-i|}\right] \\ &\leqslant 4C(\rho) \sum_{i=0}^{\infty} \left(\rho^2/\varepsilon_\rho\right)^i,\end{aligned}$$

and since $\rho^2 < \varepsilon_\rho$ for $\rho < \rho_{k,1}$, we have

$$\|\mathscr{B}^{-1}\| < C_1(\rho), \qquad \|g\| \leqslant C_1(\rho)\|q\|. \tag{5.15}$$

It follows from (3.7) that

$$|q_i| \leqslant C(\rho)\omega\left(f^{(k-1)}, \delta_n\right), \qquad \|q\| \leqslant C(\rho)\omega\left(f^{(k-1)}, \delta_n\right). \tag{5.16}$$

Using (5.15), (5.16), (3.15) and (1.2), we obtain the conclusion of Theorem 1.

THEOREM 2. *For every function $f \in C^k$ and every net Δ_n*

$$\|f^{(k)} - S_n^{(k)}\| \leqslant C\omega\left(f^{(k)}, \delta_n\right). \tag{5.17}$$

For every function $f \in W_p^k$ $(1 \leqslant p < \infty)$ and every net Δ_n that satisfies (1.3)

$$\|f^{(k)} - S_n^{(k)}\|_p \leqslant C(M, p)\omega\left(f^{(k)}, \delta_n\right)_p. \tag{5.18}$$

For every function $f \in W_p^k$ $(1 \leqslant p < \infty)$ and every net Δ_n that satisfies (1.2) *for $\rho < \rho_{k,p}$,*

$$\|f^{(k)} - S_n^{(k)}\|_p \leqslant C(\rho, p)\omega\left(f^{(k)}, \delta_n\right)_p. \tag{5.19}$$

Here C is an absolute positive constant, and $C(M, p)$ and $C(\rho, p)$ depend only on the indicated parameters.

PROOF. We first prove (5.17). We write the system (2.7) in matrix form

$$\mathscr{A}l = q, \tag{5.20}$$

where $l = (l_1, \dots, l_N)^{\mathrm{T}}$, $q = (q_1, \dots, q_N)^{\mathrm{T}}$, $q_i = u_i h_i h_{i+1}/(h_i + h_{i+1})$, and $\mathscr{A}$ is the matrix (4.1) with $\alpha_i = (-1)^{k+1} h_{i+1}/(h_i + h_{i+1})$ and $\beta_i = (-1)^{k+1} h_i/(h_i + h_{i+1})$ $(i = 1, \dots, N)$. Since $\mathscr{A}$ satisfies (4.15), we have $\|\mathscr{A}^{-1}\| \leqslant k^{-1}$ by (4.16). By (3.12) we obtain $\|q\| \leqslant C_2\omega(f^{(k)}, \delta_n)$. Taking account of (3.17) and the inequality $\|l\| < k^{-1}C_2\omega(f^{(k)}, \delta_n)$, we obtain (5.17).

We now prove (5.18). By periodicity and (2.10) we obtain the system of equations

$$\mathscr{A}\bar{l} = \bar{q},$$

where $\mathscr{A}$ is the matrix in (5.20), $\bar{l} = (\bar{l}_1, \dots, \bar{l}_N)^{\mathrm{T}}$, $\bar{q} = (\bar{q}_1, \dots, \bar{q}_N)^{\mathrm{T}}$, and $\bar{q}_i = \bar{u}_i h_i h_{i+1}/(h_i + h_{i+1})$. It is easily verified that, under (1.3), the matrix $\mathscr{A}$ satisfies (4.15) and (4.17). Taking account of (4.16), (4.18), and (4.14), we obtain

$$\|\mathscr{A}^{-1}\|_p \leqslant C(M). \tag{5.21}$$

From (3.16) we have

$$\|f^{(k)} - S_n^{(k)}\|_p \leqslant C_1(M)\omega\left(f^{(k)}, \delta_n\right)_p + C_2(M)\delta_n^{1/p}\|\bar{l}\|_p. \tag{5.22}$$

Taking account of (3.11) and (3.9), we obtain

$$\|\bar{q}\|_p = \left(\sum_{i=1}^{N} h_i^{-1}\left|h_i^{1/p}\bar{q}_i\right|^p\right)^{1/p} \leqslant h^{-1/p}C_4(M)\omega\left(f^{(k)}, \delta_n\right)_p,$$

where $h = \min_i h_i$. Since $\|\bar{l}\|_p \leqslant \|\mathscr{A}^{-1}\|_p\|\bar{q}\|_p$, inequality (5.18) follows from (5.22) and the preceding estimates.

We now prove (5.19). We multiply both sides of (2.10) by $h_i^{1/p}$ $(1 \leqslant p < \infty)$ and make the change of variables

$$\sigma_i = h_i^{1/p}\bar{l}_i.$$

Then, taking account of periodicity, we obtain the system of equations

$$\mathscr{A}_p\sigma = q, \qquad \mathscr{A}_p = \mathscr{D}_p\mathscr{A}\mathscr{D}_p^{-1}, \tag{5.23}$$

where $\sigma = (\sigma_1,\ldots,\sigma_N)^{\mathrm{T}}$, $q = (q_1,\ldots,q_N)^{\mathrm{T}}$, $q_i = u_i h_i^{1+1/p}h_{i+1}/(h_i + h_{i+1})$, the diagonal matrix $\mathscr{D}_p = \operatorname{diag}(h_1^{1/p},\ldots,h_N^{1/p})$, and $\mathscr{A}$ is the matrix (4.1) with $\alpha_i = (-1)^{k+1}h_{i+1}/(h_i + h_{i+1})$, $\beta_i = (-1)^{k+1}(1 - |\alpha_i|)$, and $\gamma_i = k + 1$ $(i = 1,\ldots,N)$.

Let us estimate $\|\mathscr{A}_p^{-1}\|_p$. For $p = \infty$ we have $\mathscr{A}_p = \mathscr{A}$ and we obtain, by the dominance relation, and taking account of (4.15),

$$\|\mathscr{A}^{-1}\| \leqslant k^{-1}. \tag{5.24}$$

Moreover, let $\mathscr{A}_p^{-1} = \{b_{ij}\}_{i,j=1}^N$ and $\mathscr{A}^{-1} = \{a_{ij}\}_{i,j=1}^N$; then from (4.13) we have

$$|b_{ij}| \leqslant (h_i/h_j)^{1/p}|a_{ij}|,$$

and the estimates (4.12) hold for a_{ij}.

By (5.8), we deduce, by an argument similar to what we used for (5.9) and (5.14), that

$$|b_{ij}| \leqslant C(\rho, p)\left[\left(\frac{\rho^{1+1/p}}{\varepsilon_\rho}\right)^{|j-i|} + \left(\frac{\rho^{1+1/p}}{\varepsilon_\rho}\right)^{N-|j-i|}\right]$$
$$\leqslant 4C(\rho, p)\sum_{m=0}^{\infty}\left(\frac{\rho^{1+1/p}}{\varepsilon_\rho}\right)^m.$$

Since $\rho^{1+1/p} < \varepsilon_p$ for $\rho < \rho_{k,p}$ $(1 \leqslant p < \infty)$, by (4.14) we obtain

$$\left\|\mathscr{A}_p^{-1}\right\|_p \leqslant C_1(\rho, p) \qquad (1 \leqslant p < \infty). \tag{5.25}$$

From (3.16) we have

$$\|f^{(k)} - S_n^{(k)}\|_p \leqslant C_1\omega\left(f^{(k)}, \delta_n\right)_p + C_3(\rho)\|\sigma\|_p,$$

and since $\|\sigma\|_p \leqslant \|\mathscr{A}_p^{-1}\|_p\|q\|_p$ we establish (5.19) by taking account of (3.14) and (3.15).

This completes the proof of the theorem.

THEOREM 3. *There exist a function* $f \in W_p^k$ $(1 \leqslant p < \infty)$ *and a sequence of nets* Δ_n, *satisfying* (1.2) *for* $\rho \geqslant \rho_{k,p}$, *for which*

$$\|f^{(l)} - S_n^{(l)}\|_q \nrightarrow 0 \qquad (n \to \infty;\, 1 \leqslant q \leqslant p),$$

where $l = 0, 1, \ldots, k$.

PROOF. Without loss of generality, we may suppose that W_p^k is the space of 2-periodic functions. Let there be given a real number $\rho \geqslant 2$. We define the sequence Δ_n in the following way:

$$\Delta_n: -1 = x^{(n)}_{-(\sigma_n+\nu_n+1)} < \cdots < x^{(n)}_{-1} < x^{(n)}_0 < \cdots < x^{(n)}_{\sigma_n+\nu_n} = 1, \quad (5.26)$$

where $x_0^{(n)} = 2^{-n}$, $x^{(n)}_{-i} = -x^{(n)}_{i-1}$, and the points $x_i^{(n)}$ $(i = 1,\dots,\sigma_n + \nu_n - 1)$ are determined by assigning the steps $h_i^{(n)} = x_i^{(n)} - x_{i-1}^{(n)}$ according to the formulas

$$\begin{aligned} h_i^{(n)} &= 2^{1-n}\rho^{-i} \qquad (i = 1,2,\dots,\sigma_n - 1),\\ h_i^{(n)} &= 2^{1-n}\rho^{-\sigma_n} \qquad (i = \sigma_n, \sigma_n + 1,\dots,\sigma_n + \nu_n - 1),\\ h^{(n)}_{\sigma_n+\nu_n} &= 2^{1-n}\rho^{-\sigma_n}d_n, \qquad 1 \leqslant d_n < 2, \end{aligned} \quad (5.27)$$

where σ_n is an odd number and d_n and ν_n are determined by the conditions $2^{-1}h_0^{(n)} + h_1^{(n)} + \cdots + h^{(n)}_{\sigma_n+\nu_n} = 1$, $1 \leqslant d_n < 2$.

Corresponding to each net Δ_n we construct a function $f_n \in W_p^k$. Set

$$\begin{aligned} f_n\left(x_{2i}^{(n)}\right) &= \left(h_{2i+1}^{(n)}\right)^{k-1/p} \qquad (i = 1,2,\dots,(\sigma_n - 1)/2),\\ f_n\left(x_{2i+1}^{(n)}\right) &= 0 \qquad (i = 0,1,\dots,(\sigma_n - 1)/2),\\ f_n^{(j)}\left(x_i^{(n)}\right) &= 0 \qquad (j = 1,2,\dots,k - 1;\ i = 1,2,\dots,\sigma_n), \end{aligned} \quad (5.28)$$

and on each interval $[x_{i-1}^{(n)}, x_i^{(n)}]$ $(i = 2,3,\dots,\sigma_n)$ let the function be an algebraic polynomial of degree $2k - 1$. For $x \in [0,1]/[x_i^{(n)}, x_{\sigma_n}^{(n)}]$ we set $f_n(x) \equiv 0$, and for $x \in [-1,0)$ we set $f_n(x) = (-1)^{k+1}f_n(-x)$ and continue it to the whole line with period 2.

From now on, the index n will be omitted from $h_i^{(n)}$ and $x_i^{(n)}$, since all estimates will be made for a given net and function f_n. As in [8], it is easily shown that the conditions

$$S_n^{(k)}(-1) = S_n^{(k)}(1) = 0, \quad \text{i.e. } S_n^{(k)}(x_i) = m_i = 0 \qquad (i = -\sigma_n - \nu_n - 1, \sigma_n + \nu_n)$$

are satisfied for the given δ_n and f_n, and that in addition $S_n^{(k)}$ is an odd function, so that

$$m_{-1} = -m_0. \quad (5.29)$$

Using (2.6) and (5.29) to determine m_i, we obtain the system of equations

$$\begin{aligned} &\left[(k + 1)(\rho + 1) + (-1)^k\right]m_0 + (-1)^{k+1}\rho m_1 = 0,\\ &(-1)^{k+1}m_{i-1} + (k + 1)(\rho + 1)m_i + (-1)^{k+1}\rho m_{i+1} = b_i \quad (i = 1,2,\dots,\sigma_n - 1),\\ &(-1)^{k+1}m_{i-1} + 2(k + 1)m_k + (-1)^{k+1}m_{i+1} = 0\\ &\qquad\qquad (i = \sigma_n, \sigma_n + 1,\dots,\sigma_n + \nu_n - 2),\\ &(-1)^{k+1}d_n m_{\sigma_n+\nu_n-2} + (1 + d_n)(k + 1)m_{\sigma_n+\nu_n-1} = 0, \end{aligned} \quad (5.30)$$

where $b_i = h_i v_i$ is calculated from (2.5) and (2.4), i.e.

$$\begin{aligned} &b_1 = k!C_{2k+1}^k h_0^{-1/p}\rho^{1-k+3/p}, \quad b_{\sigma_n-2} = (-1)^k k!C_{2k+1}^k h_0^{-1/p}\rho^{(\sigma_n-2)/p}, \quad b_{\sigma_n-1} = 0,\\ &b_i = -k!C_{2k+1}^k h_0^{-1/p}\left(1 + (-1)^k\rho^{-k-1}\right)\rho^{1+(i+1)/p} \qquad (i = 2,4,\dots,\sigma_n - 3),\\ &b_i = k!C_{2k+1}^k h_0^{-1/p}\rho^{i/p}\left(\rho^{1-k+2/p} + (-1)^k\right) \qquad (i = 3,5,\dots,\sigma_n - 4). \end{aligned} \quad (5.31)$$

We obtain $m_0 = m_0(n)$ from (5.30) by Cramer's rule. If A denotes the determinant of the system, and B is the determinant differing from A in the first column, which consists of the b_i, we then obtain

$$m_0 = B/A. \tag{5.32}$$

We introduce some auxiliary quantities in order to estimate A and B. Set

$$\gamma_0 = (k+1)(\rho+1) + (-1)^k, \qquad \beta_0 = (-1)^{k+1}\rho;$$
$$\alpha_i = (-1)^{k+1}, \qquad \beta_i = (-1)^{k+1}\rho,$$
$$\gamma_i = (k+1)(\rho+1) \qquad (1 \leqslant i \leqslant \sigma_n - 1);$$
$$\alpha_i = \beta_i = (-1)^{k+1}, \quad \gamma_i = 2(k+1), \quad (\sigma_n \leqslant i \leqslant \sigma_n + \nu_n - 2);$$
$$\alpha_{\sigma_n+\nu_n-1} = (-1)^{k+1}, \qquad \gamma_{\sigma_n+\nu_n-1} = (k+1)(1+d_n).$$

We define the following determinants by using the determinant P_{ij} introduced in (4.2):

$$F_j = F_j(\rho) = P_{1,j}, \quad M_j = P_{\sigma_n+\nu_n-j,\sigma_n+\nu_n-1}, \quad A = P_{0,\sigma_n+\nu_n-1},$$
$$D_{\sigma_n+\nu_n-i} = P_{i,\sigma_n+\nu_n-1} \qquad (0 \leqslant i \leqslant \sigma_n - 1).$$

Since all these determinants have dominant main diagonal and positive elements, they are positive.

As in [8] we obtain the estimates

$$0 < A \leqslant C_1 \varepsilon_\rho^{\sigma_n} \varepsilon_1^{\nu_n}, \qquad D_{\nu_n+j} \geqslant C_2 \varepsilon_\rho^j \varepsilon_1^{\nu_n} > 0, \tag{5.33}$$

where ε_ρ is defined above and $\varepsilon_1 = \varepsilon_{\rho|\rho=1}$.

Expanding B in terms of the b_i, we obtain

$$B = \sum_{i=1}^{\sigma_n-2} (-1)^i b_i (-1)^{(k+1)i} \rho^i D_{\sigma_n+\nu-i-1}. \tag{5.34}$$

Taking account of the values of the b_i (see (5.31)), we have for even i,

$$\begin{aligned}\kappa_i &= b_i \rho^i D_{\sigma_n+\nu_n-i-1} + (-1)^k b_{i+1} \rho^{i+1} D_{\sigma_n+\nu_n-i-2} \\ &= -k! C_R^k l2k + 1 h_0^{-1/p} \rho^{(1+1/p)(i+1)} \\ &\quad \times \left[\left(1 + (-1)^k \rho^{-k-1}\right) D_{\sigma_n+\nu_n-i-1} + (-1)^{k+1}\left((-1)^k + \rho^{1-k+2/p}\right) D_{\sigma_n+\nu_n-i-2}\right].\end{aligned}$$

In addition, we deduce by reasoning similar to that used in [8],

$$\begin{aligned}D_{\sigma_n+\nu_n-i-1} &= (k+1)(\rho+1) D_{\sigma_n+\nu_n-i-2} D_{\sigma_n+\nu_n-i-3} \\ &\geqslant k(\rho+1) D_{\sigma_n+\nu_n-i-2} + \rho\left(D_{\sigma_n+\nu_n-i-2} - D_{\sigma_n+\nu_n-i-3}\right) \\ &\geqslant k(\rho+1) D_{\sigma_n+\nu_n-i-2}.\end{aligned}$$

Evidently $2 < \rho_{k,p} \leqslant \rho$. For $k > 2$ we obtain

$$\begin{aligned}\kappa_i &\leqslant -k! C_{2k+1}^k h_0^{-1/p} \rho^{(1+1/p)(i+1)} D_{\sigma_n+\nu_n-i-2}\left[7(\rho+1)/4 + (-1)^{k+1} \rho^{1-k+2/p} - 1\right] \\ &\leqslant -k! C_{2k+1}^k h_0^{-1/p} \rho^{(1+1/p)(i+1)} D_{\sigma_n+\nu_n-i-2}.\end{aligned}$$

Since we have $\rho^2 - 1 > 0$ and $1 - \rho^{-2} > 3/4$, for $k = 1$, we deduce by taking account of (5.33) that

$$\kappa_i \leqslant -C_3 h_0^{-1/p} \rho^{(1+1/p)i} \varepsilon_\rho^{\sigma_n - i} \varepsilon_1^{\nu_n} < 0 \qquad (i = 2, 4, \ldots, \sigma_n - 3).$$

Therefore

$$B \leqslant \sum_{i=1}^{(\sigma_n-3)/2} \kappa_{2i} + (-1)^k \rho b_1 D_{\nu_n + \sigma_n - 2}$$

$$\leqslant -C_3 h_0^{-1/p} \varepsilon_\rho^{\sigma_n} \varepsilon_1^{\nu_n} \sum_{i=1}^{(\sigma_n-3)/2} \left(\rho^{1+1/p}/\varepsilon_\rho\right)^i + C_6 h_0^{-1/p} \varepsilon_\rho^{\sigma_n} \varepsilon_1^{\nu_n}.$$

Since $\rho^{1+1/p} \geqslant \varepsilon_\rho$ for $\rho \geqslant \rho_{k,p}$, we have

$$B \leqslant -C_4 h_0^{-1/p} \varepsilon_\rho^{\sigma_n} \varepsilon_1^{\nu_n} \sigma_n + C_6 h_0^{-1/p} \varepsilon_\rho^{\sigma_n} \varepsilon_1^{\nu_n},$$

and if we choose σ_n sufficiently large ($\sigma_n \to \infty$ as $n \to \infty$) we obtain

$$B \leqslant -C_5 h_0^{-1/p} \varepsilon_\rho^{\sigma_n} \varepsilon_1^{\nu_n} \sigma_n.$$

Taking account of (5.32) and (5.33), we deduce that

$$m_0 < -C_6 h_0^{-1/p} \sigma_n. \tag{5.35}$$

Hence the following estimates hold for $l = 0, 1, \ldots, k$:

$$\|S_n^{(l)}\|_q \geqslant \|S_n^{(l)}\|_{L_q[0,x_0]} \geqslant C_1 h_0^{1/q} \|S_n^{(l)}\|_{L_\infty[0,x_0]}$$

$$\geqslant C_2 h_0^{k-l+1/q} \|S_n^{(k)}\|_{L_\infty[0,x_0]} \geqslant C_2 h_0^{k-l+1/q} |m_0| \geqslant C_3 h_0^{k-l+1/q-1/p} \sigma_n.$$

Choosing $\sigma_n \geqslant n h_0^{l-k-1/q+1/p}$, we have

$$\|S_n^{(l)}\|_q \geqslant C_4 n \to \infty \qquad (n \to \infty)$$

and consequently we obtain the conclusion of Theorem 3 by the Banach-Steinhaus theorem.

§6. Convergence of derivatives of order $k + 1$

Let S_n be a multiply interpolating spline that interpolates $f \in C^k$ together with its derivatives up to order $k - 1$ on the set Δ_n.

THEOREM 4. *For every function $f \in W_p^{k+1}$ $(1 \leqslant p \leqslant \infty)$*

$$\|f^{(k+1)} - S_n^{(k+1)}\|_p \leqslant C\omega\left(f^{(k+1)}, \delta_n\right)_p, \tag{6.1}$$

where C is an absolute positive constant.

PROOF. If we multiply both sides of (2.14) for $l = k + 1$ by $h_i^{1/p}$ and make the change of variables $g_i = h_i^{1/p} d_i$, then by periodicity we obtain the system of equations

$$\mathscr{A}_p g = q \qquad (p = 1, \infty), \tag{6.2}$$

where

$$g = (g_1, \ldots, g_N)^{\mathrm{T}}, \qquad q = (q_1, \ldots, q_N)^{\mathrm{T}},$$

$$q_i = R h_i^{-1+1/p}\left[u_{i-1} h_{i-1} h_i/(h_{i-1} + h_i) - u_i h_i h_{i+1}/(h_i + h_{i+1})\right],$$

and the matrix $\mathscr{A}_p$ ($p = 1, \infty$) is the matrix $\mathscr{A}$ (see (4.1)) where when $p = 1, \infty$,

$$\gamma_i = k + 1 + (-1)^{k+1}\left[1 - h_{i-1}/(h_{i-1} + h_i) - h_{i+1}/(h_i + h_{i+1})\right];$$

when $p = 1$,

$$\alpha_i = (-1)^{k+1} h_i/(h_{i+1} + h_i), \qquad \beta_i = (-1)^{k+1} h_i/(h_i + h_{i+1});$$

when $p = \infty$,

$$\alpha_i = (-1)^{k+1} h_{i-1}/(h_{i-1} + h_i), \qquad \beta_i = (-1)^{k+1} h_{i+1}/(h_i + h_{i+1}).$$

It is evident that $\mathscr{A}_1$ satisfies (4.7) when $k \geqslant 2$, and $\mathscr{A}_\infty$ satisfies (4.15); consequently, if we take account of (4.14), we have $\|\mathscr{A}_p^{-1}\|_p < C_1$ ($p = 1, \infty$) for $k \geqslant 2$. We can establish a similar estimate for $k = 1$. In this case we have

$$\gamma_i = 1 + h_i/(h_{i-1} + h_i) + h_i/(h_i + h_{i+1}).$$

Let $\mathscr{A}_p^{-1} = \{a_{ij}\}_{i,j=1}^N$; then it is easy to show for $p = 1, \infty$ that a_{ij} satisfies (4.12).

As in [10], it is easily shown that

$$\|\mathscr{A}_1^{-1}\|_1 \leqslant C_2, \qquad \|\mathscr{A}_\infty^{-1}\|_\infty \leqslant C_2,$$

and therefore, for $p = 1, \infty$ and $k = 1, 2, \ldots$, we have

$$\|\mathscr{A}_p^{-1}\|_p \leqslant C_2. \tag{6.3}$$

By periodicity, we deduce from (3.13) and (3.9) that

$$\|q\|_p \leqslant C_3 \omega\left(f^{(k+1)}, \delta_n\right)_p \qquad (1 \leqslant p \leqslant \infty), \tag{6.4}$$

and since $\|g\|_p \leqslant \|\mathscr{A}_p^{-1}\|_p \|q\|_p$, we obtain

$$\|f^{(k+1)} - S_n^{(k+1)}\|_1 \leqslant C_4 \omega\left(f^{(k+1)}, \delta_n\right)_1 \qquad \left(f \in W_1^{k+1}\right) \tag{6.5}$$

and

$$\|f^{(k+1)} - S_n^{(k+1)}\| \leqslant C_4 \omega\left(f^{(k+1)}, \delta_n\right) \qquad \left(f \in C^{k+1}\right), \tag{6.6}$$

from (3.20) and (3.21), respectively, i.e. we have obtained (6.1) for $p = 1$ and $p = \infty$.

We construct an operator A_n: $L_1 \to L_1$ in the following way: for every $g \in L_1$ ($g \not\equiv \text{const}$) we form $\tilde{g}(x) = g(x) - \int_a^b g(t)\,dt$, and for $g(x) \equiv C_0$ we take $\tilde{g}(x) \equiv 0$; then we take the $(k+1)$th integral of $\tilde{g}$ and obtain $\tilde{g}_{k+1} \in W_1^{k+1}$. Then we construct the m.i.s. S_n for $\tilde{g}_{k+1}$ and take the $(k+1)$th derivative. Consequently we obtain A_n: $g \to S_n^{(k+1)} \in C$. It is easily shown that A_n is a bounded linear operator from L_p to L_p ($1 \leqslant p \leqslant \infty$).

It follows from (6.5) that A_n: $L_1 \to L_1$ satisfies $\|A_n\|_1 \leqslant C_1$ for all n, and by (6.6) we also obtain, for A_n: $L_\infty \to L_\infty$,

$$\|A_n\|_\infty \leqslant C_2 \quad \text{for all } n.$$

Applying the Riesz-Thorin theorem, we obtain, for A_n: $L_p \to L_p$,

$$\|A_n\|_p \leqslant C_3 \quad \text{for all } n. \tag{6.7}$$

It is known that for every $f \in W_p^{k+1}$ ($1 \leqslant p \leqslant \infty$) there is a spline $\tilde{S}_n$ of deficiency 1 with the knots of Δ_n for which

$$\|f^{(k+1)} - \tilde{S}_n^{(k+1)}\|_p \leqslant C_4 \omega\left(f^{(k+1)}, \delta_n\right)_p.$$

Hence, taking account of (6.7), we deduce that

$$\begin{aligned}
\|f^{(k+1)} - S_n^{(k+1)}\|_p &\leqslant \|f^{(k+1)} - \tilde{S}_n^{(k+1)}\|_p + \|\tilde{S}_n^{(k+1)} - S_n^{(k+1)}\|_p \\
&\leqslant \|f^{(k+1)} - \tilde{S}_n^{(k+1)}\|_p + \|A_n\left(\tilde{S}_n^{(k+1)} - f^{(k+1)}\right)\|_p \\
&\leqslant \left(1 + \|A_n\|_p\right)\|f^{(k+1)} - \tilde{S}_n^{(k+1)}\|_p \leqslant C_5 \omega\left(f^{(k+1)}, \delta_n\right)_p.
\end{aligned}$$

This completes the proof of Theorem 4.

§7. Convergence and divergence of derivatives of order $l\,(l = k+2, k+3,\dots,2k+1)$

We denote by $\rho_{l,p} > 1$ the unique root of the equation

$$(k+1)(x+1)x^{l-k-1-1/p} = 1 + x^{1+2(k-k-1-1/p)} \qquad (x \geqslant 1), \tag{7.1}$$

where $1 \leqslant p \leqslant \infty$, $1/p = 0$ for $p = \infty$, and $l = k+2, k+3,\dots,2k+1$.

It is easily shown that

$$(k+1)^{1/(l-k-1-1/p)} \leqslant \rho_{l,p} \leqslant (2k+2)^{1/(l-k-1-1/p)}. \tag{7.2}$$

THEOREM 5. *For every function $f \in W_p^l$ $(l = k+2, k+3,\dots,2k+1)$, and for every sequence of nets Δ_n satisfying* (1.3), *we have the estimate*

$$\|f^{(l)} - S_n^{(l)}\|_p \leqslant C(M)\omega(f^{(l)}, \delta_n)_p, \tag{7.3}$$

and if the sequence of nets Δ_n satisfies (1.2) *with $\rho < \rho_{l,p}$ $(1 \leqslant p \leqslant \infty)$ then*

$$\|f^{(l)} - S_n^{(l)}\|_p \leqslant C(\rho, p)\omega(f^{(l)}, \delta_n)_p. \tag{7.4}$$

In the case $l = k+2$, $p = 1$, we have

$$\|f^{(k+2)} - S_n^{(k+2)}\|_1 \leqslant C_1\omega(f^{(k+2)}, \delta_n)_1. \tag{7.5}$$

PROOF. We first prove (7.3). We multiply both sides of (2.14) by h_i^{l-k-1} and make the change of variables $g_i = h_i^{l-k-1}$. Then, taking account of the periodicity of Δ_n and the spline, we obtain the system of equations

$$\alpha_i g_{i-1} + \gamma_i g_i + \beta_i g_{i+1} = q_i \qquad (i = 1,2,\dots,N), \tag{7.6}$$
$$g_0 = g_N, \qquad g_{N+1} = g_1,$$

where

$$\alpha_i = (-1)^{k+1}h_{i-1}/(h_{i-1} + h_i), \qquad \beta_i = (-1)^{k+1}h_i/(h_i + h_{i+1}),$$
$$\gamma_i = k + 1 + (-1)^l + (-1)^{l+k}(\alpha_i + \beta_i),$$
$$q_i = R\left(\frac{h_{i-1}}{h_{i-1} + h_i}u_{i-1} + (-1)^{k+l}\frac{h_{i+1}}{h_i + h_{i+1}}u_i\right) \qquad (i = 1,2,\dots,N).$$

We write the system (7.6) in the matrix form

$$\mathscr{A}g = q, \tag{7.7}$$

where $g = (g_1,\dots,g_N)^{\mathrm{T}}$, $q = (q_1,\dots,q_N)^{\mathrm{T}}$, and $\mathscr{A}$ has the form (4.1). Since $k + 2 \leqslant l \leqslant 2k + 1$, we have $l = 3$ for $k = 1$. Using (4.16), we obtain $\|\mathscr{A}^{-1}\| \leqslant 1$, and it was shown in [10] that also $\|\mathscr{A}^{-1}\|_1 \leqslant C_1$. Consequently, by using (4.14), we find that

$$\|\mathscr{A}^{-1}\|_p \leqslant C_1 \tag{7.8}$$

for $k = 1$, $1 \leqslant p \leqslant \infty$.

It is easy to verify, taking account of (1.3), that when $k \geqslant 2$ conditions (4.15) and (4.17) are satisfied by $\mathscr{A}$. Using (4.16), (4.18), and (4.14), we deduce that

$$\|\mathscr{A}^{-1}\|_p \leqslant C_2(M). \tag{7.9}$$

It follows from (3.14) that

$$|q_i| \leqslant C_1(|u_{i-1}| + |u_i|)$$
$$\leqslant C_2(M)\delta_n^{l-k-1-1/p}(\mu_{i-1,l}^2 + \mu_{i-1,l}^3 + \mu_{i,l}^2 + \mu_{i,l}^3).$$

Hence, by (3.9) and (3.10),

$$\|q\|_p \leqslant C_3(M)\delta_n^{l-k-1-1/p}\omega\big(f^{(l)},\delta_n\big)_p \qquad (1 \leqslant p \leqslant \infty). \tag{7.10}$$

Taking account of (7.9) and (7.10), we have

$$\|g\|_p \leqslant \|\mathscr{A}^{-1}\|_p\|q\|_p \leqslant C_4(M)\delta_n^{l-k-1-1/p}\omega\big(f^{(l)},\delta_n\big)_p. \tag{7.11}$$

Let $h = \min_i h_i$; then

$$h_i^{1/p}|d_i| = h_i^{k+1-l+1/p}|g_i| \leqslant h^{k+1-l+1/p}|g_i|$$

and consequently

$$\max_{1\leqslant i\leqslant N} |d_i| \leqslant h^{k+1-l}|g| \qquad (p=\infty),$$

$$\left(\sum_{i=1}^{N} h_i|d_i|^p\right)^{1/p} \leqslant h^{k+1-l+1/p}\|g\|_p \qquad (1 \leqslant p < \infty). \tag{7.12}$$

Then (7.3) follows from (1.3), (3.22), (3.23), and (7.12).

We now prove (7.4) and (7.5). We multiply both sides of (2.14) by $h_i^{1/p}$ ($1/p = 0$ when $p=\infty$) and make the change of variables $g_i = h_i^{1/p}d_i$; then we obtain the system

$$\mathscr{B}_p g = q, \qquad \mathscr{B}_p = \mathscr{D}_p^{-1}\mathscr{A}\mathscr{D}_p, \tag{7.13}$$

where $g = (g_1,\dots,g_N)^{\mathrm{T}}$, $q = (q_1,\dots,q_N)^{\mathrm{T}}$, the diagonal matrix

$$\mathscr{D}_p = \operatorname{diag}\big(h_1^{l-k-1-1/p}, h_2^{l-k-1-1/p},\dots,h_N^{l-k-1-1/p}\big),$$

and $\mathscr{A}$ has the form (4.1) with

$$\alpha_i = (-1)^{k+1}h_{i-1}/(h_{i-1}+h_i),\quad \beta_i = (-1)^{k+1}h_{i+1}/(h_i+h_{i+1}),$$
$$\gamma_i = k+1+(-1)^l+(-1)^{l+k}(\alpha_i+\beta_i),$$
$$q_i = Rh_i^{k+1-l+1/p}\big[u_{i-1}h_{i-1}/(h_{i-1}+h_i)+(-1)^{k+l}u_ih_{i+1}/(h_i+h_{i+1})\big]$$

$(i = 1,2,\dots,N)$.

Let $\mathscr{A}^{-1} = \{a_{ij}\}_{i,j=1}^N$ and $\mathscr{B}_p^{-1} = \{b_{ij}\}_{i,j=1}^N$; since $\mathscr{D}_p^{-1} = \mathscr{D}_p^{-1} = \mathscr{D}_p^{-1}\mathscr{A}^{-1}\mathscr{D}_p$ we have

$$b_{ij} = \left(\frac{h_j}{h_i}\right)^{l-k-1-1/p} a_{ij} = (-1)^{i+j}\left(\frac{h_j}{h_i}\right)^{l-k-1-1/p} A_{ji}/A. \tag{7.14}$$

By the periodicity of Δ_n we have

$$h_j/h_i \leqslant \rho^{\min\{|i-j|,\,N-|i-j|\}}. \tag{7.15}$$

We put $\kappa = \rho^{l-k-1-1/p}$; then, taking account of (4.12), we deduce that

$$|b_{ji}| \leqslant C_1\kappa^{|j-i|}|\mathscr{K}(i+1,j)|/P_{i+1,j} \qquad (1 < i \leqslant j < N),$$

$$|b_{ji}| \leqslant C_1\kappa^{N-|j-i|}|\mathscr{L}(j,i-1)|/P_{j,i-1} \qquad (1 < j < i < N),$$

$$|b_{ji}| \leqslant C_1\left(\kappa^{N-i}\frac{|\mathscr{K}(i+1,N)|}{P_{i+1,N}} + \kappa^{i}\frac{|\mathscr{L}(1,i-1)|}{P_{1,i-1}}\right) \quad (1 \leqslant i \leqslant N, j = 1, N),$$

$$|b_{ji}| \leqslant C_1\left(\kappa^{j}\frac{|\mathscr{K}(2,j)|}{P_{2,j}} + \kappa^{N-j}\frac{|\mathscr{L}(j,N-1)|}{P_{j,N-1}}\right) \quad (1 < j < N, i = 1, N). \tag{7.16}$$

Just as in §5, we introduce the notation (5.10) and consider two cases:

$$l \text{ odd}, \qquad \gamma_i = k + 1 + |\alpha_i| - |\alpha_{i+1}|, \tag{7.17}$$

$$l \text{ even}, \qquad \gamma_i = k + 1 - |\alpha_i| + |\alpha_{i+1}|. \tag{7.18}$$

We find the minima of Φ^{-1} and Ψ^{-1} in the domains (5.11). In the case (7.17) we have

$$\Phi^{-1}(\alpha, i, j) = \begin{vmatrix} \left(\frac{k+1-|\alpha_{i+1}|}{|\alpha_i|}+1\right) & (-1)^{k+1}\frac{1-|\alpha_{i+1}|}{|\alpha_i|} & 0 & \cdots & 0 & 0 & 0 \\ (-1)^{k+1} & \left(\frac{k+1-|\alpha_{i+2}|}{|\alpha_{i+1}|}+1\right) & (-1)^{k+1}\frac{1-|\alpha_{i+2}|}{|\alpha_{i+1}|} & \cdots & 0 & 0 & 0 \\ \cdots & \cdots & \cdots & \cdots & \cdots & \cdots & \cdots \\ 0 & 0 & 0 & \cdots & (-1)^{k+1} & \left(\frac{k+1-|\alpha_j|}{|\alpha_{j-1}|}+1\right) & (-1)^{k+1}\frac{1-|\alpha_j|}{|\alpha_{j-1}|} \\ 0 & 0 & 0 & \cdots & 0 & (-1)^{k+1} & \left(\frac{k+1-|\alpha_{j+1}|}{|\alpha_j|}+1\right) \end{vmatrix}$$

As in [10], we easily obtain

$$\frac{\partial}{\partial|\alpha_m|}\Phi^{-1}(\alpha, i, j) < 0 \qquad (i \leqslant m \leqslant j+1),$$

and therefore

$$\Phi^{-1}(\alpha, i, j) \geqslant \Phi^{-1}(\alpha, i, j)\big|_{\alpha_i = \cdots = \alpha_{j+1} = \rho/(1+\rho)} \geqslant C_1\left(\frac{\varepsilon_\rho}{\rho}\right)^{j-i}, \tag{7.19}$$

where

$$\varepsilon_\rho = \left[(k+1)(\rho+1) + \sqrt{(k+1)^2(\rho+1)^2 - 4\rho}\,\right]/2.$$

In the case (7.18), taking account of the fact that $k \geqslant 2$ when l is even, we deduce that

$$\frac{|\alpha_i|}{|\alpha_{j+1}|}\Phi^{-1}(\alpha, i, j) = \frac{|\alpha_i|}{|r\alpha_{j+1}|}\frac{P_{ij}}{|\mathscr{K}(i, j)|} = \frac{P_{ij}}{|\mathscr{K}(i+1, j+1)|} \geqslant C_1\left(\frac{\varepsilon_\rho}{\rho}\right)^{j-i}. \tag{7.20}$$

Hence

$$|b_{ji}| \leqslant C_2(\rho)\left[\left(\frac{\rho\kappa}{\varepsilon_\rho}\right)^{|j-i|} + \left(\frac{\rho\kappa}{\varepsilon_\rho}\right)^{N-|j-i|}\right] \qquad (i, j = 1, 2, \ldots, N)$$

and, taking account of (4.13) and (4.14), we deduce that

$$\left\|\mathscr{B}_p^{-1}\right\|_p \leqslant C_3(\rho, p)\sum_{i=0}^{N}(\rho\kappa/\varepsilon_\rho)^i.$$

Since $\rho\kappa = \rho^{l-k-1/p} < \varepsilon_p$ for $\rho < \rho_{l,p}$, we have

$$\left\|\mathscr{B}_p^{-1}\right\|_p \leqslant C_4(\rho, p). \tag{7.21}$$

If $p = 1$ and $l = k + 2$, then $\kappa = 1$ and we hgave $|b_{ij}| = |a_{ij}|$ instead of (7.14). If we calculate the minima of Φ^{-1} and Ψ^{-1} in the domains $0 < |\alpha_i| < 1$, $0 < |\beta_i| < 1$ $(i = 1, \dots, N)$ as in [10], we obtain

$$\|\mathscr{B}_1^{-1}\|_1 \leqslant C, \tag{7.22}$$

where C is an absolute constant.

It follows from (3.14), (3.9), and (3.10) that

$$\|q\|_p \leqslant C(\rho)\omega\left(f^{(l)}, \delta_n\right)_p \qquad (1 \leqslant p \leqslant \infty), \tag{7.23}$$

and we have, for $p = 1$ and $l = k + 2$,

$$\|q\|_1 \leqslant C_1\omega\left(f^{(k+2)}, \delta_n\right)_1. \tag{7.24}$$

Taking account of (3.22), (3.23), (7.21) and (7.23), we obtain (7.4), and when $p = 1$ and $l = k + 2$ we obtain (7.5) from (7.22) and (7.24).

This completes the proof of Theorem 5.

THEOREM 6. *Let $l = k + 2, k + 3, \dots, 2k + 1$ and $1 \leqslant p \leqslant \infty$. Then for every pair (l, p) except $(k + 2, 1)$ there are a function $f \in C^l$ and a sequence of nets Δ_n satisfying (1.2) for $\rho \geqslant \rho_{l,p}$ for which*

$$\|f^{(l)} - S_n^{(l)}\|_p \nrightarrow 0 \qquad (\delta_n \to 0, n \to \infty).$$

PROOF. Without loss of generality, we may suppose that C^l is the space of 2-periodic functions. Let $\kappa = \rho^{l-k-1}$; then for $\rho \geqslant \rho_{l,\infty}$ we have, by (7.2),

$$\kappa \geqslant k + 1. \tag{7.25}$$

We define a sequence of nets Δ_n as follows:

$$\Delta_n\colon -1 = x^{(n)}_{-(\sigma_n+\nu_n+1)} < \cdots < x^{(n)}_{-1} < x^{(n)}_0 < \cdots < x^{(n)}_{\sigma_n+\nu_n} = 1, \tag{7.26}$$

where σ_n is odd, $\sigma_n \geqslant \rho^{2n}$, $x_0^{(n)} = \xi_n(2\rho^{\sigma_n+n})^{-1}$, $x_{-i}^{(n)} = -x_{i-1}^{(n)}$ $(i = 1, \dots, \sigma_n + \nu_n)$ and the points $x_i^{(n)}$ are determined by defining the steps $h_i^{(n)} = x_i^{(n)} - x_{i-1}^{(n)}$ by the formulas

$$h_i^{(n)} = \xi_n\rho^{i-n-\sigma_n} \qquad (i = 1, 2, \dots, \sigma_n - 1),$$
$$h_i^{(n)} = \xi_n\rho^{-n} \qquad (i = \sigma_n, \sigma_n + 1, \dots, \sigma_n + \nu_n),$$

and the numbers ν_n and ξ_n are determined by the conditions

$$2^{-1}h_0^{(n)} + h_1^{(n)} + \cdots + h^{(n)}_{\sigma_n+\nu_n} = 1, \qquad 1 \leqslant \xi_n < 2.$$

Evidently $h_{-i}^{(n)} = h_i^{(n)}$ for all i, and we have the conditions

$$\sup_n \max_{|i-j|=1} h_i^{(n)}/h_j^{(n)} \leqslant \rho, \qquad \lim_{n\to\infty} \delta_n = 0. \tag{7.27}$$

For each net Δ_n (7.26) we construct a function $f_n \in C^l$. For $i = 1, \dots, (\sigma_n - 3)/2$ we set

$$\varphi_{n,2i}\left(x_{2i}^{(n)}\right) = \left(h_{2i}^{(n)}\right)^l, \qquad \varphi_{n,2i}\left(x_{2i-1}^{(n)}\right) = \varphi_{n,2i}\left(x_{2i+1}^{(n)}\right) = 0,$$
$$\varphi_{n,2i}^{(j)}\left(x_{2i-1}^{(n)}\right) = \varphi_{n,2i}^{(j)}\left(x_{2i}^{(n)}\right) = \varphi_{n,2i}^{(j)}\left(x_{2i+1}^{(n)}\right) = 0 \qquad (j = 1, \dots, l).$$

On each interval $[x_{2i-1}^{(n)}, x_{2i}^{(n)}]$ or $[x_{2i}^{(n)}, x_{2i+1}^{(n)}]$ the function $\varphi_{n,2i}$ is an algebraic polynomial of degree at most $2l + 1$. On $[0,1]/[x_{2i-1}^{(n)}, x_{2i+1}^{(n)}]$ we set $\varphi_{n,2i} \equiv 0$, and for $x \in [-1, 0]$ we set $\varphi_{n,2i}(x) = (-1)^l\varphi_{n,2i}(-x)$. It is clear that $\varphi_{n,2i} \in C^l$ and

$\|\varphi_{n,2i}^{(l)}\|_p \leqslant C_1$ $(1 \leqslant p \leqslant \infty)$. For $1 \leqslant p < \infty$ we set $f_n(x) = \varphi_{n,\sigma_n-3}(x)$, and $f_n(x) = \sum_{i=1}^{(\sigma_n-3)/2} \varphi_{n,2i}(x)$ for $p = \infty$.

Since f_n is 2-periodic, and Δ_n is symmetric with respect to $x = 0$, the function $S_n^{(l)}$ is even. Consequently

$$d_{-1} = d_1, \quad d_i = S_n^{(l)}\big(y_i^{(n)}\big) - H_i^{(l)}\big(y_i^{(n)}\big), \quad y_i^{(n)} = \big(x_{i-1}^{(n)} + x_i^{(n)}\big)/2.$$

Since $H_0^{(l)}(0) = 0$ and $\|H^{(l)}(x)\|_p \leqslant C_1$, we obtain the estimates

$$\|S_n^{(l)}\|_\infty \geqslant |S_n^{(l)}(0)| = |d_0|,$$

$$\|S_n^{(l)}\|_p \geqslant \|S_n^{(l)} - H^{(l)}\|_p - \|H^{(l)}\|_p \geqslant \|S_n^{(l)} - H^{(l)}\|_p - C_1, \tag{7.28}$$

where $H(x) = H(x, f_n)$ is a Hermite spline of degree $2k + 1$.

Hence we obtain the following system of equations for determining the d_i:

$$\begin{gathered}
\big[(k+1)(\rho+1) + (-1)^l(1-\rho)\big]d_0 + (-1)^{k+1}2\rho\kappa d_1 = 0,\\
(-1)^{k+1}d_{i-1} + (k+1)(\rho+1)\kappa d_i + (-1)^{k+1}\rho\kappa^2 d_{i+1} = q_i \quad (i = 1, 2, \ldots, \sigma_n - 1),\\
(-1)^{k+1}2d_{\sigma_n-1} + \big[2(k+1)(\rho+1) + (-1)^l(\rho-1)\big]\kappa d_{\sigma_n}\\
+ (-1)^{k+1}\cdot(\rho+1)\kappa d_{\sigma_n+1} = 0,\\
(-1)^{k+1}d_{i-1} + 2(k+1)d_i + (-1)^{k+1}d_{i+1} = 0\\
(i = \sigma_n + 1, \sigma_n + 2, \ldots, \sigma_n + \nu_n - 1),\\
(-1)^{k+1}d_{\sigma_n+\nu_n-1} + \big(2k + 2 + (-1)^{k+1}\big)d_{\sigma_n+\nu_n} = 0,
\end{gathered} \tag{7.29}$$

where the q_i are calculated from (2.14), (2.4) and (2.5).

In the case when $f_n(x) = \varphi_{n,m}(x)$ (m even, $2 \leqslant m \leqslant \sigma_n - 3$), we have

$$\begin{aligned}
q_{m-1} &= Rk!C_{2k+1}^k(-1)^{k+l}\rho^{l-k}\kappa,\\
q_m &= Rk!C_{2k+1}^k\kappa\big[1 + (-1)^{k+l+1}\rho^{-k} + (-1)^{l+1}\rho\big],\\
q_{m+1} &= Rk!C_{2k+1}^k\kappa\big[-\rho^{-l} + (-1)^{k+1}\rho^{k+1-l} + (-1)^l\rho^{1-l}\big],\\
q_{m+2} &= Rk!C_{2k+1}^k\kappa(-1)^k\rho^{k+1-2l},
\end{aligned} \tag{7.30}$$

and the remaining $q_i = 0$.

From the system (7.29) we obtain $d_i = d_i(n)$ $(0 \leqslant i \leqslant \sigma_n - 3)$ by Cramer's rule. Let A denote the determinant of the system (7.29), and let B be the determinant obtained from A by replacing the $(i+1)$th column by q_j's. Then

$$d_i = B_i/A. \tag{7.31}$$

In order to estimate B_i and A, we define the following quantities. We set

$$\begin{gathered}
\gamma_0 = (k+1)(\rho+1) + (-1)^l(1-\rho), \quad \beta_0 = (-1)^{k+1}2\rho\kappa; \qquad \alpha_i = (-1)^{k+1},\\
\gamma_i = (k+1)(\rho+1)\kappa, \quad \beta_i = (-1)^{k+1}\rho\kappa^2\ (1 \leqslant i \leqslant \sigma_n - 1); \quad \alpha_{\sigma_n} = 2\cdot(-1)^{k+1},\\
\gamma_{\sigma_n} = \big[2(k+1)(\rho+1) + (-1)^l(\rho-1)\big]\kappa, \qquad \beta_{\sigma_n} = (-1)^{k+1}(\rho+1)\kappa;\\
\alpha_i = \beta_i = (-1)^{k+1}, \qquad \gamma_i = 2(k+1)(\sigma_n + 1 \leqslant i \leqslant \sigma_n + \nu_n - 1);\\
\alpha_{\sigma_n+\nu_n} = (-1)^{k+1}, \qquad \gamma_{\sigma_n+\nu_n} = 2(k+1) + (-1)^{k+1}.
\end{gathered}$$

Using the determinant P_{ij} defined in (4.2), we introduce the determinants

$$F_j = F_j(\rho) = P_{1,j}, \qquad \mathcal{K}_j = \mathcal{K}_j(\rho) = P_{0,j-1},$$

$$\mathcal{M}_1 = 2(k+1) + (-1)^{k+1}, \qquad \mathcal{M}_j = P_{\sigma_n+\nu_n-j+1,\sigma_n+\nu_n} \quad (1 \leqslant j \leqslant \nu_n + 1),$$

$$D_{\nu_n+1+\sigma_n-i} = P_{i,\nu_n+\sigma_n} \qquad (1 \leqslant i \leqslant \sigma_n - 1).$$

We also set

$$\varepsilon_\rho = \kappa\left[(k+1)(\rho+1) + \sqrt{(k+1)^2(1+\rho)^2 - 4\rho}\,\right]/2.$$

Then

$$F_j = \frac{1}{\kappa\sqrt{(k+1)^2(\rho+1)^2 - 4\rho}}\left[\varepsilon_\rho^{j+1} - \left(\rho\kappa^2/\varepsilon_\rho\right)^{j+1}\right] \qquad (j \geqslant 1)$$

and evidently

$$\frac{k(k+2)}{(k+1)^2}\,\frac{\varepsilon_\rho^{j+1}}{\kappa\sqrt{(k+1)^2(\rho+1)^2 - 4\rho}} \leqslant F_j \leqslant \frac{\varepsilon_\rho^{j+1}}{\kappa\sqrt{(k+1)^2(\rho+1)^2 - 4\rho}}. \tag{7.32}$$

Hence

$$0 < F_{j-1} < 4F_j/3\varepsilon_\rho. \tag{7.33}$$

Since $M_2 > M_1 > 0$, we have

$$\begin{aligned}
\mathcal{M}_j &= 2(k+1)\mathcal{M}_{j-1} - \mathcal{M}_{j-2} = (2k+1)\mathcal{M}_{j-1} + \left(\mathcal{M}_{j-1} - \mathcal{M}_{j-2}\right) \\
&\geqslant (2k+1)\mathcal{M}_{j-1} > 0 \qquad (2 \leqslant j \leqslant \nu_n), \\
\mathcal{M}_{\nu_n+1} &= \left[2(k+1)(1+\rho) + (-1)^l(\rho-1)\right]\kappa\mathcal{M}_{\nu_n} - (\rho+1)\kappa\mathcal{M}_{\nu_n-1} \\
&\geqslant (2k+1)(1+\rho)\kappa\mathcal{M}_{\nu_n} - [(\rho+1)\kappa/(2k+1)]\mathcal{M}_{\nu_n} \\
&= (\rho+1)\kappa\mathcal{M}_{\nu_n}(2k+1 - 1/(2k+1)) \geqslant 2k(\rho+1)\kappa\mathcal{M}_{\nu_n}.
\end{aligned}$$

We also obtain

$$\begin{aligned}
D_{\nu_n+2} &= (k+1)(\rho+1)\kappa\mathcal{M}_{\nu_n+1} - 2\rho\kappa^2\mathcal{M}_{\nu_n} \geqslant 2^{-1}(k+1)(\rho+1)\kappa\mathcal{M}_{\nu_{n+1}} \\
&\quad + k(k+1)(\rho+1)^2\kappa^2\mathcal{M}_{\nu_n} - 2\rho\kappa^2\mathcal{M}_{\nu_n} \geqslant 2^{-1}(k+1)(\rho+1)\kappa\mathcal{M}_{\nu_n+1}.
\end{aligned}$$

Now we prove by induction that

$$D_{\nu_n+1+\sigma_n-i} \geqslant \frac{(k+1)}{2}(\rho+1)\kappa D_{\nu_n+\sigma_n-i} \qquad (i \geqslant \sigma_n - 2). \tag{7.34}$$

In fact,

$$\begin{aligned}
D_{\nu_n+1+\sigma_n-i} &= (k+1)(\rho+1)\kappa D_{\nu_n+\sigma_n-i} - \rho\kappa^2 D_{\nu_n+\sigma_n-i-1} \\
&\geqslant 2^{-1}(k+1)(\rho+1)\kappa D_{\nu_n+\sigma_n-i} + D_{\nu_n+\sigma_n-i+1}\left[(k+1)^2(\rho+1)^2\kappa^2/4 - \rho\kappa^2\right] \\
&\geqslant 2^{-1}(k+1)(\rho+1)\kappa D_{\nu_n+\sigma_n-i}.
\end{aligned}$$

Since $0 < C_1\varepsilon_1^{\nu_n} \leqslant \mathcal{M}_{\nu_n} \leqslant C_2\varepsilon_1^{\nu_n}$ and

$$D_{\nu_n+\sigma_n-i+1} = F_{\sigma_n-i}\mathcal{M}_{\nu_n+1} - 2\rho\kappa^2 F_{\sigma_n-i-1}\mathcal{M}_{\nu_n},$$

we deduce, taking account of (7.32), that

$$D_{\nu_n+\sigma_n-i+1} \leqslant F_{\sigma_n-i}\mathcal{M}_{\nu_n+1} \leqslant C_3\varepsilon_1^{\nu_n}\varepsilon_\rho^{\sigma_n-i} \tag{7.35}$$

and, because $\rho\kappa/\varepsilon_\rho \leqslant 1$,

$$D_{\nu_n+\sigma_n-i} = \mathscr{M}_{\nu_n}\Big\{\big[2(k+1)(\rho+1)+(-1)^l(\rho-1)\big]\kappa F_{\sigma_n-i} - 2\rho\kappa^2 F_{\sigma_n-i-1}\Big\}$$

$$-(\rho+1)\kappa F_{\sigma_n-i}\mathscr{M}_{\nu_n-1} \geqslant F_{\sigma_n-i}\mathscr{M}_{\nu_n}\kappa\big[(2k+1)(\rho+1)+1-8\rho\kappa/3\varepsilon_\rho\big]$$

$$-(\rho+1)\kappa F_{\sigma_n-i}\mathscr{M}_{\nu_n-1} \geqslant F_{\sigma_n-i}\kappa\big[\mathscr{M}_{\nu_n}(3\rho+4-8/3)-(\rho+1)\mathscr{M}_{\nu_n-1}\big]$$

$$\geqslant 2\rho\kappa F_{\sigma_n-i}\mathscr{M}_{\nu_n} \geqslant C_4\varepsilon_1^{\nu_n}\varepsilon_\rho^{\sigma_n-i}, \tag{7.36}$$

$$\mathscr{K}_j = \big[(k+1)(\rho+1)+(-1)^l\big]F_{j-1} - 2\rho\kappa F_{j-2}$$

$$\geqslant F_{j-1}(2\rho+2-\rho-8\rho\kappa/3\varepsilon_\rho) \geqslant F_{j-1}/3 \geqslant C_5\varepsilon_\rho^i,$$

$$\mathscr{K}_j \leqslant C_6\varepsilon_\rho^j,\ 0 < A < \mathscr{K}_{\sigma_n}\mathscr{M}_{\nu_n+1} \leqslant C_7\varepsilon_1^{\nu_n}\varepsilon_\rho^{\sigma_n}. \tag{7.37}$$

Expanding B_0 in terms of the first column, we obtain

$$B_0 = \sum_{j=1}^{\sigma_n-3}(-1)^j q_j(-1)^{(k+1)j}(\rho\kappa^2)^j 2\kappa^{-1}D_{\nu_n+\sigma_n-j}$$

$$= \sum_{j=1}^{\sigma_n-3}(-1)^{kj}q_j 2\kappa^{-1}(\rho\kappa^2)^j D_{\nu_j+\sigma_n-j} = (-1)^l 2Rk!C_{2k+1}^k$$

$$\times(\rho\kappa^2)^{m-1}\Big\{\rho^{l-k}D_{\nu_n+\sigma_n-m+1} + (\rho\kappa^2)\big[(-1)^l+(-1)^{k+1}\rho^{-k}-\rho\big]$$

$$\times D_{\nu_n+\sigma_n-m} + (\rho\kappa^2)^2\big[(-1)^{k+l+1}\rho^{-l}+(-1)^k\rho^{1-l}+(-1)^{l+1}\rho^{k+1-l}\big]$$

$$\times D_{\nu_n+\sigma_n-m-1} + (\rho\kappa^2)^3(-1)^{l+k}\rho^{k+1-2l}D_{\nu_n+\sigma_n-m-2}\Big\}$$

$$= (-1)^l 2Rk!C_{2k+1}^k(\rho\kappa^2)^{m-1}\mathscr{L}. \tag{7.38}$$

Let $2 \leqslant i \leqslant \sigma_n - 3$ and $m = \sigma_n - 3$; then

$$B_i = (-1)^{ki}\mathscr{K}_i(\rho\kappa^2)^{-i}\sum_{j=\sigma_n-4}^{\sigma_n-1}(-1)^{kj}q_j(\rho\kappa^2)^j D_{\nu_n+\sigma_n-j}$$

$$= (-1)^{ki+l}\mathscr{K}_i Rk!C_{2k+1}^k(\rho\kappa^2)^{\sigma_n-1-i}\mathscr{L}. \tag{7.39}$$

We have the inequalities

$$\kappa = \rho^{l-k-1} \geqslant k+1, \qquad \rho^k = \rho^{2k-1} \geqslant \rho^{l-k+1} \geqslant k+1,$$

$$\rho^{k+1-2l} = \rho^{-k-1}\kappa^{-2}, \quad \rho^{l+k} = \rho\kappa, \qquad \rho^{k+1} \geqslant \rho(k+1). \tag{7.40}$$

We now estimate $\mathscr{D}$. We consider four cases.

1. Let k and l be odd; then

$$\mathscr{L} \geqslant \rho\kappa D_{\nu_n+\sigma_n-m+1} - (\rho+1-\rho^{-k})(\rho\kappa^2)D_{\nu_n+\sigma_n-m}$$

$$\geqslant \big[\rho\kappa 2^{-1}(k+1)(\rho+1)\kappa + (\rho+1-\rho^{-k})\rho\kappa^2\big]D_{\nu_n+\sigma_n-m}$$

$$\geqslant \rho^{k+1}\kappa^2 D_{\nu_n+\sigma_n-m}. \tag{7.41}$$

2. Let k be odd and l even; then

$$\begin{aligned}\mathscr{L} &\geqslant \rho\kappa D_{\nu_n+1+\sigma_n-m} - \rho\kappa^2(\rho-1)D_{\nu_n+\sigma_n-m} \\ &\quad -\rho^2\kappa^4\left(\rho^{1-l}+\rho^{k+1-l}\right)D_{\nu_n+\sigma_n-m-1} - \rho^{-k-1}\kappa^{-2}\left(\rho\kappa^2\right)^3 D_{\nu_n+\sigma_n-m-2} \\ &\geqslant 2\rho\kappa^2 D_{\nu_n+\sigma_n-m} - \rho^2\kappa^4\left(\rho^{1-l}+\rho^{k+1-l}\right)D_{\nu_n+\sigma_n-m-1} - \rho^2\kappa^3 D_{\nu_n+\sigma_n-m-2} \\ &\geqslant \rho\kappa^3\left[2(\rho+1)-\rho\kappa\left(\rho^{1-l}+\rho^{k+1-l}\right)\right]D_{\nu_n+\sigma_n-m-1} - \rho^2\kappa^3 D_{\nu_n+\sigma_n-m-2} \\ &\geqslant 2\rho^2\kappa^3 D_{\nu_n+\sigma_n-m-1} - \rho^2\kappa^3 D_{\nu_n+\sigma_n-m-2} \geqslant \rho^2\kappa^3 D_{\nu_n+\sigma_n-m-1}. \end{aligned} \tag{7.42}$$

3. Let k be even and l odd; then, remembering that in this case $k \geqslant 2$, we have

$$\begin{aligned}\mathscr{L} &\geqslant \rho\kappa D_{\nu_n+\sigma_n-m+1} - \rho\kappa^2\left(\rho+1+\rho^{-k}\right)D_{\nu_n+\sigma_n-m} - \rho^{-k+2}\kappa^4 D_{\nu_n+\sigma_n-m-2} \\ &\geqslant \rho\kappa^2\left[3(\rho+1)/2-\left(\rho+1+\rho^{-k}\right)\right]D_{\nu_n+\sigma_n-m} - \rho^2\kappa^3 D_{\nu_n+\sigma_n-m-2} \\ &\geqslant \left[3\rho^2\kappa^3(\rho+1)/4-\rho^2\kappa^3\right]D_{\nu_n+\sigma_n-n-2} \geqslant 2^{-1}\rho^2\kappa^3 D_{\nu_n+\sigma_n-m-2}. \end{aligned} \tag{7.43}$$

4. Let k and l be even; then

$$\begin{aligned}\mathscr{L} &\geqslant \rho\kappa D_{\nu_n+\sigma_n-m+1} - \rho^2\kappa^2 D_{\nu_n+\sigma_n-m} - \rho^2\kappa^3 D_{\nu_n+\sigma_n-m-1} \\ &\geqslant \rho\kappa^2 D_{\nu_n+\sigma_n-m} - \rho^2\kappa^3 D_{\nu_n+\sigma_n-m-1} \geqslant \rho\kappa^3 D_{\nu_n+\sigma_n-m-1}. \end{aligned} \tag{7.44}$$

Taking account of (7.41)–(7.44) and (7.36), we obtain

$$\mathscr{L} \geqslant C_8\varepsilon_1^{\nu_n}\varepsilon_\rho^{\sigma_n-m}. \tag{7.45}$$

Hence

$$|d_0| \geqslant C_9\left(\rho\kappa^2/\varepsilon_\rho\right)^m, \qquad \operatorname{sgn} d_0 = \operatorname{sgn}(-1)^l R, \tag{7.46}$$

and, when $m = \sigma_n - 3$,

$$|d_i| \geqslant C_{10}\left(\rho\kappa^2/\varepsilon_\rho\right)^{\sigma_n-i} \qquad (2 \leqslant i \leqslant \sigma_n - 3). \tag{7.47}$$

Let us establish the unboundedness of the derivatives of the spline as $n \to \infty$. We take

$$f_n(x) = \sum_{i=1}^{(\sigma_n-3)/2} \varphi_{n,2i}(x).$$

Then, recalling that in interpolating periodic spline is a linear operator and that (7.28) and (7.46) are satisfied and $\rho\kappa^2 \geqslant \varepsilon_\rho$, we deduce that

$$\left\|S_n^{(l)}\right\| \geqslant \left|S_n^{(l)}(0)\right| \geqslant C_{10}\sum_{i=1}^{(\sigma_n-3)/2}\left(\rho\kappa^2/\varepsilon_\rho\right)^{2i} \geqslant C_{11}\sigma_n \to \infty \qquad (n\to\infty).$$

Since $\rho^{1-1/p}\kappa^2/\varepsilon_\rho > 1$ for $1 \leqslant p < \infty$, we obtain, taking account of (7.28) and (7.47),

$$\begin{aligned}\left\|S_n^{(l)}\right\|_p &\geqslant C_1\left(\sum_{i=2}^{\sigma_n-3}|d_i|^p h_i\right)^{1/p} \geqslant C_2\left(\sum_{i=2}^{\sigma_n-3}\left|\frac{\rho\kappa^2}{\varepsilon_\rho}\right|^{p(\sigma_n-i)}\rho^{i-n-\sigma_n}\right)^{1/p} \\ &\geqslant \frac{C_3}{\rho^{n/p}}\left(\sum_{i=2}^{\sigma_n-3}\left(\rho^{1-1/p}\kappa^2/\varepsilon_\rho\right)^{(\sigma_n-i)p}\right)^{1/p} \geqslant C_4\sigma_n^{1/p}\rho^{-n/p} \to \infty \qquad (n\to\infty). \end{aligned}$$

Applying the Banach-Steinhaus theorem, we obtain the conclusion of Theorem 6.

Bibliography

1. J. H. Ahlberg, E. N. Nilson and J. L. Walsh, *The theory of splines and their application*, Academic Press, 1967.

2. S. B. Stechkin and Yu. N. Subbotin, *Splines in numerical mathematics*, "Nauka", Moscow, 1976. (Russian)

3. Yu. S. Zav′yalov, B. I. Kvasov and V. L. Miroshnichenko, *Methods of spline functions*, "Nauka", Moscow, 1980. (Russian)

4. A. Sharma and A. Meir, *Degree of approximation of spline interpolation*, J. Math. Mech. **15** (1966), 759–767.

5. A. Meir and A. Sharma, *On uniform approximation by cubic splines*, J. Approximation Theory **2** (1969), 270–274.

6. Martin Marsden, *Cubic spline interpolation of continuous functions*, J. Approximation Theory **10** (1974), 103–111.

7. Carl de Boor, *On cubic spline functions that vanish at all knots*, Advances in Math. **20** (1976), 1–17.

8. N. L. Zmatrakov, *Convergence of an interpolation process for parabolic and cubic splines*, Trudy Mat. Inst. Steklov. **138** (1975), 71–93; English transl. in Proc. Steklov Inst. Math. **138** (1975).

9. ______, *Uniform convergence of the third derivatives of interpolating cubic splines*, Vychisl. Sistemy Vyp. **72** (1977), 10–29. (Russian)

10. ______, *Convergence of third derivatives of interpolating cubic splines in the L_p metrics* ($1 \leqslant p \leqslant \infty$), Mat. Zametki **30** (1981), 83–99; English transl. in Math. Notes **30** (1981).

11. ______, *Divergence of the third derivatives of interpolating cubic splines in L_p metrics*, Mat. Zametki **31** (1982), 707–722; English transl. in Math. Notes **31** (1982).

12. A. K. Duĭsekov, *On interpolation by splines of fifth degree and their application in the collocation method*, Candidate's Dissertation, Inst. Mat. Mekh. Akad. Nauk Kazakh. SSR, Alma-Ata, 1976. (Russian)

13. Yu. N. Subbotin, *Multiply interpolating splines*, All-Union Sympos. Theory of Approximation of Functions in the Complex Domain, Abstracts of Reports, Bashkir. Gos Univ., Ufa, 1980, pp. 131–132. (Russian)

14. Shmuel Friedland and Charles A. Micchelli, *Bounds on the solutions of difference equations and spline interpolation at knots*, Linear Algebraic and Appl. **20** (1978), 219–251.

15. Charles A. Micchelli, *Infinite spline interpolation*, Approximation in Theorie und Praxis (Proc. Sympos., Siegen, 1979; G. Meinardus, editor), Bibliographisches Inst. Mannheim, 1979, pp. 209–238.

16. Carl de Boor, *Odd-degree spline interpolation at a biinfinite knot sequence*, Approximation Theory (Proc. Internat. Colloq., Bonn, 1976), Lecture Notes in Math., vol. 566, Springer-Verlag, 1976, pp. 30–53.

17. G. Birkhoff, M. H. Schultz and R. S. Varga, *Piecewise Hermite interpolation in one and two variables with applications to partial differential equations*, Numer. Math. **11** (1968), 232–256.

Translated by **R. P. BOAS**

Труды Матем. Инст. Стеклов.
164(1983)

Proceedings of the Steklov
Institute of Mathematics
1985, Issue 2

REPRESENTATION OF MEASURABLE FUNCTIONS BY MULTIPLE TRIGONOMETRIC SERIES

UDC 517.5

V. I. IVANOV

ABSTRACT. The following problem was posed by P. L. Ul'yanov: Does there exist a universal trigonometric series $\sum_0^\infty a_n \cos nx + b_n \sin nx$ for which

$$\sum_0^\infty \left(|a_n|^{2+\varepsilon} + |b_n|^{2+\varepsilon}\right) < \infty \quad (\forall \varepsilon > 0)? \tag{1}$$

This paper investigates questions of the representation of measurable functions by multiple trigonometric series and in a system with spectrum in a half-space, together with some of their subsystems, convergence being taken in the sense of measure and in the space $\varphi(L)$. Universal series and null-series are constructed for these systems. For the coefficients of these series (1) holds. In particular, a postive answer to Ul'yanov's question follows from the results.

Bibliography: 12 titles.

Introduction

Let $\mathbf{R}^m$ be m-dimensional Euclidean space, whose elements we shall denote by $x = (x_1, \ldots, x_m)$; let $xy = x_1y_1 + \cdots + x_my_m$ and $|x| = \sqrt{xx}$; let T^m be the m-dimensional torus $\{x \in \mathbf{R}^m | \pi \leqslant x_j < \pi, j = 1, \ldots, m\}$; let $\mathbf{Z}^m$ be the integral lattice in $\mathbf{R}^m$; for $A \subseteq \mathbf{Z}^m$, let $e(A) = \{e^{i\nu x}\}_{\nu \in A}$ be the subsystem of the trigonometric system with spectrum in A; let $\#A$ be the number of elements of the finite set A; let

$$\bar{d}_m = (A) = \varlimsup_{R \to \infty} \frac{\#\{\nu \in A | |v| \leqslant R\}}{\#\{\nu \in \mathbf{z}^m | |v| \leqslant R\}}$$

be the upper density of A; and let $|E|$ be the measure of the set $E \subseteq \mathbf{R}^m$.

We shall denote by Φ the set of all finite, even functions $\varphi(x) = \varphi(|x|)$ which are continuous at zero, nondecreasing on $[0, \infty)$ and satisfy $\varphi(0) = 0$ and $\varlimsup_{x \to \infty} \varphi(2x)/\varphi(x) < \infty$. For each function $\varphi \in \Phi$ define the space $\varphi(L(T^m)) = \varphi(L)$ as the set of all measurable, almost-everywhere finite complex-valued functions $f(x)$ on T^m whose "φ-distance" from zero

$$\|f\|_\varphi = \int_{T_m} \varphi(f(x))\, dx < \infty.$$

The assumption $\varlimsup_{x \to \infty} \varphi(2x)/\varphi(x) < \infty$ is the necessary and sufficient condition for $\varphi(L)$ to be a linear space (cf. [1]). We shall denote by $S(T^m)$ the set of all measurable complex-valued functions on T^m which can assume infinite values as well on a set of positive measure.

1980 *Mathematics Subject Classification*. Primary 42B05, 42B99, 42C30.

The system $e(A)$ is called *complete* in the space $\varphi(L)$ if it is possible to approximate any $f \in \varphi(L)$ in "φ-distance" by linear combinations of functions of $e(A)$. It is called *minimally complete* if (in addition) none of its proper subsets is complete in $\varphi(L)$.

Consider series with respect to the system $e(A)$

$$\sum_{\nu \in A} c_\nu e^{i\nu x}. \tag{1}$$

The series (1) in the system $e(A)$ is called *universal* in $S(T^m)$ (in $\varphi(L)$) in the sense of convergence almost everywhere (convergence in "φ-distance") if for any $f \in S(T^m)$ ($f \in \varphi(L)$) there exists a positive sequence $\{R_k\}$ such that almost everywhere

$$\lim_{k\to\infty} \sum_{|\nu| \leqslant R_k,\, \nu \in A} c_\nu e^{i\nu x} = f(x) \qquad \left(\lim \left\| f(x) - \sum_{|\nu| \leqslant R_k,\, \nu\in A} c_\nu e^{i\nu x} \right\| = 0\right).$$

The series (1) in the system $e(A)$ is called *universal with respect to subseries* in $S(T^m)$ (in $\varphi(L)$) in the sense of convergence in measure (in "φ-distance") if for every $f \in S(T^m)$ ($f \in \varphi(L)$) there exists a subseries $\sum_{\nu_k \in A} c_{\nu_k} e^{i\nu_k x}$ of (1) whose spherical partial sums converge to $f(x)$ in measure

$$\left(\text{respectively, } \lim_{R\to\infty} \left\| f(x) - \sum_{|\nu_k| \leqslant R,\, \nu_k \in A} c_{\nu_k} e^{i\nu_k x} \right\|_\varphi = 0\right).$$

The series (1) in system $e(A)$ is called a *null-series* in the sense of convergence in "φ-distance" if not all c are 0 yet

$$\lim_{R\to\infty} \left\| \sum_{|\nu| \leqslant R,\, \nu\in A} c_\nu e^{i\nu x} \right\|_\varphi = 0.$$

In this paper we study questions of the representation of measurable functions by series with respect to the multiple trigonometric system and with respect to a system with spectrum in a half-space, together with some of their subsystems, for convergence in the sense of measure and in the sense of "φ-distance". For these systems null-series and universal series both in the ordinary sense and with respect to subseries are constructed. For the coefficients of the series that we construct, the following condition holds:

$$\sum_{\nu\in A} |c_\nu|^2 \omega(|c_\nu|) < \infty, \tag{2}$$

where $\omega \in \varphi$ and the series depends on ω. In the cases under consideration it is not essential that the partial sums which converge be spherical. Convergence of the spherical partial sums could be replaced by convergence of the partial sums defined by an arbitrary norm in $\mathbf{R}^m$.

Many papers have been devoted to the questions of representation of measurable functions by trigonometric series and by general orthogonal series. See for example the survey articles of Talalyan [2] and Ul′yanov [1]. We shall now state results which apply directly to our investigations.

MEN'SHOV'S THEOREM [3]. *For any measurable real-valued function $f(x)$ on T, finite-valued or equal to infinity on a set of positive measure, there exists a trigonometric series converging to $f(x)$ in measure whose coefficients tend to zero.*

Talalyan [4] proved that an arbitrary complete orthonormal system is a representation system in the space $\varphi(L)$ if $\varphi(x+y) \leqslant \varphi(x) + \varphi(y)$ and $\lim_{x\to\infty} \varphi(x)/x = 0$. Ul'yanov [1] found a necessary and sufficient condition on φ for a Faber-Schauder system to be a representation system in $\varphi(L)$.

Universal series were constructed by Men'shov and Kozlov for the trigonometric system and by Talalyan [2] for complete orthonormal systems. About the coefficients of universal series it was known only that they can tend to zero. Ul'yanov posed the question of the rate at which the coefficients of universal trigonometric series can tend to zero. In particular, does there exist a (universal) trigonometric series for which the condition

$$\sum_{\nu \in A} |c_\nu|^{2+\varepsilon} < \infty \quad \forall \varepsilon > 0 \tag{3}$$

holds?

Pogosyan [5] established the existence of a series in a complete uniformly bounded orthonormal system satisfying condition (3) and universal with respect to permutations. Shaginyan [6] constructed a trigonometric series satisfying (3) which is universal with respect to subseries in the sense of summability by the methods (C, α) (for the definition cf. [2]), $\alpha > 0$.

We have obtained the following results.

THEOREM 1. *If Π is a half-space in $\mathbf{R}^m$, $A_\Pi = \Pi \cap \mathbf{Z}^m$, and $\omega \in \Phi$, then for the system $e(A_\Pi)$ there exists a series* (1) *satisfying condition* (2) *which is universal in $S(T^m)$ with respect to subseries in the sense of convergence in measure.*

COROLLARY (Sharpening of Men'shov's theorem). *For any two real-valued measurable functions $f_1(x)$ and $f_2(x)$ on T which are finite-valued or equal to infinity on a set of positivie measure there exists a trigonometric series*

$$\sum_{k=0}^{\infty} a_k \cos kx + b_k \sin kx, \tag{4}$$

which converges in measure to $f_1(x)$ while its conjugate series

$$\sum_{k=1}^{\infty} a_k \sin kx - b_k \cos kx$$

converges to $f_2(x)$. Moreover,

$$\sum_{k=0}^{\infty} \left(|a_k|^{2+\varepsilon} + |b_k|^{2+\varepsilon}\right) < \infty \quad \forall \varepsilon > 0.$$

An important problem of the theory of trigonometric series posed by N. N. Luzin in 1915 remains unsolved up to now. Does there exist, for every measurable function finite almost-everywhere or equal to infinity on a set of positive measure, a trigonometric series (4) converging to it almost everywhere? It is known (cf. for example [7], Chapter XIV, §1) that it is impossible to represent even some finite-valued functions in the sense of convergence almost everywhere by series in the system $e(\mathbf{Z}_+)$, for example functions which vanish on a set of positive measure (but are not identically zero). However, if convergence almost everywhere is replaced by convergence in measure, then, as follows from Theorem 1, it is possible to represent any measurable functions by series in the system $e(\mathbf{Z}_+)$.

THEOREM 2. *If Π is a half-space in $\mathbf{R}^m$, $A_\Pi = \Pi \cap \mathbf{Z}^m$, and $\omega \in \Phi$, then for the system $e(A_\Pi)$ there exists a series* (1) *satisfying condition* (2) *which is universal in $S(T^m)$ in the sense of convergence almost everywhere.*

By setting $\omega(x) = 1/|\ln|x||$ we obtain a positive answer to Ul'yanov's question.

THEOREM 3. *If Π is a half-space in $\mathbf{R}^m$, $A_\Pi = \Pi \cap \mathbf{Z}^m$, and ϕ, $\omega \in \Phi$, then the following conditions are equivalent*:

1) $\underline{\lim}_{x\to\infty} \varphi(x)/\ln x = 0$.

2) *The system $e(A_\Pi)$ is complete in the space $\varphi(L)$.*

3) *In $e(A_\Pi)$ there exists a series* (1) *satisfying condition* (2) *which is universal in the space $\omega(L)$ in the sense of convergence in "φ-distance".*

4) *In $e(A_\Pi)$ there exists a series* (1) *satisfying condition* (2) *which is universal with respect to subseries in the space $\varphi(L)$ in the sense of convergence in "φ-distance".*

5) *In $e(A_\Pi)$ there exists a null-series* (1) *satisfying condition* (2) *in the sense of convergence in "φ-distance".*

THEOREM 4. *If φ, $\omega \in \Phi$, then the following conditions are equivalent*:

1) $\underline{\lim}_{x\to\infty} \varphi(x)/x = 0$.

2) *The system $e(\mathbf{Z}^m)$ is not minimally complete in the space $\varphi(L)$.*

3) *In $e(\mathbf{Z}^m)$ there exists a series* (1) *satisfying condition* (2) *which is universal in the space $\varphi(L)$ in the sense of convergence in "φ-distance".*

4) *In $e(\mathbf{Z}^m)$ there exists a series* (1) *satisfying condition* (2) *which is universal with respect to subseries in the space $\varphi(L)$ in the sense of convergence in "φ-distance",*

5) *In $e(\mathbf{Z}^m)$ there exists a null-series* (1) *satisfying condition* (2) *in the sense of convergence in "φ-distance".*

Talalyan [4] has discovered that a complete orthonormal system with a finite number of functions removed remains a representation system in the space L_p, $0 < p < 1$. Price [8] has given an example of a trigonometric subsystem which is complete in measure whose spectrum has zero density. The trigonometric series constructed by Shaginyan [6] which is universal with respect to subseries in the sense of summability by the methods (C, α), $\alpha > 0$, has a spectrum of zero density. F. G. Arutyunyan has announced results on the representation of finite functions by series in certain subsystems of the multiple trigonometric system in the sense of convergence almost everywhere [9].

On the set of all subsets $A \subseteq \mathbf{Z}^m$ we define two operators

$$PA = \{n \in \mathbf{Z}^m | \forall s \in \mathbf{N}\, \exists r \in \mathbf{Z}^m\, (r \neq 0)\, \forall k \in \mathbf{Z},\, 1 \leqslant |k| \leqslant s,\, n + rk \in A\},$$

$$QA = \{n \in \mathbf{Z}^m | \forall s \in \mathbf{N}\, \exists r \in \mathbf{Z}^m\, (r \neq 0)\, \forall k \in \mathbf{N},\, 1 \leqslant k \leqslant s,\, n + rk \in A\}.$$

THEOREM 5. *If, for some $k \in \mathbf{N}$, $Q^k A = \mathbf{Z}^m$ and $\omega \in \Phi$, then in the system $e(A)$ there exists a series* (1) *satisfying condition* (2) *which is universal in $S(T^m)$ with respect to subseries in the sense of convergence in measure.*

THEOREM 6. *If* $Q^k A = \mathbf{Z}^m$ *for some* $k \in \mathbf{N}$ *and* $\omega \in \Phi$, *then in the system* $e(A)$ *there exists a series* (1) *satisfying condition* (2) *which is universal in* $S(T^m)$ *in the sense of convergence almost everywhere.*

THEOREM 7. *If* $Q^k A = \mathbf{Z}^m$ *for some* $k \in \mathbf{N}$, $\varphi, \omega \in \Phi$, *and* $\underline{\lim}_{x\to\infty} \varphi(xX)/\ln x = 0$, *then in the system* $e(A)$ *for the space* $\varphi(L)$ *there exist a null-series, a universal series, and a series universal with respect to subseries in the sense of convergence in "*φ*-distance".*

THEOREM 8. *If* $P^k A = \mathbf{Z}^m$ *for some* $k \in \mathbf{N}$, φ, $\omega \in \Phi$, *and* $\underline{\lim}_{x\to\infty} \varphi(x)/x = 0$, *then in the system* $e(A)$ *for the space* $\varphi(L)$ *there exist a null-series, a universal series and a series universal with respect to subseries in the sense of convergence in "*φ*-distance".*

We note that if $\bar{d}_m(A) = 1$, then $PA = QA = \mathbf{Z}^m$ and Theorems 5–8 hold for the system $e(A)$. On the other hand, for any $d < 1$ there exists $A \subseteq \mathbf{Z}^m$ with $\bar{d}_m(A) > d$ such that the system $e(A)$ is not complete in measure. A set A satisfying the condition $QA = \mathbf{Z}^m$ (or $PA = \mathbf{Z}^m$) can in a sense be as sparse as desired; and in particular it may have zero density. More precisely, for any sequence $\{R_k\} \uparrow \infty$ there exist two sets $A_1 = \{\nu_k\}_{k\in\mathrm{N}} \subset \mathbf{Z}^m$ and $A_2 = \{n_k, m_k\}_{k\in\mathrm{N}} \subset \mathbf{Z}^m$ for which $QA_1 = \mathbf{Z}^m$ and $PA_2 = \mathbf{Z}^m$, and

$$|\nu_{k+1}| \geqslant |\nu_k| + R_k, \qquad |m_{k+1}| \geqslant |n_{k+1}| \geqslant |m_k| + R_k.$$

Theorems 5–7 hold for the system $e(A_1)$, and Theorem 8 for $e(A_2)$.

Let B: $\mathbf{Z}^m \to \mathbf{Z}^m$ be a linear transformation, $f = \{f_1(k), \dots, f_m(k)\}$: $\mathbf{Z}^m \to \mathbf{Z}^m$ a polynomial transformation, where the $f_i(k)$ $(i = 1, \dots, m)$ are polynomials in m variables with integer coefficients, and $J_f(0) = |\det(\partial f_i/\partial k_j)(0)|$.

THEOREM 9. *If* $b \in \mathbf{Z}^m$, *then the system* $e(b + B\mathbf{Z}^m)$ *is complete in measure if and only if* $|\det B| = 1$.

THEOREM 10. *If* $J_f(0) \neq 1$, *then the system* $e(f(\mathbf{Z}^M))$ *is not complete in measure.*

This paper consists of three sections. Certain auxiliary results are proved in §1. The second section studies the representation of functions in the systems $e(\mathbf{Z}^m)$ and $e(A_{\Pi})$. The third section studies representation in certain subsystems of these two systems. The main results of the paper were announced in [10].

The author wishes to thank B. S. Kashin, S. B. Stechkin, P. L. Ul′yanov, and V. A. Yudin for helpful discussions of the results of this paper.

§1. Auxiliary results

LEMMA 1. *Suppose the function* $\varphi(x) \in \Phi$ *is convex upward for* $x \geqslant 0$ *and* $E \subseteq T^m$. *Then for any function* $f \in L(E)$

$$\frac{1}{|E|}\int_E \varphi(f(x))\,dx \leqslant \varphi\left(|E|^{-1}\int_E |f(x)|dx\right). \tag{1.1}$$

PROOF. If $f(x)$ is piecewise constant, i.e. $E = \bigcup_1^n E_i$, $E_i \cap E_j = \varnothing$, and $f(x) = c_i$ for $x \in E_i$, then by the upward convexity of $\varphi(x)$

$$\frac{1}{|E|}\int_E \varphi(f(x))\,dx = \frac{1}{|E|}\sum_{i=1}^n \varphi(c_i)|E_i| \leqslant \varphi\left(\sum_{i=1}^n |c_i|\frac{|E_i|}{|E|}\right)$$
$$= \varphi\left(\frac{1}{|E|}\int_E |f(x)|dx\right).$$

For an arbitrary $f \in \varphi(L)$ the proof concludes using approximation by piecewise-constant functions.

LEMMA 2. *Suppose $\varphi(x) \in \Phi$ and $\overline{\lim}_{x\to\infty} \varphi(x)/x^p < \infty$ $(p \geqslant 1)$. Then there exist functions $\varphi_1(x)$, $\varphi_2(x) \in \Phi$ and a constant $c_\varphi > 0$ such that*

$$\varphi(x) \leqslant \varphi_1(x) \qquad (x \geqslant 0), \tag{1.2}$$

$$\varphi(L) = \varphi_1(L), \tag{1.3}$$

$$\varphi_1(x+y) \leqslant c_\varphi\{\varphi_1(x) + \varphi_1(y)\} \qquad (x, y \geqslant 0), \tag{1.4}$$

$$\int_{T^m} \varphi(f(x))\,dx \leqslant \varphi_2(\|f\|_p). \tag{1.5}$$

PROOF. There exists $x_0 > 0$ such that

$$\varphi(x) > 0, \quad \varphi(2x) \leqslant A_\varphi \varphi(x), \quad \varphi(x) \leqslant B_\varphi x^p \qquad (x \geqslant x_0). \tag{1.6}$$

For $0 \leqslant x \leqslant x_0$ let $\varphi^*(x)$ be the upper boundary of the smallest convex region containing the region $\{0 \leqslant x \leqslant x_0,\ 0 \leqslant y \leqslant \varphi(x)\}$, and let $\varphi^*(x) = \varphi^*(x_0)$ for $x \geqslant x_0$. Then $\varphi^*(x)$ is convex upward for $x \geqslant 0$. Set

$$\varphi_1(x) = \begin{cases} \varphi^*(x), & 0 \leqslant x \leqslant x_0, \\ \varphi(x) + \varphi^*(x_0) - \varphi(x_0), & x \geqslant x_0, \end{cases}$$

$$\varphi_2(x) = (2\pi)^m \varphi^*\left(\frac{x}{(2\pi)m/p}\right) + B_\varphi |x|^p.$$

It is obvious that $\varphi_1(x)$, $\varphi_2(x) \in \Phi$, $\varphi(x) \leqslant \varphi_1(x)$, and $\varphi(L) = \varphi_1(L)$. Since $\varphi^*(x)$ is convex upward, for $x, y \leqslant x_0/2$ we have

$$\varphi_1(x+y) \leqslant \varphi_1(x) + \varphi_1(y). \tag{1.7}$$

If $x \geqslant x_0$ and $x \geqslant y$, then by (1.6)

$$\begin{aligned}\varphi_1(x+y) &\leqslant \varphi_1(2x) \leqslant \varphi(2x) + \varphi^*(x_0) - \varphi(x_0) \\ &\leqslant A_\varphi\{\varphi(x) + \varphi^*(x_0) - \varphi(x_0)\} \\ &\leqslant A_\varphi\{\varphi_1(x) + \varphi_1(y)\}.\end{aligned} \tag{1.8}$$

Finally, if $x_0/2 \leqslant x \leqslant x_0$ and $x \geqslant y$, then

$$\varphi_1(x+y) \leqslant \frac{\varphi_1(2x_0)}{\varphi_1(x_0/2)}\{\varphi_1(x) + \varphi_1(y)\}.$$

From this and from (1.7) and (1.8) we have (1.4) for

$$c_\varphi = \max\left\{A_\varphi, \frac{\varphi_1(2x_0)}{\varphi_1(x_0/2)}\right\}.$$

For the proof of (1.5) we make use of (1.1) and (1.6):

$$\begin{aligned}\int_{T^m}\varphi(f(x))\,dx &= \int_{|f(x)|\leqslant x_0}\varphi(f(x))\,dx+\int_{|f(x)|\geqslant x_0}\varphi(f(x))\,dx\\ &\leqslant \int_{T^m}\varphi^*(f(x))\,dx+B_\varphi\int_{T^m}|f(x)|^p\,dx\\ &\leqslant (2\pi)^m\varphi^*\big((2\pi)^{-m}\|f\|_1\big)+B_\varphi\|f\|_p^p\\ &\leqslant (2\pi)^m\varphi^*\big((2\pi)^{-m/p}\|f\|_p\big)+B_\varphi\|f\|_p^p=\varphi_2\big(\|f\|_p\big).\end{aligned}$$

LEMMA 3. *Suppose $\varphi(x)$, $\psi(x)\in\Phi$, and $\varphi(x)\asymp\psi(x)$ $(x\to\infty)$. The system of functions $\{f_k(x)\}_1^\infty$ is complete in $\varphi(L)$ if and only if it is complete in $\psi(L)$.*

PROOF. By assumption

$$\psi(x)\leqslant c_1\varphi(x)\quad\text{for } x\geqslant x_0,\quad\text{and}\quad \varphi(L)=\psi(L). \tag{1.9}$$

Suppose $f(x)\in\psi(L)$, $\varepsilon>0$, and $c_2=\max\{1,x_0/\varepsilon\}$. If the system $\{f_k(x)\}$ is complete in $\varphi(L)$, then its linear span contains a function $g(x)$ such that

$$\int_{T^m}\varphi(c_2f-g)\,dx\leqslant\varepsilon/c_1.$$

Set $E=\{x\in T^m\,|\,|c_2f(x)-g(x)|\leqslant x_0\}$. For $x\in E$ we have $|f(x)-g(x)/c_2|\leqslant\varepsilon$, and by (1.9)

$$\begin{aligned}\int_{T^m}\psi\left(f(x)-\frac{g(x)}{c_2}\right)dx &= \int_E\psi\left(f(x)-\frac{g(x)}{c_2}\right)dx+\int_{T^m\setminus E}\psi\left(f(x)-\frac{g(x)}{c_2}\right)dx\\ &\leqslant |T^m|\psi(\varepsilon)+\int_{T^m\setminus E}\psi(c_2f(x)-g(x))\,dx\\ &\leqslant |T^m|\psi(\varepsilon)+c_1\int_{T^m}\varphi(c_2f(x)-g(x))\,dx\\ &\leqslant |T^m|\psi(\varepsilon)+\varepsilon.\end{aligned}$$

The lemma is proved.

By Lemmas 2 and 3 we may assume henceforth that for $\varphi\in\Phi$

$$\varphi(x+y)\leqslant c_\varphi\{\varphi(x)+\varphi(y)\}\qquad(x,y\geqslant 0). \tag{1.10}$$

LEMMA 4. *If $\varphi\in\Phi$ and $\underline{\lim}_{x\to\infty}\varphi(x)/x=0$, then for any $\varepsilon>0$ there exists a polynomial $t(x)=\sum_{|k|\leqslant n}c_ke^{ikx}$ such that $c_0=1$ and $\int_0^{2\pi}\varphi(t(x))\,dx\leqslant\varepsilon$.*

PROOF. We first construct a function $h(x)\in C[-\pi,\pi]$ which is even, nondecreasing on $[0,\pi]$, and such that

$$\int_{-\pi}^{\pi}\frac{dx}{h(x)}=\infty,\qquad\int_{-\pi}^{\pi}\varphi\left(\frac{1}{h(x)}\right)dx<\infty. \tag{1.11}$$

There exists a positive sequence $\{y_k\}_1^\infty\uparrow\infty$ satisfying the conditions

$$\frac{\varphi(y_k)}{y_k}\leqslant\frac{1}{2^k},\qquad\sum_{k=1}^{\infty}\frac{1}{y_k}=s\leqslant\pi.$$

For $0 \leqslant x \leqslant \pi$ set

$$\alpha_0 = \pi, \qquad \alpha_1 = s - \frac{1}{y_1}, \dots, \alpha_k = s - \sum_{i=1}^{k} \frac{1}{y_i},$$

$$\beta_k = \alpha_k + \min\left(\frac{\alpha_{k-1} - \alpha_k}{2}, \frac{1}{2^k \varphi(y_{k+1})} \right),$$

$$h(x) = \begin{cases} 1/y_k, & \text{for } x \in [\beta_k, \alpha_{k-1}] \\ \text{linear for } x \in [\alpha_k, \beta_k] \end{cases}, \qquad h(-x) = h(x).$$

It is obvious that $h(x) \in C[-\pi, \pi]$, $h(0) = 0$, and furthermore

$$\int_0^\pi \varphi\left(\frac{1}{h(x)}\right) dx = \sum_{k=1}^{\infty} \left(\int_{\alpha_k}^{\beta_k} \varphi\left(\frac{1}{h(x)}\right) dx + \int_{\beta_k}^{\alpha_{k-1}} \varphi\left(\frac{1}{h(x)}\right) dx \right)$$
$$\leqslant \sum_{k=1}^{\infty} \left(\frac{\varphi(y_k)}{y_k} + \frac{\varphi(y_{k+1})}{2^k \varphi(y_{k+1})} \right) \leqslant \sum_{k=1}^{\infty} \frac{1}{2^k} < \infty,$$

$$\int_0^\pi \frac{dx}{h(x)} = \sum_{k=1}^{\infty} \int_{\alpha_k}^{\alpha_{k-1}} \frac{dx}{h(x)} \geqslant \sum_{k=1}^{\infty} y_k \frac{1}{y_k} = \infty.$$

Let $\varepsilon > 0$. Consider the family of continuous functions

$$h_a(x) = \begin{cases} h(x), & |x| \geqslant a \\ h(a), & |x| \leqslant a \end{cases}, \qquad a > 0.$$

Choose $a_1 > 0$ such that $\int_{|x| \leqslant a_1} \varphi(1/h(x))\, dx \leqslant \varepsilon/2$. Since by (1.11) we have

$$\lim_{a \to 0} \int_{|x| \leqslant \pi} h_a^{-1}(x)\, dx = \infty,$$

it follows that for some $a < a_1$ and $\varepsilon_a = ((1/2\pi) \int_{|x| \leqslant \pi} h_a^{-1}(x)\, dx)$ we must have

$$\varphi(\varepsilon_a / h(a_1)) \leqslant \varepsilon / 4\pi, \qquad \varepsilon_a \leqslant 1/2.$$

Approximate $h_a(x)$ by a trigonometric polynomial $t(x)$ such that $1/h_a(x) \leqslant t(x) \leqslant 2/h_a(x)$. Let

$$t^*(x) = \left(\frac{1}{2\pi} \int_{-\pi}^{\pi} t(x)\, dx \right)^{-1} t(x) = \sum_{|k| \leqslant n} c_k e^{ikx}, \qquad c_0 = 1.$$

For $|x| \leqslant a_1$

$$t^*(x) \leqslant 2\pi t(x) \Big/ \int_{-\pi}^{\pi} \frac{dx}{h_a(x)} \leqslant \frac{2\varepsilon_a}{h_a(x)} \leqslant \frac{1}{h(x)}.$$

Therefore

$$\int_{|x| \leqslant a_1} \varphi(t^*(x))\, dx \leqslant \int_{|x| \leqslant a_1} \varphi(1/h(x))\, dx \leqslant \varepsilon/2.$$

For $|x| \geqslant a_1$ we have

$$t^*(x) \leqslant 2\varepsilon_a / h_a(a_1) \leqslant 2\varepsilon_a / h(a_1),$$

and

$$\int_{|x|\geqslant a_1} \varphi(t^*(x))\,dx \leqslant \int_{-\pi}^{\pi} \varphi(2\varepsilon_a/h(a_1))\,dx \leqslant \varepsilon/2.$$

Finally

$$\int_{-\pi}^{\pi} \varphi(t^*(x))\,dx \leqslant \varepsilon.$$

LEMMA 5. *If* $\varphi \in \Phi$ *and* $\underline{\lim}_{x\to\infty} \varphi(x)/\ln x = 0$, *then for* $\varepsilon > 0$ *there exists a polynomial* $p(x) = \sum_0^n c_k e^{ikx}$ *such that* $c_0 = 1$ *and* $\int_{-\pi}^{\pi}(p(x))\,dx \leqslant \varepsilon$.

PROOF. As in the proof of Lemma 4 we construct a function $h(x) \in C[-\pi, \pi]$ which is even, nondecreasing on $[0, \pi]$, and such that

$$\int_{-\pi}^{\pi} \ln\frac{1}{h(x)}\,dx = \infty, \qquad \int_{-\pi}^{\pi} \varphi\left(\frac{1}{h(x)}\right) dx < \infty. \tag{1.12}$$

Let $\varepsilon > 0$. Choose $a_1 > 0$ such that $\int_{|x|\leqslant a_i} \varphi(1/h(x))\,dx \leqslant \varepsilon/2$. Consider the family of continuous functions

$$h_a(x) = \begin{cases} h(x), & |x| \geqslant a \\ h(a), & |x| \leqslant a \end{cases}, \qquad a > 0.$$

Denote by $G(f)$ the quantity $\exp((1/2\pi)\int_{-\pi}^{\pi}\ln|f(x)|\,dx)$. Since by (1.12) $\lim_{a\to 0} G(1/h_a) = \infty$, for some $a < a_1$ and $\varepsilon_a = (G(1/h_a))^{-1}$ we have

$$\varphi(2\varepsilon_a/h(a_1)) \leqslant \varepsilon/4\pi, \qquad \varepsilon_a \leqslant 1/2. \tag{1.13}$$

There exists a trigonometric polynomial $t(x)$ such that $1/h_a^2(x) \leqslant t(x) \leqslant 4/h_a^2(x)$. Since it is positive, $t(x) = |p(e^{ix})|^2$, where $p(z)$ is a polynomial with nonnegative spectrum having no zeros for $|z| < 1$. Set

$$p^*(x) = p(e^{ix})/\sqrt{G(t)} = \sum_{k=0}^{n} c_k e^{ikx}.$$

By Jensen's formula

$$|p(0)|^2 = \exp\left(\frac{1}{2\pi}\int_{-\pi}^{\pi} \ln|p(e^{ix})|^2\,dx\right) = G(t),$$

so that $|c_0|^2 = |p(0)|^2/G(t) = 1$. We may assume that $c_0 = 1$. We note that

$$1/\sqrt{G(t)} \leqslant 1/\sqrt{G\left(\frac{1}{h_a^2}\right)} = 1/G\left(\frac{1}{h_a}\right) = \varepsilon_a, \qquad p^*(x) \leqslant \frac{2\varepsilon_a}{h_a(x)}.$$

Hence by (1.13)

$$\begin{aligned}\int_{-\pi}^{\pi} \varphi(p^*(x))\,dx &= \int_{|x|\leqslant a_1} \varphi(p^*(x))\,dx + \int_{|x|\geqslant a_1} \varphi(p^*(x))\,dx \\ &\leqslant \int_{|x|\leqslant a_1} \varphi\left(\frac{1}{h(x)}\right) dx + \int_{|x|\geqslant a_1} \varphi\left(\frac{2\varepsilon_a}{h_a(a_1)}\right) dx \\ &\leqslant \frac{\varepsilon}{2} + 2\pi\varphi\left(\frac{2\varepsilon_a}{h(a_1)}\right) \leqslant \varepsilon.\end{aligned}$$

Let $K = \prod_1^m [\alpha_i, \beta_i]$ be a parallelepiped with sides parallel to the coordinate axes; let $\chi_K(x)$ be the characteristic function of K and $\sum_{\nu \in \mathbf{Z}^m} c_\nu e^{i\nu x}$ its Fourier series expansion.

LEMMA 6. *For every set $A \subseteq \mathbf{Z}^m$ and $p \geqslant 2$*

$$\left\| \sum_{\nu \in A} c_\nu e^{i\nu x} \right\|_p \leqslant \lambda_p |K|^{1/p}. \tag{1.14}$$

PROOF. For simplicity consider the case $m = 2$. Let

$$A = \{(\nu_1, \nu_2)\} = \bigcup_{\nu_1 \in B} \bigcup_{\nu_2 \in B_{\nu_1}} (\nu_1, \nu_2).$$

If a_{ν_1} and b_{ν_1} are the Fourier coefficients of the characteristic functions of the intervals $[\alpha_1, \beta_1]$ and $[\alpha_2, \beta_2]$, then the Fourier coefficients of the parallelepiped K are $c_\nu = a_{\nu_1} b_{\nu_2}$, $\nu = (\nu_1, \nu_2)$. Applying sequentially the Paley inequality (cf. for example [7], Chapter XII, Theorem 5.1), we obtain

$$\begin{aligned} \int_{T^2} \left| \sum_{\nu \in A} c_\nu e^{i\nu x} \right|^p dx &= \int_{-\pi}^{\pi} \int_{-\pi}^{\pi} \left| \sum_{\nu_1 \in B} a_{\nu_1} \Big(\sum_{\nu_2 \in B_{\nu_1}} b_{\nu_2} e^{i\nu_2 x_2} \Big) e^{i\nu_1 x_1} \right|^p dx_1\, dx_2 \\ &\leqslant \lambda_p \sum_{\nu_1 \in B} |\nu_1|^{p-2} |a_{\nu_1}|^p \int_{-\pi}^{\pi} \left| \sum_{\nu_2 \in B_{\nu_1}} b_{\nu_2} e^{i\nu_2 x_2} \right|^p dx_2 \\ &\leqslant \lambda_p^2 \sum_{\nu_1 \in B} |\nu_1|^{p-2} |a_{\nu_1}|^p \sum_{\nu_2 \in B_{\nu_1}} |\nu_2|^{p-2} |b_{\nu_2}|^p \\ &\leqslant \lambda_p^2 \sum_{\nu_1 \in \mathbf{Z}} |\nu_1|^{p-2} |a_{\nu_1}|^p \sum_{\nu_2 \in \mathbf{Z}} |\nu_2|^{p-2} |b_{\nu_2}|^p. \end{aligned} \tag{1.15}$$

Since

$$|a_\nu| \leqslant \left| \sin \frac{\nu(\beta_1 - \alpha_1)}{2} \right| \Big/ \pi |\nu| = \left| \sin \frac{\nu \Delta_1}{2} \right| \Big/ \pi |\nu|,$$

it follows that

$$\begin{aligned} \sum_{\nu \in \mathbf{Z}} |\nu|^{p-2} |a_\nu|^p &\leqslant \frac{1}{\pi} \sum_{\nu \in \mathbf{Z}} \left| \sin \frac{\nu \Delta_1}{2} \right|^p \Big/ \nu^2 \\ &\leqslant \frac{1}{\pi} \Bigg\{ \sum_{|\nu| \leqslant 1/\Delta_1} |\nu|^{p+2} \Delta_1^p + \sum_{|\nu| \geqslant 1/\Delta_1} \frac{1}{\nu_2} \Bigg\} \leqslant c_p \Delta_1. \end{aligned}$$

From this and from (1.15) we get

$$\int_{T^2} \left| \sum_{\nu \in A} c_\nu e^{i\nu x} \right|^p dx \leqslant \lambda_p^2 c_p \Delta_1 \Delta_2 = \lambda_p^2 c_p |K|.$$

LEMMA 7. *Let Π be a half-space in $\mathbf{R}^m$, $A_\Pi = \Pi \cap \mathbf{Z}^m$, φ, $\omega \in \Phi$, and $\underline{\lim}_{x \to \infty} \varphi(x)/\ln x = 0$. Then for any $f \in \varphi(L)$ and any $\varepsilon > 0$ and $r > 0$, there exists a polynomial*

$$t(x) = \sum_{r \leqslant |\nu| \leqslant R,\ \nu \in A_\Pi} c_\nu e^{i\nu x}$$

such that

$$\int_{T^m} \varphi(f(x) - t(x))\, dx < \varepsilon; \tag{1.16}$$

$$\int_e \varphi\Big(\sum_{r \leqslant |\nu| \leqslant R'} c_\nu e^{i\nu x} \Big)\, dx \leqslant \lambda_\varphi \int_e \varphi(f(x))\, dx + \varepsilon \tag{1.17}$$

for any $e \subseteq T^m$ *and* $r \leqslant R' \leqslant R$;

$$\int_{T^m} \varphi\Big(\sum_{\nu \in B_{R'}} c_\nu e^{i\nu x} \Big)\, dx \leqslant \varepsilon \tag{1.18}$$

for any subset $B_{R'}$ *of* $|\nu| = R$; *and*

$$\sum_{r \leqslant |\nu| \leqslant R} |c_\nu|^2 \omega(|c_\nu|) < \varepsilon. \tag{1.19}$$

PROOF. By (1.10) it suffices to prove the lemma for $f(x) = \alpha\chi_\Delta(x)$, where $\chi_\Delta(x)$ is the characteristic function of a parallelepiped $\Delta \subseteq T^m$ with sides parallel to the coordinate axes and $\alpha \in \mathbf{R}$. For simplicity assume $\alpha = 1$.

Using Lemma 5 we find a polynomial $p(x_1) = \sum_1^n b_k e^{ikx_1}$ such that

$$\int_T \varphi(1 - p(x_1))\, dx_1 < \varepsilon, \qquad \sum_1^n |b_k| \leqslant M. \tag{1.20}$$

Choose the natural number l such that

$$\varphi(M)\frac{|\Delta|}{l} \leqslant \varepsilon, \quad c_\varphi^{M-1}\varphi_2\bigg(c_p\Big(\frac{|\Delta|}{l} \Big)^{1/p} \bigg) \leqslant \varepsilon, \quad M^2 n\omega\Big(\frac{M|\Delta|}{l} \Big) \leqslant \varepsilon, \tag{1.21}$$

where $\varphi_2(x)$ and p are from Lemma 2. Partition the parallelepiped Δ into l disjoint parallelepipeds Δ_i $(i = 1, \ldots, l)$ with sides parallel to the coordinate axes and equal volumes $|\Delta_i| = |\Delta|/l$. Then $\Delta = \bigcup_1^l \Delta_i$, and $\chi_\Delta(x) = \sum_1^l \chi_{\Delta_i}(x)$. If Δ_i is the Cartesian product of the intervals Δ_i^k $(k = 1, \ldots, m)$, i.e. $\Delta_i = \prod_1^m \Delta_i^k$, then by Fejér's lemma (cf., for example, [11], Chapter I, §20) for $i = 1, \ldots, l$ and for all integers r_i with $|r_i| \geqslant r_0$ we shall have

$$\int_{\Delta_i^1} \varphi(1 - p(r_i x))\, dx \leqslant 2|\Delta_i^1| \int_T \varphi(1 - p(x_1))\, dx_1 \leqslant 2\varepsilon|\Delta_i^1|. \tag{1.22}$$

We shall use $S_R(x, f)$ to denote the partial sum of order R of the Fourier series of $f(x)$:

$$S_R(x, f) = \sum_{|\nu| \leqslant R} a_\nu(f) e^{i\nu x}.$$

For $i = 1, \ldots, l$ choose natural numbers n_i such that

$$l c_\varphi^{M+l}\varphi_2(\delta) < \varepsilon, \qquad \|R_{n_i}(x, \chi_{\Delta_i})\|_p = \|\chi_{\Delta_i}(x) - S_{n_i}(x, \chi_{\Delta_i})\|_p \leqslant \delta, \tag{1.23}$$

where c_φ and $\varphi_2(x)$ are from Lemma 2. For the desired polynomial $t(x)$ we take

$$t(x)=\sum_1^l S_{n_i}(x,\chi_{\Delta_i})p(r_ix_1)=\sum_{|\nu|\leqslant R}c_\nu e^{i\nu x},$$

where the $r_i\in\mathbf{Z}$, $|r_i|\geqslant r_0$, are chosen so that the spectrum of $t(x)$ belongs to A_{II}, $|r_{i+1}|>|r_i|$, the spectra of the polynomials $S_{n_i}(x,\chi_{\Delta_i})e^{ir_ikx_1}$ $(i=1,\dots,l;\ k=1,\dots,n)$ are disjoint, and $|\nu|\geqslant r$.

By (1.5), (1.10), (1.20), (1.22), and (1.23)

$$\begin{aligned}\int_{T^m}\varphi(\chi_\Delta(x)-t(x))\,dx&\leqslant c_\varphi\left\{\int_{T^m}\varphi\left(\chi_\Delta(x)-\sum_1^l\chi_{\Delta_i}(x)p(r_ix_1)\right)dx\right.\\&\qquad\left.+\int_{T^m}\varphi\left(\sum_1^l R_{n_i}(x,\chi_{\Delta_i})p(r_ix_1)\right)dx\right\}\\&\leqslant c_\varphi\left\{\sum_1^l\int_{\Delta_i}\varphi(1-p(r_ix_1))\,dx+\int_{T^m}\varphi\left(M\sum_1^l|R_{n_i}(x,\chi_{\Delta_i})|\right)dx\right\}\\&\leqslant c_\varphi\left\{\sum_1^l\prod_2^m|\Delta_i^k|\int_{\Delta_i^1}\varphi(1-p(r_ix_1))\,dx_1+c_\varphi^{M+l-2}\sum_1^l\varphi_2\left(\|R_{n_i}\|_p\right)\right\}\\&\leqslant 2c_\varphi\sum_1^l|\Delta_i|\varepsilon+lc_\varphi^{M+l}\varphi_2(\delta)\leqslant 2c_\varphi|\Delta|\varepsilon+\varepsilon.\end{aligned}\tag{1.24}$$

The inequality (1.16) is now established.

Since for the partial sum $S_{R'}(x,t)$

$$\begin{aligned}|S_{R'}(t)|&=\left|\sum_1^{q-1}S_{n_i}(x,\chi_{\Delta_i})p(r_ix_1)+S_{n_q}(x,\chi_{\Delta_q})\sum_1^{s-1}b_ke^{ir_qkx_1}\right.\\&\qquad\left.+S_{R'}\left(S_{n_q}(x,\chi_{\Delta_q})b_xe^{ir_qsx_1}\right)\right|\\&\leqslant\left|\sum_1^{q-1}\chi_{\Delta_i}(x)p(r_ix_1)\right|+M\sum_1^l|R_{n_i}(x,\chi_{\Delta_i})|\\&\qquad+M\left|\chi_{\Delta_q}(x)\right|+M\left|S_{R'}\left(S_{n_q}(x,\chi_{\Delta_q})e^{ir_qsx_1}\right)\right|,\end{aligned}$$

for $e\subseteq T^m$ we get

$$\begin{aligned}&\int\varphi(S_{R'}(x,t))\,dx\\&\quad\leqslant c_\varphi^3\left\{\int_e\varphi\left(\sum_1^{q-1}\chi_{\Delta_i}(x)p(r_ix_1)\right)dx+\int_{T^m}\varphi\left(M\sum_1^l|R_{n_i}(x,\chi_{\Delta_i})|\right)dx\right.\\&\qquad\left.+\int_{T^m}\varphi\left(M\chi_{\Delta_q}(x)\right)dx+\int_{T^m}\varphi\left(MS_{R'}\left(S_{n_q}(x,\chi_{\Delta_q})e^{ir_qsx_1}\right)\right)dx\right\}\\&\quad=c_\varphi^3\{J_1+J_2+J_3+J_4\}.\end{aligned}\tag{1.25}$$

As in the proof of (1.24),

$$J_1 = \int_e \varphi\left(\sum_1^{q-1} \chi_{\Delta_i}(x) p(r_i x_1)\right) dx$$
$$\leqslant c_\varphi \left\{ \int_e \varphi\left(\sum_1^{q-1} \chi_{\Delta_i}(x)\right) dx + \int_{T^m} \varphi\left(\sum_1^{l} |1 - p(r_i x_1)| \chi_{\Delta_i}(x)\right) dx \right\}$$
$$\leqslant c_\varphi \int_e \varphi(\chi_\Delta(x))\, dx + 2c_\varphi |\Delta| \varepsilon, \tag{1.26}$$

$$J_2 = \int_{T^m} \varphi\left(M \sum_1^l \left|R_{n_i}(x, \chi_{\Delta_i})\right|\right) dx \leqslant \varepsilon. \tag{1.27}$$

Furthermore,

$$J_3 = \int_{T^m} \varphi\left(M \chi_{\Delta_q}(x)\right) dx \leqslant \varphi(M) \frac{|\Delta|}{l} \leqslant \varepsilon. \tag{1.28}$$

Using (1.5) and Lemma 6 we obtain

$$J_4 = \int_{T^m} \varphi\left(M S_{R'}\left(S_{n_q}(x, \chi_{\Delta_q}) e^{i r_q s x_1}\right)\right) dx$$
$$\leqslant c_\varphi^{M-1} \varphi_2\left(\left\|S_{R'}\left(S_{n_q}(x, \chi_{\Delta_q}) e^{i r_q s x_1}\right)\right\|_p\right)$$
$$\leqslant c_\varphi^{M-1} \varphi_2\left(c_p \left(\frac{|\Delta|}{l}\right)^{1/p}\right) \leqslant \varepsilon. \tag{1.29}$$

Substituting (1.26)–(1.29) into (1.25), we obtain (1.17). Inequality (1.18) is proved exactly like (1.29).

Finally, since

$$\left|a_\nu(\chi_{\Delta_i})\right| = \frac{1}{(2\pi)^m} \left| \int_{T^m} e^{-i\nu x} \chi_{\Delta_i}(x)\, dx \right| \leqslant \frac{|\Delta|}{l},$$

it follows that

$$\sum_{r \leqslant |\nu| \leqslant R} |c_\nu|^2 \omega(|c_\nu|) = \sum_{i=1}^{l} \sum_{k=1}^{n} \sum_{|\nu| \leqslant n_i} \left|a_\nu(\chi_{\Delta_i})\right|^2 |b_k|^2 \omega\left(\left|a_\nu(\chi_{\Delta_i}) b_k\right|\right)$$
$$\times \sum_{i=1}^{l} \sum_{k=1}^{n} \sum_{|\nu| \leqslant n_i} |a_\nu|^2 M^2 \omega\left(\frac{M|\Delta|}{l}\right)$$
$$= M^2 n \omega\left(\frac{M|\Delta|}{l}\right) \sum_1^l \frac{|\Delta|}{l} = M^2 n \omega\left(\frac{M|\Delta|}{l}\right) \leqslant \varepsilon |\Delta|.$$

Inequality (1.19) and the lemma are now proved.

Analogously to Lemma 7, one can prove

LEMMA 8. *Suppose $\varphi, \omega \in \Phi$, and $\underline{\lim}_{x \to \infty} \varphi(x)/x = 0$. Then for any $f \in \varphi(L)$ and any $\varepsilon > 0$ and $r > 0$, there exists a polynomial*

$$t(x) = \sum_{r \leqslant |\nu| \leqslant R} c_\nu e^{i\nu x}$$

such that

$$\int_{T^m} \varphi(f(x) - t(x))\, dx \leqslant \varepsilon; \tag{1.30}$$

$$\int_e \varphi\Big(\sum_{r \leqslant |\nu| \leqslant R'} c_\nu e^{i\nu x}\Big)\, dx \leqslant c_\varphi \int_e \varphi(f(x))\, dx + \varepsilon, \tag{1.31}$$

for any $e \subseteq T^m$ *and* $r \leqslant R' \leqslant R$;

$$\int_{T^m} \varphi\Big(\sum_{\nu \in B_{R'}} c_\nu e^{i\nu x}\Big)\, dx \leqslant \varepsilon \tag{1.32}$$

for any subset $B_{R'}$ *of* $|\nu| = R'$; *and*

$$\sum_{r \leqslant |\nu| \leqslant R} |c_\nu|^2 \omega(|c_\nu|) \leqslant \varepsilon. \tag{1.33}$$

We shall need certain properties of linear transformations in $\mathbf{Z}^m$.

LEMMA 9. *Let* $m \geqslant 2$ *and* $e = (1, 0, \dots, 0) \in \mathbf{Z}^m$. *Then for any* $a = (a_1, \dots, a_m) \in \mathbf{Z}^m$ *with* $\gcd(a_1, \dots, a_m) = 1$ *there exists a linear transformation* $C\colon \mathbf{Z}^m \to \mathbf{Z}^m$ *such that*

$$\det C = 1, \quad Ce = a, \quad C^{-1}a = e. \tag{1.34}$$

PROOF. Induction of m. If $\gcd(a_1, a_2) = 1$, then there exist $\lambda_1, \lambda_2 \in \mathbf{Z}$ such that $a_1\lambda_1 - a_2\lambda_2 = 1$. Thus for $m = 2$ one can set

$$C = \begin{pmatrix} a_1 & \lambda_1 \\ a_2 & \lambda_2 \end{pmatrix}.$$

Now suppose $\gcd(a_1, \dots, a_{m-1}) = r$. By the induction hypothesis there exists a matrix

$$\overline{C} = \begin{Vmatrix} a_1 & \cdots \\ \vdots & \\ a_{m-1} & \cdots \end{Vmatrix}, \qquad \det \overline{C} = r.$$

Since $\gcd(r, a_m) = 1$, for some $\mu_1, \mu_2 \in \mathbf{Z}$

$$\mu_1 r + \mu_2 a_m = 1. \tag{1.35}$$

Consider the matrix

$$C = \left\Vert \begin{array}{c|c} & \pm \mu_2 a_1/r \\ \overline{C} & \vdots \\ & \pm \mu_2(a_{m-1})/r \\ \hline a_m\, 0 \cdots 0 & \mu_1 \end{array} \right\Vert.$$

The signs of the elements of the last column are chosen so that the algebraic cofactor of a_m equals μ_2. By (1.35) this matrix is the one desired.

LEMMA 10. *For any linear transformation* $B\colon \mathbf{Z}^m \to \mathbf{Z}^m$ *there exist linear transformations* $c_i\colon \mathbf{Z}^m \to \mathbf{Z}^m$ $(i = 1, 2, 3)$ *such that* $\det C_i = 1$, $C_1 B$ *is a triangular matrix, and* $C_2 B C_3$ *is a diagonal matrix.*

PROOF. In reducing a matrix to triangular form we may add to a row another row multiplied by an integer, interchange two rows, or multiply a row by −1. In reducing a matrix to diagonal form we may do analogous operations on the columns as well. For definiteness we shall show how to reduce a matrix to diagonal form. Multiplying the matrix by −1 if necessary, we can assure that at least one element is positive. Choose the smallest positive element, and, interchanging rows and columns, move it to the place of b_{11}. Subtracting or adding the first row and the first column to the remaining ones the necessary number of times, we make all the elements of the first row and the first column nonnegative and less than b_{11}, then move the smallest of these to the place of b_{11}. After a finite number of such steps all elements of the first row and first column will be zero except b_{11}. Next analogous operations are carried out with the remaining rows and columns. As a result the matrix B will be reduced to diagonal form by use of matrices C_2 and C_3, for which $\det \overline{C}_i = \pm 1$. Multiplying them, if necessary, by −1, we obtain the desired matrices C_2 and C_3.

§2. Representation of functions by series in the systems $e(\mathbf{Z}^m)$ and $e(A_\Pi)$

THEOREM 1′. *Suppose Π is a half-space in $\mathbf{R}^m$, $A_\Pi = \Pi \cap \mathbf{Z}^m$, $\varphi, \omega \in \Phi$, and*

$$\underline{\lim}_{x\to\infty} \frac{(x)}{\ln x} = 0, \qquad \lim_{x\to\infty} \varphi(x) = \infty.$$

Then in system $e(A_\Pi)$ there exists a series (1) satisfying (2) which is universal with respect to subseries in $S(T^m)$ in the sense of convergence in measure and universal in $\varphi(L)$ in the sense of convergence in "φ-distance".

PROOF. Let $\{F_i(x)\}_1^\infty$ be the sequence of all complex functions whose real and imaginary parts assume rational values on intervals with rational endpoints. By Lemma 7 there exists a sequence of polynomials

$$t_i(x) = \sum_{r_i \leqslant |\nu| \leqslant R_i,\ \nu \in A_n} c_\nu e^{i\nu x} \qquad (i = 1, 2, \ldots;\ r_i < R_i < r_{i+1}),$$

satisfying the conditions

1) $$\int_{T^m} \varphi(F_i(x) - t_i(x))\, dx \leqslant \frac{1}{2^i};$$

2) $$\int_e \varphi\Bigg(\sum_{r_i \leqslant |\nu| \leqslant R} c_\nu\text{-}e^{i\nu x}\Bigg)\, dx \leqslant \lambda_\varphi \int_e \varphi(F_i)\, dx + \frac{1}{2^i}$$

for $e \subseteq T^m$ and $r_i \leqslant R \leqslant R_i$;

3) $$\sum_{r_i \leqslant |\nu| \leqslant R_i} |c_\nu|^2 \omega(|c_\nu|) \leqslant \frac{1}{2^i}.$$

We shall show that the series

$$\sum_{i=1}^\infty t_i(x) = \sum_{i=1}^\infty \sum_{r_i \leqslant |\nu| \leqslant R_i,\ \nu \in A_\Pi} c_\nu e^{i\nu x} \tag{2.1}$$

is the one sought. Condition (2) for the coefficients follows from 3).

Universality of (2.1) *with respect to subseries in* $\varphi(L)$. Let $f \in \varphi(L)$. By induction define sequences of functions

$$\{f_n\}, \quad \{F_{i_n}\}, \quad \{t_{i_n}\} \qquad (n = 1, 2 \dots)$$

by the conditions

$$f_1(x) = f(x), \quad f_n(x) = f_{n-1}(x) - t_{i_{n-1}}(x), \quad \int_{T^m} \varphi(f_n(x) - F_{i_n}(x))\, dx \leqslant \frac{1}{2^n}. \tag{2.2}$$

Consider the series $\sum_{n=1}^{\infty} t_{i_n}$, which is a subseries of (2.1). Since

$$F_{i_n} = (F_{i_n} - f_n) + (f_{n-1} - F_{i_{n-1}}) + (F_{i_{n-1}} - t_{i_{n-1}}),$$

it follows from 1), 2), and (2.2) that for the partial sum of the subseries

$$\begin{aligned} &\int_{T^m} \varphi\left(f(x) - \sum_{k=1}^{n-1} t_{i_k}(x) - \sum_{r_{i_n} \leqslant |\nu| \leqslant R} c_\nu e^{i\nu x} \right) dx \\ &\quad \leqslant c_\varphi^2 \left\{ \int_{T^m} \varphi(f_{n-1}(x) - t_{i_{n-1}}(x))\, dx + \lambda_\varphi \int_{T^m} \varphi(F_{i_n})\, dx + \frac{1}{2^n} \right\} \\ &\quad \leqslant 2\lambda_\varphi c_\varphi^4 \left\{ \int_{T^m} \varphi(f_{n-1}(x) - F_{i_{n-1}}(x))\, dx + \int_{T^m} \varphi(F_{i_{n-1}}(x) - t_{i_{n-1}}(x))\, dx \right. \\ &\qquad\qquad \left. + \int_{T^m} \varphi(f_n(x) - F_{i_n}(x))\, dx + \frac{1}{2^n} \right\} \\ &\quad \leqslant \bar{\lambda}_\varphi \frac{1}{2^n}. \end{aligned}$$

From this we conclude that

$$\lim_{R \to \infty} \int_{T^m} \varphi\left(f(x) - \sum_{k=1}^{n-1} t_{i_k}(x) - \sum_{r_{i_n} \leqslant |\nu| \leqslant R} c_\nu e^{i\nu x} \right) dx = 0.$$

Universality of (2.1) *with respect to subseries in the sense of convergence in measure.* Let

$$\begin{aligned} f = f_1 + if_2 &\in S(T^m), & A &= \{x \mid |f| < \infty\}, \\ A_{11} &= \{x \mid f_1 = +\infty, f_2 = +\infty\}, & A_{12} &= \{x \mid f_1 = +\infty, f_2 = -\infty\}, \\ A_{21} &= \{x \mid f_1 = -\infty, f_2 = +\infty\}, & A_{22} &= \{x \mid f_1 = -\infty, f_2 = -\infty\}, \\ B_{11} &= \{x \mid f_1 = +\infty, |f_2| < \infty\}, & B_{21} &= \{x \mid f_1 = -\infty, |f_2| < \infty\}, \\ B_{12} &= \{x \mid |f_1 < \infty, f_2 = +\infty\}, & B_{22} &= \{x \mid |f_1| < \infty, f_2 = -\infty\}. \end{aligned} \tag{2.3}$$

By induction define sequences of functions $\{f_n\}$, $\{F_{i_n}\}$, and $\{t_{i_n}\}$ from the conditions

$$f_n(x) = f_n^1 + if_n^2 = \begin{cases} f, & x \in A, \\ \pm n \pm in, & x \in A_{11}, A_{12}, A_{21}, A_{22}, \\ \pm n + if_2, & x \in B_{11}, B_{21}, \\ f_1 \pm in, & x \in B_{12}, B_{22}, \end{cases} \tag{2.4}$$

$$\int_{T^m} \varphi\left(f_n(x) - \sum_{k=1}^{n-1} t_{i_k}(x) - F_{i_n}(x) \right) dx \leqslant \frac{1}{2^n}.$$

Consider the series

$$\sum_{n=1}^{\infty} t_{i_n}(x) = \sum_{n=1}^{\infty} \sum_{r_{i_n} \leqslant |\nu| \leqslant R_{i_n}} c_\nu e^{i\nu x}.$$

Let us break it up into the sum of two real series:

$$\sum_{n=1}^{\infty} t_{i_n}(x) = \sum_{n=1}^{\infty} \sum_{r_{i_n} \leqslant |\nu| \leqslant R_{i_n}} (a_\nu \cos \nu x + b_\nu \sin \nu x)$$
$$+ i \sum_{n=1}^{\infty} \sum_{r_{i_n} \leqslant |\nu| \leqslant R_{i_n}} (a_\nu \sin \nu x - b_\nu \cos \nu x).$$

We shall prove that the first series converges in measure to $f_1(x)$ and the second to $f_2(x)$. For definiteness we shall prove this for the series

$$\sum_{n=1}^{\infty} \sum_{r_{i_n} \leqslant |\nu| \leqslant R_{i_n}} (a_\nu \cos \nu x + b_\nu \sin \nu x) = \sum_{n=1}^{\infty} \sum_{r_{i_n} \leqslant |\nu| \leqslant R_{i_n}} A_\nu(x) = \sum_{n=1}^{\infty} p_{i_n}(x). \tag{2.5}$$

With minor modifications the proof will be carried out analogously to the proof of the corresponding assertion in [2]. For $x \in A \cup B_{12} \cup B_{22} = A'$ we have $f_n^1(x) = f_1(x)$. Hence, as in the proof of universality in $\varphi(L)$, by (2.4)

$$\int_{A'} \varphi\left(f_1(x) - \sum_{k=1}^{n-1} p_{i_k}(x) - \sum_{r_{i_n} \leqslant |\nu| \leqslant R} A_\nu(x) \right) dx \leqslant \lambda_\varphi \frac{1}{2^n}.$$

Hence for any $\varepsilon > 0$

$$\lim_{R\to\infty} \left| \left\{ x \in A' \left| \left| f_1(x) - \sum_{k=1}^{n-1} p_{i_k}(x) - \sum_{r_{i_n} \leqslant |\nu| \leqslant R} A_\nu(x) \right| \geqslant \varepsilon \right. \right\} \right| = 0.$$

We shall show that for any $M > 0$ and $x \in B = A_{11} \cup A_{12} \cup B_{11}$

$$\lim_{R\to\infty} \left| \left\{ x \in B \left| \sum_{k=1}^{n-1} p_{i_k}(x) + \sum_{r_{i_n} \leqslant |\nu| \leqslant R} A_\nu(x) \leqslant M \right. \right\} \right| = 0. \tag{2.6}$$

Let $\varepsilon > 0$. By 1) and (2.4), for $n \geqslant n_0$

$$\left| \left\{ x \in B \left| \sum_{k=1}^{n-1} p_{i_k}(x) < \frac{n-1}{2} \right. \right\} \right|$$
$$\leqslant \frac{1}{\varphi((n-1)/2)} \int_B \varphi\left(n - 1 - \sum_{k=1}^{n-1} p_{i_k}(x) \right) dx$$
$$\leqslant \frac{c_\varphi}{\varphi((n-1)/2)} \left\{ \int_{T^m} \varphi\left(f_{n-1}(x) - \sum_{k=1}^{n-2} t_{i_k}(x) - F_{i_{n-1}}(x) \right) dx \right.$$
$$\left. + \int_{T^m} \varphi\left(F_{i_{n-1}}(x) - t_{i_{n-1}}(x) \right) dx \right\}$$
$$\leqslant \frac{\bar{c}_\varphi}{\varphi((n-1)/2)} \frac{1}{2^{n-1}} \leqslant \varepsilon. \tag{2.7}$$

For $x \in A_{11} \cup A_{12}$ we have $f_n = f_{n-1} + 1 + i$ and

$$F_{i_n} = 1 \pm i + \left(f_{n-1} - \sum_{k=1}^{n-2} t_{i_k} - F_{i_n} \right) - \left(f_n - \sum_{k=1}^{n-1} t_{i_k} - F_{i_n} \right) + \left({}_{i_{n-1}} + t_{i_{n-1}} \right).$$

Analogously, for $x \in B_{11}$ we have $f_n = f_{n-1} + 1$ and

$$F_{i_n} = 1 + \left(f_{n-1} - \sum_{k=1}^{n-2} t_{i_k} - F_{i_n} \right) - \left(f_n - \sum_{k=1}^{n-1} t_{i_k} - F_{i_n} \right) + \left(F_{i_{n-1}} - t_{i_{n-1}} \right).$$

Therefore

$$\int_B \varphi\left(F_{i_n}(x)\right) dx \leqslant c_\varphi^3 \left\{ \int_{T^m} \varphi(1+i)\, dx + \int_{T^m} \varphi\left(f_{n-1} - \sum_{k=1}^{n-2} t_{i_k} - F_{i_{n-1}} \right) dx \right.$$

$$\left. + \int_{T^m} \varphi\left(f_n - \sum_{k=1}^{n-1} t_{i_k} - F_{i_n} \right) dx + \int_{T^m} \varphi\left(F_{i_{n-1}} - t_{i_{n-1}}\right) dx \right\} \leqslant \lambda_\varphi.$$

Hence by 2) for any n and $r_{i_n} \leqslant R \leqslant R_{i_n}$

$$\int_B \varphi\left(\sum_{r_{i_n} \leqslant |\nu| \leqslant R} A_\nu(x) \right) dx \leqslant \int_B \varphi\left(\sum_{r_{i_n} \leqslant |\nu| \leqslant R} c_\nu e^{i\nu x} \right) dx$$

$$\leqslant \lambda_\varphi \int_B \varphi\left(F_{i_n}\right) dx + \frac{1}{2^n} \leqslant \bar{\lambda}_\varphi.$$

Therefore, noting the hypothesis that $\lim_{x \to \infty} \varphi(x) = \infty$, there exists $\alpha = \alpha(\varepsilon) > 0$ such that

$$\left| \left\{ x \in B \,\middle|\, \sum_{r_{i_n} \leqslant |\nu| \leqslant R} A_\nu(x) < -\alpha \right\} \right| \leqslant \frac{1}{\varphi(\alpha)} \int_B \varphi\left(\sum_{r_{i_n} \leqslant |\nu| \leqslant R} A_\nu(x) \right) dx$$

$$\leqslant \frac{\bar{\lambda}_\varphi}{\varphi(\alpha)} < \varepsilon. \tag{2.8}$$

If now $(n-1)/2 - \alpha > M$, then for $n \geqslant n_0$ it follows from (2.7) and (2.8) that

$$\left| \left\{ x \in B \,\middle|\, \sum_{k=1}^{n-1} p_{i_k}(x) + \sum_{r_{i_n} \leqslant |\nu| \leqslant R} A_\nu(x) \leqslant M \right\} \right| \leqslant 2\varepsilon.$$

Relation (2.6) follows from this. Analogously one can verify that for any $M > 0$ and $C = A_{21} \cup A_{22} \cup B_{21}$

$$\lim_{R \to \infty} \left| \left\{ x \in C \,\middle|\, \sum_{k=1}^{n-1} p_{i_k}(x) + \sum_{r_{i_n} \leqslant |\nu| \leqslant R} A_\nu(x) \geqslant -M \right\} \right| = 0.$$

Hence, (2.5) converges to $f_1(x)$ in measure. The theorem is proved.

THEOREM 2′. *Let Π be a half-space in $\mathbf{R}^m$, $A_\Pi = \Pi \cap \mathbf{Z}^m$, φ, $\omega \in \Phi$, and*

$$\underline{\lim}_{x \to \infty} \frac{\varphi(x)}{\ln x} = 0, \qquad \lim_{x \to \infty} \varphi(x) = \infty.$$

Then in the system $e(A_\Pi)$ there exists a series (1) satisfying condition (2) which is universal in $S(T^m)$ in the sense of convergence almost everywhere and universal in $\varphi(L)$ in the sense of convergence in "φ-distance".

PROOF. Let $\{F_i(x)\}$ be as in Theorem 1′. Applying Lemma 7, we construct a sequence of polynomials

$$t_i(x) = \sum_{r_i \leqslant |\nu| \leqslant R_i,\ \nu \in A_\Pi} c_\nu e^{i\nu x} \qquad (i = 1, 2, \ldots;\ r_i < R_i < r_{i+1}),$$

possessing the properties

1) $$\int_{T^m} \varphi\left(F_i(x) - \sum_1^i t_j(x) \right) dx \leqslant \frac{1}{2^i} \varphi\left(\frac{1}{2^i} \right),$$

2) $$\sum_{r_i \leqslant |\nu| \leqslant R_i} |c_\nu|^2 \omega(|c_\nu|) < \frac{1}{2^i}.$$

We shall show that the series

$$\sum_1^\infty t_i(x) = \sum_{i=1}^\infty \sum_{r_i \leqslant |\nu| \leqslant R_i} c_\nu e^{i\nu x}$$

is the one sought. The condition for the coefficients of (2) follows from 2). Since by 1)

$$\left| \left\{ x \,\middle|\, \left| F_i(x) - \sum_1^i t_j(x) \right| \geqslant \frac{1}{2^i} \right\} \right| \leqslant \frac{1}{\varphi(1/2i)} \int_{T^m} \varphi\left(F_i(x) - \sum_1^i t_j(x) \right) dx \leqslant \frac{1}{2^i},$$

the proof of universality in $S(T^m)$ in the sense of convergence almost everywhere is carried out in the usual manner (cf., for example [12], p. 382).

We shall verify the universality in $\varphi(L)$. For any $f \in \varphi(L)$ there exists a sequence $\{F_{n_k}\}$ such that

$$\int_{T^m} \varphi(f(x) - F_{n_k}(x))\, dx \leqslant \frac{1}{2^k} \varphi\left(\frac{1}{2^k} \right).$$

From this and from 1) we get

$$\begin{aligned} &\int_{T^m} \varphi\left(f(x) - \sum_1^{n_k} t_j(X) \right) dx \\ &\qquad \leqslant c_\varphi \left\{ \int_{T^m} \varphi(f(x) - F_{n_k}(x))\, dx + \int_{T^m} \varphi\left(F_{nk}(x) - \sum_1^{n_k} t_j(x) \right) dx \right\} \\ &\qquad \leqslant \frac{c_\varphi}{2^{k-1}} \varphi\left(\frac{1}{2^k} \right). \end{aligned}$$

The theorem is proved.

Theorems 1 and 2 are corollaries of Theorems 1′ and 2′. For the proof of Theorem 3 it remains to show that if

$$\varlimsup_{x \to \infty} \frac{\varphi(x)}{\ln x} > 0,$$

then the system $e(A_\Pi)$ is not complete in $\varphi(L)$; moreover by Lemma 3 it suffices to consider the case $\varphi(x) = \ln(1 + |x|)$. We may assume that $\Pi = \{x \in \mathbf{R}^m | ax > 0\}$, where $a \in \mathbf{R}^m$. First suppose that $a = (1, 0, \ldots, 0)$ and that $K_R = \{x |\, |x_i| \leqslant R\}$ is a cube in $\mathbf{R}^m$. Then

$$J = \inf_{c_\nu} \int_{T^m} \ln\left(1 + \left|2 - \sum_{\nu \in \Pi \cap K_R} c_\nu e^{i\nu x}\right|\right) dx$$

$$\geqslant \inf_{c_\nu} \int_{T^m} \ln\left|2 + \sum_{\nu_1 = 1}^{R} \left(\sum_{|\nu_i| \leqslant R} c_\nu e^{i(\nu_2 x_2 + \cdots + \nu_m x_m)}\right) e^{i\nu_1 x_1}\right| dx. \tag{2.9}$$

Since by Jensen's formula we have for functions of Nevanlinna class

$$f(z) = \sum_0^\infty a_k z^k, \qquad f(0) \neq 0,$$

$$\int_0^{2\pi} \ln|f(e^{ix})|dx \geqslant 2\pi \ln|f(0)|,$$

it follows that for almost all $x_2, \ldots, x_m$

$$\int_0^{2\pi} \ln\left|2 - \sum_{\nu_1 = 1}^{R} c_{\nu_1}(x_2, \ldots, x_m) e^{i\nu_1 x_1}\right| dx_1 \geqslant 2\pi \ln 2.$$

From this and from (2.9) we get

$$J \geqslant (2\pi)^m \ln 2.$$

If $a = (a_1, \ldots, a_m) \in \mathbf{Z}^m$, $a_1 > 0$ and $C\colon \mathbf{Z}^m \to \mathbf{Z}^m$ is a linear transformation, as in Lemma 9, i.e. $\det C = 1$, $Ca = (1, 0, \ldots, 0)$, then we form a linear transformation of $\mathbf{R}^m$ via the formula $x = C^* y$. We shall have

$$ax = aC^* y = Cay = y_1,$$

and Π will map onto the half-space $\Pi' = \{y | y_1 > 0\}$. The general case $a \in \mathbf{R}^m$ reduces to the preceding, since for any cube K_R there exists $a' \in \mathbf{R}^m$ with rational coordinates such that there are no points of $\mathbf{Z}^m$ in K_R between the hyperplanes $ax = 0$ and $a'x = 0$.

The proof of Theorem 4 is carried out analogously to the proof of Theorem 3 only with the use of Lemma 8.

§3. Representation of functions by series in subsystems of the systems $e(\mathbf{Z}^m)$ and $e(A_\Pi)$

Let $A \subseteq \mathbf{Z}^m$, and let PA and QA be the sets defined in the Introduction. These sets possess, for example, the property

$$QA = \{n \in \mathbf{Z}^m | \forall s \in \not< \forall R > 0\, \exists r \in \mathbf{Z}^m,\, |r| \geqslant R\, \forall k \in \mathbf{N},$$
$$1 \leqslant k \leqslant s, n + rk \in A\}. \tag{3.1}$$

Indeed, if $|r| \leqslant R_0$ for some $n \in \mathbf{Z}^m$ and all $s \in \mathbf{N}$, then the set of such r is finite and there exists r_0 with $|r_0| \leqslant R_0$ such that

$$\{n + kr_0\}_{k \in \mathbf{N}} \subseteq A.$$

Thus for r we may take $r = tr_0$ for any $t \in N$. We have now reached a contradiction.

LEMMA 11 (extension). *Suppose* $A \subseteq \mathbf{Z}^m$, φ, $\omega \in \Phi$, $\underline{\lim}_{x\to\infty} \varphi(x)/\ln x = 0$, $f \in \varphi(L)$, *and* ε *and* r *are positive. If there exists a polynomial in the system* $e(QA)$ *for which assertions* (1.16)–(1.19) *of Lemma* 7 *hold, then there exists a polynomial in the system* $e(A)$ *for which the same assertions hold.*

PROOF. By hypothesis, for any $\varepsilon > 0$, $r > 0$, and any $f \in \varphi(L)$ there exists a polynomial

$$t(x) = \sum_{r\leqslant|\nu|\leqslant R,\, \nu\in QA} a_\nu e^{i\nu x}, \tag{3.2}$$

for which assertions (1.16)–(1.19) of Lemma 7 hold. Let us enumerate in order of increasing modulus those ν for which $a_\nu \neq 0$: $a_{\nu_1}, \ldots, a_{\nu_l}$. Applying Lemma 7, we obtain polynomials

$$T_j^*(x) = \sum_1^{s_j} c_k e^{ikx} \qquad (j = 1, \ldots, l),$$

for which

$$\int_0^{2\pi} \varphi\left(a_{\nu_j} - T_j^*(x)\right) dx \leqslant \frac{\varepsilon}{(2\pi)^m c_\varphi^l}; \tag{3.3}$$

$$\int_0^{2\pi} \varphi\left(\sum_{k=1}^{q} c_k e^{ikx}\right) dx \leqslant \lambda_\varphi \varphi(a_{\nu_j}) + \frac{\varepsilon}{(2\pi)^m c_\varphi^l} \tag{3.4}$$

for $1 \leqslant q \leqslant s_j$, and

$$\varphi(c_k) \leqslant \varepsilon/(2\pi)^m, \qquad k = 1, \ldots, s_j; \tag{3.5}$$

$$\sum_1^{s_j} |c_k|^2 \omega(c_k) \leqslant \varepsilon/l. \tag{3.6}$$

Using (3.1), choose $r_j \in \mathbf{Z}^m$ $(j = 1, \ldots, l)$ so that

$$\{\nu_j + kr_j\}_{k=1}^{s_j} \subset A, \tag{3.7}$$

$$\alpha_1 = \min_{1\leqslant k\leqslant s_1} |\nu_1 + kr_1| \geqslant r,$$

$$\alpha_{j+1} \min_{1\leqslant k\leqslant s_{j+1}} |\nu_{j+1} + kr_{j+1}| \geqslant \max_{1\leqslant k\leqslant s_j} |\nu_j + kr_j| + 1 = \beta_j + 1. \tag{3.8}$$

Form the polynomials

$$T_j(x) = \sum_1^{s_j} c_k e^{i(\nu_j + kr_j)x}, \qquad T(x) = \sum_1^{l} T_j(x).$$

The spectrum of the polynomial $T(x)$ is in A by (3.7). Inequality (1.19) holds for the coefficients of $T(x)$ by (3.6). Making a change of variable via the formulas

$$y_1 = r_j x = r_j^1 x_1 + \cdots + r_j^m x_m, \qquad r_j^1 > 0,$$
$$y_i = x_i \qquad (i = 1, \ldots, m),$$

we obtain

$$\int_{T^m} \varphi\left(a_{\nu_j} - \sum_{k=1}^{s_j} c_k e^{ikr_jx}\right) dx = \frac{1}{r_j^1} \int_{T^{m-1}} \int_0^{2\pi r_j^1} \varphi\left(a_{\nu_j} - T_j^*(y_1)\right) dy$$
$$= (2\pi)^{m-1} \int_0^{2\pi} \varphi\left(a_{\nu_j} - T_j^*(y_1)\right) dy_1 \leqslant \frac{\varepsilon}{2\pi c_\varphi^l}. \quad (3.9)$$

Hence

$$\int_{T^m} \varphi(f(x) - T(x))\, dx$$
$$\leqslant c_\varphi \left\{ \int_{T^m} \varphi(f(x) - t(x))\, dx + c_\varphi^{l-1} \sum_{j=1}^{l} \int_{T^m} \varphi\left(a_{\nu_j} - \sum_1^{s_j} c_k e^{ikr_jx}\right) dx \right\}$$
$$\leqslant \varepsilon. \quad (3.10)$$

Inequality (1.16) is now established.

Let $\alpha_s \leqslant R \leqslant \beta_s$. Since for the partial sum we have

$$S_R(T) = \sum_1^{s-1} T_j + S_R(T_s),$$

it follows that for $e \subseteq T^m$

$$\int_e \varphi(S_R(T))\, dx \leqslant c_\varphi^2 \left\{ \int_{T^m} \varphi\left(\sum_1^{s-1} \left(a_{\nu_j} e^{i\nu_j x} - T_j(x)\right)\right) dx \right.$$
$$\left. + \int_e \varphi\left(\sum_1^{s-1} a_{\nu_j} e^{i\nu_j x}\right) dx + \int_{T^m} \varphi(S_R(T_s))\, dx \right\}. \quad (3.11)$$

As in the proof of (3.10),

$$\int_{T^m} \varphi\left(\sum_{j=1}^{s-1} \left(a_{\nu_j} e^{i\nu_j x} - T_j(x)\right)\right) dx \leqslant \varepsilon. \quad (3.12)$$

Furthermore,

$$\sum_{j=1}^{s-1} a_{\nu_j} e^{i\nu_j x} = \sum_{|\nu| \leqslant R'} a_\nu e^{i\nu x} - \sum_{\nu \in B_{R'}} a_\nu e^{i\nu x},$$

$$\int_e \varphi\left(\sum_{j=1}^{s-1} a_{\nu_j} e^{i\nu_j x}\right) dx \leqslant c_\varphi \left\{ \int_e \varphi\left(\sum_{|\nu| \leqslant R'} a_\nu e^{i\nu x}\right) dx + \int_{T^m} \varphi\left(\sum_{\nu \in BR'} a_\nu e^{i\nu x}\right) dx \right\}$$
$$\leqslant c_\varphi \left\{ \lambda_\varphi \int_e \varphi(f(x))\, dx + \varepsilon \right\}. \quad (3.13)$$

Since

$$S_R(T_s) = \sum_{|\nu_s + kr_s| \leqslant R} c_k e^{i(\nu_s + kr_s)x} = \sum_{k=k_1+1}^{k_r} c_k e^{i(\nu_s + kr_s)x}$$
$$= \sum_1^{k_s} c_k e^{i(\nu_s + kr_s)x} - \sum_1^{k_1} c_k e^{i(\nu_s + kr_s)x},$$

it follows from (3.4) and (1.18), as in the proof of (3.9), that

$$\int_{T^m} \varphi(S_R(T_s))\,dx \leqslant c_\varphi\left\{\int_{T^m} \varphi\left(\sum_1^{k_s} c_k e^{i(\nu_s+kr_s)x}\right) dx + \int_{T^m} \varphi\left(\sum_1^{k_1} c_k e^{i(\nu_s+kr_s)x}\right) dx\right\}$$

$$\leqslant c_\varphi\left\{(2\pi)^{m-1}\lambda_\varphi \varphi(a_{\nu_s}) + (2\pi)^{m-1}\lambda_\varphi \varphi(a_{\nu_s})\right\} \leqslant \bar{\lambda}_\varphi \varepsilon. \qquad (3.14)$$

Taking account of (3.11)–(3.14), we obtain (1.17). If B_R is an arbitrary subset of $|\nu| = R$, we have by (3.5)

$$\int_{T^m} \varphi\left(\sum_{\nu_s+kr_s \in B_R} c_k e^{i(\nu_s+kr_s)x}\right) dx = \int_{T^m} \varphi(c_{k_2})\,dx \leqslant \varepsilon.$$

The lemma is proved.

Analogously one can prove

LEMMA 12 (extension). *Suppose* $A \subseteq \mathbf{Z}^m$, $\varphi, \omega \in \Phi$, $\underline{\lim}_{x\to\infty} \varphi(x)/x = 0$, $f \in \varphi(L)$, *and* ε *and* r *are positive. If there exists a polynomial in the system* $e(\mathscr{P}A)$ *for which assertions* (1.30)–(1.33) *of Lemma* 8 *hold, then there exists a polynomial in the system* $e(A)$ *for which the same assertions hold.*

Theorems 5–8 are proved using Lemmas 11 and 12 analogously to the way in which Theorems 1–4 were proved.

LEMMA 13. *If* $A \subseteq \mathbf{Z}^m$ *and* $\bar{d}_m(A) = 1$, *then* $\mathscr{P}A = QA = \mathbf{Z}^m$.

PROOF. Since $\mathscr{P}A \subseteq QA$, we shall prove the assertion of the lemma for the set $\mathscr{P}A$. Suppose the contrary, i.e.

$$\exists n \in \mathbf{Z}^m\, \exists s \in \mathbf{N}\, \forall r \in \mathbf{Z}^m\, \exists k \in \mathbf{Z},\ 0 \leqslant |k| \leqslant s,\ n + rk \notin A.$$

Denote by $B_s \subset \mathbf{Z}^m$ the following set:

$$\forall r \in \mathbf{Z}^m\, \exists k \in \mathbf{Z},\ 0 \leqslant |k| \leqslant s,\ n + rk \in B_s.^*$$

By assumption A does not contain B_s. We shall show that $\underline{d}_m(B_s) > 0$. Since density is not changed by shifting by a fixed vector, we may assume $n = 0$. For an arbitrary $R > 0$ consider the spherical layer

$$\Gamma = \left\{x \in \mathbf{R}^m \,\middle|\, \frac{R}{s}\left(1 - \frac{1}{3s}\right) \leqslant |x| \leqslant \frac{R}{s}\right\}.$$

Through all points of $\Gamma \cap \mathbf{Z}^m$ construct lines passing through the origin. An arbitrary line will intersect this layer in two intervals. Let $[x_1, x_2]$ be one of these. Since $|(k+1)x_1| < |kx_2|$ for $k = \pm 1, \ldots, \pm s$, it follows that for these k the intervals $[kx_1, kx_2]$ are disjoint. But if $m \in [x_1, x_2] \cap \mathbf{Z}^m$, then either it or one of the points km ($k = \pm 1, \ldots, \pm s$) lies in the ball $|x| \leqslant R$ and B_s. Therefore the number of points

---

* *Translator's note.* This display is reproduced as it appears in the original, but the author may have intended A in place of B_s.

of B_s lying in the ball $|x| \leqslant R$ is not less than half the points of $\Gamma \cap \mathbf{Z}^m$. But this number is proportional to the volume of this layer. Therefore $\underline{d}_m(B_s) > 0$. Hence $\bar{d}_m(A) < 1$. We have now obtained a contradiction. It follows that $\mathcal{P}A = \mathbf{Z}^m$.

LEMMA 14. *For any positive sequence* $\{R_k\} \uparrow \infty$ *there exist two sets*

$$A_1 = \{n_k\}_{k \in \mathbf{N}} \subset \mathbf{Z}^m, \qquad A_2 = \{\nu_k, m_k\}_{k \in \mathbf{N}} \subset \mathbf{Z}^m,$$

for which $QA_1 = \mathbf{Z}^m, \mathcal{P}A_2 = \mathbf{Z}^m$ *and*

$$|n_{k+1}| \geqslant |n_k| + R_k, \qquad |m_{k+1}| \geqslant |\nu_{k+1}| \geqslant |m_k| + R_i.$$

PROOF. For definiteness we shall indicate how A_1 iks constructed. Enumerate $\mathbf{Z}^m = \{\nu_k\}_{k \in \mathbf{N}}$. Let

$$n_1 = \nu_1, n_2 = \nu_1 + r_{12}, n_3 = \nu_1 + 2r_{12}, n_4 = \nu_2 + r_{22}, n_5 = \nu_2 + 2r_{22},$$

where r_{12} and r_{22} are chosen so that ν_1 and $\nu_1 + r_{12}$, ν_2 and $\nu_2 + r_{22}$, lie on lines passing through the origin and

$$|\nu_1 + r_{12}| = |\nu_1| + |r_{12}| \geqslant |\nu_1| + R_1,$$
$$|\nu_1 + 2r_{12}| = |\nu_1| + |\nu_1 + r_{12}| + R_2,$$
$$|\nu_2 + r_{22}| \geqslant |n_3| + R_3,$$
$$|\nu_2 + 2r_{22}| = |\nu_2 + r_{22}| + |r_{22}| \geqslant |\nu_2 + r_{22}| + R_4.$$

Then construct analogously elements

$$n_{5+j} = \nu_1 + jr_{13}, \quad n_{8+j} = \nu_2 + jr_{23}, \quad n_{11+j} = \nu_3 + jr_{33} \qquad (j = 1, 2, 3)$$

etc.

The distinction between the systems $e(\mathbf{Z}^m)$ and $e(A_\Pi)$ in the problems under consideration is well seen in the following example.

Let a and x_0 be in $\mathbf{R}^m$; let $0 < \alpha \leqslant \pi/2$; and let

$$K_\alpha = \{x \in \mathbf{R}^m | a(x - x_0) \geqslant |a_0|\,|x - x_0| \cos \alpha\}$$

be the cone with vertex angle α. Finally let $K_{\pi/2}$ be a half-space and

$$A_\alpha = K_\alpha \cap \mathbf{Z}^m \quad (0 < \alpha \leqslant \pi/2), \qquad A_\alpha = (\mathbf{R}^m \setminus K_{\pi-\alpha}) \cap \mathbf{Z}^m \quad (\pi/2 < \alpha < \pi).$$

If $\underline{\lim}_{x\to\infty} \varphi(x)/\ln x > 0$, then $e(A_\alpha)$ is complete in $\varphi(L)$ for $\pi/2 < \alpha < \pi$ and not complete for $\alpha \leqslant \pi/2$. But if $\underline{\lim}_{x\to\infty} \varphi(x)/\ln x = 0$, then $e(A_\alpha)$ is complete in $\varphi(L)$ for any $\alpha > 0$.

Instead of Theorem 9 we shall prove a still more general assertion. We first show that with any linear nondegenerate transformation

$$B\colon \mathbf{Z}^m \to \mathbf{Z}^m, \qquad |\det B| = r$$

certain subspaces of functions on T^m are associated.

Let B^* be the matrix transpose of B. Consider the system

$$B^*\alpha = 0 \pmod{2\pi}. \tag{3.15}$$

By Lemma 10 there exist integer matrices C_1 and C_2 such that $\det C_1 = \det C_2 = 1$ and

$$C_1 B^* C_2 = \begin{Vmatrix} n_1 & & 0 \\ & \ddots & \\ 0 & & n_m \end{Vmatrix}$$

is a diagonal matrix with $|n_1 \cdots n_m| = r$. The systems (3.15) and

$$C_1 B^* \alpha = 0 \pmod{2\pi} \tag{3.16}$$

are equivalent. If $\alpha = C_2\beta$, then (3.16) assumes the form

$$n_1\beta = 0 \pmod{2\pi}, \dots, n_m\beta_m = 0 \pmod{2\pi},$$

$$\beta_1 = \frac{2\pi k_1}{|n_1|}\ (0 \leqslant k \leqslant |n_1| - 1), \dots, \beta_m = \frac{2\pi k_m}{|n_m|}\ (0 \leqslant k_m \leqslant |n_m| - 1).$$

Altogether we obtain r vectors

$$\beta = \left(\frac{2\pi k_1}{|n_1|}, \dots, \frac{2\pi k_m}{|n_m|}\right).$$

Then (3.15) also has precisely the r solutions $\alpha^1, \dots, \alpha^r$, where $\alpha^k = C_2\beta_k$.

We shall describe the factor-group $\mathbf{Z}^m/B\mathbf{Z}^m$. Since $C_1^*\mathbf{Z}^m = \mathbf{Z}^m$, we have $\nu \in B\mathbf{Z}^m$ if and only if $C_2^*\nu \in C_2^*BC_1^*\mathbf{Z}^m$. But the matrix $C_2^*BC_1^* = (C_1B^*C_2)^*$ is diagonal. The factor-group $\mathbf{Z}^m/C_2^*BC_1^*\mathbf{Z}^m = \{a_1, \dots, a_r\}$, where

$$a_l = (k_1^l, \dots, k_m^l), \qquad 0 \leqslant k_i^l \leqslant |n_i| - 1.$$

Therefore $\mathbf{Z}^m/B\mathbf{Z}^m = \{b_1, \dots, b_r\}$, where $b_l = (C_2^*)^{-1}a_l$, and moreover $b_1 = 0$. For measurable functions on T^m consider the operators

$$F_l(f) = \frac{1}{r}\sum_1^r e^{-i(b_l, \alpha^k)} f(x + \alpha^k) \qquad (l = 1, \dots, r).$$

They have the properties

$$F_l(e^{i\nu x}) = \begin{cases} e^{i\nu x}, & \nu \in b_l + B\mathbf{Z}^m, \\ 0, & \nu \notin b_l + B\mathbf{Z}^m. \end{cases}$$

Indeed, for this it suffices to verify that

$$\frac{1}{r}\sum_1^r e^{i(b_l, \alpha^k)} = \begin{cases} 1, & l = 1, \\ 0, & l \neq 1. \end{cases}$$

For $l = 1$ this is obvious. If $l \neq 1$, then

$$\sum_1^r e^{i(b_l, \alpha^k)} = \sum_1^r e^{i\left((C_2^*)^{-1} a_l,\, \alpha^k\right)} = \sum_1^r e^{i(a_l, C_2^{-1}\alpha^k)} = \sum_1^r e^{i(a_l, \beta^k)}$$

$$= \sum_{k_1=0}^{|n_1|-1} e^{iq_1 2\pi k_1/|n_1|} \cdots \sum_{k_m=0}^{|n_m|-1} e^{iq_m^l 2\pi k_m/|n_m|} = 0.$$

since at least one of the $q_k^l \neq 0$, and so one of the factors is 0.

We can now introduce the subspaces connected with the transformation B. Set

$$H_{B, b_l} = \left\{ f(x) \,\middle|\, f(x) = \frac{1}{r}\sum_1^r e^{-i(b_l, \alpha^k)} f(s + \alpha^k) \text{ almost everywhere} \right\},$$

where $r = |\det B|$, the α^k are the solutions of (3.15), and $\mathbf{Z}^m/B\mathbf{Z}^m = \{b_1, \dots, b_r\}$.

THEOREM 9′. *The system* $e(A)$ *is complete in* $\varphi(L)$ *if and only if for any linear transformation* B: $\mathbf{Z}^m \to \mathbf{Z}^m$ *with* $|\det B| = r$ $(r \neq 0, 1)$ *and any* $b \in \mathbf{Z}^m/B\mathbf{Z}^m$ *the systems* $e(A \cap (b + B\mathbf{Z}^m))$ *are complete in* $\varphi(L) \cap H_{B,b}$.

PROOF. Since for any $\nu \in \mathbf{Z}^m$ there exists $b \in \mathbf{Z}^m/B\mathbf{Z}^m$ such that $\nu \in b + B\mathbf{Z}^m$, the sufficiency is obvious. Let us prove the necessity.

If $f \in \varphi(L) \cap H_{B,b}$, then for any polynomial $t(x)$

$$\int_{T^m} \varphi(f(x) - t(x))\, dx = \frac{1}{r}\sum_1^r \int_{T^m} \varphi(f(x + \alpha^k) - t(x + \alpha^k))\, dx$$

$$\geqslant \frac{1}{rC_\varphi^{r-1}} \int_{T^m} \varphi\left(\frac{1}{r}\sum_1^r e^{-i(b,\alpha^k)} f(x + \alpha^k) - \frac{1}{r}\sum_1^r e^{-i(b,\alpha^k)} t(x + \alpha^k)\right) dx \geqslant c,$$

provided for some $b \in \mathbf{Z}^m/B\mathbf{Z}^m$ the system $e(A \cap (b + B\mathbf{Z}^m))$ is not complete in $\varphi(L) \cap H_{B,b}$.

The theorem is proved.

It is obvious that Theorem 9 is an immediate corollary of Theorem 9′, since if $A = b + B\mathbf{Z}^m$ and $|\det B| \neq 1$, then there exists $b_1 \in \mathbf{Z}^m$ such that $A \cap (b_1 + dB\mathbf{Z}^m) = \varnothing$. We note that convergence in measure is equivalent to convergence in $\varphi(L)$ for

$$\varphi(x) = \frac{|x|}{1 + |x|}.$$

PROOF OF THEOREM 10. Let

$$A = n + B\mathbf{Z}^m + f(\mathbf{Z}^m),$$

where $n \in \mathbf{Z}^m$, B is a linear transformation of $\mathbf{Z}^m$,

$$f(k) = \{f_1(k), \dots, f_m(k)\}, \qquad f_i(k) = \sum_{2 \leqslant |\alpha| \leqslant n_i} a_\alpha K^\alpha,$$

$\alpha = (\alpha_1, \dots, \alpha_m)$ is a multi-index, $|\alpha| = \alpha_2 + \cdots + \alpha_m$ and $a_\alpha \in \mathbf{Z}$. Without loss of generality we may assume $n = 0$.

We shall show that A fails to contain some set $\{n_1 + rk\}_{k \in \mathbf{Z}^m}$, $n_1 \in \mathbf{Z}^m$, $r \in \mathbf{N}$, if $|\det B| \neq 1$. Hence by Theorem 9 the incompleteness of $e(A)$ in measure will follow.

By Lemma 10 we may consider the matrix B diagonal, i.e.

$$A = \{\gamma_1 k_1 + f_1(k), \dots, \gamma_m k_m + f_m(k)\}_{k \in \mathbf{Z}^m},$$

where all the $\gamma_i \geqslant 0$. Consider two cases.

Case 1. Suppose zero is among the γ_i, for example $\gamma_1 = 0$. For $j = 1, \dots, m$ we have

$$f_j(0, 0, \dots, 0) = 0 \pmod 4, \qquad f_j(2, 0, \dots, 0) = 0 \pmod 4.$$

If we examine the vectors

$$\{(\gamma_1 k_1 + f_1(k)) \pmod 4, \dots, (\gamma_m k_m + f_m(k)) \pmod 4\}$$

for all $k = (k_1, \dots, k_m)$, $0 \leqslant k_i \leqslant 3$, at least two of them will be the same (zero).

Hence there exists a vector $n_0 = (n_1,\ldots,n_m)$, $0 \leqslant n_i \leqslant 3$, such that A fails to contain the set

$$\{n_0 + 4k\}_{k\in\mathbf{Z}^m}.$$

Case 2. Suppose $|\det B| \geqslant 2$ and that the largest of the γ_i is $\gamma_1 = q \geqslant 2$. If $r = q^2$, then

$$f_j(0,q,\ldots,q)\,(\operatorname{mod} r) = 0, \quad f_j(q,\ldots,q)\,(\operatorname{mod} r) = 0 \qquad (j = 1,\ldots,m),$$
$$\gamma_1 q = 0\,(\operatorname{mod} r), \quad \gamma_i 0 = 0\,(\operatorname{mod} r), \quad \gamma_i q\,(\operatorname{mod} r) = \gamma_i q \qquad (j = 2,\ldots,m).$$

Therefore again among the vectors

$$\{(\gamma_1 k_1 + f_1(k))\,(\operatorname{mod} r),\ldots,(\gamma_m k_m + f_m(k))\,(\operatorname{mod} r)\}$$

$(0 \leqslant k_i \leqslant r-1)$ two are the same. Hence A fails to contain some set

$$\{n_0 + rk\}_{k\in,\mathbf{Z}^m}.$$

The theorem is proved.

BIBLIOGRAPHY

1. P. L. Ul'yanov, *Representation of functions by series, and the classes* $\varphi(L)$, Uspekhi Mat. Nauk **27** (1972), no. 2 (164), 3–52; English transl. in Russian Math. Surveys **27** (1972).

2. A. A. Talalyan, *The representation of measurable functions by series*, Uspekhi Mat. Nauk **15** (1960), no. 5 (95), 77–141; English transl. in Russian Math. Surveys **15** (1960).

3. D. E. Men'shov, *On convergence in measure of trigonometric series*, Trudy Mat. Inst. Steklov. **32** (1950); English transl. in Amer. Math. Soc. Transl. (1) **3** (1962).

4. A. A. Talalyan, *The representation of functions in the classes* $L_p[0,1]$, $0 < p < 1$, *by orthogonal series*, Acta Math. Acad. Sci. Hungar. **21** (1970), 1–9. (Russian)

5. N. B. Pogosyan, *Representation of measurable functions by orthogonal series*, Mat. Sb. **98 (140)** (1975), 102–112; English transl. in Math. USSR Sb. **27** (1975).

6. L. A. Shaginyan, *On the summability to infinity of trigonometric series and series in the Walsh system*, Mat. Sb. **108 (150)** (1979), 457–470; English transl. in Math. USSR Sb. **36** (1980).

7. A. Zygmund, *Trigonometric series*, 2nd rev. ed., vol. 2, Cambridge Univ. Press, 1959.

8. J. J. Price, *Sparse subsets of orthonormal systems*, Proc. Amer. Math. Soc. **35** (1972), 161–164.

9. F. G. Arutyunyan, *Representation of functions by multiple series*, Akad. Nauk Armyan. SSR Dokl. **64** (1977), 72–76. (Russian)

10. V. I. Ivanov, *Representation of measurable functions by multiple trigonometric series*, Dokl. Akad. Nauk SSSR **259** (1981), 279–282; English transl. in Soviet Math. Dokl. **24** (1981).

11. N. K. Bari, *Trigonometric series*, Fizmatgiz, Moscow, 1961; English transl., *A treatise on trigonometric series*. Vols. I, II, Pergamon Press, Oxford, and Macmillan, New York, 1964.

12. R. S. Guter and P. L. Ul'yanov, *On new results in the theory of orthogonal series*, Appendix to the Russian transl. of S. Kaczmarz and H. Steinhaus, *Theorie der Orthogonalreihen*, Fizmatgiz, Moscow, 1958, pp. 333–456. (Russian)

Translated by R. L. COOKE

Труды Матем. Инст. Стеклов.
164(1983)

Proceedings of the Steklov
Institute of Mathematics
1985, Issue 2

ON LUZIN'S C-PROPERTY FOR A CONJUGATE FUNCTION

UDC 517.5

K. I. OSKOLKOV

ABSTRACT. In this paper are obtained quantitative estimates of N. N. Luzin's C-property, i.e. estimates for the differences of the conjugate function $\tilde{f}(x)$ of the form

$$|\tilde{f}(x)-\tilde{f}(y)| \leqslant \left(C_f(x)+C_f(y)\right)w\left(|x-y|\right).$$

Here $w(\delta)$ is a function which is nonnegative for $\delta>0$ and tends monotonically to zero as $\delta \downarrow 0$, while the coefficient $C_f(x)$ is nonnegative and finite for almost all x. It is assumed as well that the original function $f(x)$ for which $\tilde{f}(x)$ is trigonometrically conjugate is continuous and that a majorant $w(\delta)$ of its uniform modulus of continuity is known. It is established that for the function $w(\delta)=w(\omega,\delta)$ one can take the functions

$$w(\delta)=\omega(\delta)\log\log\min\left(\frac{1}{\omega(\delta)},\frac{\omega(\delta)}{\delta}\right), \qquad w(\delta)=\omega(\delta)\log\log(1/\delta),$$

and it is shown that for certain moduli of continuity (near the class Lip 1 and also the class of essentially bounded functions) the presence of coefficients of $w(\delta)$ tending to infinity as $\delta\to 0$ is, in general, necessary. In a series of cases, the order of growth of these coefficients is elucidated.

Bibliography: 9 titles.

§1. Introduction

Let $f(x)$ be a real-valued function, measurable and finite almost everywhere on some interval $[a,b]\subset(-\infty,\infty)$. Then according to a classical theorem of Luzin ([1], p. 65) this function possesses the C-property, i.e. for any arbitrarily small positive number ε there exists a perfect set $F=F_\varepsilon(f)$ with Lebesgue measure meas $F \geqslant (b-a)-\varepsilon$ such that, considered only at points $x\in F$, the function $f(x)$ is continuous on this set.

In view of this theorem the question arises of a quantitative characterization of the continuity of a measurable almost-everywhere finite function on subsets of its interval of definition having positive Lebesgue measure. One of the natural methods for such a characterization consists of examining the uniform modulus of continuity of a function on subsets of measure arbitrarily close to a full measure. To be precise we shall examine here difference properties of measurable almost-everywhere finite functions which are expressed by estimates of the form

$$|f(x)-f(y)| \leqslant (C(x)+C(y))w\left(|x-y|\right), \tag{1.1}$$

1980 *Mathematics Subject Classification*. Primary 42A50.

where $w(\delta)$ is a function defined for all sufficiently small positive δ and possessing the properties

$$0 \leqslant w(\delta_2) \leqslant w(\delta_1) \quad (0 \leqslant \delta_2 \leqslant \delta_1); \qquad w(\delta) \to 0 \quad (\delta \to 0), \tag{1.2}$$

while $C(x) = C(f, w, x)$ is a certain measurable function which is nonnegative and finite for almost every $x \in [a, b]$.

In view of the C-property of Luzin it is natural to call each function w with the properties (1.2) for which estimate (1.1) holds with some almost-everywhere finite function $C(x)$ a modulus of C-continuity of the given measurable almost-everywhere finite function $f(x)$. It is clear that a modulus of C-continuity is not unique and that for a given function $f(x)$ there exists, in general, a whole class of different moduli of C-continuity, including some which are nonequivalent in the sense of order of vanishing as $\delta \to 0$. We shall denote this class by $W_C(f)$.

In the present paper we examine the moduli of C-continuity of function $\tilde{f}(x)$:

$$\tilde{f}(x) = \mathrm{PV}\, \frac{1}{2\pi} \int_{-\pi}^{\pi} f(t) \cot \frac{t-x}{2}\, dt,$$

which are trigonometrically conjugate to continuous 2π-periodic functions $f(t)$ with prescribed ordinary uniform moduli of continuity

$$\omega(f, \delta) = \max\{|f(x) - f(y)| : |x - y| \leqslant \delta\}.$$

In other words, we shall study functions $\tilde{f}(x)$ which are conjugate to functions $f(x)$ of the given 2π-periodic class H_ω,

$$H_\omega = \{f \colon \omega(f, \delta) \leqslant \omega(\delta)\ (0 \leqslant \delta \leqslant \pi)\},$$

where $\omega(\delta)$ is a certain modulus of continuity (cf. [2]), and obtain estimates of the moduli of C-continuity $w(\tilde{f}, \delta)$ expressed with the aid of $\omega(\delta)$. The sharpness of these estimates will also be studied.

Our purpose is thus to find functions $w(\delta) = w(\omega, \delta)$ for a given modulus of continuity $\omega(\delta)$ which decrease as fast as possible as $\delta \to 0$ and are such that for all f from the class H_ω the following inclusion holds:

$$w(\delta) \in W_C(\tilde{f}). \tag{1.3}$$

Let us exhibit some consequences of our results and point out their connection with known estimates of the uniform modulus of continuity of conjugate functions. First of all, it will be proved that for any modulus of continuity $\omega(\delta) \not\equiv 0$

$$w(\delta) = \omega(\delta) u(\delta) \in W_C(\tilde{f}) \qquad (f \in H_\omega), \tag{1.4}$$

where

$$u(\delta) = \log\log \min\left(\frac{\omega(1)}{\omega(\delta)}, \frac{\omega(\delta)}{\delta\omega(1)}\right). \tag{1.5}$$

In particular, the following cruder estimate holds:

$$\omega(f, \delta) \log\log(1/\delta) \in W_C(f). \tag{1.6}$$

We note that in (1.4) $\omega(\delta)u(\delta) \to 0$ as $\delta \to 0$ for any modulus of continuity, and that this estimate is characterized by the presence of a coefficient $u(\delta)$ of $\omega(\delta)$ which tends to $+\infty$ as $\delta \to 0$ provided $\omega(\delta)/\delta \to \infty$ as $\delta \to 0$. Hence (1.4) contains for the classes Lip α $(0 < \alpha < 1)$ an obviously high, nonoptimal estimate of the moduli of C-continuity of conjugate functions, since according to a theorem of Privalov [3] if $f \in \operatorname{Lip} \alpha$ and $0 < \alpha < 1$, then $\omega(\tilde{f}, \delta) = O(\delta^\alpha)$ as $\delta \to 0$.

The coefficients $u(\delta)$ tending to $+\infty$ as $\delta \to 0$ are superfluous for these classes. Nevertheless we shall prove that in the general case of classes H_ω, in particular for classes H_ω close to Lip 1 or to Lip 0 (the essentially bounded functions), for example the classes H_ω, where

$$\omega(\delta) = \delta \exp\left(\alpha \frac{\log(1/\delta)}{\log\log(1/\delta)}\right) \quad \text{or} \quad \omega(\delta) = \exp\left(-\alpha \frac{\log(1/\delta)}{\log\log(1/\delta)}\right) \tag{1.7}$$

($\delta \leqslant \delta_0$; α a positive constant), the existence of such coefficients in the estimate (1.4) is necessary, and without them the estimate ceases to be true. In particular for the classes H_ω with modulus of continuity of the form (1.7) even the coefficient $\log\log(1/\delta)$ in (1.6) cannot in general be replaced by any other tending to infinity more slowly.

Further, for the uniform modulus of continuity $\omega(\tilde{f}, \delta)$ of functions $\tilde{f}$ conjugate to functions of the given class H_ω, Bari and Stechkin [4] established the estimate

$$\omega(\tilde{f}, \delta) \leqslant c\beta(\omega, \delta), \tag{1.8}$$

where

$$\beta(\omega, \delta) = \int_0^\delta \frac{\omega(t)}{t}\, dt + \delta \int_\delta^\pi \frac{\omega(t)}{t^2}\, dt, \tag{1.9}$$

while c here and in what follows denotes absolute positive constants, in general not always the same. Also in [4] the sharpness of (1.8) and (1.9) is established: for each modulus of continuity $\omega(\delta)$ for which

$$\int_0^\pi \frac{\omega(t)}{t}\, dt < \infty, \tag{1.10}$$

there exists a function $f \in H_\omega$ such that

$$\omega(\tilde{f}, \delta) \neq o(\beta(\omega, \delta)) \qquad (\delta \to 0),$$

and if the integral in (1.10) diverges, then there exists a function $f \in H_\omega$ for which the conjugate function is not essentially bounded.

In particular, it follows from this that if

$$\beta(\omega, \delta) \neq O(\omega(\delta)) \qquad (\delta \to 0), \tag{1.11}$$

then there exists a function $f \in H_\omega$ such that

$$\omega(\tilde{f}, \delta) \neq O(\omega(\delta)) \qquad (\delta \to 0). \tag{1.12}$$

Here it will be established (Theorem 2) that in case (1.11) there exists a function $f \in H_\omega$ such that the conjugate function $\tilde{f}$ not only (cf. (1.12)) fails to inherit the uniform modulus of continuity of f on the entire period, but also does not inherit it on any subset of it of positive measure, i.e. as a C-modulus. In other words, for this function

$$\omega(\delta) \notin W_C(\tilde{f}), \tag{1.13}$$

and if the function $g(x)$ is continuous while $\operatorname{meas}\{x\colon g(x) = \tilde{f}(x)\} > 0$, then

$$\omega(g, \delta) \neq O(\omega(\delta)) \qquad (\delta \to 0).$$

Here and throughout what follows we suppose that all sets are 2π-periodic and $\operatorname{meas} E = \operatorname{meas}\{E \cap (0, 2\pi]\}$.

Moreover, under certain additional hypotheses on the modulus of continuity $\omega(\delta)$ one can succeed in finding (Theorem 1) coefficients $u(\delta) = u(\omega, \delta)$ which guarantee the imbedding (1.4) for all $f \in H_\omega$ and are also sharp (Theorem 3) for the class H_ω with respect to the order of tending to ∞ as $\delta \to 0$.

This entire paper is closely related in its results and methods of obtaining those results to the author's previous papers [5] and [6], where quantitative estimates of Luzin's C-property for the classes $H[\omega, p]$ (sets of functions with prescribed majorant of the integral moduli of continuity) were studied.

§2. An estimate of the modulus of C-continuity of the conjugate function

We shall assume that the modulus of continuity $\omega(\delta)$ satisfies the condition

$$\omega(\delta)/\delta \to \infty \qquad (\delta \to 0), \tag{2.1}$$

i.e. the corresponding class H_ω is not equivalent to Lip 1. This assumption is natural and does not limit the generality, since if $f \in \mathrm{Lip}\,1$, the conjugate function is absolutely continuous and, consequently, $\omega(\delta) \equiv \delta \in W_C(\tilde{f})$ for $f \in \mathrm{Lip}\,1$.

Further, as in [5] and [6], consider the sequence $\Delta = \Delta(\omega) = \{\delta_k(\omega)\}_0^\infty$ and the auxiliary function $\Omega(\delta) = \Omega(\omega, \delta)$, which make it possible to tell how the quantities $\omega(\delta)$ and $\delta/\omega(\delta)$ tend to zero as $\delta \to 0$. The definition and properties of the sequence Δ and the function Ω are contained in the following lemma (for the proof cf. [6]).

LEMMA 1. *Suppose the modulus of continuity $\omega(\delta)$ satisfies condition* (2.1) *and the sequence $\Delta(\omega) = \{\delta_k\}_0^\infty$ is defined by induction as follows*:

$$\delta_0 = \pi; \quad \delta_{k+1} = \min\left\{\delta \colon \max\left(\frac{\omega(\delta)}{\omega(\delta_k)}, \frac{\delta\omega(\delta_k)}{\delta_k\omega(\delta)}\right) = \frac{1}{2}\right\} \qquad (k \geqslant 0). \tag{2.2}$$

Set

$$\Omega(\omega, \delta) = 2^{1-k} \qquad (\delta \in (\delta_{k+1}, \delta_k], k = 0, 1, \ldots). \tag{2.3}$$

Then

$$\delta_{k+1}/\delta_k \leqslant 1/4 \qquad (k = 0, 1, \ldots), \tag{2.4}$$

$$\frac{1}{c}\omega(\delta) \leqslant \sum_{k=0}^{\infty} \omega(\delta_k) \min\left(1, \frac{\delta}{\delta_k}\right) \leqslant c\omega(\delta) \qquad (\delta \geqslant 0), \tag{2.5}$$

$$\Omega(\delta) \geqslant \max\left(\frac{\omega(\delta)}{\omega(\pi)}, \frac{\omega(\pi)\delta}{\omega(\delta)}\right) \qquad (\delta > 0). \tag{2.6}$$

At this juncture it will be useful for what follows to make a remark: considering the entire class of moduli of continuity with condition (2.1), the sequences $\Delta(\omega)$ defined by (2.2) are, to a significant degree, arbitrary, and, in general, do not process any additional properties beyond (2.4). This is clear from the following lemma.

LEMMA 2. *Let $\Delta = \{\delta_k\}_0^\infty$ be an arbitrary sequence of numbers satisfying the conditions*

$$\delta_0 = \pi, \quad \delta_k > 0, \quad 4\delta_{k+1}/\delta_k \leqslant 1 \qquad (k = 0, 1, \ldots). \tag{2.7}$$

Then there exists a modulus of continuity $\omega(\delta)$ such that $\Delta = \Delta(\omega)$, i.e. for all $k = 0, 1, \ldots$ the relations (2.2) *hold.*

PROOF. Fix some sequence $\theta = \{\theta_k\}_0^\infty$ consisting of zeros and ones, i.e. for any $k \geqslant 0$ either $\theta_k = 0$ or $\theta_k = 1$, and set

$$\omega(0) = 0, \quad \omega_0 = a, \quad \omega_k = \omega_k(\theta) = a2^{-k}\prod_{l=0}^{k-1}\left(4\frac{\delta_{l+1}}{\delta_l}\right)^{\theta_l} \qquad (k = 1, 2, \ldots), \tag{2.8}$$

$$\omega(\delta) = \omega(\theta, \delta) = \frac{(\delta - \delta_{k+1})\omega_k + (\delta_k - \delta)\omega_{k+1}}{\delta_k + \delta_{k+1}} \qquad (\delta \in (\delta_{k+1}, \delta_k], k \geqslant 0), \tag{2.9}$$

where a is an arbitrary positive constant.

We verify that $\omega(\delta)$ is a convex function on $(0, \pi]$ and, in particular, is a modulus of continuity; and that relations (2.2) hold. First of all, if for a given $k \geqslant 0$ we have $\theta_k = 0$, then

$$\omega(\delta_{k+1}) = 0.5\omega(\delta_k); \qquad \frac{\omega(\delta_{k+1})}{\delta_{k+1}} = 0.5\frac{\omega(\delta_k)}{\delta_k}\frac{\delta_k}{\delta_{k+1}} \geqslant 2\frac{\omega(\delta_k)}{\delta_k}, \tag{2.10}$$

while if $\theta_k = 1$, then

$$\frac{\omega(\delta_{k+1})}{\delta_{k+1}} = 2\frac{\omega(\delta_k)}{\delta_k}; \qquad \omega(\delta_{k+1}) = 2\omega(\delta_k)\frac{\delta_{k+1}}{\delta_k} \leqslant 0.5\omega(\delta_k). \tag{2.11}$$

From (2.10) and (2.11) it follows that to complete the proof of the lemma it remains only to verify that for any sequence θ the sequence

$$\frac{\omega_k - \omega_{k+1}}{\delta_k - \delta_{k+1}} = \frac{\omega(\delta_k)}{\delta_k}\frac{1 - 0.5(4\gamma_k)^{\theta_k}}{1 - \gamma_k} = \omega'(\delta) \qquad (\delta \in (\delta_{k+1}, \delta_k], k \geqslant 0)$$

is increasing; here we have set $\gamma_k = \delta_{k+1}/\delta_k$. From (2.10) and (2.11) it is clear that to do this it suffices to establish that

$$2\frac{1 - 0.5(4\gamma_{k+1})^{\theta_{k+1}}}{1 - \gamma_{k+1}} \geqslant \frac{1 - 0.5(4\gamma_k)^{\theta_k}}{1 - \gamma_k}.$$

But this relation is obviously satisfied since if $0 < \gamma < 0.25$ and $\theta = 0$ or $\theta = 1$, then

$$1 > \frac{1 - 0.5(4\gamma)^\theta}{1 - \gamma} \geqslant 0.5.$$

REMARK 1. From (2.8) it follows that if the set of natural numbers k for which the strict inequality $\delta_{k+1}/\delta_k < 1/4$ holds is infinite, then there exists an uncountable set of pairwise distinct moduli of continuity $\omega(\delta)$, normed, for example, by the condition $\omega(\pi) = 1$, and such that $\Delta = \{\delta_k\} = \delta(\omega)$.

REMARK 2. Suppose (cf. (1.9))

$$\beta(\omega, \delta) = \int_0^\delta \frac{\omega(t)}{t}dt + \delta\int_\delta^\pi \frac{\omega(t)}{t^2}dt,$$

and the sequence $\Delta(\omega) = \{\delta_k\}_0^\infty$ is defined by (2.2). The condition

$$\delta_k/\delta_{k+1} = O(1) \qquad (k \to \infty) \tag{2.12}$$

is necessary and sufficient for the validity of the estimate

$$\beta(\omega, \delta) = O(\omega(\delta)) \qquad (\delta \to 0). \tag{2.13}$$

Indeed, let $l \geqslant 0$ and $\delta \in (\delta_{l+1}, \delta_l]$; then by (2.2)

$$\beta(\omega, \delta) \leqslant 2\delta \sum_{k=0}^{l-1} \frac{\omega(\delta_{k+1})}{\delta_{k+1}} \log \frac{\delta_k}{\delta_{k+1}} + 2\omega(\delta) \log \frac{\delta_l}{\delta_{l+1}} + \omega(\delta) \log \frac{\delta_l}{\delta_{l+1}}$$

$$+ \sum_{k=l+1}^{\infty} \omega(\delta_k) \log \frac{\delta_k}{\delta_{k+1}} \leqslant c\left(\delta \frac{\omega(\delta_l)}{\delta_l} + \omega(\delta) + \omega(\delta_{l+1})\right) \sup_k \log \frac{\delta_k}{\delta_{k+1}},$$

whence, taking account of (2.5), we see that (2.13) follows from (2.12). On the other hand, it is obvious that for any $l \geqslant 0$

$$\beta(\omega, \delta) \geqslant \int_{\delta_{l+1}}^{\delta_l} \frac{\omega(t)}{t} dt, \qquad \beta(\omega, \delta_{l+1}) \geqslant \delta_{l+1} \int_{\delta_{l+1}}^{\delta_l} \frac{\omega(t)}{t^2} dt,$$

and since at least one of the relations

$$\text{a)} \quad \omega(\delta_{l+1}) = 0.5\omega(\delta_l) \quad \text{or} \quad \text{b)} \quad \omega(\delta_l)/\delta_l = 0.5\omega(\delta_{l+1})/\delta_{l+1} \tag{2.14}$$

holds, (2.12) follows from (2.13).

REMARK 3. Suppose the modulus of continuity $\omega(\delta)$ has the form (1.7). Then for sufficently large k the sequence $\{\delta_k\} = \Delta(\omega)$ satisfies

$$k^{-A} \leqslant \delta_{k+1}/\delta_k \leqslant k^{-B}, \tag{2.15}$$

where $A = A(\alpha)$ and $B = B(\alpha)$ are positive constants.

THEOREM 1. *Suppose the modulus of continuity $\omega(\delta)$ satisfies condition* (2.1) *and the function $\Omega(\omega, \delta)$ is defined by* (2.2) *and* (2.3). *Set*

$$w(\delta) = w(\omega, \delta) = \omega(\delta) \log\log 2/\Omega(\omega, \delta). \tag{2.16}$$

Then

$$w(\delta) \in W_C(\tilde{f}) \qquad (f \in H_\omega), \tag{2.17}$$

i.e. $w(\delta)$ is a modulus of C-continuity of the functions $\tilde{f}$ conjugate to functions from the class H_ω. Also

$$|\tilde{f}(x) - \tilde{f}(y)| \leqslant (C_f(x) + C_f(y)) w(|x-y|) \qquad (|x-y| \leqslant \pi), \tag{2.18}$$

where the exponential estimate

$$\operatorname{meas}\{x\colon C_f(x) > z\} \leqslant c \exp(-z/c) \qquad (z > 0). \tag{2.19}$$

holds for the distribution function of the coefficient $C_f(x)$.

PROOF. Let $0 < \delta \leqslant 2\pi$; let $g(x)$ be a summable 2π-periodic function; and let $s_\delta(g, x)$ and $h_\delta(g, x)$ denote respectively the Steklov mean and the Hilbert transform of the function g with parameter δ, i.e.

$$s_\delta(g, x) = \frac{1}{2\delta} \int_{|x-t| \leqslant \delta} g(t)\, dt, \qquad h_\delta(g, x) = \mathrm{PV} \int_{|x-t| \leqslant \delta} \frac{g(t)}{t - x} dt.$$

Further set

$$h_\delta^{(1)}(g, x) = \int_{2\pi \geqslant |x-t| \geqslant \delta} \frac{g(t) - g(x)}{(t-x)^2} dt, \tag{2.20}$$

and denote by $M(g, x)$ and $H(g, x)$ respectively the Hardy-Littlewood maximal function and the maximal Hilbert transform of the function g, i.e.

$$M(g, x) = \sup_{\delta \in (0, 2\pi]} s_\delta(|g|, x), \qquad H(g, x) = \sup_{\delta \in (0, 2\pi]} |h_\delta(g, x)| \tag{2.21}$$

(cf. [7], Chapter I, §13 and Chapter VII, §7).

It is obvious that for almost all x

$$|h_\delta(g,x)| \leqslant H(g,x), \qquad |h_\delta^{(1)}(g,x)| \leqslant c\delta^{-1}M(g,x). \tag{2.22}$$

Further, if $g(x)$ is absolutely continuous, then

$$h_\delta^{(1)}(g,x) = \frac{1}{\delta}\int_{|t-x|\leqslant\delta} g'(t)\,\mathrm{sgn}(t-x)\,dt + \int_{2\pi\geqslant|t-x|\geqslant\delta} \frac{g'(t)}{t-x}\,dt,$$

and hence, for absolutely continuous $g(x)$,

$$|h_\delta(g,x)| \leqslant c\delta M(g',x), \qquad |h_\delta^{(1)}(g,x)| \leqslant c(M(g',x) + H(g',x)). \tag{2.23}$$

The main point in the proof of Theorem 1 consists in using the exponential estimate of the distribution of the maximal Hilbert transform for bounded functions. This estimate was established by Carleson ([9], p. 139, Lemma 4): let $g(x)$ be an essentially bounded function and $\|g\|_\infty = \mathrm{ess\,sup}_x|g(x)|$. Then

$$\mathrm{meas}\{x\colon H(g,x) > z\|g\|_\infty\} \leqslant \exp(-z/c) \qquad (z>0). \tag{2.24}$$

Now suppose $f \in H_\omega$, $x, y \in (0, 2\pi]$, and $\delta = x - y > 0$. Applying transformations of a well-known type (cf., for example [7], Chapter III, §13) we obtain

$$|\pi(\tilde f(x) - \tilde f(y)) - (h_\delta(f,x) - h_\delta(f,y) + \delta h_\delta^{(1)}(f,x))| \leqslant c\omega(\delta). \tag{2.25}$$

To estimate h, we introduce the function $f_k(x)$, setting (cf. also (2.14))

$$f_k(x) = \begin{cases} s_{\delta_k}(f,x) & \text{if } \omega(\delta_{k+1}) = 0.5\omega(\delta_k) \\ s_{\delta_{k+1}}(f,x) & \text{if } \omega(\delta_{k+1}) < 0.5\omega(\delta_k). \end{cases}$$

These functions possess the following easily verifiable properties:

$$\|f - f_k\|_\infty \leqslant c\omega(\delta_{k+1}), \quad \|f_k'\|_\infty \leqslant c\omega(\delta_k)/\delta_k \qquad (f \in H_\omega). \tag{2.26}$$

Taking account of (2.22) and (2.23), we have

$$\begin{aligned} |h_\delta(f,x)| &\leqslant |h_\delta(f - f_l, x)| + |h_\delta(f_l, x)| \\ &\leqslant H(f - f_l, x) + c\delta M(f_l', x) \\ &\leqslant \|f - f_l\|_\infty \varphi_l(x) + c\delta\|f_l'\|_\infty, \\ \delta|h_\delta^{(1)}(f,x)| &\leqslant \delta|h_\delta^{(1)}(f - f_l, x)| + \delta|h_\delta^{(1)}(f_l, x)| \\ &\leqslant c(M(f - f_l, x) + \delta(M(f_l', x) + H(f_l', x))) \\ &\leqslant c(\|f - f_l\|_\infty + \delta\|f_l'\|_\infty \varphi_l^{(1)}(x)), \end{aligned}$$

where

$$\varphi_l(x) = H\left(\frac{f - f_l}{\|f - f_l\|_\infty}, x\right), \qquad \varphi_l^{(1)}(x) = H\left(\frac{f_l'}{\|f_l'\|_\infty}, x\right) + 1.$$

From this and (2.26) it follows that

$$|h_\delta(f,x)| \leqslant c\left(\omega(\delta_{l+1})\varphi_l(x) + \delta\frac{\omega(\delta_l)}{\delta_l}\right),$$

$$\delta|h_\delta^{(1)}(f,x)| \leqslant c\left(\omega(\delta_{l+1}) + \delta\frac{\omega(\delta_l)}{\delta_l}\varphi_l^{(1)}(x)\right),$$

and since $\delta_{l+1} < \delta \leqslant \delta_l$, by (2.5) we have

$$|\tilde{f}(x) - \tilde{f}(y)| \leqslant c\omega(\delta)(\psi_l(x) + \psi_l(y)), \tag{2.27}$$

where $\psi_l(x) = \max(\varphi_l(x), \varphi_l^{(1)}(x))$.

Suppose that

$$D_f(x) = \sup_{k \geqslant 0} \frac{\psi_k(x)}{\log(k+2)}.$$

In view of (2.24)

$$\begin{aligned} \operatorname{meas}\{x: D_f(x) > z\} &\leqslant \sum_{k=0}^{\infty} \operatorname{meas}\{x: \psi_k(x) > z\log(k+2)\} \\ &\leqslant c\sum_{k=0}^{\infty} (k+2)^{-z/c} \leqslant c\exp\left(-\frac{z}{c}\right) \qquad (z \geqslant z_0). \end{aligned}$$

From this and (2.27), taking account of the definition (2.3) of $\Omega(\omega, \delta)$, we obtain (2.18) and (2.19). The theorem is proved.

REMARK 4. It is not difficult to establish with the aid of the same reasoning the following estimate, which is sharper than (2.16)–(2.19), although more cumbersome: if $f \in H_\omega$ and $x, y \in [0, 2\pi]$ with $|x - y| = \delta$, $\delta_{l+1} < \delta \leqslant \delta_l$, then

$$\begin{aligned} |\tilde{f}(x) - \tilde{f}(y)| \leqslant c\Bigg(&\sum_{k=l+1}^{\infty} \omega(\delta_{k+1}) \min\left(\psi_k(x) + \psi_k(y), \log\frac{\delta_k}{\delta_{k+1}}\right) \\ &+ \omega(\delta_{l+1}) \min\left(\psi_l(x) + \psi_l(y), \log\frac{\delta}{\delta_{l+1}}\right) \\ &+ \delta\frac{\omega(\delta_l)}{\delta_l} \min\left(\psi_l(x), \log\frac{\delta_l}{\delta}\right) \\ &+ \delta\sum_{k=0}^{l-1} \frac{\omega(\delta_k)}{\delta_k} \min\left(\psi_k(x), \log\frac{\delta_k}{\delta_{k+1}}\right)\Bigg), \end{aligned} \tag{2.28}$$

where the following estimate holds for the distributions of the functions $\psi_k(x)$:

$$\operatorname{meas}\{x: \psi_k(x) > z\} \leqslant c\exp(-z/c) \qquad (z > 0, k = 0, 1, \dots). \tag{2.29}$$

It is clear that the uniform estimate (1.8), (1.9) also follows from (2.28), since for $0 < \delta \leqslant \pi$

$$\begin{aligned} \frac{1}{c}\beta(\omega, \delta) \leqslant &\sum_{k=l+1}^{\infty} \omega(\delta_{k+1}) \log\frac{\delta_k}{\delta_{k+1}} + \omega(\delta_{l+1}) \log e\frac{\delta}{\delta_{l+1}} \\ &+ \delta\frac{\omega(\delta_l)}{\delta_l} \log e\frac{\delta_l}{\delta} + \delta\sum_{k=0}^{l-1} \frac{\omega(\delta_k)}{\delta_k} \log\frac{\delta_k}{\delta_{k+1}} \leqslant c\beta(\omega, \delta). \end{aligned} \tag{2.30}$$

§3. On the sharpness of the estimates of the moduli of C-continuity

The following assertions shows that if condition (2.1) holds and

$$\beta(\omega, \delta) = \int_0^\delta \frac{\omega(t)}{t}\,dt + \delta\int_\delta^\pi \frac{\omega(t)}{t^2}\,dt \neq O(\omega(\delta)) \qquad (\delta \to 0), \tag{3.1}$$

i.e. the class of functions

$$\tilde{H}_\omega = \{g(x): g(x) + \tilde{f}(x), f \in H_\omega\}$$

is not imbedded in H_ω in the uniform sense, then this class is not imbedded in H_ω in the sense of moduli of C-continuity either (cf. (1.13)).

THEOREM 2. *Suppose conditions* (2.1) *and* (3.1) *hold for the modulus of continuity* $\omega(\delta)$. *Then there exists a function* $f \in H_\omega$ *such that* $\omega(\delta) \notin W_C(\tilde{f})$. *Moreover, for any measurable set* $F \subset (0, 2\pi]$ *of positive measure, the following relation holds at almost every point* $x \in F$:

$$\limsup_{y \to x, y \in F} \frac{|\tilde{f}(y) - \tilde{f}(x)|}{\omega(|y - x|)} = +\infty. \tag{3.2}$$

Further, as we noted in the Introduction, under additional conditions on the modulus of continuity $\omega(\delta)$ one can prove that the coefficient $\log\log(1/\Omega(\omega, \delta))$ is optimal with respect to order of growth as $\delta \to 0$. This can be seen from the following assertion.

THEOREM 3. *Suppose the sequence* $\Delta(\omega) = \{\delta_k\}_1^\infty$ *and the function* $\Omega(\omega, \delta)$ *are defined by* (2.2) *and* (2.3) *and that there exists an increasing sequence of natural numbers* $\{k_s\}_0^\infty$ *such that*

$$\log s = O\left(\log \frac{\delta_{d_s}}{\delta_{k_s+1}}\right), \quad k_s = O(\log s) \qquad (s \to \infty). \tag{3.3}$$

Then there exists a function $f(x) \in H_\omega$ *such that if*

$$w(\delta) = o\left(\omega(\delta) \log\log \frac{1}{\Omega(\omega, \delta)}\right) \qquad (\delta \to 0),$$

it follows that $w(\delta) \notin W_C(\tilde{f})$. *Moreover, for an arbitrary measurable set* F *of positive measure the following relation holds at almost every point* $x \in F$

$$\limsup_{y \to x, y \in F} \frac{|\tilde{f}(y) - \tilde{f}(x)|}{\omega(|y - x|) \log\log(2/\Omega(\omega, |y - x|))} > 0. \tag{3.4}$$

Before proving this theorem we remark that condition (3.3) is obviously satisfied if there exists a positive number A such that for all sufficiently large natural numbers k $\delta_{k+1}/\delta_k \leqslant k^{-A}$.

Therefore by Remark 3 (cf. (2.15)) to Theorem 3, in particular, the assertion made in the Introduction about the sharpness of the coefficient $\log\log(1/\delta)$ in (1.5) follows for the modulus of C-continuity of functions $\tilde{f}$ from the class $\tilde{H}_\omega$, where the modulus of continuity has the form (1.7). These classes are extremal in the sense that for them the coefficient $u(\delta)$ has the greatest possible growth.

We note also that in view of Lemma 2 it appears difficult to give any sufficiently simply formulated and at the same time sharp estimates of the moduli of C-continuity of functions $\tilde{f}$ ($f \in H_\omega$) for an arbitrary class H_ω without imposing any limitations on the modulus of continuity $\omega(\delta)$ (cf. also (2.28)).

PROOF OF THEOREMS 2 AND 3. We divide the set of all nonnegative whole numbers into two sequences K^a and K^b (cf. also (2.2) and (2.14); one of the sequences K^a or K^b may be empty) by setting

$$\begin{aligned} K^a &= \{k : \omega(\delta_{k+1}) = 0.5\omega(\delta_k)\}, \\ K^b &= \{k : \omega(\delta_{k+1}) < 0.5\omega(\delta_k), \omega(\delta_k)/\delta_k = 0.5\omega(\delta_{k+1})/\delta_{k+1}\}. \end{aligned} \tag{3.5}$$

We shall find the functions $f(x)$ whose existence is asserted in Theorems 2 and 3 in the form of sums of series in a suitably "smoothed" system of Rademacher type. To be precise let $[z]$ denote the integer part of the number $z \geqslant 0$, and put

$$r(x) = \frac{\pi}{4}\operatorname{sgn}\sin x, \quad n_k = [\pi/8\delta_k], \quad h_k = \pi/2n_k, \quad r_k(x) = r(n_k x),$$

$$g_k^a(x) = s_{h_{k+1}}(r_k, x), \qquad g_k^b(x) = s_4 l h_k(r_k, x - h_k). \tag{3.6}$$

We remark that in view of (2.4)

$$4\delta_k \leqslant h_k \leqslant 5\delta_k \qquad (k \geqslant 3). \tag{3.7}$$

Further, the uniform moduli of continuity of the functions g_k satisfy the estimates

$$\omega(g_k^a, \delta) \leqslant c\min(1, \delta/\delta_{k+1}); \qquad \omega(g_k^b, \delta) \leqslant c\min(1, \delta/\delta_k). \tag{3.8}$$

We note that (cf. [7], Chapter I, §2)

$$\tilde{r}(x) = \log\left|\cot\frac{x}{2}\right|. \tag{3.9}$$

We first estimate $\tilde{g}_k^a(x)$. We have

$$\tilde{g}_k^a(x) = \tilde{s}_{h_{k+1}}(r_k, x) = s_{h_{k+1}}(\tilde{r}_k, x) = s_{n_k h_{k+1}}(\tilde{r}, n_k x). \tag{3.10}$$

Suppose $k \geqslant 3$ and $x \in (\delta_{k+1}, \delta_k]$. Then taking account of (2.4) and (3.7), we see that

$$n_k(x + h_{k+1})/2 < \pi/4,$$

and so

$$\log\left|\cot\frac{t}{2}\right| \geqslant \log\frac{\delta_k}{x} - c \quad \text{when } |t - n_k x| \leqslant n_k h_{k+1}.$$

From this and from (3.9) and (3.10) it follows that

$$\tilde{g}_k^a(x) = \frac{1}{2n_k h_{k+1}}\int_{|t - n_k x| \leqslant n_k h_{k+1}} \log\left|\cot\frac{t}{2}\right| dt \geqslant \log\frac{\delta_k}{x} - c \qquad (x \in (\delta_{k+1}, \delta_k]). \tag{3.11}$$

On the other hand, if $y \in [2\delta_k, 4\delta_k]$, where $k \geqslant 4$, then, as is easy to see $|\tilde{g}_k^a(y)| \leqslant c$. From this estimate and from (3.11), using (3.10), we conclude that for the measure of the sets

$$G_k^a(z) = \left\{x\colon \min_{\delta \in (2\delta_k, 3\delta_k)} |\tilde{g}_k^a(x) - \tilde{g}_k^a(x + \delta)| > z\right\} \qquad (z > 0)$$

the following estimate holds:

$$m_k^a(z) = \operatorname{meas} G_k^a(z) \geqslant e^{-z-c} \qquad \left(0 < z \leqslant \log\frac{\delta_k}{\delta_{k+1}}\right). \tag{3.12}$$

We now estimate the differences of the functions $\tilde{g}_k^b(x)$. To do this we note that

$$h_k\, dg_k^b(x)/dx = r_k(x),$$

and consequently, if $x \in (\delta_{k+1}, \delta_k]$, $\delta \in (2\delta_{k+1}, 3\delta_{k+1})$, then

$$h_k\frac{d\tilde{g}_k^b(x)}{dx} = \log\left|\cot\frac{n_k x}{2}\right| \leqslant h_k\frac{d\tilde{g}_k^b(t)}{dt}$$

$$\leqslant h_k\frac{d\tilde{g}_k^b(x + \delta)}{dx} \leqslant -\log\frac{\delta_k}{x} - c \qquad (x \leqslant t \leqslant x + \delta).$$

Therefore

$$h_k\left(\tilde{g}_k^b(x) - \tilde{g}_k^b(x+\delta)\right) \geqslant \delta\left(\log\frac{\delta_k}{x} - c\right) \qquad \left(x \in (\delta_{k+1}, \delta_k], \delta \in (2\delta_{k+1}, 3\delta_{k+1}\right). \tag{3.13}$$

From (3.13) and (3.7) it follows that for the measures of the sets

$$G_k^b(z) = \left\{x: \min_{\delta\in(\delta_{k+1}, 3\delta_{k+1})} \delta^{-1}\left|\tilde{g}_k^b(x) - \tilde{g}_k^b(x+\delta)\right| > z/5\delta_k\right\}$$

an estimate analogous to (3.12) holds:

$$m_k^b(z) = \operatorname{meas} G_k^b(z) \geqslant e^{-z-c} \qquad \left(0 < z \leqslant \log\frac{\delta_k}{\delta_{k+1}}\right). \tag{3.14}$$

We now set (cf. (3.5))

$$f_k(x) = \omega(\delta_k)\begin{cases} g_k^a(x) & \text{if } k \in K^a, \\ g_k^b(x) & \text{if } k \in K^b. \end{cases}$$

Further, in the proof of Theorem 3 it will be convenient for us to assume that the sequence $K = \{k_s\}_1^\infty$ in the hypothesis of the theorem satisfies the relations

$$\text{a)}\quad \log\frac{\delta_{k_s}}{\delta_{k_s+1}} \geqslant \log s \quad\text{and}\quad \text{b)}\quad k_{s+1} - k_s \geqslant m \quad (s = 1, 2, \ldots) \tag{3.15}$$

where inequalities (3.15a) hold instead of the first of the estimates (3.3) and the second of estimates (3.3) continues to hold; m is a fixed sufficiently large natural number ($m \leqslant c$) whose actual value will be shown below (cf. (3.21) and (3.22)). It is easy to see that these assumptions do not restrict the generality. Now for the proof of Theorem 2 we choose and consider fixed in what follows a certain sequence $K = \{k_s\}_1^\infty$ with conditions (3.15)–such sequences do exist, since by (3.1) and Remark 2

$$\limsup_{k\to\infty} \frac{\delta_k}{\delta_{k+1}} = +\infty.$$

Suppose further that $\xi = \{\xi_k\}_{k\in K}$ is some sequence of real numbers and

$$f(x) = f_\xi(x) = \sum_{k\in K} f_k(x - \xi_k), \tag{3.16}$$

$$F_l(x) = \sum_{k\in K,\, k>l} f_k(x-\xi_k), \quad \Phi_l(x) = \sum_{k\in K,\, k<l} f_k(x-\xi_k) \qquad (l \in K), \tag{3.17}$$

where for a given set E and a given number ξ the shift of the set E by ξ is denoted by $E + \xi$, and

$$E + \xi = \{x: x = y + \xi, y \in G\};$$

$$G_\xi = \left\{\limsup_{k\to\infty,\, k\in K^a\cap K} G_k^a\right\} \cup \left\{\limsup_{k\to\infty,\, k\in K^a\cap K} G_k^0\right\}.$$

Since by (3.12), (3.14), (3.15a), and (3.17)

$$\sum_{k\in K^a\cap K} \operatorname{meas} G_k^a(z_k) + \sum_{k\in K^b\cap K} \operatorname{meas} G_k^b(z_k) \geqslant \sum_{k\in K} e^{-z_k-c} = e^{-c}\sum_{s=1}^\infty \frac{1}{s} = +\infty,$$

it follows from the Borel-Cantelli lemma (cf. [8], Chapter XIII, Lemma 1.24) that there exists a sequence of shifts ξ such that almost every point $x \in (0, 2\pi]$ belongs to G_ξ, i.e.

$$\operatorname{meas} G_\xi = 2\pi. \tag{3.18}$$

Throughout what follows we assume that the sequence ξ with property (3.18) is fixed and that the natural numbers k and l belong to the sequence K.

Further, in view of (2.5), (3.5), (3.8), and (3.15b) the following estimates hold:

$$\omega(f, \delta) \leqslant c\omega(\delta) \qquad (\delta > 0), \tag{3.19}$$

$$\|F_l\|_\infty \leqslant c2^{-m}\omega(\delta_{l+1}), \qquad \|\Phi_l'\|_\infty \leqslant c2^{-m}\omega(\delta_l)/\delta_l. \tag{3.20}$$

Let $\kappa = 1/40$, and

$$E_k = \left\{x\colon |\tilde{F}_k(x)| > \frac{\kappa}{2} z_k \omega(\delta_{k+1})\right\} \cup \left\{x\colon M(\tilde{\Phi}_k', x) > \frac{\kappa}{2} z_k \frac{\omega(\delta_k)}{\delta_k}\right\}.$$

Then by (3.20) and (2.24)

$$\operatorname{meas} E_k \leqslant c\exp(-\kappa z_k 2^m/c), \tag{3.21}$$

and, consequently (cf. (3.17)),

$$\sum_{k \in K} \operatorname{meas} E_k < \infty. \tag{3.22}$$

From (3.22) it follows that for the sequence of functions

$$\Psi_k(x) = \max\left(\frac{\tilde{F}_k(x)}{\kappa z_k \omega(\delta_k)}, \frac{M(\tilde{\Phi}_k', x)\delta_k}{\kappa z_k \omega(\delta_k)}\right)$$

the following estimate holds for almost all x:

$$\limsup_{k \to \infty} \Psi_k(x) \leqslant 0.5.$$

From this it follows that for an arbitrary number $\varepsilon > 0$ there exist a set E_ε and a number $k(\varepsilon) \in K$ such that

$$\operatorname{meas} E_\varepsilon \geqslant 2\pi - \varepsilon; \qquad \Psi_k(x) \leqslant 1 \qquad (x \in E_\varepsilon, k \geqslant k(\varepsilon)). \tag{3.23}$$

Now suppose that F is some set of positive measure, and for $\varepsilon < \operatorname{meas} F$ let F_ε denote the set of points of density of the intersection $E_\varepsilon \cap F \cap G$. In view of (3.23) and (3.18)

$$\operatorname{meas} F_\varepsilon \geqslant \operatorname{meas} F - \varepsilon, \tag{3.24}$$

but since ε is an arbitrary small positive number, to complete the proof of Theorems 2 and 3, respectively, it suffices to establish that (3.2) and (3.4) hold at each point $x \in F_\varepsilon$.

Thus, let us examine an arbitrary point $x \in F_\varepsilon$. Then by the definition of the set G at least one of the two sets $K^a \cap K$ or $K^b \cap K$ contains a subsequence L^a or L^b respectively tending to $+\infty$ and such that

$$\text{either a) } x \in G_l^a \quad (l \in L^a) \quad \text{or} \quad \text{b) } x \in G_l^b \quad (l \in L^b). \tag{3.25}$$

Suppose first that (3.25a) holds. Since x is a point of density of $E_\varepsilon \cap F \cap G$, for all sufficiently large $l \in L^a$ there exists a point $y = y_l$ such that

$$y \in (x + 2\delta_l, x + 3\delta_l) \cap E_\varepsilon \cap F. \tag{3.26}$$

But then by the definition of G_l^a and f_l^a

$$|\tilde{f}(x)-\tilde{f}(y)|>\omega(\delta_l)z_l-|\tilde{F}_l(x)|-|\tilde{F}_l(y)|-M(\tilde{\Phi}_l',x)|x-y|$$
$$\geqslant \omega(\delta_l)z_l\big(1-\kappa(\Psi_l(x)+\Psi_l(y))(\omega(\delta_{l+1})/\omega(\delta_l)+|x-y|/\delta_l)\big),$$

whence, taking account of (3.23) and (3.26) and the definition of κ ($\kappa=1/40$), we obtain that for $l\geqslant k(\varepsilon)$

$$|\tilde{f}(x)-\tilde{f}(y)|\geqslant 0.2\omega(|x-y|)z_l. \tag{3.27}$$

Suppose now that (3.25b) holds. Then for all sufficiently large $l\in L^b$ there exists a point $y=y_l$ such that

$$y\in(x+2\delta_{l+1},y+3\delta_{l+1})\cap F\cap E_\varepsilon. \tag{3.28}$$

Then by the definition G_l^b and f_l^b

$$|\tilde{f}(x)-\tilde{f}(y)|\geqslant\frac{\omega(\delta_l)}{5\delta_l}|x-y|z_l-|\tilde{F}_l(x)|-|\tilde{F}_l(y)|-M(\tilde{\Phi}_l',x)|x-y|$$
$$\geqslant\frac{\omega(\delta_l)}{5\delta_l}|x-y|z_l\left(1-5x(\Psi_l(x)+\Psi_l(y))\left(\frac{\omega(\delta_{l+1})}{|x-y|}\frac{\delta_l}{\omega(\delta_l)}+1\right)\right),$$

whence, taking account of (3.23), (3.28), and the definition of κ, we obtain that for $l\geqslant k(\varepsilon)$

$$|\tilde{f}(x)-\tilde{f}(y)|\geqslant 0.03\omega(|x-y|)z_l. \tag{3.29}$$

From (3.24), (3.27), and (3.29), taking account of (3.17), (3.26), and (3.23), we obtain (3.2), which completes the proof of Theorem 2. But to finish Theorem 3, it suffices to use the definition (2.2), (2.3) of the function $\Omega(\delta)$, relation (3.17), and the second of the estimates (3.3).

BIBLIOGRAPHY

1. N. N. Luzin, *The integral, and trigonometric series*, GITTL, Moscow, 1951. (Russian)

2. S. Nikolsky [S. M. Nikol'skiĭ], *La série de Fourier d'une fonction dont le module de continuité est donné*, C. R. (Dokl.) Acad. Sci. URSS **52** (1946), 191–194.

3. I. Priwaloff [I. I. Privalov], *Sur les fonctions conjuguées*, Bull. Soc. Math. France **44** (1916), 100–103.

4. N. K. Bari and S. B. Stechkin, *Best approximations and differential properties of two conjugate functions*, Trudy Moskov. Mat. Obshch. **5** (1956), 483–522. (Russian)

5. K. I. Oskolkov, *Uniform modulus of continuity of summable functions on sets of positive measure*, Dokl. Akad. Nauk SSSR **229** (1976), 304–306; English transl. in Soviet Math. Dokl. **17** (1976).

6. ______, *Approximation properties of summable functions on sets of full measure*, Mat. Sb. **103** (**145**) (1977), 563–589; English transl. in Math. USSR Sb. **32** (1977).

7. A. Zygmund, *Trigonometric series*, 2nd rev. ed., vol. 1, Cambridge Univ. Press, 1959.

8. ______, *Trigonometric series*, 2nd rev. ed., vol. 2, Cambridge Univ. Press, 1959.

9. Lennart Carleson, *On convergence and growth of partial sums of Fourier series*, Acta Math. **116** (1966), 135–157.

Translated by R. L. COOKE

Труды Матем. Инст. Стеклов.
164(1983)

Proceedings of the Steklov
Institute of Mathematics
1985, Issue 2

AVERAGED MODULI OF SMOOTHNESS FOR FUNCTIONS OF SEVERAL VARIABLES, AND THE FUNCTION SPACES GENERATED BY THEM

UDC 517.5

V. A. POPOV AND V. KH. KHRISTOV

ABSTRACT. Averaged moduli of smoothness are introduced for functions of several variables, and their properties are investigated. The function spaces generated by the averaged moduli are introduced, and various norms are proved to be equivalent. An imbedding theorem is formulated for these function spaces, and the connection with Besov spaces is explained.

Bibliography: 8 titles.

§1. Introduction

Let the points in m-dimensional Euclidean space $\mathbf{R}^m$ be denoted by x, y, t, h, $x = (x_1,\dots,x_m)$ with the norm $|x| = \max\{|x_i|\colon 1 \leqslant i \leqslant m\}$. If α is the multi-index $(\alpha_1,\dots,\alpha_m)$, the α_i being nonnegative integers, then we use the notation $|\alpha| = \sum_1^m|\alpha_i|$ and $D^\alpha = \partial^{|\alpha|}/\partial x_1^{\alpha_1} \cdots \partial x_m^{\alpha_n}$. The collection of 2π-periodic bounded measurable functions on $\mathbf{R}^m$ is denoted by M.

For any $x \in \mathbf{R}^m$ and any positive integer k the kth-order local modulus of smoothness at x is defined for a function $f \in M$ by the expression ($\delta \geqslant 0$)

$$\omega_k(f, x; \delta) = \sup\left\{\left|\Delta_h^k f(t)\right|\colon t, t + kh \in \Omega_{k\delta/2}(x)\right\}, \tag{1}$$

where

$$\Delta_h^k f(t) = \sum_{j=0}^{k} (-1)^{j+k}\binom{k}{j} f(t + jh) \quad \text{and} \quad \Omega_{k\delta/2}(x) = \{y\colon |x - y| \leqslant k\delta/2\}.$$

For an $f \in M$ the kth averaged modulus of smoothness in L_p ($1 \leqslant p \leqslant \infty$) is defined by the expression

$$\tau_k(f; \delta)_p = \|\omega_k(f, \cdot\,; \delta)\|_p, \tag{2}$$

where

$$\|g\|_p = \|g\|_{L_p(G)} = \left\{\int\int\cdots\int_G |g(x)|^p\,dx\right\}^{1/p}, \quad 1 \leqslant p < \infty,$$

$$\|g\|_\infty = \|g\|_{L_\infty(G)} = \sup\{|g(x)|\colon x \in G\},$$

and $G\{x\colon 0 \leqslant x_i \leqslant 2\pi, i = 1, 2, \dots, m\}$.

In the one-dimensional case the definitions (1) and (2) coincide with the definitions of the kth averaged modulus in [1]–[4] (see [1]–[3] about the history and about some applications of this modulus).

1980 *Mathematics Subject Classification*. Primary 46E35.

In §2 we investigate the properties of the modulus $\tau_k(f;\delta)_p$. In §3 we introduce the function space A^{θ}_{pq} generated by $\tau_k(f;\delta)_p$ and prove that various norms are equivalent. Some imbedding theorems for these spaces A^{θ}_{pq} are formulated in §4.

In our view the averaged moduli of smoothness $\tau_k(f;\delta)_p$ and the function spaces A^{θ}_{pq} they generate (see §3) can be used in the theory of error estimates in numerical methods for solving problems in mathematical physics, numerical integration, and other areas of numerical analysis.

§2. Properties of the averaged modulus

The modulus $\tau_k(f;\delta)_p$ satisfies the following inequalities (cf. [1]–[3] in the case $m-1$) $(f, g \in M)$:

$$\begin{aligned} &\tau_k(f;\delta)_p \leqslant \tau_k(f;\delta')_p \quad \text{if } \delta \leqslant \delta', \\ &\tau_k(f+g;\delta)_p \leqslant \tau_k(f;\delta)_p + \tau_k(g;\delta)_p, \\ &\tau_k(f;\delta)_p \leqslant 2\tau_{k-1}(f;k\delta/(k-1))_p. \end{aligned} \tag{3}$$

LEMMA 1. *If* $D^{\alpha}f \in M$, $\alpha_i = 0, 1,$, *then*

$$\tau_k(f;\delta)_p \leqslant \delta\sqrt{m} \sum_{\substack{\alpha,\,|\alpha|=1\\ \alpha_i=0,1}} \tau_{k-1}\left(D^{\alpha}f;\frac{k\delta}{k-1}\right)_p. \tag{4}$$

PROOF. Let t and h be such that $t, t+kh \in \Omega_{k\delta/2}(x)$. Consequently, $|h| \leqslant \delta$. Using the formula for increments of a function in the direction h (see [5], §4.6), we have that

$$\begin{aligned} \left|\Delta^k_h f(t)\right| &\leqslant \sqrt{m}\,|h| \left|\int_0^1 \Delta^{k-1}_h f'_h(t+uh)\,du\right| \leqslant \delta\sqrt{m}\int_0^1 \sum_{i=1}^m \left|\Delta^{k-1}_h f'_{x_i}(t+uh)\right| du \\ &\leqslant \delta\sqrt{m}\sum_{i=1}^m \sup\left\{\left|\Delta^{k-1}_h f'_{x_i}(t+uh)\right| : t, t+kh \in \Omega_{k\delta/2}(x), 0 \leqslant u \leqslant 1\right\} \\ &\leqslant \delta\sqrt{m}\sum_{i=1}^m \omega_{k-1}\left(f'_{x_i}, x; k\delta/(k-1)\right), \end{aligned}$$

i.e.,

$$\omega_k(f;x;\delta) \leqslant \delta\sqrt{m}\sum_{i=1}^m \omega_{k-1}\left(f'_{x_i}, x, k\delta/(k-1)\right).$$

The inequality (4) is obtained by taking the L_p-norm on both sides of the last inequality and using Minkowski's inequality on the right-hand side.

LEMMA 12 (see [6]).

$$\tau_k(f;\lambda\delta)_p \leqslant (2(\lambda+1))^{k+m}\tau_k(f;\delta)_p \qquad (\lambda \geqslant). \tag{5}$$

PROOF. We have that

$$\left|\Delta^k_{2nh}f(t)\right| \leqslant \sum_{i=0}^{(2n-1)k} A_i^{2n,k}\left|\Delta^k_h f(t+ih)\right|,$$

where the numbers $A_i^{2n,k}$ are determined from the equalities

$$(1+t+t^2+\cdots+t^{2n-1})^k = \sum_{i=0}^{(2n-1)k} A_i^{2n,k} t^i.$$

If t, $t+2nh \in \Omega_{kn\delta}(x)$, then for $0 \leqslant i \leqslant (2n-1)k$ the points $t+ih$, $t+ih+kh$ belong to some neighborhood of the form $\Omega_{k\delta/2}(y)$, where y is a point of the form $x+\sum_{s=1}^{m} j_s k\delta l_s/2$, with l_s the sth coordinate unit vector, and $|j_s| \leqslant n-1$, $s=1,2,\dots,m$. Consequently,

$$|\Delta_{2nh}^k f(t)| \leqslant \sum_{i=0}^{(2n-1)k} A_i^{2n,k} \sum_{j_1=1}^{n-1} \sum_{j_2=1-n}^{n-1} \cdots \sum_{j_m=1-n}^{n-1} \omega_k\left(f, x+\sum_{s=1}^{m} j_s \frac{k\delta}{2} l_s; \delta\right),$$

$$\omega_k(f, x; n\delta) \leqslant (2n)^k (2n-1)^m \omega_k\left(f, x+\sum_{s=1}^{m} j_s \frac{k\delta}{2} l_s; \delta\right),$$

$$\tau_k(f; n\delta)_p \leqslant (2n)^{k+m} \tau_k(f; \delta)_p.$$

The last inequality gives us (5), by (3).

LEMMA 3. *If $f \in M$ and $D^\alpha f \in L_p$ for all α such that $\alpha_i = 0$ or 1 for $i=1,\dots,m$, then*

$$\tau_1(f; \delta)_p \leqslant 2 \sum_{\substack{\alpha,\, |\alpha| \geqslant 1 \\ \alpha_i = 0,1}} \delta^{|\alpha|} \|D^\alpha f\|_p. \tag{6}$$

The proof is based on the following formula (proved by induction on the number of variables):

$$f(t) - f(x) = \sum_{\substack{\alpha,\, |\alpha| \geqslant 1 \\ \alpha_i = 0,1}} \int_{x^{(\alpha)}}^{t^{(\alpha)}} D^\alpha f(x^{(1-\alpha)} + u^{(\alpha)})\, du^{(\alpha)}, \tag{7}$$

where the following symbolic notation is employed (x, t, and u are m-dimensional vectors, and β is a multi-index with $\beta_i = 0$ or 1 for $i=1,\dots,m$):

$$(u^{(\beta)})_i = \begin{cases} u_i & \text{if } \beta_i = 1, \\ 0 & \text{if } \beta_i = 0, \end{cases}$$

$$\int_{x^{(\alpha)}}^{t^{(\alpha)}} f\, du^{(\alpha)} = \prod_{\beta_i = 1} \int_{x_i}^{t_i} f\, du_i, \qquad i = 1,2,\dots,m.$$

Formula (7) gives us that

$$\omega_1(f, x; \delta) \leqslant 2 \sum_{\alpha,\, |\alpha| \geqslant 1} \int_{x^{(\alpha)} - \delta^{(\alpha)}/2}^{x^{(\alpha)} + \delta^{(\alpha)}/2} |D^\alpha f(x^{(1-\alpha)} + u^{(\alpha)})| du^{(\alpha)}$$
$$= 2 \sum_{\alpha,\, |\alpha| \geqslant 1} \int_{-\delta^{(\alpha)}/2}^{\delta^{(\alpha)}/2} |D^{(\alpha)} f(x^{(\alpha)} + u^{(\alpha)})| du^{(\alpha)}.$$

We get (6) by taking the L_p-norm of both sides of the last inequality and using the generalized Minkowski inequalities (see [5], §1.3.1). Lemmas 1 and 3 at once imply

LEMMA 4. *If $f \in M$ and $D^\alpha f \in L_p$, $0 \leqslant \alpha_i \leqslant k$, then*

$$\tau_k(f; \delta)_p \leqslant c(k, m) \sum_{\substack{\alpha,\, |\alpha| \geqslant k \\ 0 \leqslant \alpha_i \leqslant k}} \delta^{(\alpha)} \|D^\alpha f\|_p.$$

LEMMA 5. *Suppose that f, f^+, $f^- \in M$ and $f^-(x) \leqslant f(x) \leqslant f^+(x)$ for all $x \in \mathbf{R}^m$. Then*

$$\tau_k(f; \delta)_p \leqslant \tau_k(f^+, \delta)_p + \tau_k(f^-; \delta)_0 + c(k)\{\tau_1(f^+ - f^-; \delta)_p + \|f^- f^-\|_p\}.$$

The last inequality was proved in [7] using different notation.

§3. The function spaces generated by the modulus $\tau_k(f;\delta)_p$

Let $1 \leqslant p \leqslant \infty$ and let T_N be the collection of trigonometric polynomials in m variables of order $\leqslant N$ in each variable. Then we define the best one-sided approximation of an $f \in M$ by trigonometric polynomials of order $0 \leqslant A < \infty$ in the L_p-metric to be

$$\tilde{E}_A(f)_p = \inf\{\|P_A - Q_A\|_p, P_A, Q_A \in T_{[A]}, Q_A(x) \leqslant f(x) \leqslant P_A(x)\}.$$

For $\theta > 0$, k a positive integer ($k > \theta$), and $1 \leqslant p, q \leqslant \infty$ we define the space ${}^iA^\theta_{pq}$ ($i = 1, 2, 3$) to be the collection of all functions $f \in M$ such that the corresponding norm ${}^i\|f\|$ is finite.

Let ($a > 1$)

$${}^1\|f\|_{A^\theta_{pq}(\mathbf{R}^m)} = {}^1\|f\|_p + \left(\int_0^\infty \left(t^{-\theta}\tau_k(f;t)_p\right)^q \frac{dt}{t}\right)^{1/q},$$

$${}^2\|f\|_{A^\theta_{pq}(\mathbf{R}^m)} = {}^2\|f\| = \|f\|_p + \tilde{E}_0(f)_p + \left(\sum_{l=0}^\infty \left(a^{\theta l}\tilde{E}_{a^l}(f)_p\right)^q\right)^{1/q},$$

$${}^3\|f\|_{A^\theta_{pq}(\mathbf{R}^m)} = {}^3\|f\| = \inf\max\left\{\|P_0\|_p + \left(\sum_{l=0}^\infty \left(a^{\theta l}\|P_{a^l}\|_p\right)^q\right)^{1/q},\right.$$
$$\left.\|Q_0\|_p + \left(\sum_{l=0}^\infty \left(a^{\theta l}\|Q_{a^l}\|_p\right)^q\right)^{1/q}\right\},$$

where the infimum is taken over all representations of f (in the L_p-sense) in the form

$$f(x) = P_0 + \sum_{l=0}^\infty P_{a^l}(x) = Q_0 + \sum_{l=0}^\infty Q_{a^l}(x), \tag{8}$$

where $P_0, Q_0 \in T_0$ and $P_{a^l}, Q_{a^l} \in T_{[a^l]}$, are such that

$$Q_0 + \sum_{l=0}^N Q_{a^l}(x) \leqslant f(x) \leqslant P_0 + \sum_{l=0}^N P_{a^l}(x) \tag{9}$$

for all $x \in \mathbf{R}^m$ and N is a positive integer.

THEOREM 1. *The norms* ${}^i\|\ \|_{A^\theta_{pq}}$, $i = 1, 2, 3$, *are equivalent for 2π-periodic bounded functions and for given $\theta > 0$, ($k > \theta$), $1 \leqslant p, q \leqslant \infty$, and $a > 1$.*

PROOF. With use of the direct and converse theorems obtained in [8] for one-sided approximations of periodic functions by trigonometric polynomials in $\mathbf{R}^m$, the proof that ${}^1\|\ \|$ and ${}^2\|\ \|$ are equivalent for arbitrary m is analogous to the proof in [7] that they are equivalent for $m = 1$.

We prove that if $f \in M$ and ${}^2\|f\|_{A^\theta_{pq}} < \infty$,, then ${}^3\|f\|_{A^\theta_{pq}} \leqslant c\,{}^2\|f\|_{A^\theta_{pq}}$ (see [5], §5.6). Suppose that the trigonometric polynomials $g^-_{a^l}(x)$ and $g^{a_l+}(x) \in T_{[a^l]}$, $l = 0, 1, \ldots$, are such that $g^-_{a^l}(x) \leqslant f(x) \leqslant g^+_{a^l}(x)$ and $\|g^+_{a^l} - g^-_{a^l}\|_p \leqslant 2\tilde{E}_{a^l}(f)_p$. Let

$$P_{a^l} = g^+_{a^l} - g^+_{a^{l-1}}, \qquad Q_{a^l} = g^-_{a^l} - g^-_{a^{l-1}},\ l = 1, 2, \ldots,$$
$$P_1(x) = g^+_1(x) - \sup_t f(t), \quad Q_1(x) = g^-_1(x) - \inf_t f(t),$$
$$P_0 = \sup\{f(t) : t \in \mathbf{R}^m\}, \qquad Q_0 = \inf\{f(t) : t \in \mathbf{R}^m\}.$$

Since the finiteness of ${}^2\|f\|$ implies that $\tilde{E}_{a^l}(f)_p \to 0$ ($l \to \infty$), (8) and (9) are satisfied in the L_p-sense for any x and N.

Moreover,

$$\|Q_0\|_p = \left\|\inf_t f(t)\right\|_p \leqslant \|f\|_p + \left\|f - \inf_t f(t)\right\|_p \leqslant \|f\|_p + \tilde{E}_0(f)_p,$$

$$\|Q_1\|_p = \left\|g_1^- - \inf_t f(t)\right\|_p \leqslant \|g_1^- - f\|_p + \left\|f - \inf_t f(t)\right\|_p$$
$$\leqslant 2\tilde{E}_1(f)_p + \tilde{E}_0(f)_p \leqslant 3\tilde{E}_0(f)_p,$$

$$\|Q_{a^l}\|_p \leqslant \|g_{a^l}^- - f\|_p + \|g_{a^{l-1}}^- - f\|_p \leqslant \|g_{a^l}^- - g_{a^l}^+\|_p + \|G_{a^{l-1}}^- - g_{a^{l-1}}^+\|_p$$
$$\leqslant 4\tilde{E}_{a^{l-1}}(f)_p, \qquad l = 1, 2, \ldots$$

Analogous inequalities are valid also for the polynomials P_0, P_1, and P_{a^l}. Consequently,

$$^3\|f\|_{A^\theta_{pq}} \leqslant c\left\{\|f\|_p + \tilde{E}_0(f)_p + \left(3\tilde{E}_0^q(f)_p + \sum_{l=1}^{\infty}\left(a^{\theta l}\tilde{E}_{a^{l-1}}(f)_p\right)^q\right)^{1/q}\right\}$$
$$\leqslant c\left\{\|f\|_p + \tilde{E}_a(f)_p + \left(\sum_{l=0}^{\infty}\left(a^{\theta l}\tilde{E}_{a^l}(f)_p\right)^q\right)^{1/q}\right\} = {}^2\|f\|_{A^\theta_{pq}}.$$

Let us prove that if $f \in M$ and $^3\|f\|_{A^\theta_{pq}} < \infty$, then $^1\|f\| \leqslant c^3\|f\|$. First note that if

$$f_N^+(x) = P_0 + \sum_{l=0}^{N} P_{a^l}(x) \quad \text{and} \quad f_N^-(x) = Q_0 + \sum_{l=0}^{N} Q_{a^l}(x) \qquad \left(f_N^+, f_N^- \in T_{[a^N]}\right),$$

then we have from Lemma 5, Lemma 4, and Bernstein's inequality that (cf. the analogous inequality in [5], §5.6)

$$\tau_k(f; a^{-N})_p \leqslant \tau_k\left(f_N^+; a^{-N}\right)_p + \tau_k\left(f_N^-; a^{-N}\right)_p$$
$$+ c(k)\left\{\tau_1\left(f_N^+ - f_N^-; a^{-N}\right)_p + \|f_N^+ - f_N^-\|_p\right\}$$
$$\leqslant c(k)\left\{\sum_{\substack{\beta, |\beta| \geqslant k \\ 0 \leqslant \beta \leqslant k}} a^{-N|\beta|}\left(\|D^\beta f_N^+\|_p + \|D^\beta f_N^-\|_p\right)\right.$$
$$\left. + \sum_{\substack{\alpha, |\alpha| \geqslant 1 \\ \alpha_1 = 0, 1}} a^{-N|\alpha|}\|D^\alpha(f_N^+ - f_N^-)\|_p + \|f_N^+ - f_N^-\|_p\right\}$$
$$\leqslant c(k)\left\{\sum_{\beta, |\beta| = k} a^{-Nk}\left(\|D^\beta f_N^+\|_p + \|D^\beta f_N^-\|_p\right) + \|f_N^+ - f_N^-\|_p\right\}$$
$$\leqslant c(k)\left\{a^{-Nk}\sum_{l=0}^{N} a^{lk}\left(\|P_{a^l}\|_p + \|Q_{a^l}\|_p\right) + \sum_{l=N+1}^{\infty}\left(\|P_{a^l}\|_p + \|Q_{a^l}\|_p\right)\right\}.$$

By using the last estimate the inequality $^1\|f\| \leqslant c^3\|f\|$ can be proved according to the scheme of proof of the analogous inequality between the norms $^1\|\ \|$ and $^6\|\ \|$ for a Besov space (see [5], §5.6). It is only necessary to observe (see [7]) that for $t \geqslant 4\pi$

$$\tau_k(f; t) = \tau_k(f; 4\pi)_p \leqslant c(k)\tilde{E}_0(f)_p.$$

§4. Imbedding theorems for the spaces A^θ_{pq}

We can obtain some imbedding theorems for the spaces A^θ_{pq} by starting from the fact that the space A^θ_{pq} has an equivalent polynomial norm $^3\|\ \|$ and by using inequalities connecting the norms of polynomials in different metrics and in spaces with different dimensions (see [5], §8.10.4). These theorems are analogous to the imbedding theorems for Besov spaces. For example:

THEOREM 2. (cf. [5], §6.3). *If* $\theta > 0$, $1 \leqslant p, q < \infty$, $p \leqslant p' \leqslant \infty$ *and* $\theta' = \theta - m(1/p - 1/p') > 0$, *then*

$$A^{\theta}_{pq}(\mathbf{R}^m) \subset A^{\theta'}_{p'q}(\mathbf{R}^m),$$

i.e., if $f \in A^{\theta}_{pq}(\mathbf{R}^m)$, *then* $f \in A^{\theta'}_{p'q}$ *and* $\|f\|_{A^{\theta'}_{p'q}} \leqslant c\|f\|_{A^{\theta}_{pq}}$.

The proof is analogous to that of the corresponding assertion for Besov spaces ([5], §6.3).

If $B^{\theta}_{pq}(\mathbf{R}^m)$ denotes the Besov space with parameters $\theta > 0$, $1 \leqslant p, q \leqslant \infty$, then the spaces A^{θ}_{pq} and B^{θ}_{pq} are connected as follows:

THEOREM 3. *If* $\theta > m/p$, *then* $A^{\theta}_{pq}(\mathbf{R}^m) = B^{\theta}_{pq}(\mathbf{R}^m)$ (*with equivalent norms*).

A proof is given for $m = 1$ in [7]. For arbitrary m a proof can be obtained either by interpolation between A^{θ}_{1q} and $A^{\theta}_{\theta q}$ or by the connections between best approximations and best one-sided approximations of a function f in the case when there exist fractional derivatives of a given order α (the α_i can also be nonintegers). In view of its awkwardness, we shall present the proof later.

BIBLIOGRAPHY

[1] A. Andreev, V. A. Popov and Bl. Sendov, *Theorems of Jackson type for best one-sided approximations by trigonometric polynomials and splines*, Mat. Zametki **26** (1979), 791–804; English transl. in Math. Notes **26** (1979).

[2] V. A. Popov, *Converse theorem for the onesided trigonometrical approximations*, C. R. Acad. Bulgare Sci. **30** (1977), 1529–1532.

[3] ______, *Direct and converse theorems for oneside approximation*, Linear Spaces and Approximation (Proc. Conf., Oberwolfach, 1977), Internat. Schriftenreihe Numer. Math., Vol. 40, Birkhäuser, 1980, pp. 449–458.

[4] V. A. Popov and A. Andreev, *Stechkin's type theorems for one-sided trigonometrical and spline approximation*, C. R. Acad. Bulgare Sci. **34** (1978), 151–154.

[5] S. M. Nikol'skiĭ, *Approximation of functions of several variables and imbedding theorems*, 2nd rev. aug. ed., "Nauka", Moscow, 1977; English transl. of 1st ed., Springer-Verlag, 1974.

[6] A. S. Andreev, V. A. Popov and Bl. Sendov, *Error estimates in numerical solution of ordinary differential equations*, Zh. Vychisl. Mat. i Mat. Fiz. **21** (1981), 635–650; English transl. in USSR Comput. Math. and Math. Phys. **21** (1981).

[7] V. A. Popov, *Function spaces, generated by the averaged moduli of smoothness*, PLISKA Stud. Math. Bulgar. **5** (1983), 132–143.

[8] ______, *One-sided approximation of periodic functions of several variables*, C. R. Acad. Bulgare Sci. **35** (1982), 1639–1642.

Translated by H. H. McFADEN

Труды Матем. Инст. Стеклов.
164(1983)

Proceedings of the Steklov
Institute of Mathematics
1985, Issue 2

AN ANALOGUE OF A. A. MARKOV'S INEQUALITY. APPLICATION TO INTERPOLATION AND FOURIER SERIES

UDC 517.518

A. A. PRIVALOV

ABSTRACT. An analgoue of A. A. Markov's inequality is proved by the use of sets of positive capacity. A class of perfect compact sets F is introduced, for each of which there are a matrix $\mathfrak{M}$ of interpolation knots and a weight function σ such that both the Lagrange interpolation process formed for f by means of this matrix, and the Fourier series of f in the system of polynomials orthogonal on F with weight function σ, converge uniformly to f on F, provided that f is continuous, $f \in C(F)$. It is shown that the capacity of every compact set in the class considered here is zero.

Bibliography: 22 titles.

Introduction

Let F be an infinite closed set, $F \subset [0; 1]$; let $C(F)$ be the space of real functions f that are continuous on $[0; 1]$ with the seminorm

$$\|f\|_{C(F)} = \max_{x \in F} |f(x)|,$$

and let H_n $(n = 0, 1, 2, \dots)$ be the set of all polynomials of degree at most n with real coefficients.

Let $\mathfrak{M} = \{x_{kn}\}$, $0 \leqslant x_{1,n} < x_{2,n} < \cdots < x_{n,n} \leqslant 1$, $n = 1, 2, 3, \dots$, be a matrix of interpolation knots. For every $f \in C(F)$ and every $n \in \mathbf{N}$, set

$$\mathscr{L}_n(\mathfrak{M}, f, x) = \sum_{k=1}^{n} f(x_{kn}) l_{kn}(\mathfrak{M}, x), \tag{1}$$

$$l_{kn}(\mathfrak{M}, x) = \frac{\omega_n(x)}{\omega_n(x_{kn})(x - x_{kn})}, \qquad \omega_n(x) = \prod_{i=1}^{n} (x - x_{in}).$$

Denote by

$$\mathscr{L}_n(\mathfrak{M}, F) = \|\mathscr{L}_n(\mathfrak{M}, x)\|_{C(F)}, \qquad \mathscr{L}_n(\mathfrak{M}, x) = \sum_{k=1}^{n} |l_{kn}(\mathfrak{M}, x)| \tag{2}$$

the norm of the Lagrange operator (1) acting from $C[0; 1]$ to $C(F)$, and introduce the numbers

$$\mathscr{L}(\mathfrak{M}, F) = \varlimsup_{n \to \infty} \mathscr{L}_n(\mathfrak{M}, F), \qquad \mathscr{L}(F) = \inf_{\mathfrak{M}} \mathscr{L}(\mathfrak{M}, F), \tag{3}$$

which play a fundamental role in interpolation theory.

1980 *Mathematics Subject Classification*. Primary 26D05, 26D10, 41A05, 42C10, 42C15; Secondary 31A15.

Indeed, if $\mathscr{L}(F) < \infty$, there is a matrix $\mathfrak{M}_0$ such that $\mathscr{L}(\mathfrak{M}_0, F) < \infty$, and therefore, by the Banach-Steinhaus theorem, the sequence $\{\mathscr{L}_n(\mathfrak{M}_0, f, x)\}$ of Lagrange interpolation polynomials converges uniformly on F to f for every $f \in C(F)$; but if $\mathscr{L}(F) = \infty$, then for every matrix $\mathfrak{M}$ there is an $f \in F$ such that

$$\varlimsup_{n\to\infty} \|\mathscr{L}_n(\mathfrak{M}, f, x)\|_{C(F)} = \infty.$$

From the point of view of interpolation theory it is natural to separate the closed sets into two classes K and K^c, putting the closed sets F for which $\mathscr{L}(F) < \infty$ into class K, and all others into K^c.

That K is not empty was shown by Korovkin [1] and Mergelyan [2], and it follows from Faber [3], Bernstein [4] and Erdös [5] that if F contains an interval or has positive Lebesgue measure then $F \in K^c$, so that if $F \in K$ then meas $F = 0$. We also note that there are closed sets of measure zero that belong to K^c (see [6] and [7]).

Now let Σ_F be the set of functions $\sigma(x)$, nondecreasing on $[0;\ 1]$, whose restrictions to F have infinitely many points of increase. Take any $\sigma \in \Sigma_F$; orthogonalizing, by the Gram-Schmidt process, the set of powers $\{x^n\}_0^\infty$, we obtain a system of polynomials $\{P_n(\sigma, x)\}_0^\infty$ orthonormal on F with weight function σ: $\int_F P_m(\sigma, t)P_n(\sigma, t)\, d\sigma(t) = \delta_{mn}$, $m, n = 1, 2, 3, \ldots$, where δ_{mn} is the Kronecker symbol (see [8], Chapter VIII, §§6, 7 and §2,s Theorem 3). Now every function $f \in C(F)$ can be formally expanded in a Fourier series with respect to the system of polynomials that are orthogonal on F with weight function σ, i.e.

$$f(\dot{x}) \sim S_\sigma(f, x) = \sum_{k=0}^{\infty} a_k P_k(\sigma, x), \tag{4}$$

where

$$a_k = \int_F f(t) P_k(\sigma, t)\, d\sigma(t), \qquad k = 0, 1, 2, \ldots.$$

By analogy with interpolation we separate the closed sets $F \subset [0;\ 1]$ into two classes S and S^c, putting into S the closed sets F for each of which there is a function $\sigma \in \Sigma_F$ such that the Fourier series $S_\sigma(f, x)$ of every function $f \in C(F)$ converges to it, uniformly on F, and putting all other sets into S^c.

A fundamental problem of the theory of approximation of functions on sets is to find criteria for sets to belong to these classes.

In this paper we single out the class $\mathfrak{N}$ of infinite closed subsets of K and S, and obtain metric characterizations of the closed sets $F \in \mathfrak{N}$ in terms of capacity (see [9], Chapter V, §2 or [10], Chapter VII, §3).

In solving the problem of the metric characterization of sets in $\mathfrak{N}$, as in many other problems of approximation theory, there is special value in inequalities in which the derivative of a polynomial is estimated in terms of the values of the polynomial. Consequently in the first part of the paper we investigate the behavior of

$$\varphi(n, F) = \sup_{P_n \in H_n,\ \|P_n\|_{C(F)} \neq 0} \frac{\|P_n'(x)\|_{C(F)}}{\|P_n(x)\|_{C(F)}} \tag{5}$$

and for sets F of positive capacity [9], [10] we show that

$$\lim_{n\to\infty} \sqrt[n]{\varphi(n, F)} = 1.$$

In the second part we find conditions on F under which it belongs to the class $\mathfrak{N} \subset K \cap S$. Moreover, for sets $F \in \mathfrak{N}$ we establish the inequalities $\varphi(n, F) \geqslant q^n$, where $q > 1$ and is independent of n, and by the same token show that the capacity of every set $F \in \mathfrak{N}$ is zero.

§1. Analogue of A. A. Markov's inequality

Markov's inequality [11], in its classical form, is an inequality for the derivative of a polynomial on an interval $[a; b]$ if we know the maximum of its modulus of this interval: for every $n \in \mathbf{N}$, the inequality is

$$\sup_{P_n \in H_n,\, \|P_n\|_{C[a;\, b]} \neq 0} \frac{(b-a)\|P_n'(x)\|_{C[a;\, b]}}{2n^2\|P_n(x)\|_{C[a;\, b]}} = 1. \tag{6}$$

In 1916, I. I. Privalov [12]–[14] generalized the statement of the problem of estimating the derivative of a trigonometric polynomial, given a bound for the polynomial not on the whole interval $[-\pi, \pi]$ but only on some subset of positive measure (see also [15]). As Stechkin has remarked, it is natural to generalize Markov's problem in the same direction, by considering polynomials on any infinite closed set F.

It can be shown that in the general case, under the transition from $[a, b]$ to a set of positive measure, and replacement of $(b - a)$ in (6) by the measure of F, the equation (6) is not preserved. Hence the problem must be formulated as follows: if F is an infinite closed set, what condition on a positive nondecreasing function $\varphi(n, F)$ with integral argument will guarantee that

$$\sup_{P_n \in H_n,\, \|P_n\|_{C[a;\, b]} \neq 0} \frac{\|P_n'(x)\|_{C(F)}}{\varphi(n, F)\|P_n(x)\|_{C(F)}} = 1? \tag{7}$$

We shall investigate the function $\varphi(n, F)$ under the assumptions that the closed set F is perfect, of positive capacity, and without irregular points: all points of F are to be regular in Wiener's sense [16]. A point $x \in F$ is said to be regular if the series $\sum 2^n \Delta(F_n)$ diverges, where $\Delta(F_n)$ is the capacity, or equivalently the transfinite diameter, of the set F_n, the intersection of F with the interval

$$1/2^n \leqslant |t - x| \leqslant 1/2^{n-1}.$$

This condition indicates, first, that the complement F^c with respect to the extended complex plane is connected and regular, i.e. that F^c has a Green function $G(z)$ with logarithmic pole at infinity, and therefore if $H(z)$ is the harmonic conjugate of $G(z)$ in F^c then the function $w = \varphi(z) = \exp\{G(z) + iH(z)\}$ maps F^c conformally, but not necessarily one-to-one, on the exterior $|w| > 1$ of the unit disk, with $\varphi(\infty) = \infty$; second, if $R > 1$ and $\mathscr{L}_R$ is a level line of F^c, $\mathscr{L}_R = \{z \in \mathbf{C}\colon\ G(z) = \ln R$ or $|\varphi(z)| = R\}$, and if the distance $\rho(A, B)$ between the sets A and B is defined by

$$\rho(A, B) = \inf_{z \in A,\, t \in B} |z - t|, \tag{8}$$

then the distance $\rho(\mathcal{L}_R, F)$ from the level line $\mathcal{L}_R$ to the closed set F is a strictly positive function of R and F. This is equivalent to F's being of positive capacity and reducible, i.e. if the closure of every nonempty portion of F (the intersection of F with an interval) has positive capacity, then F contains no irregular points (see Keldysh [17], also [9] and [18]).

THEOREM 1.1. *If the perfect set F is of positive capacity and reducible, then*

$$\lim_{n\to\infty} \sqrt[n]{\varphi(n, F)} = 1, \tag{9}$$

where $\varphi(n, F)$ is defined by (5).

THEOREM 1.2. *If the positive nondecreasing function $\varphi(n)$ on the integers satisfies* (9) *and*

$$1/\varphi(n) = O(1/n^2), \tag{10}$$

then there is a reducible perfect set F of positive capacity such that

$$\overline{\lim_{n\to\infty}} \sup_{P_n \in H_n,\, \|P_n\|_{C(F)} \neq 0} \frac{\|P_n'(x)\|_{C(F)}}{\varphi(n)\|P_n(x)\|_{C(F)}} > 0. \tag{11}$$

Before we verify these propositions, we introduce some concepts that will be needed later, and prove some lemmas.

If $x_0, x_1, \dots, x_n$ are $n+1$ points of F, the number

$$V(x_0, x_1, x_2, \dots, x_n) = \prod_{0 \leqslant k < l \leqslant n} (x_l - x_k), \qquad n \geqslant 1,$$

is equal to the Vandermonde determinant of $x_0, x_1, \dots, x_n$ and depends continuously on its arguments. From this and the closure of F there follows the existence of $V_{n+1}(F)$, the maximum modulus of $|V(x_0, x_1, \dots, x_n)|$, when $x_0, x_1, \dots, x_n$ run through all sets of $n+1$ points of F. Let

$$t_{0,n}, t_{1,n}, t_{2,n}, \dots, t_{n,n} \tag{12}$$

denote points of F such that

$$V_{n+1}(F) = |F(t_{0,n}, t_{1,n}, t_{2,n}, \dots, t_{n,n})|. \tag{13}$$

Let $T_n(z, F) = z^n + a_1^{(n)} z^{n-1} + \cdots + a_n^{(n)}$ be the polynomial deviating least from zero on F among all polynomials $P_n(z) = z^n + b_1^{(n)} z^{n-1} + \cdots + b_n^{(n)}$ that have zeros only on F, and

$$m_n(F) = \max_F |T_n(z, F)|.$$

As Fekete [19] showed, the limit

$$\Delta(F) = \lim_{n\to\infty} \sqrt[n]{m_n(F)} \tag{14}$$

exists. This limit is equal to the capacity of F.

LEMMA 1. *If F is an infinite set, the following inequalities are satisfied for all $n \in \mathbf{N}$, $n \geqslant 2$:*

$$m_n(F) \leqslant V_{n+1}(F)/V_n(F) \leqslant (n+1)m_n(F). \tag{15}$$

For a proof of (15) see Goluzin [10] (Chapter VII, §1, inequality (8)).

LEMMA 2. *Let F be a perfect reducible set of positive capacity. Then if $P_n(z)$ is a polynomial of degree at most n, we have, for points $z \in F^c$,*

$$|P_n(z)| \leqslant \|P_n\|_{C(F)} |\varphi(z)|^n, \tag{16}$$

and for points z lying on or inside the level line $\mathscr{L}_R$, $R > 1$, we have

$$|P_n(z)| \leqslant \|P_n\|_{C(F)} R^n. \tag{17}$$

Inequality (17) was proved by Bernstein [20] when F is an interval, and by Walsh ([18], p. 77) in the general case.

LEMMA 3. *Let F be a perfect reducible set of positive capacity, $\Delta(F) > 0$. Then every sequence $\{P_n(z)\}$ of polynomials that satisfies*

$$\overline{\lim_{n\to\infty}} \|P_n\|_{C(F)}^{1/n} \leqslant \Delta(F), \tag{18}$$

also satisfies

$$\overline{\lim_{n\to\infty}} \|P_n'\|_{C(F)}^{1/n} \leqslant \Delta(F). \tag{19}$$

PROOF. Consider a decreasing sequence $\{R_k\}$ of real numbers with limit 1. Choose any $k \in \mathbf{N}$ and let $\mathscr{L}_{R_k} = \{z \in \mathbf{C}: |\varphi(z)| = R_k\}$, a level line of F^c. Then by Lemma 2 (see (17)) we have, on and inside $\mathscr{L}_{R_k}$, for every polynomial $P_n \in H_n$, the inequality

$$|P_n(z)| \leqslant \|P_n\|_{C(F)} R_k^n. \tag{20}$$

Denote by ρ_k the distance from the level line $\mathscr{L}_{R_k}$ to F. It is evident that $\rho_k > 0$. For every point $x \in F$ the disk $|\xi - x| \leqslant \rho_k$ is a subset of the closed domain bounded by $\mathscr{L}_{R_k}$. Hence we obtain, by (20) and Cauchy's integral formula,

$$|P_n'(x)| = \left| \frac{1}{2\pi i} \int_{|\xi - x| = \rho_k} \frac{P_n(\xi)\, d\xi}{(\xi - x)^2} \right| \leqslant \frac{\|P_n\|_{C(F)} R_k^n}{\rho_k}$$

and

$$\|P_n'\|_{C(F)}^{1/n} \leqslant \|P_n\|_{C(F)}^{1/n} R_k \rho_k^{-1/n}.$$

By (18) we have, for every $k \in \mathbf{N}$,

$$\overline{\lim_{n\to\infty}} \|P_n'\|_{C(F)}^{1/n} \leqslant \Delta(F) R_k.$$

Letting $k \to \infty$, we obtain (19).

This completes the proof of the lemma.

LEMMA 4 (see [18], §7.7). *Let F be a perfect reducible set of positive capacity, $\Delta(F) > 0$; let $P_n(z, F)$ be the polynomial $\prod_{k=0}^n (z - t_{k,n})$ with zeros at the points* (12). *Then*

$$\lim_{n\to\infty} \left(\|P_n(z, F)\|_{C(F)}\right)^{1/n+1} = \Delta(F). \tag{21}$$

PROOF OF THEOREM 1.1. Choose any polynomial $P_n(x) \in H_n$ with $\|P_n\|_{C(F)} \neq 0$. The Lagrange interpolation formula with knots at $t_{0,n}, t_{1,n}, \dots, t_{n,n}$ (see (12)) and the recurrence relation for the Vandermonde determinant let us write the identity

$$\begin{aligned} P_n(x) &= \sum_{k=0}^{n} P_n(t_{kn}) \frac{V(t_{0,n}, \dots, t_{k-1,n}, x, t_{k+1,n}, \dots, t_{n,n})}{V(t_{0,n}, t_{1,n}, \dots, t_{n,n})} \\ &= \sum_{k=0}^{n} P_n(t_{kn})(-1)^{n-k} \frac{V(t_{0,n}, \dots, t_{k-1,n}, t_{k+1,n}, \dots, t_{n,n}) P_{kn}(x, F)}{V(t_{0,n}, t_{1,n}, \dots, t_{n,n})}, \end{aligned} \tag{22}$$

where

$$P_{kn}(x, F) = \prod_{\substack{i=0 \\ i \neq k}}^{n} (x - t_{i,n}).$$

Differentiating (22) with respect to x and using (13), we obtain

$$\|P'_{n(x)}\|_{C(F)} \leqslant \|P_n(x)\|_{C(F)} \frac{(n+1)V_n(F) \max_{0 \leqslant k \leqslant n} \|P'_{kn}(x, F)\|_{C(F)}}{V_{n+1}(F)}. \tag{23}$$

For $n \geqslant 1$ we set

$$\psi(n, F) = \frac{(n+1)V_n(F) \max_{0 \leqslant k \leqslant n} \|P'_{kn}(x, F)\|_{C(F)}}{V_{n+1}(F)}, \tag{24}$$

and $\psi(0, F) = 1$ for $n = 0$.

We may suppose without loss of generality that $\psi(n, F)$ is a nondecreasing function of n; if not, we set

$$\psi(n, F) = \sup_{1 \leqslant m \leqslant n} \frac{(n+1)V_m(F) \max_{0 \leqslant k \leqslant m} \|P'_{km}(x, F)\|_{C(F)}}{V_{m+1}(F)}.$$

By (23) and the definition (24) of $\psi(n, F)$, we have, for every polynomial $P_n \in H_n$ with $\|P_n\|_{C(F)} \neq 0$,

$$\|P'_n\|_{C(F)} / \|P_n\|_{C(F)} \leqslant \psi(n, F),$$

or, equivalently (see (5)), $\varphi(n, F) \leqslant \psi(n, F)$. From this and the inequality

$$\|P'_n(x)\|_{C(F)} = \|P_1(x)\|_{C(F)}$$

for the polynomial $P_1(x) = x + 1 - \beta$, where $\beta = \sup F$, it follows that

$$1 \leqslant \varphi(n, F) \leqslant \psi(n, F). \tag{25}$$

Moreover by (13) we have, for every integer k, $0 \leqslant k \leqslant n$, and every point $x \in F$,

$$|P_{kn}(x, F)| \cdot |V(t_{0,n}, \dots, t_{k-1,n}, t_{k+1,n}, \dots, t_{n,n})| \leqslant V_{n,+1}(F).$$

The inequality becomes an equality for $x = t_{k_n}$, and consequently

$$\|P_{kn}(x, F)\|_{C(F)} = \prod_{\substack{i=0 \\ i \neq k}}^{n} |t_{kn} - t_{in}| = |P'_n(x, F)|_{x = t_{kn}},$$

where $P_n(x, F)$ is a polynomial of degree $n + 1$ whose zeros are the points (12).

Applying Lemmas 4 and 3, we deduce the relations

$$\varlimsup_{n\to\infty} \max_{0\leqslant k\leqslant n} \|P_{kn}(x, F)\|_{C(F)}^{1/(n+1)} = \varlimsup_{n\to\infty} \max_{0\leqslant k\leqslant n} |P_n'(t_{kn}, F)|^{1/(n+1)} \leqslant \Delta(F).$$

Again applying Lemma 3, and using the continuity of the power functions, we have

$$\varlimsup_{n\to\infty} \max_{0\leqslant k\leqslant n} \|P_{kn}'(x, F)\|_{C(F)}^{1/n} \leqslant \Delta(F).$$

Hence by (14) and (15), and the definition (24) of ψ, we obtain

$$\begin{aligned} \varlimsup_{n\to\infty} \sqrt[n]{\psi(n, F)} &= \varlimsup_{n\to\infty} \left\{ \frac{V_n(F)\max_{0\leqslant k\leqslant n}\|P_{kn}'(x, F)\|_{C(F)}}{(n+1)^{-1}V_{n+1}(F)} \right\}^{1/n} \\ &\leqslant \varlimsup_{n\to\infty} \left\{ \frac{\max_{0\leqslant k\leqslant n}\|P_{kn}'(x, F)\|_{C(F)}}{(n+1)^{-1}m_n(F)} \right\}^{1/n} = 1. \qquad (26) \end{aligned}$$

Then (9) follows from (25) and (26).

This completes the proof of Theorem 1.1.

We are going to need the Cantor perfect sets defined by the following simple construction. Let $\{p_n\}$, $n = 0, 1, 2, \ldots$, be a nondecreasing sequence of real numbers, with $p_0 = 1$ and $p_n > 1$ for $n \geqslant 1$. From the unit interval $\delta = [0; 1]$ we delete an interval of length $(\operatorname{meas}\delta)(1 - 1/p_1) = 1 - 1/p_1$ so that the two remaining intervals δ_{11} and δ_{12} have lengths $(\operatorname{meas}\delta)/2p_1 = 1/p_1$. We say for short that δ_{11} and δ_{12} are obtained from δ by the operation (p_1). Let E_1 denote the set consisting of δ_{11} and δ_{12}. Applying the operation (p_2) to δ_{11} and δ_{12}, we obtain a subset E_2 of E_1 that consists of four congruent intervals, the sum of whose lengths is $1/p_1p_2$. Continuing this process indefinitely, we define an infinite sequence $\{E_n\}$, $n = 0, 1, 2, \ldots, E_0 = \delta$, of nested sets E_n, each consisting of 2^n congruent disjoint intervals, the sum of whose lengths is $1/p_0p_1 \cdots p_n$. The intersection of the sets E_n gives us the perfect set $F = F(p_0p_1\ldots)$ that we shall need.

LEMMA 5. *The Cantor perfect set $F = F(p_0p_1p_2\ldots)$ has positive capacity if and only if the series*

$$\sum_{\nu=1}^{\infty} \frac{\ln p_\nu}{2^\nu} < \infty$$

(see [9], Chapter V, §6, Theorem 3 and the proposition on German p. 117, English p. 124).

PROOF OF THEOREM 1.2. If a positive nondecreasing function $\varphi(n)$ on the integers satisfies conditions (9) and (10), there are two possible cases: the function $\varphi_1(n) = \varphi(n)n^{-2}$ is bounded, or $\varphi_1(n)$ is unbounded. In the first case the validity of (11) follows from the fact that for every interval $[a; b]$, $a < b$, we have (6), for which the extremal function is the Tchebycheff polynomial of degree n, which deviates least from zero on $[a; b]$. In the second case we have

$$\varphi(n) = \exp\{o(n)\}. \qquad (27)$$

Choose any convergent series $\sum_1^\infty a_k$ with strictly decreasing positive terms, such that $a_n \geqslant n^{-2}$, $n = 1, 2, \ldots$. Set

$$p_n = \exp\{2^n a_n\}, \qquad n = 1, 2, 3, \ldots \tag{28}$$

We may suppose without loss of generality that $p_n \geqslant 2^{n+1}$, $n = 1, 2, \ldots$. We construct the Cantor perfect set $F = F(p_0 p_1 \cdots)$ with these numbers p_n. By Lemma 5, the set F has positive capacity and is reducible.

On the set F we consider the following function:

$$\binom{m}{2}^{-1} \sum_{0 \leqslant i < j \leqslant m-1} \ln \frac{1}{|x_i - x_j|};$$

it attains its minimum at the points (12): $t_{0,m-1}, t_{1,m-1}, \ldots, t_{m-1,m-1}$. It is easy to see that since the p_n increase and satisfy the inequalities $p_n \geqslant 2^{n+1}$, $n = 1, 2, \ldots$, then for every $m = 2^{n+1}$ each component of E_n contains at least one point $t_{k,m-1}$, $0 \leqslant k \leqslant m - 1$, $m = 2^{n+1}$. On the other hand we have, by (12), (13), and the fact that the Lagrange polynomials are fundamental,

$$\left\| \frac{P_m(x, F)}{P'_m(t_{k,m-1})(x - t_{k,m-1})} \right\|_{C(F)} = 1, \qquad k = 0, 1, 2, \ldots, m-1, \tag{29}$$

where the $P_m(x, F)$ are the polynomials from Lemma 4.

Now let $x_0 \in F$ be a point where the polynomial $P_m(x, F)$, of degree $m = 2^{n+1}$, attains its absolute maximum on F, and let i be the index such that

$$\min_{0 \leqslant k \leqslant m-1} |x_0 - t_{k,m-1}| = |x_0 - t_{i,m-1}|.$$

Then (see (28))

$$|x_0 - t_{i,m-1}| \leqslant \frac{1}{2^n p_0 p_1 p_2 \cdots p_n} = \exp\left\{-2^n \sum_{k=1}^n \frac{a_k}{2^{n-k}} - n \ln 2\right\}$$

and consequently from this and (29) we obtain

$$\frac{\|P_m(x, F)\|_{C(F)}}{\|P'_m(x, F)\|_{C(F)}} \geqslant \exp\left\{2^n \sum_{k=1}^n \frac{a_k}{2^{n-k}} + n \ln 2\right\}. \tag{30}$$

But by the choice of $\sum a_k$ and the fact that $\varphi(n)$ satisfies (27), it is always possible to satisfy the inequalities

$$\varphi(2^{n+1}) \leqslant \exp\left\{2^n \sum_{k=1}^n \frac{a_k}{2^{n-k}} + n \ln 2\right\}, \qquad n \geqslant n_0.$$

Observing also that

$$\lim_{n \to \infty} \left(\exp\left\{2^n \sum_{k=1}^n \frac{a_k}{2^{n-k}} + n \ln 2\right\}\right)^{1/2^{n+1}} = 1,$$

we find from (30) that (11) is also valid in the case when $\varphi_1(n) = \varphi(n) n^{-2}$ is unbounded.

This completes the proof of Theorem 1.2.

REMARK 1. V. I. Belyĭ conjectured in conversation that the distance from our set F to the level line $\mathscr{L}_R$, $R = 1/n$, never exceeds c/n^2, where the positive constant c depends only on F. We will show that in general Belyĭ's conjecture is not correct. Let

$\rho(\mathcal{L}_R, x)$ be the distance from $x \in F$ to the level line $\mathcal{L}_R$ and let $\rho(\mathcal{L}_R, f) = \inf_{x\in F}\rho(\mathcal{L}_R, x)$. Since $\varphi(n, F)$ is defined by (5) and the inequality

$$|P_n'(x)| = \left|\frac{1}{2\pi i}\int_{|\xi - x| = \rho(\mathcal{L}_R, x)} \frac{P_n(\xi)\,d\xi}{(\xi - x)^2}\right| \leqslant \frac{\|P_n\|_{C(F)}R^n}{\rho(\mathcal{L}_R, x)} \leqslant \frac{\|P_n\|_{C(F)}R^n}{\rho(\mathcal{L}_R, F)},$$

holds for every polynomial $P_n \in H_n$ and every point $x \in F$, by Lemma 2 (see (17)), we have

$$\rho(\mathcal{L}_R, F) \leqslant R^n/\varphi(n, F).$$

But for every $\varphi(n) > 0$ that satisfies (9) there is a set F, of positive capacity and reducible, such that $\varphi(n, F) \geqslant \varphi(n)$ for $n = n_1, n_2, \dots$ (Theorem 1.2), and consequently

$$\rho(\mathcal{L}_R, F) \leqslant 3\varphi(n), \qquad R = 1 + 1/n, n = n_1, n_2, n_3, \dots.$$

REMARK 2. It is evident that the conclusions of Theorems 1.1 and 1.2 are valid for compact subsets F of the complex plane provided that F and F^c have the same boundary.

§2. Interpolation and Fourier series

We shall find conditions on a set $F \subset [0; 1]$ that guarantee that it belongs to class K or S.

THEOREM 2.1. *Let there be given two sequences $\{\varepsilon_n\}$ and $\{\eta_n\}$ of positive numbers such that*

$$\eta_n \leqslant 1/n, \qquad \varepsilon_n = O\{(n\eta_n/6)^n\}, \tag{31}$$

and let the closed set $F \subset [0; 1]$ be representable, for every $n \geqslant n_0$, as the union of n disjoint nonempty closed sets $\Phi_1^{(n)}, \dots, \Phi_n^{(n)}$, each of diameter at most ε_n, and such that

$$\min_{1\leqslant i\neq j\leqslant n} \rho\left(\Phi_i^{(n)}, \Phi_j^{(n)}\right) \geqslant \eta_n, \tag{32}$$

where the distance ρ between two sets is defined by (8). *Then $F \in K$.*

PROOF. Let the closed set F satisfy the hypotheses of the theorem. Choose any integer $n \geqslant n_0$ and represent F in the form

$$F = \bigcup_{i=1}^{n} \Phi_i^{(n)}, \quad \Phi_i^{(n)} \neq \varnothing, \quad \Phi_i^{(n)} \cap \Phi_j^{(n)} = \varnothing, \qquad 1 \leqslant i \neq j \leqslant n,$$

where the distance between two of the component closed sets, $\Phi_i^{(n)}$ and $\Phi_j^{(n)}$, $i \neq j$, is at least η_n and the diameter of each is at most ε_n. From each set $\Phi_i^{(n)}$ we select arbitrarily a point x_{in}, $1 \leqslant i \leqslant n$, to form the nth row of a matrix $\mathfrak{M}$. We have thus defined all the rows of $\mathfrak{M}$ with index $n \geqslant n_0$. We determine the remaining $n_0 - 1$ rows of $\mathfrak{M}$ arbitrarily from F. Let us show that this matrix $\mathfrak{M}$ of interpolation knots is the matrix of a Lagrange interpolation process that converges uniformly on F; that is, $\mathcal{L}(\mathfrak{M}, f) < \infty$ (see (3)). Choose any point $x \in F$ and integer $n \geqslant n_0$. Then the nth row contains a knot x_{in} such that

$$|x - x_{n}| \leqslant \varepsilon_n. \tag{33}$$

Also observe that by (32) and the construction of the nth row of $\mathfrak{M}$ we have

$$\min_{1\leqslant i\leqslant n-1}|x_{i+1,n}-x_{i,n}|\geqslant\eta_n.$$

It is easy to verify the inequalities

$$\min_{1\leqslant k\leqslant n}\left|\prod_{i=1,i\neq k}^{n}(x_{kn}-x_{in})\right|\geqslant(k-1)!(n-k)!\eta_n^{n-1}>\frac{(n-1)!\eta_n^{n-1}}{2^{n-1}}.\qquad(34)$$

From (31), (33), (34), the Lagrange formula, the Markov inequality (6), and Stirling's formula, we have, for every continuous function f,

$$\begin{aligned}|f(x)-\mathscr{L}_n(\mathfrak{M},f,x)|&\leqslant|\mathscr{L}_n(\mathfrak{M},f,x)-\mathscr{L}_n(\mathfrak{M},f,x_{in})|+|f(x_{in})-f(x)|\\&\leqslant\sum_{k=1}^{n}|f(x_{kn})|\,|l_{kn}(\mathfrak{M},x)-l_{kn}(\mathfrak{M},x_{in})|+\omega(f,\varepsilon_n)\\&\leqslant\|f\|_{C(F)}n(n-1)^2\varepsilon_n\frac{2^n}{(n-1)!\eta_n^{n-1}}+\omega(f,\varepsilon_n)\\&=O(n^{7/2}3^{-n}e^n\eta_n)+\omega(f,\varepsilon_n)=O(\eta_n)+\omega(f,\varepsilon_n),\end{aligned}\qquad(35)$$

where $\omega(f,\delta)$ is the modulus of continuity of f on [0; 1]. Thus if $n\geqslant n_0$ we have

$$\|f(x)-\mathscr{L}_n(\mathfrak{M},f,x)\|_{C(F)}=O(\eta_n)+\omega(f,\varepsilon_n),$$

i.e. the sequence $\{\mathscr{L}_n(\mathfrak{M},f,x)\}$ of Lagrange interpolation polynomials converges uniformly to $f\in C(F)$ on F, and consequently $F\in K$.

This completes the proof of the theorem.

REMARK 1. We call the component closed sets $\Phi_i^{(n)}$, $1\leqslant i\leqslant n$, $n\geqslant n_0$, in the hypothesis of Theorem 2.1 *sets of rank n*. We may suppose without loss of generality that each component closed set $\Phi_j^{(n+1)}$, $1\leqslant j\leqslant n+1$, of rank $n+1$ is contained in a set of rank n. From this and the construction of $\mathfrak{M}$ that guarantees the uniform convergence of the Lagrange interpolation process on F for every function $f\in C(F)$, it follows that $\mathfrak{M}$ can be constructed so that each successive row is obtained from its predecessor by adding a new knot. Therefore if we observe that the numbering of the rows has no effect on the behavior of the interpolation process, we may set

$$x_{k,n+1}=x_{k,n},\qquad 1\leqslant k\leqslant n,\ n=n_0,n_0+1,n_0+2,\dots.$$

REMARK 2. Let the polynomial $P_{n-1}(x)\in H_{n-1}$ of degree $\leqslant n-1$ have absolute value no greater than M at the knots x_{kn}, $1\leqslant k\leqslant n$, of the nth row of $\mathfrak{M}$, as constructed in Theorem 2.1 or Remark 1; then for every $x\in F$ there is a knot x_{in} such that $|x-x_{in}|\leqslant\varepsilon_n$, and therefore (cf. (35))

$$\begin{aligned}|P_{n-1}(x)-P_{n-1}(x_{in})|&=|\mathscr{L}_n(\mathfrak{M},P_{n-1},x)-\mathscr{L}_n(\mathfrak{M},P_{n-1},x_{in})|\\&\leqslant\sum_{k=1}^{n}|P_{n-1}(x_{kn})|\,|l_{kn}(\mathfrak{M},x)-l_{kn}(\mathfrak{M},x_{in})|\\&=MO(\eta_n).\end{aligned}\qquad(36)$$

Here $O(\eta_n)$ is a function, the same for all polynomials of degree at most $n-1$, that tends to zero as $n\to\infty$.

THEOREM 2.2. *Let the closed set F satisfy the hypotheses of Theorem* 2.1. *Then there are a function $\sigma \in \Sigma_F$ and a system of polynomials $\{P_n(\sigma, x)\}_0^\infty$, orthogonal on F with weight function σ, such that the Fourier series* (4) *in this system of every continuous function $f \in C(F)$ converges uniformly on F to f.*

PROOF. By Remark 1 to Theorem 2.1 there exists a matrix $\mathfrak{M}$ such that $x_{k,n+1} = x_{k,n}$, $n = n_0, n_0 + 1, n_0 + 2, \ldots$, for which the sequence $\{\mathscr{L}_n(\mathfrak{M}, f, x)\}$ of Lagrange interpolation polynomials converges uniformly to f on F, provided only that $f \in C(F)$. Set $x_{11} = x_{1,n_0}, x_{12} = x_{11}, x_{22} = x_{2,n_0}, \ldots, x_{n_0-1,n_0-1} = x_{n_0-1,n_0}$. Then we have defined all the rows of $\mathfrak{M}$. Without loss of generality we may suppose that the point $x = 0 \in F$ and that $x_{11} = 0$. Consider the function

$$\sigma(x) = \sum_{x_{kk}^k < x} \{\sigma(x_{kk} + 0) - \sigma(x_{kk} - 0)\}, \qquad \sigma(0) = 1/2, 0 \leqslant x \leqslant 1,$$

which has jumps at the points x_{kk}, $k = 1, 2, \ldots$, of amount $d_k = \sigma(x_{kk} + 0) - \sigma(x_{kk} - 0) = (1/2)^k$. It is clear that $\sigma \in \Sigma_F$. Let $\{P_n(\sigma, x)\}$ be the system of polynomials that are orthogonal on F with weight function σ, obtained by orthogonalizing $\{x^n\}_0^\infty$ by the Gram-Schmidt method, and let

$$S_{\sigma,n}(f, x) = \sum_{k=0}^{n-1} a_k P_k(\sigma, x), \qquad a_k = \int_F f(t) P_k(\sigma, t)\, d\sigma(t)$$

be the partial sums of the Fourier series of f in the system $\{P_n(\sigma, x)\}$. By Toepler's theorem, the properties of $\mathfrak{M}$, the definition of σ, and (35), we have, for every $f \in C(F)$,

$$\int_F |f(x) - S_{\sigma,n}(f, x)|^2\, d\sigma(x) \leqslant \int_F |f(x) - \mathscr{L}_n(\mathfrak{M}, f, x)|^2\, d\sigma(x)$$
$$\leqslant 2^{-n}\{O(\eta_n) + \omega(f, \varepsilon_n)\}^2. \tag{37}$$

Suppose the contrary, namely that there is a function $f \in C(F)$ with $\|f\|_{C(F)} = 1$, whose Fourier series (4) does not converge uniformly on F. Then there are a positive number ε and an infinite sequence $\{n_i\}$ of integers such that

$$\left\|S_{\sigma,n_i}(f, x) - f(x)\right\|_{C(F)} \geqslant \varepsilon, \qquad i = 1, 2, 3, \ldots \tag{38}$$

Let us show that in this case we can find for every i a knot x_{k_i,n_i}, $1 \leqslant k_i \leqslant n_i$, among the knots x_{k,n_i}, $1 \leqslant k \leqslant n_i$, of the ith row of $\mathfrak{M}$, such that

$$\left|S_{\sigma,n_i}(f, x_{k_i,n_i}) - f(x_{k_i,n_i})\right| \geqslant \frac{\varepsilon}{2\mathscr{L}_{n_i}(\mathfrak{M}, F)}, \qquad i = 1, 2, 3, \ldots \tag{39}$$

In fact, if (39) were not valid, then all the knots in the ith row of $\mathfrak{M}$ would satisfy

$$\left|S_{\sigma,n}(f, x_{k,n_i})\right| \leqslant \|f\|_{C(F)} + \frac{\varepsilon}{2\mathscr{L}_{n_i}(\mathfrak{M}, F)}, \qquad 1 \leqslant k \leqslant n_i. \tag{40}$$

Hence by Remark 2 to Theorem 2.1 (see (36)) and the fact that each polynomial $P_k(\sigma, x)$, $0 \leqslant k \leqslant n - 1$, is a linear combination of $1, x, x^2, \ldots, x^k$, $0 \leqslant k \leqslant n - 1$, i.e. that $S_{\sigma,n_i}(f, x)$ is a polynomial of degree at most $n_i - 1$, we have

$$\left|S_{\sigma,n_i}(f, x) - S_{\sigma,n_i}(f, x_{k,n_i})\right| = \left(\|f\|_{C(F)} + \frac{\varepsilon}{2\mathscr{L}_{n_i}(\mathfrak{M}, F)}\right) O(\eta_n) \tag{41}$$

everywhere on F. Here x_{k,n_i} satisfies $|x - x_{k,n_i}| \leqslant \varepsilon_{n_i}$. Let x_0 be determined by the equation

$$\|f(x) - S_{\sigma,n_i}(f, x)\|_{C(F)} = |f(x_0) - S_{\sigma,n_i}(f, x_0)|$$

and let x_{k,n_i} be the knot in the n_ith row of $\mathfrak{M}$, closest x_0. Then by (40), (41), and the assumption that (39) is false, we obtain

$$\begin{aligned}\|f(x) - S_{\sigma,n_i}(f, x)\|_{C(F)} &\leqslant |S_{\sigma,n_i}(f, x_0) - S_{\sigma,n_i}(f, x_{k,n_i})| \\ &\quad + |f(x_0) - f(x_{k,n_i})| + |f(x_{k,n_i}) - S_{\sigma,n_i}(f, x_{k,n_i})| \\ &= \left(\|f\|_{C(F)} + \varepsilon/2\mathscr{L}_{n_i}(\mathfrak{M}, F)\right)O(\eta_{n_i}) \\ &\quad + \varepsilon/2\mathscr{L}_{n_i}(\mathfrak{M}, F) + \omega(f, \varepsilon_{n_i}).\end{aligned}$$

But $\omega(f, \varepsilon_{n_i}) \to 0$ and $O(\eta_{n_i}) \to 0$ as $i \to \infty$. Moreover, $\mathscr{L}_{n_i}(\mathfrak{M}, F) \geqslant 1$. Therefore

$$\|f(x) - S_{\sigma,n_i}(f, x)\|_{C(F)} < \varepsilon, \qquad i \geqslant i_0. \tag{42}$$

Inequality (42) contradicts (38), and therefore (39) is valid for some k_i, $1 \leqslant k_i \leqslant n_i$, $i \geqslant i_0$. Now we have

$$\begin{aligned}\int_F |f(x) - S_{\sigma,n_i}(f, x)|^2 d\sigma(x) &\geqslant |f(x_{k_i,n_i}) - S_{\sigma,n_i}(f, x_{k_i,n_i})|^2 \\ &\quad \times \left[\sigma(x_{k_i,n_i} + 0) - \sigma(x_{k_i,n_i} - 0)\right] \\ &= |f(x_{k_i,n_i}) - S_{\sigma,n_i}(f, x_{k_i,n_i})|^2 \left[\sigma(x_{k_i,k_i} - 0)\right] \\ &\geqslant \frac{\varepsilon^2}{4\mathscr{L}_{n_i}^2(\mathfrak{M}, F)} \frac{1}{2^{k_i}} \geqslant \frac{\varepsilon^2}{2^{n_i+2}\mathscr{L}_{n_i}^2(\mathfrak{M}, F)}.\end{aligned}$$

Combining this inequality with (37), we obtain

$$\varepsilon^2/4\mathscr{L}_{n_i}^2(\mathfrak{M}, F) \leqslant \left\{O(\eta_n) + \omega(f, \varepsilon_{n_i})\right\}^2. \tag{43}$$

By the Banach-Steinhaus theorem, $\sup \mathscr{L}_n(\mathfrak{M}, F) < \infty$, and $O(\eta_{n_i}) + \omega(f, \varepsilon_{n_i}) \to 0$ as $i \to \infty$. Hence it follows that (43) is false for sufficiently large i, and consequently (38) does not hold for large i.

This completes the proof of the theorem.

REMARK. The method of proof of Theorems 2.1 and 2.2 is a development of ideas of Korovkin [1], [21] and Mergelyan [2].

THEOREM 2.3. *Let a closed set F satisfy the hypotheses of Theorem* 2.1. *Then there are a number $q > 1$ and a sequence of polynomials $\{P_n\}_1^\infty$, $P_n \in H_n$, such that*

$$\|P_n'(x)\|_{C(F)} > q^n \|P_n(x)\|_{C(F)}, \qquad n \geqslant n_0.$$

In fact, let the matrix $\mathfrak{M}$ of interpolation knots be as in Theorem 2.1, and let the point $x_n \in F$ be such that

$$\|\omega_n(x)\|_{C(F)} = \left\|\prod_{i=1}^n (x - x_{in})\right\|_{C(F)} = |\omega_n(x_n)|.$$

Let x_{kn} be the knot closest to x_n. Then $0 < |x_n - x_{kn}| \leqslant \varepsilon_n$, $n > n_0$. On the other hand, by the Banach-Steinhaus theorem we have

$$\frac{\|\omega_n(x)\|_{C(F)}}{|\omega_n'(x_{kn})|\varepsilon_n} \leqslant \frac{|\omega_n(x_n)|}{|\omega_n'(x_{kn})|\,|x_n - x_{kn}|} = |l_{k_n}(\mathfrak{M}, x_n)|$$
$$\leqslant \|l_{kn}(\mathfrak{M}, x)\|_{C(F)} \leqslant \mathscr{L}_n(\mathfrak{M}, F) \leqslant \mathscr{L}(\mathfrak{M}, F) < \infty,$$

where

$$\mathscr{L}(\mathfrak{M}, F) \overset{\text{def}}{=} \sup_n \mathscr{L}_n(\mathfrak{M}, F).$$

From this and (31) we obtain

$$\frac{\|\omega_n'(x)\|_{C(F)}}{\|\omega_n(x)\|_{C(F)}} \geqslant \frac{1}{\varepsilon_n \mathscr{L}_n(\mathfrak{M}, F)} \geqslant c\left(\frac{6}{n\eta_n}\right)^n.$$

Here $c > 0$ is independent of n. But by (31) the numbers η_n must satisfy $\eta_n \leqslant 1/n$, and therefore we obtain

$$\frac{\|\omega_n'(x)\|_{C(F)}}{\|\omega_n(x)\|_{C(F)}} \geqslant c6^n = q^n, \qquad n \geqslant n_0, q > 1. \tag{44}$$

This completes the proof of the theorem.

COROLLARY. *If the set F satisfies the hypotheses of Theorem* 2.1, *then $F \in \mathfrak{N} \subset K \cap S$, and the capacity of F is zero.*

In fact, suppose that the capacity of $F \in \mathfrak{N}$ is not zero. Then by Theorem 1.1 we would have $\varphi(n, F) \leqslant (1 + \varepsilon)^n$, where $0 < \varepsilon < 1$ and $n \geqslant n_1(\varepsilon)$; but this contradicts inequality (44): $\varphi(n, F) \geqslant c \cdot 6^n$. Therefore $\Delta(F) = 0$.

I take this opportunity to thank S. B. Stechkin for discussions of the results of this paper and for remarks directed toward their improvement.

BIBLIOGRAPHY

1. P. P. Korovkin, *On the closure of systems of Tchebycheff functions*, Dokl. Akad. Nauk SSSR **78** (1951), 853–855. (Russian)

2. S. N. Mergelyan, *Some questions of the constructive theory of functions*, Trudy Mat. Inst. Steklov. **37** (1951). (Russian)

3. Georg Faber, *Über die interpolatorische Darstellung stetiger Funktionen*, Jber. Deutsch. Math.-Verein. **23** (1914), 192–210.

4. Serge Bernstein, *Sur la limitation des valeurs d'un polynome $P_n(x)$ de degré n sur tout un segment par ses valeurs en $n + 1$ points du segment*, Izv. Akad. Nauk SSSR Otdel. Mat. Estestv. Nauk **1931**, 1025–1050.

5. P. Erdös, *Problems and results on the theory of interpolation*. I, Acta Math. Acad. Sci. Hungar. **9** (1958), 381–388.

6. A. A. Privalov, *On interpolation on sets of measure zero*, Moskov. Gos. Ped. Inst. Uchen. Zap. No. 188 (1962), 163–169. (Russian)

7. ______, *On interpolation on sets of measure zero*, Mat. Sb. **60(102)** (1963), 29–38. (Russian)

8. I. P. Natanson, *Theory of functions of a real variable*, 2nd rev. ed., GITTL, Moscow, 1957; English transl., Vols. I, II, Ungar, New York, 1955, 1961.

9. Rolf Nevanlinna, *Eindeutige analytische Funktionen*, Springer-Verlag, 1936; English transl. of the 2nd ed., 1970.

10. G. M. Goluzin, *Geometric theory of functions of a complex variable*, 2nd rev. ed., "Nauka", Moscow, 1966; English transl., Amer. Math. Soc., Providence, R. I., 1969.

11. A. A. Markov, *On a problem of D. I. Mendeleev*, Zap. Imp. Akad. Nauk St. Petersburg **62** (1890), 1–24; reprinted in his *Selected papers on continued fractions and the theory of functions deviating least from zero*, OGIZ, Moscow, 1948, pp. 51–75. (Russian)

12. J. Priwaloff, [I. I. Privalov], *Sur la convergence des séries trigonométriques conjuguées*, C. R. Acad. Sci. Paris **162** (1916), 123–126.

13. ______, *The Cauchy integral*, Izv. Saratov. Univ. Fiz.-Mat. Fak. **2** (1918), no. 1 = Thesis, Saratov Univ., Saratov, 1919. (Russian)

14. ______, *Sur la convergence des séries trigonométriques conjuguées*, Mat. Sb. **32** (1924/25), 357–363. (Russian; French summary)

15. V. I. Ivanov, *Some extremal properties of polynomials and inverse inequalities of function theory*, Trudy Mat. Inst. Steklov. **145** (1980), 79–110; English transl. in Proc. Steklov Inst. Math. **1981**, no. 1(145).

16. Norbert Wiener, *The Dirichlet problem*, J. Math. and Phys. **3** (1924), 127–146.

17. M. V. Keldysh, *On the solvability and stability of the Dirichlet problm*, Uspekhi Mat. Nauk **8** (1941), 171–231; English transl. in Amer. Math. Soc. Transl. (2) **51** (1966).

18. J. L. Walsh, *Interpolation and approximation by rational functions in the complex domain*, 2nd ed., Amer. Math. Soc., Providence, R. I., 1956.

19. M. Fekete, *Über die Verteilung der Wurzeln bei gewissen algebraischen Gleichungen mit ganzzahligen Koeffizienten*, Math. Z. **17** (1923), 228–249.

20. S. N. Bernstein, *On best approximation of continuous functions by polynomials of given degree*, Soobshch. Khar′kov. Mat. Obshch. (2) **13** (1912), 47–194; rev. reprint in his *Collected Works*. Vol. 1, Izdat. Akad. Nauk SSSR, Moscow, 1952; abriged French transl., Acad. Roy. Belg. Cl. Sci. Mém (2) **4** (1912), fasc. 1.

21. P. P. Korovkin, *On the growth of polynomials on a set*, Dokl. Akad. Nauk SSSR **61** (1948), 781–784. (Russian)

22. B. Ya. Levin, *Distribution of zeros of entire functions*, GITTL, Moscow, 1956; English transl., Amer. Math. Soc., Providence, R. I., 1964.

Translated by R. P. BOAS

Труды Матем. Инст. Стеклов.
164(1983)

Proceedings of the Steklov
Institute of Mathematics
1985, Issue 2

ON THE LOCAL CHARACTER OF SOME PROPERTIES OF COMPLETE ORTHONORMAL SYSTEMS

UDC 517.52

A. A. TALALYAN

ABSTRACT. This paper considers orthonormal systems $\{\varphi_n(x)\}$ on $[0,1]$ having the following property:

For some set $E \subset [0,1]$, $\mu(E) > 0$, there exists for each function $f(x) \in L_2(E)$ a series $\Sigma\, a_n\varphi_n(x)$, $\Sigma\, a_n^2 < +\infty$, the sum of which coincides with $f(x)$ almost everywhere on E.

For such systems we prove the following statements:

1) There exist a permutation $\{\nu_n\}$ of the natural numbers and a sequence $\{a_n\}$, $\Sigma\, a_n^2 < +\infty$, such that $\Sigma\, a_n\varphi_{\nu n}(x)$ diverges almost everywhere on E.

2) $\Sigma[\varphi_n^-(x)]^2$ and $\Sigma[\varphi_n^+(x)]^2$ diverge almost everywhere on E.

The first statement is a generalization of a known result by Ul'yanov and Olevskiĭ on the almost everywhere divergence of permutations of L_2-series with respect to complete orthonormal systems. The second statement generalizes a result by Kozlov on the distribution of positive and negative values of the functions of complete orthonormal system.

Bibliography: 12 titles.

I. Many important investigations in the theorry of orthogonal series have been stimulated by a conjecture raised by N. N. Luzin [1] in 1915 to the effect that L_2-series with respect to the trigonometric system converge almost everywhere. This conjecture was confirmed only in 1966 by Carleson [2], but many important results obtained earlier in the theory of the trigonometric series, and even in the theory of general orthogonal series, are in one way or other connected with this conjecture of Luzin's. To these results belongs also a theorem stated by Kolmogorov [3] in 1927 to the effect that L_2-series with respect to a permutation of the trigonometric systems $\{\cos \nu_n x, \sin \nu_n x\}$ may diverge almost everywhere.

In praxis, the validity of this theorem, given without proof in [3], remained contested and only in 1960 did Zahorski [4] outline a multiple scheme for its proof. Later it turned out that this result is not specific for the trigonometric system only. Indeed, Ul'yanov [5] and Olevskiĭ [6] established an important theorem according to which, $\{\varphi_n(x)\}$ being an arbitrary complete orthonormal system in $L_2[0,1]$, there exists a permutation $\{\nu_n\}$ of the natural numbers such that L_2 series with respect to the system $\{\varphi_{\nu n}(x)\}$ may diverge almost everywhere.

In the present paper we consider orthonormal system (o.n. systems) in $L_2[0,1]$ which are not complete in $L_2[0,1]$ but possess the property of completeness on suitable subsets $E \subset [0,1]$, $\mu(E) > 0$, in the sense of the following definition.

1980 *Mathematics Subject Classification*. Primary 42C15; Secondary 42C20.

DEFINITION 1. The o.n. system $\{\varphi_n(x)\}$ is said to be *complete with respect to L_2-functions on the set* $E \subset [0,1]$, $\mu(E) > 0$, if every function defined on E and belonging to $L_2(E)$ is in the $L_2(E)$-metric the sum of a series $\sum a_n \varphi_n(x)$, with $\sum a_n^2 < \infty$.

It turns out that some important theorems on o.n. systems complete in $L_2[0,1]$, as for instance the above-mentioned theorem of Ul′yanov and Olevskiĭ on the divergence almost everywhere of permuted L_2-series with respect to complete systems, and likewise V. Ya. Kozlov's theorem [7] (cf. also [8]) on the almost everywhere divergence of the series $\sum|\varphi_n^+(x)|^2$ and $\sum|\varphi_n^-(x)|^2$ extend to o.n. systems which are complete for L_2 series on sets of positive measure. The following theorems are true:

THEOREM 1. *Let $\{\varphi_n(x)\}$ be an o.n. system which is complete for L_2-series on a set $E \subset [0,1]$, $\mu(E) > 0$. Then there exists a permutation $\{\nu_n\}$ of the natural numbers and a sequence $\{a_n\}$, $\sum a_n^2 < +\infty$, such that the series*

$$\sum a_n \varphi_{\nu_n}(x) \tag{1}$$

diverges unboundedly almost everywhere on E.

THEOREM 2. *If $\{\varphi_n(x)\}$ is an o.n. system which is complete for L_2-series on a set $E[0,1]$, $\mu(E) > 0$, then the series*

$$\sum_{n=1}^{\infty} [\varphi_n^-(x)]^2, \qquad \sum_{n=1}^{\infty} [\varphi_n^+(x)]^2 \tag{2}$$

diverges almost everywhere on the set E, where

$$\varphi_n^+(x) = \begin{cases} \varphi_n(x), & \text{if } \varphi_n(x) \geqslant 0, \\ 0, & \text{if } \varphi_n(x) < 0, \end{cases} \tag{3}$$

$$\varphi_n^-(x) = \varphi_n(x) - \varphi_n^+(x). \tag{4}$$

Theorem 1, being a generalization of the Ul′yanov-Olevskiĭ theorem, also shows that the phenomenon of almost everywhere divergence of permuted L_2-series with respect to complete systems has a local character, because for noncomplete o.n. systems this phenomenon is preserved on completeness sets for L_2-series with respect to these systems.

As is readily seen, for each o.n. system $\{\varphi_n(x)\}$ on $[0,1]$ which is complete on some set $E \subset [0,1]$, $\mu(E) > 0$, one can write the interval $[0,1]$ in a unique way (up to sets of measure zero) as a union of two sets A, B, $A \cap B = \varnothing$, $A \cup B = [0,1]$, with the following properties:

1) The system $\{\varphi_n(x)\}$ is not complete on any subset $E \subset A$, $\mu(E) > 0$.

2) For each measurable subset $E \subset B$, $\mu(E) > 0$, there exists a measurable subset $F \subset E$, $\mu(F) > 0$, such that the system $\{\varphi_n(x)\}$ is complete for L_2-series on F.

For convenience, let us call B a set of everywhere completeness of L_2-series for the system $\{\varphi_n(x)\}$.

From Theorems 1 and 2 and from the definition of the set B we easily get the following more general theorems.

THEOREM 3. *Let $\{\varphi_n(x)\}$ be an o.n. system on $[0,1]$ and let $B \subset [0,1]$ be the set of everywhere completeness for L_2-series with respect to the system $\{\varphi_n(x)\}$. Then there exist a permutation $\{\nu_n\}$ of the natural numbers and a sequence $\{a_n\}$, $\sum a_n^2 < +\infty$, such that the series*

$$\sum_{n=1}^{\infty} a_n \varphi_{\nu_n}(x) \tag{5}$$

diverges unboundedly almost everywhere on B.

THEOREM 4. *If $\{\varphi_n(x)\}$ is an o.n. system on $[0,1]$ and B the set of everywhere completeness for L_2-series with respect to that system, then the series*

$$\sum_{n=1}^{\infty} [\varphi_n^+(x)]^2, \qquad \sum_{n=1}^{\infty} [\varphi_n^-(x)]^2 \tag{6}$$

diverge almost everywhere on B.

In Theorems 1 and 2 it is not possible to replace the completeness with respect to L_2-series of the system $\{\varphi_n(x)\}$ by completeness of this system on E in the usual sense, i.e. by the everywhere density in $L_2(E)$ of the set of finite linear combinations of the system $\{\varphi_n(x)\}$ restricted to E.

This is seen by the example of an o.n. system constructed in [9] (formula (7.2)), having the following form:

$$\varphi_n(x) = b_n x^{n-1}, \quad x \in \left[\frac{1}{n+1}, 1\right], \qquad n = 1, 2, \ldots, \tag{7}$$

where $b_n \neq 0$, $n = 1, 2, \ldots$.

This system is complete in the ordinary sense on each interval $[\delta, 1]$, $0 < \delta < 1$, and, as is readily seen, every series $\sum a_n \varphi_n(x)$, $\sum a_n^2 < +\infty$, with respect to this system converges absolutely on $(0,1]$. It is clear that the conclusion of Theorem 1 is not true for the system (7). Since in (7) one can take all the b_n positive (upon multiplying suitable b_n by -1), also the conclusion of Theorem 2 cannot hold for the system (7), because

$$\sum_{n=1}^{\infty} [\varphi_n^-(x)]^2 < +\infty, \qquad x \in (0,1]. \tag{8}$$

For the sake of simplicity let us call a system $\{\varphi_n(x)\}$ *everywhere noncomplete* if it is not complete in the ordinary sense on any set $E \subset [0,1]$, $\mu(E) > 0$, and *everywhere complete* if for any $E \subset [0,1]$, $\mu(E) > 0$, there exists $F \subset E$, $\mu(F) > 0$, such that the system $\{\varphi_n(x)\}$ is complete on F in the ordinary sense.

In the survey article by Price [10] the following question is posed (p. 608):

Is the condition of everywhere noncompleteness of the o.n. system $\{\varphi_n(x)\}$ a necessary and sufficient condition for $\{\varphi_n(x)\}$ to be a system of unconditional convergence, i.e. does every series $\sum a_n \varphi_{\nu n}(x)$, $\sum a_n^2 < +\infty$, where $\{\nu_n\}$ is an arbitrary permutation of the natural numbers, have to converge almost everywhere on $[0,1]$?

The above example (7) also shows that the first part of this question has a negative answer. In the final section of the present paper we establish the existence of a large class of everywhere noncomplete o.n. systems which are not systems of unconditional convergence.

II. Here we give the proof of the lemmata with the aid of which we shall establish our theorems. Let us denote

$$\|f\|_E = \left(\int_E |f(x)|^2\,dx\right)^{1/2}, \tag{9}$$

if $E \subset [0,1]$, $\mu(E) > 0$. When $E = [0,1]$ also write $\|f\| = \|f\|_{[0,1]}$.

LEMMA 1. *Let $\{\varphi_n(x)\}$ be an o.n. system which is complete for L_2-series on the set $E \subset [0,1]$, $\mu(E) > 0$. Then there exists a constant M, depending only on $\{\varphi_n(x)\}$ and E, such that for each function $f \in L_2(E)$ there exists a series*

$$\sum_{n=1}^{\infty} a_n\varphi_n(x), \qquad \sum a_n^2 < +\infty, \tag{10}$$

which converges to $f(x)$ in the metric of $L_2(E)$ and is such that

$$\left\| \sum_{n=1}^{m} a_n\varphi_n(x) \right\| \leqslant M\|f\|_E, \qquad m = 1,2,\dots \tag{11}$$

PROOF. Let us denote by Φ the set of functions which are sums in the $L_2[0,1]$-metric of series with respect to the system $\{\varphi_n(x)\}$, and for a function $f(x) \in L_2(E)$ let us introduce the norm $\|f\|_E^*$ defined by

$$\|f\|_E^* = \inf_{F \in \Phi(f)} \|F\|, \tag{12}$$

where $\Phi(f)$ is the set of those functions $F(x) \in \Phi$ which are equal to $f(x)$ almost everywhere on E.

It is not hard to see that the set $L_2(E)$ with the form $\|f\|_E^*$ is a complete Banach space.

In fact, the equivalence $\|f\|_E^* = 0 \Leftrightarrow f(x) = 0$ is trivial. The identity

$$\|cf\|_E^* = |c|\,\|f\|_E^*$$

holds, because

$$\|cf\|_E^* = \inf_{F\in\Phi(cf)} \|F\| = \inf_{F\in\Phi(f)} \|cF\| = |c| \inf_{F\in\Phi(f)} \|F\| = |c|\,\|f\|_E^*.$$

In order to verify the triangle inequality we remark that

$$\Phi(f+g) \supset \{F(x) + G(x); \qquad F \in \Phi(f), G \in \Phi(g)\},$$

so that

$$\begin{aligned} \|f+g\|_E^* &\leqslant \inf_{F\in\Phi(f),\, G\in\Phi(g)} \|F+G\| \\ &\leqslant \inf_{F\in\Phi(f)} \|F\| + \inf_{G\in\Phi(g)} \|G\| \leqslant \|f\|_E^* + \|g\|_E^*. \end{aligned}$$

Let us check the completeness. Consider a fundamental sequence $f_n(x) \in L_2(E)$, $n = 1,2,\dots$; that is,

$$\|f_n - f_m\|_E^* \to 0 \quad \text{as } n, m \to \infty. \tag{13}$$

Pick a sequence of natural numbers $n_i \uparrow +\infty$ such that

$$\|f_n - f_{n_i}\|_E^* < 1/2^{i+1}, \qquad n > n_i. \tag{14}$$

With the aid of the inequalities

$$\|f_{n_{i-1}} - f_{n_i}\|_E^* < 1/2^i, \qquad i = 1, 2, \dots, \tag{15}$$

we shall show that the sequence $\{f_{n_i}(x)\}$ converges in the metric $\|\cdot\|_E^*$ to some function $f(x) \in L_2(E)$. Let us remark that if $g_1(x)$ and $g_2(x)$ are any two functions in $L_2(E)$ and if $F_1 \in \Phi(g_1)$ is fixed then

$$\Phi(g_1 - g_2) = \{F_1 - F_2, F_2 \in \Phi(g_2)\}. \tag{16}$$

Hence, taking an arbitrary function $F_{n_1}(x) \in \Phi(f_{n_1})$ and using (16), we get

$$\|f_{n_1} - f_{n_2}\| = \inf_{F \in \Phi(f_{n_2})} \|F_{n_1} - F\|. \tag{17}$$

It follows from (17) and (15), with $i = 1, 2$, that there exists a function $F_{n_2}(x) \in \Phi(f_{n_2})$ such that $\|F_{n_1} - F_{n_2}\| < 1/2$. Upon applying this reasoning to all differences $f_{n_{i-1}} - f_{n_i}$ we can define functions $F_{n_i} \in \Phi(f_{n_i})$, $i = 1, 2, \dots$, such that

$$\|F_{n_{i-1}} - F_{n_i}\| < 1/2^i, \qquad i = 1, 2, \dots. \tag{18}$$

Let $F(x)$ be the limit in the $L_2[0,1]$ metric of the sequence $\{F_{n_i}(x)\}$ and set $f(x) = F(x)$ for $x \in E$. It is plain that $F(x) \in \Phi$ and that moreover $F \in \Phi(f)$ and $F_{n_i} - F \in \Phi(f_{n_i} - f)$. Therefore it follows from the definition of the norm $\|\cdot\|_E^*$ (cf. (12)) that

$$\|f - f_{n_i}\|_E^* \leqslant \|F - F_{n_i}\| \to 0 \quad \text{as } i \to \infty. \tag{19}$$

Again from (13) and (19) it follows that $\|f_n - f\|_E^* \to 0$ as $n \to \infty$. This proves the completeness.

The identity operator $Af = f$, mapping our Banach space with the norm $\|\cdot\|_E^*$ into L_2 with the norm $\|\cdot\|_E$, is continuous, because $\|f\|_E^* \geqslant \|f\|_E$ for each $f \in L_2(E)$. In view of Banach's inverse mapping theorem this shows that there exists a number $\alpha > 0$ such that $\|f\|_E^* \leqslant \alpha\|f\|_E$ for each $f \in L_2(E)$. Taking $M = 2\alpha$ for a function $f(x)$ which differs from zero, we get $\|f\|_E^* < \frac{1}{2}M\|f\|_E$, and therefore there exists an $F \in \Phi(f)$ such that $\|F\| < M\|f\|_E$. Then in view of Bessel's inequality the series $\sum a_n\varphi_n(x)$, which converges to $F(x)$ in the $L_2[0,1]$ metric, must satisfy (11). In the case when $f(x) = 0$ for $x \in E$ we satisfy (11) by the zero series. Thus Lemma 1 is proved.

LEMMA 2. *Let $\{\varphi_n(x)\}$ be an o.n. system which is complete for L_2-series on the set $E \subset [0,1]$, $\mu(E) > 0$. Then for each measurable set $A \subset E$, $\mu(A) > 0$, for each natural number N and $\eta > 0$ there exists a function $\psi(x)$ and a linear combination*

$$\sum_{i=N+1}^{m} a_i\varphi_i(x), \qquad m > N, \tag{20}$$

satisfying the conditions

$$\psi(x) = 1, x \in A_1, \psi(x) = -1, x \in A_2, \psi(x) = 0, x \in E \setminus (A_1 \cup A_2), \tag{21}$$

where

$$A_1 \cup A_2 \subset A, \quad A_1 \cap A_2 = \varnothing, \quad \mu(A_1) = \mu(A_2) = \tfrac{1}{4}\mu(A), \tag{22}$$

$$\left\| \sum_{i=N+1}^{m} a_i\varphi_i(x) - \psi(x) \right\|_E < \eta, \tag{23}$$

$$\left\| \sum_{i=N+1}^{n} a_i\varphi_i(x) \right\| \leqslant \beta \|\psi(x)\|_E, \qquad N+1 \leqslant n \leqslant m, \tag{24}$$

where β is a constant depending only on the set E and the system $\{\varphi_n(x)\}$.

PROOF. Let us consider a sequence $\{r_n(A,x)\}$ of functions defined on the set A and having properties analogous to the Rademacher functions $r_n(x)$. This means that we consider a sequence of decompositions $A_1^{(n)}, \dots, A_{2^n}^{(n)}$, $n = 1, 2, \dots$, of the set A such that

$$\bigcup_{i=1}^{2n} A_i^{(n)} = A, \quad A_i^{(n)} \cap A_j^{(n)} = \varnothing, \qquad i \neq j, \qquad \mu\left(A_i^{(n)}\right) = \bar{2}^n\mu(A), \tag{25}$$

$$A_{2i-1}^{(n)} \cup A_{2i}^{(n)} = A_i^{(n-1)}, \quad \mu\left(A_{2i-1}^{(n)}\right) = \mu\left(A_{2i}^{(n)}\right) = \tfrac{1}{2}\mu\left(A_i^{(n-1)}\right), \qquad 1 \leqslant i \leqslant 2^{n-1}, \tag{26}$$

setting

$$r_n(A,x) = 1, \quad x \in A_{2i+1}^{(n)}, \qquad r_n(A,x) = -1, \quad x \in A_{2i}^{(n)}, 1 \leqslant i \leqslant 2^{n-1},$$
$$r_n(A,x) = 0, \qquad x \in E \setminus A. \tag{27}$$

By Lemma 1 we can find series $\sum a_i^{(n)}\varphi_i(x)$, where $\sum_1^\infty (a_i^{(n)})^2 < +\infty$, such that

$$\sum_{i=1}^{\infty} a_i^{(n)}\varphi_i(x) = \frac{1}{2} r_n(A,x), \qquad x \in A, \tag{28}$$

$$\left\| \sum_{i=1}^{m} a_i^{(n)}\varphi_i(x) \right\| \leqslant M \left\| \frac{1}{2} r_n(A,x) \right\|_E, \qquad m = 1, 2, \dots \tag{29}$$

Here (28) is taken in the sense of the metric of $L_2(E)$.

It follows from (29) that the sequences $\{a_i^{(n)}\}_{n=1}^\infty$ are bounded for each i, so that for a suitable sequence $\{n_j\}$ we get

$$\lim_{j\to\infty} \left(a_i^{(n_j)} - a_i^{(n_{j+1})}\right) = 0, \qquad 1 \leqslant i \leqslant N. \tag{30}$$

Again it follows from (30) that we must have

$$\left\| \sum_{i=1}^{N} \left(a_i^{(n_j)} - a_i^{(n_{j+1})}\right)\varphi_i(x) \right\| < \frac{\eta}{2}. \tag{31}$$

It is readily seen that the function

$$\psi(x) = \tfrac{1}{2} r_{n_j}(A,x) - \tfrac{1}{2} r_{n_{j+1}}(A,x) \tag{32}$$

and the linear combination

$$\sum_{i=N+1}^{m} a_i\varphi_i(x) = \sum_{i=N+1}^{m} \left(a_i^{(n_j)} - a_i^{(n_{j+1})}\right)\varphi_i(x), \tag{33}$$

where $m \geqslant N+1$ is chosen so that (cf. (28))

$$\left\| \sum_{i=1}^{m} \left(a_i^{(n_j)} - a_i^{(n_{j+1})}\right)\varphi_i(x) - \psi(x) \right\|_E < \frac{\eta}{2}, \tag{34}$$

satisfy the requirements of Lemma 2 with the constant $\beta = 2^{1/2}M$. In fact, that condition (23) is fulfilled follows from (31), (33) and (34). Condition (21) follows at once from (32) and from the definition of the functions $r_n(A, x)$. Finally, it follows from (33) and (34) that

$$\left\| \sum_{i=N+1}^{n} a_i\varphi_i(x) \right\| \leqslant M\left\| \frac{1}{2} r_{n_j}(x, A) \right\|_E + M\left\| \frac{1}{2} r_{n_{j+1}}(A, x) \right\|_E$$
$$= M(\mu(A))^{1/2} = 2^{1/2}M\|\psi(x)\|_E, \qquad N+1 \leqslant n \leqslant m.$$

Lemma 2 is proved.

LEMMA 3. *Let $\{\varphi_n(x)\}$ be an o.n. system which is complete for L_2-series on a set $E \subset [0,1]$, $\mu(E) > 0$. Then for any measurable subset $A \subset E$, $\mu(A) > 0$, for any natural number N and $\eta > 0$ there exist a function $f(x)$ and a linear combination of the form* (20) *possessing the following properties*:

$$f(x) = 1, x \in A_1, \quad f(x) = -1, x \in A_2, \quad f(x) = 0, x \in E\setminus A, \tag{35}$$

$$A_1 \cup A_2 = A, \quad A_1 \cap A_2 = \varnothing, \quad \mu(A_1) = \mu(A_2) = \tfrac{1}{2}\mu(A), \tag{36}$$

$$\left\| \sum_{i=N+1}^{m} a_i\varphi_i(x) - f(x) \right\|_E < \eta, \tag{37}$$

$$\left\| \sum_{i=N+1}^{n} a_i\varphi_i(x) \right\| \leqslant M\|f(x)\|_E, \qquad N+1 \leqslant n \leqslant m, \tag{38}$$

where M is a constant depending only on the system $\{\varphi_n(x)\}$ and the set E.

PROOF. Applying Lemma 2, we define a function $\psi_1(x)$ and a linear combination

$$\sum_{i=N+1}^{m_1} a_i\varphi_i(x), \qquad m_1 > N, \tag{39}$$

with the properties

$$\psi_1(x) = 1, \quad x \in A_1^{(1)}, \qquad \psi_1(x) = -1, \quad x \in A_2^{(1)},$$
$$\psi_1(x) = 0, \qquad x \in E\setminus\left(A_1^{(1)} \cup A_2^{(1)}\right), \tag{40}$$

$$A_1^{(1)} \cap A_2^{(1)} = \varnothing, \quad A_1^{(1)} \cup A_2^{(1)} \subset A, \quad \mu\left(A_1^{(1)}\right) = \mu\left(A_2^{(1)}\right) = \tfrac{1}{4}\mu(A), \tag{41}$$

$$\left\| \sum_{i=N+1}^{m_1} a_i\varphi_i(x) - \psi(x) \right\| \leqslant \frac{\eta}{2}, \qquad \left\| \sum_{i=N+1}^{m_1} a_i\varphi_i(x) \right\| \leqslant \beta\|\psi_1(x)\|_E. \tag{42}$$

Assume that we have defined the polynomials

$$\sum_{i=m_{j-1}+1}^{m_j} a_i\varphi_i(x), \qquad 1 \leqslant j \leqslant k-1, N = m_0 < m_1 < \cdots < m_{k-1} \tag{43}$$

and the functions $\psi_j(x)$,

$$\begin{aligned} &\psi_j(x) = 1, \quad x \in A_1^{(j)}, \qquad \psi_j(x) = -1, \quad x \in A_2^{(j)}, \\ &\psi_j(x) = 0, \qquad x \in E\setminus\left(A_1^{(j)} \cup A_2^{(j)}\right). \end{aligned} \tag{44}$$

Applying Lemma 2 with $N = m_{l-1}$, the positive number $\eta/2^{k+1}$ and the set

$$B_{k-1} = A \Big\backslash \bigcup_{j=1}^{k-1} \left(A_1^{(j)} \cup A_2^{(j)}\right), \tag{45}$$

we define a function $\psi_k(x)$ and a polynomial

$$\sum_{i=m_{k-1}+1}^{m_k} a_i\varphi_i(x), \tag{46}$$

having the properties

$$\begin{gathered}\psi_k(x)=1,\quad x\in A_1^{(k)},\qquad \psi_k(x)=-1,\quad x\in A_2^{(k)},\\ \psi_k(x)=0,\qquad x\in E\setminus\left(A_1^{(k)}\cup A_2^{(k)}\right),\end{gathered} \tag{47}$$

$$A_1^{(k)}\cup A_2^{(k)}\subset B_{k-1},\quad A_1^{(k)}\cap A_2^{(k)}=\varnothing,\quad \mu\left(A_1^{(k)}\right)=\mu\left(A_2^{(k)}\right)=\tfrac14\mu(B_{k-1}), \tag{48}$$

$$\left\|\sum_{i=m_{k-1}+1}^{m_k} a_i\varphi_i(x)-\psi_k(x)\right\|_E<\frac{\eta}{2^{k+1}}, \tag{49}$$

$$\left\|\sum_{i=m_{k-1}+1}^{n} a_i\varphi_i(x)\right\|\leqslant\beta\|\psi_k(x)\|_E,\qquad m_{k-1}<n\leqslant m_k. \tag{50}$$

Thus by induction we may pick sequences of functions (47) and polynomials (46) satisfying conditions (47)–(50) for all $k=1,2\ldots$ and such that

$$N=m_0<m_1<\cdots<m_k<\cdots. \tag{51}$$

Let us set

$$A_1=\bigcup_{k=1}^{\infty}A_1^{(k)},\qquad A_2=\bigcup_{k=1}^{\infty}A_2^{(k)}. \tag{52}$$

It follows from (41), (45) and (48) that

$$A_1\cap A_2=\varnothing,\quad A_1\cup A_2\subset A,\quad \mu(A_1)=\mu(A_2)=\tfrac12\mu(A). \tag{53}$$

The function $f(x)$ defined by (35) with the sets A_1 and A_2 defined by (52) satisfies conditions (35) and (36) of Lemma 3.

It follows from (45), (47) and (48) that

$$f(x)=\sum_{k=1}^{\infty}\psi_k(x),\qquad x\in E, \tag{54}$$

where clearly the series (54) converges also in then $L_2(E)$ metric.

Choose ν such that

$$\left\|\sum_{k=1}^{\nu}\psi_k(x)-f(x)\right\|_E<\eta/4. \tag{55}$$

Then from (49), (51) and (55) we get

$$\left\|\sum_{i=N+1}^{m_\nu}a_i\varphi_i(x)-f(x)\right\|_E<\eta. \tag{56}$$

On the other hand, it follows from (50) that

$$\left\|\sum_{i=1}^{n}a_i\varphi_i(x)\right\|\leqslant\sum_{k=1}^{\nu}\beta\|\psi_k(x)\|_E,\qquad N+1\leqslant n\leqslant m_\nu, \tag{57}$$

where, as is readily read off from (45), (47) and (49),

$$\|\psi_k(x)\|_E = \left(\frac{\mu(A)}{2^k}\right)^{1/2}. \tag{58}$$

As (cf. (54)) $\|f\|_E = (\mu(A))^{1/2}$ and

$$\sum_{k=1}^{\infty}\left(\frac{\mu(A)}{2^k}\right)^{1/2} = \frac{1}{2^{1/2}-1}(\mu(A))^{1/2} = \frac{1}{2^{1/2}-1}\|f\|_E, \tag{59}$$

we get by (57) and (58)

$$\left\|\sum_{i=N+1}^{n} a_i\varphi_i(x)\right\| \leqslant \beta\frac{1}{2^{1/2}-1}\|f\|_E, \qquad N+1 \leqslant n \leqslant m_\nu. \tag{60}$$

Inequalities (56) and (60) mean that the polynomials

$$\sum_{i=N+1}^{m} a_i\varphi_i(x) = \sum_{i=N+1}^{m_\nu} a_i\varphi_i(x), \qquad m_\nu = m,$$

satisfy conditions (37) and (38) of Lemma 3, provided that for M we take the number $\beta/(2^{1/2}-1)$.

Lemma 3 is proved.

III. Now we prove Theorems 1 and 2. As will be seen from the following discussion, with the aid of Lemma 3 the proof of the theorem reduces to Ul'yanov's theorem [5] on the existence of L_2 series with respect to the Haar system which diverge almost everywhere after a suitable permutation of the terms. We remark that this result by Ul'yanov likewise is the basis of the proof of the aforementioned theorems of Ul'yanov [5] and Olevskiĭ [6].

PROOF OF THEOREM 1. Let $\{\varphi_n(x)\}$ be an orthonormal system which is complete for L_2-series on the set $E \subset [0,1]$, $\mu(E) > 0$. Applying Lemma 3 we define a function $f_1(x)$ and a polynomial (with respect to the same system)

$$H_1(x) = \sum_{i=1}^{n_1} a_i\varphi_i(x), \tag{61}$$

possessing the properties

$$f_1(x) = (\mu(E))^{-1/2}, \quad x \in E_1^{(1)}, \qquad f_1(x) = -(\mu(E))^{-1/2}, \quad x \in E_2^{(1)},$$
$$f_1(x) = 0, \qquad x \in [0,1]\setminus E, \tag{62}$$

$$E_1^{(1)} \cap E_2^{(1)} = \varnothing, \quad E_1^{(1)} \cup E_2^{(1)} \subset E, \quad \mu\left(E_1^{(1)}\right) = \mu\left(E_2^{(1)}\right) = \tfrac{1}{2}\mu(E), \tag{63}$$

$$\|H_1(x) - f(x)\|_E < \tfrac{1}{2}, \qquad \|H_1(x)\| \leqslant M\|f(x)\|_E. \tag{64}$$

Assume that we have already defined the functions $f_j(x)$, $1 \leqslant j \leqslant 2^k - 1$, and the polynomials

$$H_j(x) = \sum_{i=n_{j-1}+1}^{n_j} a_i\varphi_i(x), \qquad 1 \leqslant j \leqslant 2^{k-1}, k > 2, \tag{65}$$

where $n_j > n_{j-1}$, $2 \leqslant j \leqslant 2^k - 1$, and where for $1 \leqslant \nu \leqslant 2^{k-1}$ we have

$$f_{2^{k-1}+\nu-1}(x) = \left(2^{1-k}\mu(E)\right)^{-1/2}, \qquad x \in E_{2\nu-1}^{(k)};$$
$$f_{2^{k-1}+\nu-1}(x) = -\left(2^{1-k}\mu(A)\right)^{-1/2}, \qquad x \in E_{2\nu}^{(k)};$$
$$f_{2^{k-1}+\nu-1}(x) = 0, \qquad x \in E \setminus \left(E_{2\nu-1}^{(k)} \cup E_{2\nu}^{(k)}\right), \tag{66}$$

$$E_p^{(k)} \cap E_q^{(k)} = \varnothing, \quad p \neq q, \qquad \bigcup_{p=1}^{2^k} E_p^{(k)} = E, \tag{67}$$

$$\mu\left(E_p^{(k)}\right) = 2^{-k}\mu(E), \qquad 1 \leqslant p \leqslant 2^k. \tag{68}$$

Applying Lemma 3 to the sets $A = E_p^{(k)}$, $1 \leqslant p \leqslant 2^k$, we may define functions $f_j(x)$ and polynomials $H_j(x)$ of the form (65) for $2^k \leqslant j \leqslant 2^{k+1} - 1$ which satisfy the conditions

$$f_j(x) = \left(2^{-k}\mu(E)\right)^{-1/2}, \qquad x \in E_{2p-1}^{(k+1)};$$
$$f_j(x) = -\left(2^{-k}\mu(E)\right)^{-1/2}, \qquad x \in E_{2p}^{(k+1)};$$
$$f_j(x) = 0, \qquad x \in E \setminus \left(E_{2p-1}^{(k+1)} \cup E_{2p}^{k+1}\right), \tag{69}$$

where

$$j = 2^k + p - 1, \quad 1 \leqslant p \leqslant 2^k, \qquad E_{2p-1}^{(k+1)} \cup E_{2p}^{(k+1)} \subset E_p^{(k)}, \quad 1 \leqslant p \leqslant 2^k, \tag{70}$$

$$E_{2p-1}^{(k+1)} \cap E_{2p}^{(k+1)} = \varnothing, \qquad \mu\left(E_{2p-1}^{(k+1)}\right) = \mu\left(E_{2p}^{(k+1)}\right) = \tfrac{1}{2}\mu\left(E_p^{(k)}\right) \tag{71}$$

$$\left\|H_j(x) - f_j(x)\right\|_E < 1/2^j, \quad \left\|H_j(x)\right\| \leqslant M\left\|f_j(x)\right\|_E, \qquad 2^k \leqslant j \leqslant 2^{k+1}, \tag{72}$$

and likewise $n_j > n_{j-1}$ (cf. (65)) for all $j = 1, 2, \ldots$

In this way we define a sequence of polynomials with respect to the system $\{\varphi_n(x)\}$ of the form (65) and functions $f_j(x)$ of the form (69) which satisfy conditions (70), (71) and (72) for all $j \geqslant 1$. As is seen from (62), (70) and (71), the functions $f_j(x)$, $j = 1, 2, \ldots$, constitute an orthonormal system on the set E and are sort of analogous of the Haar functions $\mathrm{X}_j(x)$, $j \geqslant 1$, defined on $[0, 1]$. By the theorem of Ul′yanov [5] on the existence of almost everywhere unboundedly divergent L_2-series of the form

$$\sum_{j=1}^{\infty} c_j X_{\nu_j}(x), \qquad \sum c_j^2 < +\infty, \tag{73}$$

where $\{\nu_j\}$ is a permutation of the numbers $j = 1, 2, \ldots$, and by the definition of the functions $f_j(x)$ it follows (cf. [6]) that there exists a series

$$\sum_{j=1}^{\infty} c_j f_{\nu_j}(x), \qquad \sum_{j=1}^{\infty} c_j^2 < +\infty, \tag{74}$$

which is unboundedly divergent almost everywhere on E.

From the second inequality (72), which holds for all $j \geqslant 1$, we see that the series

$$\sum_{j=1}^{\infty} c_j H_{\nu_j}(x) = \sum_{j=1}^{\infty} \sum_{i=n_{\nu_{j-1}}+1}^{n_{\nu_j}} c_j a_i \varphi_i(x) \tag{75}$$

is a permutation of an L_2-series with respect to the system $\{\varphi_n(x)\}$.

On the other hand, from the first inequality (72), which holds for all j, and from the unbounded divergence almost everywhere on E of the series (74) it follows that the series (75) is unboundedly divergent almost everywhere on E.

Theorem 1 is proved.

Next we prove Theorem 2. Assume that the theorem is not true; that is, that one of the series (2), for example, the series

$$\sum_{n=1}^{\infty} [\varphi_n^-(x)]^2, \tag{76}$$

converges on a set $A \subset E$ with $\mu(A) > 0$. From this assumption it readily follows that the series

$$\sum_{n=1}^{\infty} a_n \varphi_n(x) \tag{77}$$

converges absolutely almost everywhere on E for every sequence $\{a_n\}$ such that $\sum a_n^2 < +\infty$. In fact,

$$\sum_{n=1}^{\infty} |a_n \varphi_n^-(x)|^2 \leqslant \left(\sum_{n=1}^{\infty} a_n^2\right)^{1/2} \left(\sum_{n=1}^{\infty} [\varphi_n^-(x)]^2\right)^{1/2} < +\infty \tag{78}$$

almost everywhere on A.

Since $\sum |a_n| \varphi_n(x)$ converges in the L_2-metric on A, it follows from (79) that for any $\varepsilon > 0$ there exists a set $A_0 \subset A$, $\mu(A_0) > \mu(A) - \varepsilon$, such that the series

$$\sum |a_n| \varphi_n^+(x) \tag{79}$$

converges in the L_2-metric on A_0. Consequently, (79) has to converge almost everywhere on A_0. It follows from this and from (78) that the series (77) converges almost everywhere on A, for any sequence $\{a_n\}$, $\sum a_n^2 < +\infty$. But the system $\{\varphi_n(x)\}$ is complete for L_2-series on E. This contradicts Theorem 1, according to which for some sequence $\{a_n\}$, $\sum a_n^2 < +\infty$, the series (77) diverges almost everywhere on E after a suitable permutation.

Theorem 2 is proved.

Let us also give a simpler proof of Theorem 2 which does not rely on Theorem 1. It follows at once from Lemma 3 that there exist sequences of functions $\{r_j(x)\}$ and polynomials $\{H_j(x)\}$,

$$H_j(x) = \sum_{i=n_{j-1}+1}^{n_j} a_i \varphi_k(x), \qquad n_j > n_{j-1}, \tag{80}$$

satisfying the conditions

$$\begin{gathered} r_j(x) = 1, \quad x \in A_1^{(j)}, \qquad r_j(x) = -1, \quad x \in A_2^{(j)}, \\ r_j(x) = 0, \qquad x \in [0,1] \setminus E, \end{gathered} \tag{81}$$

where

$$A_1^{(j)} \cap A_2^{(j)} = \varnothing, \quad A_1^{(j)} \cup A_2^{(j)} = E, \quad \mu\left(A_1^{(j)}\right) = \mu\left(A_2^{(j)}\right) = \tfrac{1}{2}\mu(E), \tag{82}$$

$$\left\| H_j(x) - \frac{1}{j} r_j(x) \right\|_E \leqslant \frac{1}{2uj}, \qquad \|H_j(x)\| < \frac{M}{j} \|r_j(x)\|_E. \tag{83}$$

It follows immediately from the second inequalities (83) and from (80) that

$$\sum_{j=1}^{\infty} \|H_j(x)\|^2 = \sum_{i=n_0-1}^{\infty} a_i^2 < M^2 \sum_{j=1}^{\infty} \frac{1}{j^2} < +\infty. \tag{84}$$

On the other hand, since everywhere on E

$$\sum_{j=1}^{\infty} \frac{1}{j} |r_j(x)| = +\infty, \tag{85}$$

we get from the first inequalities (83) that

$$\sum_{j=1}^{\infty} |H_j(x)| = \sum_{j=1}^{\infty} \left| \sum_{i=n_{j-1}+1}^{n_j} a_i \varphi_i(x) \right| = +\infty \tag{86}$$

almost everywhere on E. According to what we said above it follows from (84) and (86) that both series (2) diverge almost everywhere on E, because in the opposite case it would follow that each L_2-series with respect to the system $\{\varphi_n(x)\}$ had to converge on some subset $A \subset E$, $\mu(A) > 0$.

Theorems 3 and 4 now easily follow from Theorems 1 and 2.

Let $\{\varphi_n(x)\}$ be an o.n. system on $[0,1]$ which is complete for L_2-series on some subset $E \subset [0, \text{k}1]$, $\mu(E) > 0$. Let us consider the family of all subsets of $[0,1]$ for which the system $\{\varphi_n(x)\}$ is complete for L_2-series and, denoting by α_1 the upper bound of the measures of these sets, let us pick an arbitrary subset belonging to this family, the measure of which exceeds $\alpha_1/2$.

Assuming that the sets E_i, $1 \leqslant i \leqslant n+1$, have been chosen, let us denote by α_{n+1} the upper bound of the masures of the sets lying in $[0,1] \setminus \bigcup E_i$ for which the system $\{\varphi_n(x)\}$ is complete for L_2-series, and let us choose from them an arbitrary set E_{n+1} whose measure exceeds $\alpha_{n+1}/2$.

It is clear that in this way there is defined a finite or infinite sequence of sets E_i such that the set $A = [0,1] \setminus \bigcup E_i$ does not contain a subset of positive measure on which the system $\{\varphi_n(x)\}$ is complete for L_2-series.

It is clear that on the set $B = \bigcup E_i$ both series (6) diverge almost everywhere, because by Theorem 2 they diverge almost everywhere on each of the sets E_i.

Theorem 4 is proved.

For the proof of Theorem 3 we remark that if the system $\{\varphi_n(x)\}$ is complete for L_2-series on a set E, $\mu(E) > 0$, then for each natural number N and arbitrary positive numbers Q, α and ε there exists a polynomial of the form

$$H(x) = \sum_{i=p}^{q} a_i \varphi_{\nu_i}(x), \tag{87}$$

having the properties

$$\nu_j \neq \nu_i, \quad i \neq j, \quad \nu_i > N, \quad p \leqslant i \leqslant q, \tag{88}$$

$$\sum_{i=p}^{q} a_i^2 < \alpha, \tag{89}$$

$$\mu\left(E \cap \left\{ \sup_{p \leqslant n \leqslant q} \left| \sum_{i=p}^{n} a_i \varphi_{\nu_i}(x) \right| > Q \right\}\right) > \mu(E) - \varepsilon. \tag{90}$$

The existence of such a polynomial (87) follows immediately from Theorem 1. It is easy to see that, upon applying the same reasoning to each of the sets E_i infinitely many times, with sequences $Q_j \to +\infty$, $\varepsilon_j \to 0$ and α_j, $\sum \alpha_j < +\infty$, prescribed in advance, we may define polynomials

$$H_j(x) = \sum_{i=p_j}^{q_j} a_i \varphi_{\nu_j}(x), \qquad j = 1, 2, \ldots, \tag{91}$$

having the following properties:

1) Different polynomials H_j and $H_{j'}$, $j \neq j'$, do not contain functions $\varphi_{\nu_i}(x)$ with the same index ν_i and

$$\sum_{i=p_j}^{q_j} a_i^2 < \alpha_j. \tag{92}$$

2) For each set E_m one can select from the sequence $\{H_j(x)\}$ a subsequence $\{H_{m_j}(x)\}_{j=1}^{\infty}$ such that

$$\mu\left(E \cap \left\{x;\ \sup_{p_{m_j} \leqslant n \leqslant q_{m_j}} \left|\sum_{i=p_{m_j}}^{n} a_i \varphi_{\nu_i}(x)\right| > Q_m\right\}\right) > \mu(E_m) - \varepsilon_{m_j}. \tag{93}$$

It is clear that the series

$$\sum_{j=1}^{\infty} H_j(x) = \sum_{j=1}^{\infty} \sum_{ip_j}^{q_j} a_i \varphi_{\nu_i}(x) \tag{94}$$

with the inner sum opened up is an L_2-series which diverges almost everywhere on the set B.

Theorem 3 is proved.

This theorem shows that the phenomenon of almost everywhere divergence of permutations of L_2-series for complete systems remains in force likewise for non-complete o.n. systems possessing the property that for each set $E \subset [0, 1]$, $\mu(E) > 0$, there exists an $F \subset E$, $\mu(F) > 0$, on which the system is complete for L_2-series. This follows from the fact that for such systems the measure of the set B equals the measure of the interval of orthogonality.

IV. In this section we consider the question of the existence of o.n. systems $\{\varphi_n(x)\}$ which are not complete on any subset E and possess the property that appropriate L_2-series with respect to these systems diverge almost everywhere after a suitable permutation.

The solution to this problem given below will be based on the abovementioned theorem by Ul'yanov and the following theorem in [11] (cf. also [9], §6, Theorem 12):

THEOREM 1°. *If the system $\{f_n(x)\}$ of everywhere finite on E measurable functions is complete on E in the sense of convergence in measure, i.e., for any measurable function $f(x)$ on E there exists a sequence of linear combinations*

$$H_j(x) = \sum_{i=1}^{n_j} a_i^{(j)} f_i(x), \qquad j = 1, 2, \ldots, \tag{95}$$

converging to $f(x)$ *in measure on* E, *then all subsystems of the form* $\{f_n(x)\}_{N+1}^{\infty}$ (N *an arbitrary number*) *are complete on* E *in the sense of convergence in measure.*

With the aid of this theorem it is easy to prove the following general result:

THEOREM 5. *Let* $\{f_n(x)\}$ *be a sequence of almost everywhere finite measurable functions on* $[0,1]$, *and continue them periodically with period* 1 *to the entire real axis. If the series*

$$\sum_{n=1}^{\infty} a_n f_n(x), \qquad x \in [0,1] \tag{96}$$

with respect to that system diverges on some set $E \subset [0,1]$, $\mu(E) > 0$, *then one can find numbers* p_i *and* q_i, $p_i < q_i < p_{i+1}$, $i = 1,2,\ldots$, *such that the series*

$$\sum_{i=1}^{\infty} \sum_{n=p_i}^{q_i} a_n f_n(2^i x) \tag{97}$$

has the following properties:

1) *The system* $f_n(2^i x)$, $p_i \leqslant n \leqslant q_i$, $i = 1,2,\ldots$, *is not complete on any subset* $E \subset [0,1]$, $\mu(E) > 0$.

2) *For any subsequence* $\{i_k\}$ *the series*

$$\sum_{k=1}^{\infty} \sum_{n=p_{i_k}}^{q_{i_k}} a_n f_n(2^{i_k} x), \tag{98}$$

expanded with respect to the inner sum, diverges almost everywhere on $[0,1]$.

PROOF. Denote

$$F_N(x) = \sup_{N \leqslant p \leqslant q} \left| \sum_{n=p}^{q} a_n f_n(x) \right|, \tag{99}$$

$$F_{N,m}(x) = \sup_{N \leqslant p \leqslant q \leqslant m} \left| \sum_{n=p}^{q} a_n f_n(x) \right|, \qquad m > N. \tag{100}$$

We remark that

$$\lim_{m \to \infty} F_{N,m}(x) = F_N(x), \tag{101}$$

and that the sequence $F_N(x)$ is nonincreasing. Also, if we denote

$$F(x) = \lim_{N \to \infty} F_N(x),$$

then in view of the fact that (96) converges on E we must have([1])

$$F_N(x) \geqslant F(x) > 0, \qquad x \in E. \tag{102}$$

It follows from (102) that there exist a positive number ε_0 and a set $E_0 \subset E$, $\mu(E_0) > 0$, such that

$$F_N(x) < 2\varepsilon_0, \qquad x \in E_0,\ N = 1,2,\ldots \tag{103}$$

Again from (101) and (103) it follows that for any natural number N and $\eta > 0$ there exist a number $m > N$ and a set $E_{N,m}$ such that

$$E_{N,m} \subset E_0, \qquad \mu(E_{N,m}) > \mu(E_0) - \eta, \tag{104}$$

$$\sup_{N \leqslant p \leqslant q \leqslant m} \left| \sum_{n=p}^{q} a_n f_n(x) \right| > \varepsilon_0, \qquad x \in E_{N,m}. \tag{105}$$

([1])In this argument the functions $F_N(x)$ and $F(x)$ may take the value $+\infty$ on sets of positive measure.

This entails the existence of numbers p_i and q_i, $p_i < q_i < p_{i+1}$, and a set $A \subset E_0$, $\mu(A) > 0$, for which

$$\sup_{p_i \leqslant j_1 \leqslant j_2 \leqslant q_i} \left| \sum_{n=j_1}^{j_2} a_n f_n(x) \right| > \varepsilon_0, \qquad x \in A,\ i = 1, 2, \ldots \tag{106}$$

Now note that by virtue of (106) there exist sets $A_\nu^{(i)}$, $1 \leqslant \nu \leqslant 2^i$, such that

$$A_\nu^{(i)} \subset \left[\frac{\nu - 1}{2^i}, \frac{\nu}{2^i} \right], \qquad \mu\left(A_\nu^{(i)}\right) = \frac{1}{2^i} \mu(A), \tag{107}$$

$$\sup_{p_i \leqslant j_1 \leqslant j_2 \leqslant q_i} \left| \sum_{n=j_1}^{j_2} a_n f_n(2^i x) \right| > \varepsilon_0, \qquad x \in \bigcup_{\nu=1}^{2^i} A_\nu^{(i)},\ i = 1, 2, \ldots \tag{108}$$

That condition 2) is fulfilled now follows easily from (107) and (108). Let us check that 1) too is fulfilled. Indeed, we shall show that the system $f_n(2^i x)$, $p_i \leqslant n \leqslant q_i$, $i = 1, 2, \ldots$, is not complete in measure on any set $E \subset [0, 1]$, $\mu(E) > 0$.

Assume the contrary; that is, that the system is complete in measure on some set E_0. Then, for suitable i_0 and j_0, $1 \leqslant j_0 < 2^{i0}$, writing for short

$$\Delta' = \left(\frac{j_0 - 1}{2^{i_0}}, \frac{j_0}{2^{i_0}} \right), \quad \Delta'' = \left(\frac{j_0}{2^{i_0}}, \frac{(j_0 + 1)}{2^{i_0}} \right), \quad \Delta = \Delta' \cup \Delta'', \tag{109}$$

we have

$$\mu(\Delta' \cap E_0) < \mu(\Delta') - \tfrac{1}{8}\mu(\Delta'), \qquad \mu(\Delta'' \cap E_0) > \mu(\Delta'') - \tfrac{1}{8}\mu(\Delta''), \tag{110}$$

where by Theorem 1° the system $f_n(2^i x)$, $p_i \leqslant n \leqslant q_i$, $i \geqslant i_0$, has to be complete in measure on $E_0 \cap \Delta$ and, in particular, a suitable sequence of linear combinations $H_\nu(x)$ of the form

$$H_\nu(x) = \sum_{i=i_0}^{i_\nu} \sum_{n=p_i}^{q_i} c_n^{(\nu)} f_n(2^i x), \qquad \nu = 1, 2, \ldots, \tag{111}$$

has to converge in measure on E_0 to the function

$$f(x) = 1, \quad x \in E_0 \cap \Delta', \qquad f(x) = 0, \quad x \in E_0 \cap \Delta''. \tag{112}$$

For sufficiently large values of ν we then have

$$\mu\left(\left\{x;\ |H_\nu(x)| > \tfrac{1}{2}\right\} \cap E_0 \cap \Delta'\right) > \mu\left(E_0 \cap \Delta'\right) - \tfrac{1}{8}\mu(\Delta'), \tag{113}$$

$$\mu\left(\left\{x;\ |H_\nu(x)| > \tfrac{1}{2}\right\} \cap E_0 \cap \Delta''\right) < \tfrac{1}{8}\mu(\Delta'') \tag{114}$$

and from (109) and (110) we get

$$\mu\left(\left\{x;\ |H_\nu(x)| > \tfrac{1}{2}\right\} \cap \Delta'\right) < \tfrac{6}{8}\mu(\Delta') = \tfrac{3}{8}\mu(\Delta), \tag{115}$$

$$\mu\left(\left\{x;\ |H_\nu(x)| < \tfrac{1}{2}\right\} \cap \Delta''\right) < \tfrac{1}{8}\mu(\Delta'') = \tfrac{1}{8}\mu(\Delta). \tag{116}$$

On the other hand, denoting

$$h_\nu(x) = \sum_{i=i_0}^{i_\nu} \sum_{n=p_i}^{q_i} c_n^{(\nu)} f_n(2^{i-i_0} x), \tag{117}$$

we have $H_\nu(x) = h_\nu(2^{i_0} x)$. Therefore the following inequality must be true:

$$\mu\left(\left\{x;\ |H_\nu(x)| > \tfrac{1}{2}\right\} \cap \Delta'\right) = \mu\left(\left\{x;\ |H_\nu(x)| > \tfrac{1}{2}\right\} \cap \Delta''\right). \tag{118}$$

This contradiction completes the proof of Theorem 5.

With the aid of this theorem one can establish the existence of a large class of everywhere noncomplete o.n. which are not systems of unconditional convergence. In fact, by the Ul'yanov-Olevskiĭ theorem we may take as the series (96), for instance, an everywhere divergent L_2-series with respect to a rearrangement of the Haar system or a rearrangement of the trigonometric system. Then, clearly, for suitable subsequences i_k the corresponding systems $f_n(2^{i_k}x)$, $p_{i_k} \leqslant n \leqslant q_{i_k}$, $k = 1, 2, \ldots$, will be everywhere noncomplete o.n. systems while the corresponding series (98) turn out to be almost everywhere divergent L_2-series with respect to these systems. Let us further remark that in the second case, when $\{f_n(x)\}$ thus is a rearrangement of the trigonometric system, the corresponding system $\{f_n(2^{i_k}x)\}$ $p_{i_k} \leqslant n \leqslant q_{i_k}$, $k \geqslant 1$, will consist of functions belonging to the trigonometric system.

Let us note that an explicit concrete example of an everywhere noncomplete o.n. system with divergence is displayed in [12], where besides the methods set forth here one uses some long-known estimates for the oscillation of the partial sums of rearrangements of Haar polynomials.

Bibliography

1. N. N. Luzin, *The integral, and trigonometric series*, GITTL, Moscow, 1951. (Russian)

2. Lennart Carleson, *On the convergence and growth of partial sums of Fourier series*, Acta Math. **116** (1966), 135–157.

3. A. Kolmogoroff [A. N. Kolmogorov] and D. Menchoff [D. E. Men'shov], *Sur la convergence des séries de fonctions orthogonales*, Math. Z. **26** (1927), 432–441.

4. Zygmunt Zahorski, *Une série de Fourier permutée d'une fonction de classe L^2 divergente presque partout*, C. R. Acad. Sci. Paris **251** (1960), 501–503.

5. P. L. Ul'yanov, *Divergent Fourier series*, Uspekhi Mat. Nauk **16** (1961), no. 3(99), 61–142; English transl. in Russian Math. Surveys **16** (1961).

6. A. M. Olevskiĭ, *Divergent series in L^2 with respect to complete systems*, Dokl. Akad. Nauk SSSR **138** (1916), 545–548; English transl. in Soviet Math. Dokl. **2** (1961).

7. V. Ya. Kozlov, *On the distribution of positive and negative values of normal orthogonal functions forming a complete system*, Mat. Sb. **23(65)** (1948), 475–480. (Russian)

8. F. G. Arutyunyan, *Distribution of positive and negative values of functions forming an unconditional basis in the space $L_p(0,1)$*, Izv. Akad. Nauk Armyan. SSR Ser. Mat. **1** (1966), 79–87. (Russian)

9. A. A. Talalyan, *The representation of measurable functions by series*, Uspekhi Mat. Nauk **15** (1960), no. 5(95), 77–141; English transl. in Russian Math. Surveys **15** (1960).

10. J. J. Price, *Topics in orthogonal functions*, Amer. Math. Monthly **82** (1975), 594–609; erratum, **83** (1976), 546.

11. A. A. Talalyan, *On the convergence almost everywhere of subsequences of partial sums of general orthogonal series*, Izv. Akad. Nauk Armyan. SSR Ser. Fiz.-Mat. Nauk **10** (1957), no. 3, 17–34. (Russian)

12. ______, *On the connection between the properties of completeness and unconditional convergence for orthogonal systems*, Mat. Sb. **117(159)** 516–522; English transl. in Math. USSR Sb. **45** (1983).

Translated by J. PEETRE

Труды Матем. Инст. Стеклов.
164(1983)

Proceedings of the Steklov
Institute of Mathematics
1985, Issue 2

ON BOUNDED ORTHONORMAL SYSTEMS OF FUNCTIONS

UDC 517.5

K. TANDORI

ABSTRACT. For $1 \leqslant K \leqslant \infty$ let $\Omega(K)$ denote the class of all orthonormal systems of functions $\{\varphi_k(x)\}_0^\infty$ on $(0,1)$ which satisfy the estimate

$$|\varphi_k(x)| \leqslant K \qquad (x \in (0,1), k = 0,1,\dots).$$

Also let T be a regular Toeplitz method of summation, and let $M^T(K)$ denote the set of those sequences $a = \{a_k\}_0^\infty$ for which the series

$$\sum_{k=0}^{\infty} a_k\varphi_k(x)$$

are T-summable almost everywhere on $(0,1)$ for all systems $\{\varphi_k(x)\}_0^\infty$ of class $\Omega(K)$. It is shown that $M^T(K) = M^T(1)$ for $1 < K \leqslant \infty$. By this means the author is able to generalize, to arbitrary regular Toeplitz methods of summation, his earlier result in which he considered, instead of summability, convergence of series in bounded orthonormal systems.

Bibliography: 4 titles.

1. For a given number K, $1 \leqslant K \leqslant \infty$, we denote by $\Omega(K)$ the class of orthonormal systems $\varphi = \{\varphi_k(x)\}_0^\infty$ on $(0,1)$ for which

$$|\varphi_k(x)| \leqslant K \qquad (x \in (0,1);\ k = 0,1,\dots).$$

In particular, $\Omega(\infty)$ denotes the class of all orthonormal systems on $(0,1)$; also in the case $\varphi \in \Omega(1)$ we have $|\varphi_k(x)| = 1$ $(k = 0,1,\dots)$ a.e. on $(0,1)$. We call this type of system a *system of signs*.

For given K, $1 \leqslant K \leqslant \infty$, we denote by $M(K)$ the set of sequences of real numbers $a = \{a_k\}_0^\infty$ for which the series

$$\sum_{k=0}^{\infty} a_k\varphi_k(x) \tag{1}$$

converges almost everywhere for any system $\varphi \in \Omega(K)$.

It was proved in the author's earlier paper [1] that $M(1) = M(K)$ $(1 < K < \infty)$. We shall extend this result to general linear methods of summation.

2. Let $T = \|t_{nk}\|_{n,k=0}^\infty$ be a regular Toeplitz matrix; that is,

(i)

$$\lim_{n\to\infty} t_{nk} = 0 \qquad (k = 0,1,\dots),$$

(ii)

$$\lim_{n\to\infty} \sum_{k=0}^{\infty} t_{nk} = 1,$$

1980 *Mathematics Subject Classification*. Primary 42C05, 42A24.

(iii)

$$\sum_{k=0}^{\infty} |t_{nk}| \leqslant C \qquad (n = 0, 1, \ldots),$$

where C is a positive constant. The series $\sum_0^\infty c_k$ will be called *T-summable* if there exists an index n_0 such that the series $\sum_{k=0}^\infty t_{nk}(\sum_0^k c_l)_k$ $(n \geqslant n_0)$ are convergent and the sequence of sums $t_n = \sum_{k=0}^\infty t_{nk}(\sum_0^k c_l)$ tends to a limit $\lim_{n\to\infty} t_n$.

For a given K, $1 \leqslant K \leqslant \infty$, we denote by $M^T(K)$ the class of sequences of real numbers a such that the series (1) is T-summable almost everywhere on $(0, 1)$ for any system $\varphi \in \Omega(K)$. It follows from the definition that

$$M^T(K_1) \supseteq M^T(K_2) \qquad (1 \leqslant K_1 < K_2 \leqslant \infty). \tag{2}$$

We prove the following theorem:

THEOREM 1. *$M^T(1) = M^T(K)$ for $1 < K < \infty$.*

It is an open question whether this relation holds for $K = \infty$.

For the proof of Theorem 1 we shall require some notions from the papers by Kashin [2] and Moricz [3].

Since by (2), $M^T(1) \supseteq M^T(K)$ $(1 < K)$, in order to prove Theorem 1 it suffices to establish that $a \notin M^T(K)$ implies $a \notin M^T(1)$. Now by a well-known theorem (see, for example, [4], p. 123) there exists for each sequence $a \in l^2$ a set of positive measure over which the series in Rademacher functions $\sum_0^\infty a_k r_k(x)$ is not T-summable, so that $l^2 \supseteq M^T(1)$.

3. We introduce some notation. For a given sequence $a \in l^2$ and system $\varphi \in \Omega(\infty)$, the series

$$\sum_{k=0}^{\infty} t_{nk}\left(\sum_{l=0}^{k} a_l\varphi_l(x)\right) = \sum_{l=0}^{\infty} a_l\varphi_l(x)\left(\sum_{k=l}^{\infty} t_{nk}\right)$$

converges in the $L^2(0, 1)$ metric; we denote its sum, which is uniquely defined almost everywhere and belongs to $L^2(0, 1)$, by $t_n(a, \varphi, x)$ $(n = 0, 1, \ldots)$.

For a given K, $1 \leqslant K \leqslant \infty$, and sequence $a \in l^2$, we set

$$\|a, K, T\| = \sup_{\varphi\in\Omega(K)} \left\{\int_0^1 \sup_n t_n^2(a, \varphi, x)\, dx\right\}^{1/2},$$

$$D(a, K, T, N_1, N_2) = \sup_{\varphi\in\Omega(K)} \left\{\int_0^1 \sup_{\substack{n_1, n_2 \\ N_1 < n_1 \leqslant n_2 \leqslant N_2}} \left(t_{n_2}(a, \varphi, x) - t_{n_1}(a, \varphi, x)\right)^2 dx\right\}^{1/2}$$

$$(N_1 < N_j2),$$

$$D(a, K, T, N, \infty) = \sup_{\varphi\in\Omega(K)} \left\{\int_0^1 \sup_{\substack{n_1, n_2 \\ N < n_1 \leqslant n_2}} \left(t_{n_2}(a, \varphi, x) - t_{n_1}(a, \varphi, x)\right)^2 dx\right\}^{1/2}.$$

The next two results follow from Theorem 1.

THEOREM 2. *Let $1 \leqslant K \leqslant \infty$ and $a \in l^2$. If*

$$\lim_{N\to\infty} D(a, K, T, N, \infty) = 0, \tag{3}$$

then $a \in M^T(K)$.

THEOREM 3. *Let* $1 < K < \infty$ *and* $a \in l^2$. *If*

$$\lim_{N\to\infty} D(a, K, T, N, \infty) \neq 0, \tag{4}$$

then there exists a system $\varphi \in \Omega(1)$ *such that the series* (1) *is not T-summable almost everywhere on* $(0, 1)$.

4. PROOF OF THEOREM 2. Let $\varphi \in \Omega(K)$, and set

$$F_N(\varphi, T, x) = \sup_{\substack{n_1, n_2 \\ N < n_1 \leqslant n_2}} \left| t_{n_2}(a, \varphi, x) - t_{n_1}(a, \varphi, x) \right|.$$

Then

$$F_N(\varphi, T, x) \geqslant F_{N+1}(\varphi, T, x) \qquad (x \in (0,1), N = 0, 1, \ldots),$$

$$\int_0^1 F_N^2(\varphi, T, x)\, dx \leqslant D^2(a, K, T, N, \infty) \to 0 \qquad (N \to \infty),$$

whence it follows that $\lim_{N\to\infty} F_N(\varphi, T, x) = 0$ a.e. on $(0, 1)$.

REMARK. For $a \in l^2$ it follows from (3) that $\|a, K, T\| < \infty$. Moricz [3] has proved that for $a \in l^2$ the condition $\|a, \infty, T\| < \infty$ implies also that

$$\lim_{N\to\infty} D(a, \infty, T, N, \infty) = 0.$$

It is an open question whether this is true for the case $1 \leqslant K < \infty$.

5. For the proof of Theorem 3 we shall need a couple of lemmas.

LEMMA 1. *Let* $1 < K < \infty$. *For any sequence* $c = \{0, \ldots, 0, c_{M_1+1}, \ldots, c_{M_2}, 0, \ldots\}$ *and any natural numbers* $N_1 < N_2$ *such that* $D(c, K, T, N, N) > 0$, *there exists a system* $\psi \in \Omega(K)$ *of step functions* $\psi_k(x)$ $(k = M_1 + 1, \ldots, M_2)$ *for which*

$$\int_0^1 \max_{\substack{n_1, n_2 \\ N_1 < n_1 \leqslant n_2 \leqslant N_2}} \left(t_{n_2}(c, \psi, x) - t_{n_1}(c, \psi, x)\right)^2 dx > \frac{1}{2} D^2(c, K, T, N_1, N_2).$$

PROOF. Let ε be a positive number such that $\varepsilon < \frac{1}{2} D^2(c, K, T, N_1, N_2)$. Then by the definition of $D(c, K, T, N_1, N_2)$ there exists a system $\varphi = \{\varphi_k(x)\}_{M_1+1}^{M_2}$ with the properties

$$|\varphi_k(x)| \leqslant K \qquad (x \in (0,1), k = M_1 + 1, \ldots, M_2) \tag{5}$$

and

$$\int_0^1 \max_{\substack{n_1, n_2 \\ N_1 < n_1 \leqslant n_2 \leqslant N_2}} \left(t_{n_2}(c, \varphi, x) - t_{n_1}(c, \varphi, x)\right)^2 dx \geqslant D^2(c, K, T, N_1, N_2) - \frac{\varepsilon}{2}. \tag{6}$$

Let χ be a system of step functions $\chi_k(x)$ $(k = M_1 + 1, \ldots, M_2)$ for which

$$\int_0^1 \left(\varphi_k(x) - \chi_k(x)\right)^2 dx < \eta \qquad (k = M_1 + 1, \ldots, M_2)$$

and

$$|\chi_k(x)| \leqslant |\varphi_k(x)| \qquad (x \in (0,1), k = M_1 + 1, \ldots, M_2). \tag{7}$$

We set

$$\alpha_{ij} = \int_0^1 \chi_i(x)\chi_j(x)\, dx \qquad (M_1 < i, j \leqslant M_2).$$

Let the number a ($0 < a < 1$) be chosen such that

$$\left(1-(1-a^2)^2(1-a)\right)/a < K^2, \tag{8}$$

$$(1-a)(1-a^2)\int_0^1 \max_{\substack{n_1, n_2\\ N_1 < n_1 \leqslant n_2 \leqslant N_2}} \left(t_{n_2}(c,\varphi,k) - t_{n_2}(c,\varphi,x)\right)^2 dx$$

$$\geqslant \int_0^1 \max_{\substack{n_1, n_2\\ N_1 < n_1 < n_2 \leqslant N_2}} \left(t_{n_2}(c,\varphi,x) - t_{n_1}(c,\varphi,x)\right)^2 dx - \frac{\varepsilon}{4}.$$

We divide the interval $(1-a^2, 1)$ into $(M_2 - M_1)(M_2 - M_1 - 1)$ mutually disjoint subintervals I_{ij} ($1 \leqslant i, j \leqslant M_2 - M_1$, $i \neq j$) of common length meas $I_{ij} = a^2/(M_2 - M_1)(M_2 - M_1 - 1)$. Then, provided that η is sufficiently small ($\eta < \eta_1$), we have

$$\int_0^1 \max_{\substack{n_1, n_2\\ N_1 < n_1 \leqslant n_2 \leqslant N_2}} \left(t_{n_2}(c,\chi,x) - t_{n_1}(c,\chi,x)\right)^2 dx$$

$$\geqslant \int_0^1 \max_{\substack{n_1, n_2\\ N_1 < n_1 \leqslant n_2 \leqslant N_2}} \left(t_{n_2}(c,\varphi,x) - t_{n_1}(c,\varphi,x)\right)^2 dx - \frac{\varepsilon}{4}, \tag{10}$$

$$\frac{M_2 - M_1}{2a^2} \max_{\substack{i, j\\ 1 \leqslant i, j \leqslant M_2 - M_1, i \neq j}} |\alpha_{ij}| \leqslant K^2, \tag{11}$$

$$\alpha_{kk} > 1 - a^2 \qquad (k = M_1 + 1, \ldots, M_2). \tag{12}$$

We set

$$\bar{\psi}_k(x) = \begin{cases} \chi_k\left(\dfrac{x}{1-a^2}\right) & \left(x \in (0, 1-a^2)\right), \\ \sqrt{\dfrac{M_2 - M_1}{2a^2}|\alpha_{ki}|} & \left(x \in I_{ki}, M_1 < i \leqslant M_2, i \neq k\right), \\ -\sqrt{\dfrac{M_2 - M_1}{2a^2}|\alpha_{ki}|}\ \operatorname{sgn} \alpha_{ki} & \left(x \in I_{ik}, M_1 < i \leqslant M_2, i \neq k\right), \\ 0 & \text{otherwise} \end{cases}$$

$$(k = M_1 + 1, \ldots, M_2).$$

It is clear that the step functions $\bar{\psi}_k(x)$ ($k = M_1 + 1, \ldots, M_2$) form an orthogonal system on $(0,1)$. Furthermore, in view of (5), (7), (11), (12) and the definition of the $\bar{\psi}_k(x)$, we have

$$\left|\bar{\psi}_k(x)\right| \leqslant K \qquad \left(x \in (0,1), k = M_1 + 1, \ldots, M_2\right), \tag{13}$$

$$\int_0^1 \max_{\substack{n_1, n_2\\ N_1 < n_1 \leqslant n_2 \leqslant N_2}} \left(t_{n_2}(c,\bar{\psi},x) - t_{n_1}(c,\bar{\psi},x)\right)^2 dx$$

$$\geqslant \int_0^{1-a^2} \max_{\substack{n_1, n_2\\ N_1 < n_1 \leqslant n_2 \leqslant N_2}} \left(t_{n_2}(c,\bar{\psi},x) - t_{n_1}(c,\bar{\psi},x)\right)^2 dx$$

$$= (1-a^2)\int_0^1 \max_{\substack{n_1, n_2\\ N_1 < n_1 \leqslant n_2 \leqslant N_2}} \left(t_{n_2}(c,\chi,x) - t_{n_1}(c,\chi,x)\right)^2 dx, \tag{14}$$

$$\int_0^1 \bar{\psi}_k^2(x)\,dx \geqslant \int_0^{1-a^2} \bar{\psi}_k^2(x)\,dx = (1-a^2)\alpha_{kk} \geqslant (1-a^2)^2$$
$$(k = M_1+1,\ldots,M_2). \quad (15)$$

Finally, let

$$\psi_k(x) = \begin{cases} \bar{\psi}_k\left(\dfrac{x}{1-a}\right) & (x \in (0, 1-a)), \\ m_k r_k\left(\dfrac{x-1+a}{a}\right) & (x \in (1-a, 1)), \end{cases} \qquad (k = M_1+1,\ldots,M_2),$$

where $r_k(x) = \operatorname{sgn} \sin 2^k \pi x$ is the kth Rademacher function and the m_k are chosen so as to normalize the $\psi_k(x)$; that is,

$$(1-a)\int_0^1 \bar{\psi}_k^2(x)\,dx + m_k^2 a = 1 \qquad (k = M_1+1,\ldots,M_2). \quad (16)$$

It now follows from (8), (15) and (16) that

$$m_k \leqslant \sqrt{\frac{1-(1-a^2)^2(1-a)}{a}} \leqslant K \qquad (k = M_1+1,\ldots,M_2).$$

Hence we see from (13) that $|\psi_k(x)| \leqslant K$ ($x \in (0,1)$; $k = M_1+1,\ldots,M_2$). Clearly the step functions $\psi_k(x)$ ($k = M_1+1,\ldots,M_2$) form an orthonormal system on $(0,1)$. Furthermore,

$$\int_0^1 \max_{\substack{n_1, n_2 \\ N_1 < n_1 \leqslant n_2 \leqslant N_2}} \left(t_{n_2}(c,\psi,x) - t_{n_1}(c,\psi,x)\right)^2 dx$$
$$\geqslant \int_0^{1-a} \max_{\substack{n_1, n_2 \\ N_1 < n_1 \leqslant n_2 \leqslant N_2}} \left(t_{n_2}(c,\bar{\psi},x) - t_{n_1}(c,\bar{\psi},x)\right)^2 dx$$
$$= (1-a)\int_0^1 \max_{\substack{n_1, n_2 \\ N_1 < n_1 \leqslant n_2 \leqslant N_2}} \left(t_{n_2}(c,\bar{\psi},x) - t_{n_1}(c,\bar{\psi},x)\right)^2 dx.$$

Finally it follows from (6), (9), (10), (14) and (17) that

$$\int_0^1 \max_{\substack{n_1, n_2 \\ N_1 < n_1 \leqslant n_2 \leqslant N_2}} \left(t_{n_2}(c,\psi,x) - t_{n_1}(c,\psi,x)\right)^2 dx \geqslant D^2(c,K,T,N_1,N_2) - \varepsilon,$$

from which the estimate of Lemma 1 follows.

LEMMA 2. *Let* $1 < K < \infty$ *and* $c = \{0,\ldots,0,c_{M_1+1},\ldots,c_{M_2},0,\ldots\}$, *and let* $N_1 < N_2$ *be natural numbers such that*

$$D^2(c,K,T,N_1,N_2) > 512C^2K^2 \sum_{k=M_1+1}^{M_2} c_k^2,$$

where C *is the constant in condition* (iii). *Then there exists a system* $\psi \in \Omega(1)$ *of step functions* $\psi_k(x)$ ($k = M_1+1,\ldots,M_2$) *such that*

$$\max_{\substack{n_1, n_2 \\ N_1 < n_1 \leqslant n_2 \leqslant N_2}} \left|t_{n_2}(c,\psi,x) - t_{n_1}(c,\psi,x)\right| \geqslant \frac{1}{4K} D(c,K,T,N_1,N_2)$$

on a simple set E ($\subseteq (0,1)$) *with measure* meas $E \geqslant 1/4$.

A set is said to be *simple* if it consists of a finite number of intervals.

PROOF. We suppose without loss of generality that $D(c, K, T, N_1, N_2) = 2$. Applying Lemma 1, we obtain a system $\bar{\psi} \in \Omega(K)$ of step functions $\bar{\psi}_k(x)$ $(k = M_1 + 1, \ldots, M_2)$ such that

$$\int_0^1 \max_{\substack{n_1, n_2 \\ N_1 < n_1 \leqslant n_2 \leqslant N_2}} \left(t_{n_2}(c, \bar{\psi}, x) - t_{n_1}(c, \bar{\psi}, x)\right)^2 dx > \frac{1}{2} D^2(c, K, T, N_1, N_2) = 2. \tag{18}$$

Let $I_r = (\alpha_r, \beta_r)$ $(r = 1, \ldots, \rho)$ be a partition of $(0, 1)$ into mutually disjoint subintervals such that on any subinterval I_r each function $\bar{\psi}_k(x)$ is constant. We denote the values of the function

$$\max_{\substack{n_1, n_2 \\ N_1 < n_1 \leqslant n_2 \leqslant N_2}} \left|t_{n_2}(c, \bar{\psi}, x) - t_{n_1}(c, \bar{\psi}, x)\right|$$

on the interval I_r by w_r. By (18) we have

$$4 = D^2(c, K, T, N_1, N_2) \geqslant \sum_{r=1}^{\rho} w_r^2 \operatorname{meas} I_r > \frac{1}{2} D^2(c, K, T, N_1, N_2) = 2. \tag{19}$$

Let $1 \leqslant r_1, \ldots, r_k \leqslant \rho$ be those indices r for which $w_r \geqslant 1$; we relabel the indices r which are distinct from the $r_1, \ldots, r_\kappa$ by the sequence $s_1, \ldots, s_{\rho-\kappa}$, in increasing order. It follows from (19) that

$$4 \geqslant \sum_{k=1}^{\kappa} w_{r_k}^2 \operatorname{meas} I_{r_k} > 1. \tag{20}$$

We set

$$\gamma = \sum_{k=1}^{\kappa} w_{r_k}^2 \operatorname{meas} I_{r_k}, \qquad \delta = \sum_{k=1}^{\rho-\kappa} \operatorname{meas} I_{s_k};$$

clearly $\delta \leqslant 1$. Let $J_k' = (a_k', b_k')$ $(k = 1, \ldots, \kappa)$ be mutually disjoint intervals on $(0, \gamma)$ with measure $\operatorname{meas} J_k' = w_{r_k}^2 \operatorname{meas} I_{r_k}$ $(k = 1, \ldots, \kappa)$; and let $J_k'' = (a_k'', b_k'')$ $(k = 1, \ldots, \rho - \kappa)$ be mutually disjoint intervals on $(\gamma, \gamma + \delta)$ with measure $\operatorname{meas} J_k'' = \operatorname{meas} I_{s_k}$ $(k = 1, \ldots, \rho - \kappa)$. We set

$$\tilde{\psi}_l(x) = \begin{cases} \bar{\psi}_l\left(\dfrac{x - a_{s_k}''}{b_{s_k}'' - a_{s_k}''}\left(\beta_{s_k} - \alpha_{s_k}\right) + \alpha_{s_k}\right) & \left(x \in J_k'', k = 1, \ldots, \rho - \kappa\right), \\ \dfrac{1}{w_{r_k}} \bar{\psi}_l\left(\dfrac{x - a_{r_k}'}{b_{r_k}' - a_{r_k}'}\left(\beta_{r_k} - \alpha_{r_k}\right) + \alpha_{r_k}\right) & \left(x \in J_k', k = 1, \ldots, \kappa\right) \end{cases}$$

$$(l = M_1 + 1, \ldots, M_2).$$

Also let

$$\psi_l^*(x) = \tilde{\psi}_l((\gamma + \delta)x)/K \qquad (l = M_1 + 1, \ldots, M_2).$$

It is clear that the step functions $\psi_k^*(x)$ $(k = M_1 + 1, \ldots, M_2)$ form an orthogonal system on $(0, 1)$ such that

$$\left|\psi_k^*(x)\right| \leqslant 1 \qquad (x \in (0,1), k = M_1 + 1, \ldots, M_2). \tag{21}$$

Let $\bar{E}$ be the image of $(0, \gamma)$ under the linear map $y = x/(\gamma + \delta)$. It follows from (20) that

$$\operatorname{meas} \bar{E} \geqslant 1/2. \tag{22}$$

Furthermore, according to the definition of the $\psi_k^*(x)$ we have

$$\max_{\substack{n_1, n_2 \\ N_1 < n_1 \leqslant N_2 \leqslant N_2}} \left| t_{n_2}(c, \psi^*, x) - t_{n_1}(c, \psi^*, x) \right| = \frac{1}{K}$$

$$= \frac{1}{2K} D(c, K, T, N_1, N_2) \qquad (x \in \bar{E}). \tag{23}$$

Let $I_s^* = (\gamma_s, \delta_s)$ $(s = 1, \dots, \sigma)$ be a partition of $(0, 1)$ into mutually disjoint intervals such that each $\psi_k^*(x)$ is constant over each I_s^* and $\bar{E}$ is the union of some of the I_s^*. We denote the value of $\psi_k^*(x)$ on I_s^* by $\rho_s^{(k)}$. For each s $(1 \leqslant s \leqslant \sigma)$ let $\{\chi_s^k(x)\}_{k=M_1+1}^{M_2}$ be a sequence of step functions on $(0, 1)$ such that

$$\int_0^1 \chi_s^{(k)}(x)\, dx = 0 \qquad (k = M_1 + 1, \dots, M_2),$$

$$\int_0^1 \chi_s^{(k)}(x) \chi_s^{(l)}(x)\, dx = 0 \qquad (k, l = M_1 + 1, \dots, M_2, k \neq l),$$

and any of the $\chi_s^{(k)}(x)$ takes the values $1 - \rho_s^{(k)}$ and $-1 - \rho_s^{(k)}$. (In the case when $\rho_s^{(k)} = 1$, we set $\chi_s^{(k)}(x) \equiv 0$.) It is not too difficult to construct such a system of functions. We set

$$\chi_s^{(k)}(I_s^*, x) = \begin{cases} \chi_s^{(k)}\left(\dfrac{x - \gamma_s}{\delta_s - \gamma_s}\right) & (x \in I_s^*), \\ 0 \quad \text{otherwise} \end{cases}$$

$(k = M_1 + 1, \dots, M_2;\ s = 1, \dots, \sigma)$. We denote the system $\{\chi_s^{(k)}(I_s^*, x)\}_{k=M_1+1}^{M_2}$ by χ_s $(s = 1, \dots, \sigma)$. It follows from (21) that

$$\left|\chi_s^{(k)}(I_s^*, x)\right| \leqslant 2 \qquad \left(x \in I_s^*, k = M_1 + 1, \dots, M_2, s = 1, \dots, \sigma\right), \tag{24}$$

and, in addition,

$$\int_{I_s^*} \chi_s^{(k)}(I_s^*, x)\, dx = 0 \qquad (k = M_1 + 1, \dots, M_2, s = 1, \dots, \sigma),$$

$$\int_{I_s^*} \chi_s^{(k)}(I_s^*, x) \chi_s^{(l)}(I_s^*, x)\, dx = 0 \qquad (k, l = M_1 + 1, \dots, M_2, k \neq l, s = 1, \dots, \sigma).$$

Let

$$\psi_k(x) = \psi_k^*(x) + \sum_{s=1}^{\sigma} \chi_s^{(k)}(I_s^*, x) \qquad (k = M_1 + 1, \dots, M_2).$$

It is easy to see that the step functions $\psi_k(x)$ form an orthogonal system on $(0, 1)$ and that

$$|\psi_k(x)| = 1 \qquad (x \in (0, 1), k = M_1 + 1, \dots, M_2);$$

in particular, $\psi = \{\psi_k(x)\}_{M_1+1}^{M_2} \in \Omega(1)$.

Let $I^* \subseteq E$, and let $n_1(s)$, $n_2(s)$ be numbers such that $N_1 < n_1(s) \leqslant n_2(s) \leqslant N_2$ and

$$\max_{\substack{n_1, n_2\\ N_1 < n_1 \leqslant n_2 \leqslant N_2}} \left|t_{n_2}(c, \psi^*, x) - t_{n_1}(c, \psi^*, x)\right| = \left|t_{n_2(s)}(c, \psi^*, x) - t_{n_1(s)}(c, \psi^*, x)\right| \quad \left(x \in I_s^*\right).$$

From (23) we have

$$\begin{aligned}
&\operatorname{meas}\left\{ x \in I_s^*: \max_{\substack{n_1, n_2\\ N_1 < n_1 \leqslant n_2 \leqslant N_2}} \left|t_{n_2}(c, \psi, x) - t_{n_1}(c, \psi, x)\right| \geqslant \frac{1}{4K} D(c, K, T, N_1, N_2)\right\}\\
&= \operatorname{meas}\left\{ x \in I_s^*: \left|t_{n_2(s)}(c, \psi, x) - t_{n_1(s)}(c, \psi, x)\right| \geqslant \frac{1}{4K} D(c, K, T, N_1, N_2)\right\}\\
&\geqslant \operatorname{meas}\left\{ x \in I_s^*: \left|t_{n_2(s)}(c, \psi^*, x) - t_{n_1(s)}(c, \psi^*, x)\right| \geqslant \frac{1}{2K} D(c, K, T, N_1, N_2)\right\}\\
&\quad - \operatorname{meas}\left\{ x \in I_s^*: \left|t_{n_2(s)}(c, \chi_s, x) - t_{n_1(s)}(c, \chi_s, x)\right| \geqslant \frac{1}{4K} D(c, K, T, N_1, N_2)\right\}\\
&= \operatorname{meas}\left\{ x \in I_s^*: \max_{\substack{n_1, n_2\\ N_1 < n_1 \leqslant n_2 \leqslant N_2}} \left|t_{n_2}(c, \psi^*, x) - t_{n_1}(c, \psi^*, x)\right| \geqslant \frac{1}{2K} D(c, K, T, N_1, N_2)\right\}\\
&\quad - \operatorname{meas}\left\{ x \in I_s^*: \left|t_{n_2(s)}(c, \chi_s, x) - t_{n_1(s)}(c, \chi_s, x)\right| \geqslant \frac{1}{4K} D(c, K, T, N_1, N_2)\right\}.
\end{aligned} \tag{25}$$

From (24) and the hypothesis of the lemma, we obtain the following estimate via Tchebycheff's inequality:

$$\begin{aligned}
&\operatorname{meas}\left\{ x \in I_s^*: \left|t_{n_2(s)}(c, \chi_s, x) - t_{n_1(s)}(c, \chi_s, x)\right| \geqslant \frac{1}{4K} D(c, K, T, N_1, N_2)\right\}\\
&\leqslant 16K^2 \sum_{k=M_1+1}^{M_2} \left(\sum_{l=k}^{\infty} \left(t_{n_2(s),l} - t_{n_1(s),l}\right)\right)^2 c_k^2\\
&\quad \times \int_{I_s^*} \left(\chi_s^{(k)}(I_s^*, x)\right)^2 dx / D^2(c, K, T, N_1, N_2)\\
&\leqslant \operatorname{meas} I_s^* \cdot 256K^2C^2 \sum_{k=M_1+1}^{M_2} c_k^2 / D^2(c, K, T, N_1, N_2) \leqslant \operatorname{meas} I_s^*/2.
\end{aligned}$$

It follows from this and (23), (25) that if $I_s^* \subseteq \overline{E}$, then

$$\operatorname{meas}\left\{x \in I_s^*: \max_{\substack{n_1, n_2 \\ N_1 < n_1 \leqslant n_2 \leqslant N_2}} \left|t_{n_2}(c, \psi, x) - t_{n_1}(c, \psi, x)\right| \geqslant \frac{1}{4K} D(c, K, T, N_1, N_2)\right\} \geqslant \operatorname{meas} I_s^*/2. \tag{26}$$

Let

$$E = \bigcup_{\substack{s \\ I_s^* \subseteq \overline{E}}} \left\{x \in I_s^*: \max_{\substack{n_1, n_2 \\ N_1 < n_1 \leqslant n_2 \leqslant N_2}} \left|t_{n_2}(c, \psi, x) - t_{n_1}(c, \psi, x)\right| \geqslant \frac{1}{4K} D(c, K, T, N_1, N_2)\right\},$$

where E is a simple set and, by (22) and (26), meas $E \geqslant 1/4$.

This completes the proof of Lemma 2.

6. PROOF OF THEOREM 3. By (4) and the definition of $D(a, K, T, N, \infty)$ there exists a positive number ρ such that

$$D(a, K, T, N, \infty) \downarrow \rho \qquad (N \uparrow \infty).$$

Now by the definition of $D(a, K, T, N, \infty)$ and $D(a, K, T, N_1, N_2)$ we have, for each number N_1,

$$D(a, K, T, N_1, N_2) \uparrow D(a, K, T, N_1, \infty) \qquad (N_2 \uparrow \infty).$$

It follows from these relations that there exist for arbitrary N_0 numbers N_1 and N_2 $(N_0 < N_1 < N_2)$ such that

$$D(a, K, T, N_1, N_2) > \rho.$$

We now define three sequences of numbers $\{M_i\}_0^\infty$, $\{N_1(i)\}_1^\infty$, and $\{N_2(i)\}_1^\infty$ with the following properties:

We have the estimates

$$(-1 =) \quad M_0 < M_1 < \cdots < M_i < \cdots,$$

$$N_1(i) < N_2(i) < N_1(i + d1) < N_2(i + 1) \qquad (i = 1, 2, \ldots);$$

and, in addition, for each i $(i = 1, 2, \ldots)$

$$D(a, K, T, N_1(i), N_2(i)) > \rho, \tag{27}$$

$$\left(\frac{\rho}{3}\right)^2 \geqslant 512 C^2 K^2 \sum_{k=M_i+1}^{M_{i+1}} a_k^2, \tag{28}$$

$$2K\left(\sum_{k=0}^{M_i} |a_k| \left(2 \max_{\substack{n \\ N_1(i) < n \leqslant N_2(i)}} \sum_{k=0}^{M_i} |t_{nk}|\right) + \max_{\substack{n \\ N_1(i) < n \leqslant N_2(i)}} \left|1 - \sum_{k=0}^{\infty} t_{nk}\right|\right) \leqslant \frac{\rho}{24K}, \tag{29}$$

$$4C^2 (N_2(i) - N_1(i))^2 \left(\sum_{k=M_{i+1}+1}^{\infty} a_k^2\right) \leqslant \left(\frac{\rho}{3}\right)^2 \frac{1}{2^i}, \tag{30}$$

where C is the constant in (iii).

Let M_1 be a number such that

$$\sum_{k=M_1+1}^{\infty} a_k^2 < \left(\frac{\rho}{3}\right)^2/512C^2K^2; \tag{31}$$

such a number exists since $a \in l^2$. We then choose $N_1(1)$ so large that when $n > N_1(1)$

$$2K\left(\sum_{k=0}^{M_1} |a_k|\left(2\sum_{k=0}^{M_1} |t_{nk}| + \left|1 - \sum_{k=0}^{\infty} t_{nk}\right|\right)\right) \leqslant \frac{\rho}{24K};$$

such a number $N_1(1)$ exists by (i) and (ii). Next let $N_2(1) > N_1(1)$ be such that (27) holds when $i = 1$. For this $N_2(1)$, when $i = 1$, (29) holds as well. Finally we choose M_2 such that (30) holds when $i = 1$; this can be done since $a \in l^2$. It follows from (31) that (28) holds when $i = 1$. We define the sequences $\{M_i\}_0^\infty$, $\{N_1(i)\}_1^\infty$, and $\{N_2(i)\}_1^\infty$ by induction.

Note that, by virtue of (29), for each $\varphi \in \Omega(K)$ and sequence $a_1(i) = \{a_0, \ldots, a_{M_i}, 0, \ldots\}$ the following estimate holds throughout $(0, 1)$:

$$\max_{n_1, n_2} \left|t_{n_2}(a_1(i), \varphi, x) - t_{n_1}(a_1(i), \varphi, x)\right|$$

$$\leqslant \max_{\substack{n_1, n_2 \\ N_1(i) < n_1 \leqslant n_2 \leqslant N_2(i)}} \left|\sum_{k=0}^{M_i} t_{n_2k}\left(\sum_{l=0}^{k} a_l\varphi_l(x)\right) - \sum_{k=0}^{M_i} t_{n_1k}\left(\sum_{l=0}^{k} a_l\varphi_l(x)\right)\right|$$

$$+ \max_{\substack{n_1, n_2 \\ N_1(i) < n_1 \leqslant n_2 \leqslant N_2(i)}} \left|\sum_{l=0}^{M_i} a_l\varphi_l(x)\right| \left|\sum_{k=M_i=1}^{\infty} t_{n_2k} - \sum_{k=M_i+1}^{\infty} t_{n_1k}\right|$$

$$\leqslant 2K\sum_{k=0}^{M_i} |a_k| \max_{\substack{n \\ N_1(i) < n \leqslant N_2(i)}} \sum_{k=0}^{M_i} |t_{nk}| + 2K\sum_{l=0}^{M_i} |a_l| \max_{\substack{n \\ N_1(i) < n \leqslant N_2(i)}} \left|1 - \sum_{k=M_i+1}^{\infty} t_{nk}\right|$$

$$\leqslant 2K\sum_{k=0}^{M_i} |a_k|\left(\max_{\substack{n \\ N_1(i) < n \leqslant N_2(i)}} \sum_{k=0}^{M_i} |t_{nk}| + \max_{\substack{n \\ N_1(i) < n \leqslant N_2(i)}} \left(\sum_{k=0}^{M_i} |t_{nk}| + \left|1 - \sum_{k=0}^{\infty} t_{nk}\right|\right)\right)$$

$$= 2K\sum_{k=0}^{M_i} |a_k|\left(2\max_{\substack{n \\ N_1(i) < n \leqslant N_2(i)}} \sum_{k=0}^{M_i} |t_{nk}| + \max_{\substack{n \\ N_1(i) < n \leqslant N_2(i)}} \left|1 - \sum_{k=0}^{\infty} t_{nk}\right|\right) < \frac{\rho}{24K}, \tag{32}$$

and hence

$$\left\{\int_0^1 \max_{\substack{n_1, n_2 \\ N_1(i) < n_1 \leqslant n_2 \leqslant N_2(i)}} \left(t_{n_2}(a_1(i), \varphi, x) - t_{n_1}(a_1(i), \rho, x)\right)^2 dx\right\}^{1/2} < \frac{\rho}{3}$$

$$(\varphi \in \Omega(K)). \tag{33}$$

Furthermore, from (30) and in view of (iii), we have the following estimate for each $\varphi \in \Omega(K)$ and even for each $\varphi \in \Omega(\infty)$, and for the sequence $a_2(i) = \{0,\dots,0, a_{M_{i+1}+1},\dots\}$:

$$\left\{\int_0^1 \max_{\substack{n_1, n_2\\ N_1(i)<n_1\leqslant n_2\leqslant N_2(i)}} \left(t_{n_2}(a_2(i),\varphi,x) - t_{n_1}(a_2(i),\varphi,x)\right)^2 dx\right\}^{1/2}$$

$$\leqslant 2\left\{\int_0^1 \max_{\substack{n\\ N_1(i)<n\leqslant N_2(i)}} \left(t_n(a_2(i),\varphi,x)\right)^2 dx\right\}^{1/2}$$

$$= 2\left\{\int_0^1 \max_{\substack{n\\ N_1(i)<n_1\leqslant n_2\leqslant N_2(i)}} \left(\sum_{k=M_{i+1}+1}^{\infty} t_{nk}\left(\sum_{l=M_{i+1}+1}^{k} a_l\varphi_l(x)\right)\right)^2 dx\right\}^{1/2}$$

$$\leqslant 2\sum_{n=N_1(i)+1}^{N_2(i)} \left\{\int_0^1 \left(\sum_{k=M_{i+1}+1}^{\infty} |t_{nk}|\left(\sum_{l=M_{i+1}+1}^{k} a_l\varphi_l(x)\right)\right)^2 dx\right\}^{1/2}$$

$$\leqslant 2\sum_{n=N_1(i)+1}^{N_2(i)} \left\{\sum_{k=M_{i+1}+1}^{\infty} |t_{nk}|\right\}^{1/2} \left\{\sum_{k=M_{i+1}+1}^{\infty} |t_{nk}| \int_0^1 \left(\sum_{l=M_{i+1}+1}^{k} a_l\varphi_l(x)\right)^2 dx\right\}^{1/2}$$

$$\leqslant 2\sqrt{C}\sum_{n=N_1(i)+1}^{N_2(i)} \left\{\sum_{k=M_{i+1}+1}^{\infty} |t_{nk}|\right\}^{1/2} \left\{\sum_{l=M_{i+1}+1}^{\infty} a_l^2\right\}^{1/2}$$

$$\leqslant 2C(N_2(i) - N_1(i))\left\{\sum_{l=M_{i+1}+1}^{\infty} a_l^2\right\}^{1/2} \leqslant \frac{\rho}{3}\frac{1}{\sqrt{2^i}} \quad \left(< \frac{\rho}{3}\right). \tag{34}$$

It follows from (33) and (34) that for each $\varphi \in \Omega(K)$ and for the sequence $a(i) = \{0,\dots,0, a_{M_i+1},\dots,a_{M_{i+1}}, 0,\dots\}$ we have the estimates

$$\left\{\int_0^1 \max_{\substack{n_1, n_2\\ N_1(i)<n_1\leqslant n_2\leqslant N_2(i)}} \left(t_{n_2}(a(i),\varphi,x) - t_{n_1}(a(i),\varphi,x)\right)^2 dx\right\}^{1/2}$$

$$\geqslant \left\{\int_0^1 \max_{\substack{n_1, n_2\\ N_1(i)<n_1\leqslant n_2\leqslant N_2(i)}} \left(t_{n_2}(a,\varphi,x) - t_{n_1}(a,\varphi,x)\right)^2 dx\right\}^{1/2}$$

$$-\left\{\int_0^1 \max_{\substack{n_1, n_2\\ N_1(i)<n_1\leqslant n_2\leqslant N_2(i)}} \left(t_{n_2}(a_1(i),\varphi,x) - t_{n_1}(a_1(i),\varphi,x)\right)^2 dx\right\}^{1/2}$$

$$-\left\{\int_0^1 \max_{\substack{n_1, n_2\\ N_1(i)<n_1\leqslant n_2\leqslant N_2(i)}} \left(t_{n_2}(a_2(i),\varphi,x) - t_{n_1}(a_2(i),\varphi,x)\right)^2 dx\right\}^{1/2}$$

$$> \left\{\int_0^1 \max_{\substack{n_1, n_2\\ N_1(i)<n_1\leqslant n_2\leqslant N_2(i)}} \left(t_{n_2}(a,\varphi,x) - t_{n_1}(a,\varphi,x)\right)^2 dx\right\}^{1/2} - 2\frac{\rho}{3},$$

whence, by (27) and the definition of $D(a, K, T, N_1, N_2)$,

$$D(a(i), K, T, N_1(i), N_2(i)) > \rho/3 \qquad (i = 1, 2, \dots). \tag{35}$$

We note further that from (34) we see that for any $\varphi \in \Omega(K)$

$$\sum_{i=1}^{\infty} \int_0^1 \max_{\substack{n_1, n_2 \\ N_1(i) \leqslant n_1 \leqslant n_2 \leqslant N_2(i)}} \left(t_{n_2}(a_2(i), \varphi, x) - t_{n_1}(a_2(i), \varphi, x)\right)^2 dx \leqslant \frac{\rho^2}{9} \sum_{i=1}^{\infty} \frac{1}{2^i} < \infty,$$

whence it follows that for each system $\varphi \in \Omega(K)$

$$\lim_{i \to \infty} \max_{\substack{n_1, n_2 \\ N_1(i) < n_1 \leqslant n_2 \leqslant N_2(i)}} \left|t_{n_2}(a_2(i), \varphi, x) - t_{n_1}(a_2(i), \varphi, x)\right| = 0 \tag{36}$$

a.e. on $(0, 1)$.

The proof of Theorem 3 is now completed by the application of simple and well-known arguments. We define by induction a system $\varphi \in \Omega(1)$ of step functions and a stochastically independent sequence $\{E_i\}_1^{\infty}$ of simple sets on $(0, 1)$ such that the following conditions hold:

a) meas $E_i \geqslant 1/4$ for each $i = 1, 2, \dots$.

b) For each $i = 1, 2, \dots$,

$$\max_{\substack{n_1, n_2 \\ N_1(i) < n_1 \leqslant n_2 \leqslant N_2(i)}} \left|t_{n_2}(a(i), \varphi, x) - t_{n_1}(a(i), \varphi, x)\right| \geqslant \frac{1}{12K}\rho \qquad (x \in E_i).$$

Suppose that $\varphi_k(x) = r_k(x)$ for $k = 0, \dots, M_1$ $(x \in (0, 1))$. Let i_0 be a natural number. Suppose that we have already constructed a system of signs, which are step functions $\varphi_0(x), \dots, \varphi_{M_{i_0}}(x)$, and simple sets $E_1, \dots, E_{i_0 - 1}$ such that these functions form an orthonormal system on $(0, 1)$, the sets $E_1, \dots, E_{i_0-1}$ are stochastically independent, and, in addition, that conditions a) and b) hold for $i = 1, \dots, i_0 - 1$. Since by (28) and (35) the condition of Lemma 2 is satisfied for $c = a(i_0)$, we can apply Lemma 2 to $c = a(i_0)$; we denote the system and set so obtained by $\psi^{(i_0)} = \{\psi_k^{(i_0)}(x)\}_{M_{i_0+1}}^{M^{i_0+1}}$ and E_{i_0} respectively. There exists by hypothesis a partition of $(0, 1)$ into mutually disjoint intervals I_r $(r = 1, \dots, \rho)$ such that on any interval I_r each of the functions $\varphi_0(x), \dots, \varphi_{M_{i_0}}(x)$ is constant and each of the sets $E_1, \dots, E_{i_0-1}$ is a union of some of the I_r. We denote the left and right halves of I_r by I_r' and I_r'' respectively. Let $I = (a, b)$ be a subinterval of $(0, 1)$. For the function $f(x)$ defined on $(0, 1)$ we set

$$f(I, x) = \begin{cases} f\left(\dfrac{x - a}{b - a}\right) & (x \in I), \\ 0 \quad \text{otherwise.} \end{cases}$$

Next, we denote by $H(I)$, where $H \subseteq (0, 1)$ is a given set, the image of H under the transformation $y = (b - a)x + a$.

We set

$$\varphi_k(x) = \sum_{r=1}^{\rho} \psi_k^{(i_0)}(I_r', x) - \sum_{r=1}^{\rho} \psi_k^{(i_0)}(I_r''; x) \qquad (k = M_{i_0} + 1, \dots, M_{i_0+1})$$

and

$$E_{i_0} = \bigcup_{r=1}^{\rho} \left(\bar{E}_{i_0}(I'_r) \cup \bar{E}_{i_0}(I''_r) \right).$$

It is clear that the step functions $\varphi_0(x),\ldots,\varphi_{M_{i_0+1}}(x)$ form an orthonormal system of signs on $(0,1)$ and that the simple sets $E_1,\ldots,E_{i_0}$ are stochastically independent. By Lemma 2, the relations a) and b) hold also for the case $i = i_0$. We construct the system φ and the sequence $\{E_i\}_1^\infty$ with the required properties by induction.

From b) and (32) we obtain

$$\max_{\substack{n_1, n_2 \\ N_1(i) < n_1 \leqslant n_2 \leqslant N_2(i)}} \left| t_{n_2}(a_1(i) + a(i), \varphi, x) - t_{n_1}(a_1(i) + a(i), \varphi, x) \right| > \frac{1}{24K}\rho \qquad (x \in E_i) \quad (37)$$

for all $i = 1, 2, \ldots$.

Since the E_i $(i = 1, 2, \ldots)$ are stochastically independent and a) holds for each i, we obtain, on application of the second Borel-Cantelli Lemma,

$$\operatorname{meas}\left(\overline{\lim_{i\to\infty}} E_i \right) = 1.$$

If $x \in \overline{\lim}_{i\to\infty} E_i$, then (37) holds for infinitely many i. Let $\bar{E}$ denote the set of all points x for which $x \in E$ and (36) holds. Then meas $\bar{E} = 1$, and for $x \in \bar{E}$

$$\begin{aligned}
&\max_{\substack{n_1, n_2 \\ N_1(i) < n_1 \leqslant n_2 \leqslant N_2(i)}} \left| t_{n_2}(a, \varphi, x) - t_{n_1}(a, \varphi, x) \right| \\
&\quad \geqslant \max_{\substack{n_1, n_2 \\ N_1(i) < n_1 \leqslant n_2 \leqslant N_2(i)}} \left| t_{n_2}(a_1(i) + a(i), \varphi, x) - t_{n_1}(a_1(i) + a(i), \varphi, x) \right| \\
&\quad\quad - \max_{\substack{n_1, n_2 \\ N_1(i) < n_1 \leqslant n_2 \leqslant N_2(i)}} \left| t_{n_2}(a_2(i), \varphi, x) - t_{n_1}(a_2(i), \varphi, x) \right| \\
&\quad \geqslant \frac{\rho}{24K} - \frac{\rho}{48K} = \frac{\rho}{48K} > 0
\end{aligned}$$

for infinitely many i; that is, the sequence $\{t_n(a, \varphi, x)\}_0^\infty$ diverges at x.

Theorem 3 is now completely proved.

REMARK. Moricz [3] has proved the following theorem. If $a \in l^2$ and $\lim_{N\to\infty} D(a, \infty, T, N, \infty) \neq 0$, then there exists a system $\varphi \in \Omega(1)$ such that the series (1) is non-T-summable a.e. on $(0,1)$. It is an open question whether this holds for a system $\varphi \in \Omega(1)$.

BIBLIOGRAPHY

1. K. Tandori, *Über beschränkte orthonormierte Systems*, Acta Math. Acad. Sci. Hungar. **31** (1978), 279–285.

2. B. S. Kašin [Kashin], *On Weyl's multipliers for almost everywhere convergence of orthogonal series*, Anal. Math. **2** (1976), 249–266.

3. Ferenc Móricz, *On the T-summation of orthogonal series*, Acta Sci. Math. (Szeged) **30** (1969), 49–67.

4. A. Zygmund, *Trigonometrical series*, PWN, Warsaw, 1935.

Translated by G. G. GOULD

Труды Матем. Инст. Стеклов.
164(1983)

Proceedings of the Steklov
Institute of Mathematics
1985, Issue 2

ON CONDITIONS FOR INTEGRABILITY OF MULTIPLE TRIGONOMETRIC SERIES

UDC 517.5

S. A. TELYAKOVSKIĬ

ABSTRACT. Sufficient conditions for integrability of the sums of multiple trigonometric series are established, and corresponding estimates of integrals are obtained.

These conditions are expressed in terms of convergence of certain numerical series constructed on the basis of first or second differences of the coefficients of the original trigonometric series or the series obtained from it by termwise integration.

Bibliography: 10 titles.

Many variants are known of integrability conditions for one-variable trigonometric series with coefficients which decrease sufficiently regularly (integrability of the sums of these series is intended).

This paper is devoted to the proof of many-variable analogues of integrability conditions from the author's papers [1] and [2] relating to the one-variable case.

The integrability of both one-variable and several-variable trigonometric series was studied by the author in [3]. In particular, we shall obtain a simpler form of integrability conditions for multiple series.

It is known that sine and cosine series behave differently in regard to the matter under consideration of integrability of their sums. The integrability conditions for the series are themselves different, as are the methods of proof. In fact the proof of theorems for cosine series is comparatively simple, while sine series have required finer reasoning.

Kano [4] has represented a sine series as a termwise-differentiated cosine series to prove theorems on the integrability of one-variable series. By this method he has obtained, in particular, a simple proof of one result of [3]. This question is discussed in [2].

In this paper we apply the indicated procedure to study the integrability of multiple series.

Nosenko [5] has extended the integrability conditions of [1] to double series. For the case of cosine series our reasoning repeats his proof.

We shall consider a trigonometric series of m variables of the form

$$\sum_{\substack{k_1,\dots,k_\nu=1\\ k_{\nu+1},\dots,k_m=0}}^{\infty}{}' a_{k_1,\dots,k_m} \sin k_1x_1 \cdots \sin k_\nu x_\nu \cos k_{\nu+1}x_{\nu+1} \cdots \cos k_m x_m, \qquad (1)$$

1980 *Mathematics Subject Classification*. Primary 42B05.

where $0 \leqslant \nu \leqslant m$ and the prime on the summation sign means that if i numbers among $k_{\nu+1}, \ldots, k_m$ are equal to zero, the corresponding term must be divided by 2^i.

We shall establish some notation for differences of sequences of numbers $c_{k_1,\ldots,k_m}$ depending on m indices. Let Δ_i denote the first difference on the index k_i and $\bar{\Delta} = \Delta_1 \cdots \Delta_m$. For example for $m = 2$ we have

$$\Delta_1 c_{k_1,k_2} = c_{k_1,k_2} - c_{k_1+1,k_2}$$

and

$$\bar{\Delta} c_{k_1,k_2} = c_{k_1,k_2} - c_{k_1+1,k_2} - c_{k_1,k_2+1} + c_{k_1+1,k_2+1}.$$

It is clear that the operators Δ_i and Δ_j commute.

The notation for second differences $\Delta_i^2 = \Delta_i \Delta_i$ and $\bar{\Delta}^2 = \bar{\Delta}\bar{\Delta}$ is analogous.

We shall now note some properties of numerical sequences.

Suppose the numbers $c_{k_1,\ldots,k_m}$ satisfy the condition

$$c_{k_1,\ldots,k_m} \to 0 \quad \text{as } k_1 + \cdots + k_m \to \infty \tag{2}$$

and the series $\sum_{k=0}^{\infty} |\bar{\Delta} c_{k_1,\ldots,k_m}|$ converges. (The notation means that the summation is extended over all indices $k_1, \ldots, k_m$ from 0 to ∞.) By (2),

$$c_{k_1,k_2,\ldots,k_m} = \sum_{i_1=k_1}^{\infty} \Delta_1 c_{i_1,k_2,\ldots,k_m}. \tag{3}$$

We now pass to differences on the second index

$$c_{k_1,k_2,\ldots,k_m} = \sum_{i_1=k_1}^{\infty} \left(\sum_{i_2=k_2}^{\infty} \Delta_1 \Delta_2 c_{i_1,i_2,k_2,\ldots,k_m} \right)$$

and continue sequentially through all the remaining indices. Then we obtain

$$c_{k_1,\ldots,k_m} = \sum_{\substack{i_1=k_1 \\ \cdots \\ i_m=k_m}}^{\infty} \bar{\Delta} c_{i_1,\ldots,i_m}, \tag{4}$$

and this series converges absolutely.

LEMMA 1. *If condition* (2) *is satisfied and for some* ν, $0 \leqslant \nu \leqslant m$,

$$\sum_{k=0}^{\infty} k_1 \cdots k_\nu (k_1 + 1) \cdots (k_m + 1) \left|\bar{\Delta} c_{k_1,\ldots,k_m}\right| < \infty, \tag{5}$$

then the series

$$\sum_{k=0}^{\infty} k_1 \cdots k_\nu c_{k_1,\ldots,k_m} \tag{6}$$

converges absolutely, and

$$\sum_{k=0}^{\infty} k_1 \cdots k_\nu c_{k_1,\ldots,k_m} = 2^{-\nu} \sum_{k=0}^{\infty} k_1 \cdots k_\nu (k_1 + 1) \cdots (k_m + 1) \bar{\Delta} c_{k_1,\ldots,k_m}. \tag{7}$$

PROOF. Let us substitute the numbers $c_{k_1,\dots,k_m}$ from (4) into (6) and change the order of summation. We then obtain

$$\sum_{k=0}^{\infty} k_1 \cdots k_\nu c_{k_1,\dots,k_m} = \sum_{k=0}^{\infty} k_1 \cdots k_\nu \sum_{\substack{i_1=k_1\\ \cdots \\ i_m=k_m}}^{\infty} \bar{\Delta} c_{i_1,\dots,i_m}$$

$$= \sum_{i=0}^{\infty} \bar{\Delta} c_{i_1,\dots,i_m} \sum_{k_1=0}^{i_1} \cdots \sum_{k_m=0}^{i_m} k_1 \cdots k_\nu$$

$$= 2^{-\nu} \sum_{i=0}^{\infty} i_1 \cdots i_\nu (i_1+1) \cdots (i_m+1) \bar{\Delta} c_{i_1,\dots,i_m}.$$

The admissibility of changing the order of summation and the absolute convergence of (6) follow from (5). The lemma is proved.

We need an estimate of the integral of the polynomial

$$H_k(x) = \sum_{i_1=0}^{k_1} \cdots \sum_{i_m=0}^{k_m} \alpha_{i_1,\dots,i_m} D_{i_1}(x_1) \cdots D_{i_m}(x_m),$$

where D_j is the Dirichlet kernel of order j.

LEMMA 2. *If the numbers $\alpha_{i_1,\dots,i_m}$ satisfy the condition $|\alpha_{i_1,\dots,i_m}| \leqslant 1$, then*

$$\int_0^\pi \cdots \int_0^\pi |H_k(x)|\, dx_1 \cdots dx_m < 2^m (k_1+1) \cdots (k_m+1). \tag{8}$$

For $m = 1$ such an estimate was obtained by Sidon [6] with the aid of a theorem on strong summability of Fourier series. G. A. Fomin [7] has given a direct proof of a more general one-variable estimate which A. A. Fomin [8] has extended to the two-variable case. In none of these papers was the numerical coefficient on the right-hand side of (8) estimated.

PROOF OF LEMMA 2. We shall estimate the integral

$$I = \int_0^{2\pi} \cdots \int_0^{2\pi} |H_k(x)|\, dx_1 \cdots dx_m = 2^m \int_0^\pi \cdots \int_0^\pi |H_k(x)|\, dx_1 \cdots dx_m.$$

Let E_j denote the set of x_j belonging to the interval $[1/(k_j+1), 2\pi - 1/(k_j+1)]$, and F_j the complement of E_j in $[0, 2\pi]$. Let us represent I in the form of a sum of 2^m integrals by breaking the interval of integration on each variable x_j into the sets E_j and F_j.

Let us examine one of these integrals.

For simplicity of notation we shall suppose that for the first q variables ($0 \leqslant q \leqslant m$) the integration is extended over the sets F_j and for the remaining variables over E_j:

$$J_q \equiv \int_{F_1} \cdots \int_{F_q} \left\{ \int_{E_{q+1}} \cdots \int_{E_m} |H_k(x)|\, dx_{q+1} \cdots dx_m \right\} dx_1 \cdots dx_q.$$

We shall estimate the integral in braces using the Cauchy-Schwarz-Bunyakovskiĭ inequality. Suppose

$$h_{i_{q+1},\dots,i_m} = \sum_{i_1=0}^{k_1} \cdots \sum_{i_q=0}^{k_q} \alpha_{i_1,\dots,i_m} D_{i_1}(x_1) \cdots D_{i_q}(x_q).$$

Then

$$H_k(x) = \sum_{i_{q+1}=0}^{k_q+1} \cdots \sum_{i_m=0}^{k_m} h_{i_{q+1},\dots,i_m} D_{i_{q+1}}(x_{q+1}) \cdots D_m(x_m)$$

and

$$\{\ \}^2 \leqslant \int_{E_{q+1}} \cdots \int_{E_m} \frac{1}{4\sin^2(x_{q+1}/2)} \cdots \frac{1}{4\sin^2(x_m/2)} dx_{q+1} \cdots dx_m$$

$$\times \int_{E_{q+1}} \cdots \int_{E_m} \left[\sum_{i_{q+1}=0}^{k_{q+1}} \cdots \sum_{i_m=0}^{k_m} h_{i_{q+1},\dots,i_m} \sin\left(i_{q+1} + \frac{1}{2}\right) x_{q+1} \cdots \sin\left(i_m + \frac{1}{2}\right) x_m \right]^2 \times dx_{q+1} \cdots dx_m$$

$$\leqslant \int_{E_{q+1}} \frac{dx_{q+1}}{4\sin^2(x_{q+1}/2)} \cdots \int_{E_m} \frac{dx_m}{4\sin^2(x_m/2)}$$

$$\times 2^{m-q} \int_0^{\pi} \cdots \int_0^{\pi} \left[\sum_{i_{q+1}=0}^{k_{q+1}} \cdots \sum_{i_m=0}^{k_m} h_{i_{q+1},\dots,i_m} \sin(2i_{q+1}+1)x_{q+1} \cdots \sin(2i_m+1)x_m \right]^2 dx_{q+1} \cdots dx_m. \tag{9}$$

From (9), due to the estimate

$$\int_{E_j} \frac{dx_j}{4\sin^2(x_j/2)} = \cot \frac{1}{2(k_j+1)} \leqslant 2(k_j+1)$$

and the orthogonality of the sine system it follows that

$$\{\ \}^2 \leqslant 2^{m-q}(k_{q+1}-1) \cdots (k_m+1)\pi^{m-q} \sum_{i_{q+1}=0}^{k_{q+1}} \cdots \sum_{i_m=0}^{k_m} h^2_{i_{q+1},\dots,i_m}. \tag{10}$$

Furthermore,

$$|h_{i_{q+1},\dots,i_m}| \leqslant \sum_{i_1=0}^{k_1} \cdots \sum_{i_q=0}^{k_q} \left(i_1 + \frac{1}{2}\right) \cdots \left(i_q + \frac{1}{2}\right)$$

$$= 2^{-q}(k_1+1)^2 \cdots (k_q+1)^2.$$

Substituting this estimate in (10), we find that

$$\{\ \} \leqslant (2\pi)^{(m-q)/2} 2^{-q} (k_{q+1}+1) \cdots (k_m+1)(k_1+1)^2 \cdots (k_q+1)^2.$$

Hence we obtain

$$J_q \leqslant (2\pi)^{(m-q)/2}(k_1+1) \cdots (k_m+1). \tag{11}$$

The integral I was represented in the form of a sum of 2^m integrals. For q variables the integration over the sets F_j is carried out in C_q^m of these integrals. Estimate (11) holds for each of them. Therefore

$$I \leqslant \sum_{q=0}^{m} C_m^q (2\pi)^{(m-q)/2}(k_1+1) \cdots (k_m+1)$$

$$= (1+\sqrt{2\pi})^m (k_1+1) \cdots (k_m+1) < 4^m (k_1+1) \cdots (k_m+1).$$

The lemma is proved.

We now pass to the study of the integrability of the series (1). Everywhere in what follows convergence of (1) is understood as convergence of its rectangular partial sums of order n_i in the ith variable, $i = 1,\dots,m$, as $\min(n_1,\dots,n_m) \to \infty$.

Set

$$b_{k_1,\dots,k_m} = \frac{1}{k_1 \cdots k_\nu} a_{k_1,\dots,k_m},$$

if all the numbers $k_1,\dots,k_\nu$ are different from zero. If, however, at least one of the numbers $k_1,\dots,k_\nu$ equals zero, the numbers $b_{k_1,\dots,k_m}$ may be taken arbitrarily.

THEOREM 1. *Suppose the coefficients of the series* (1) *satisfy the condition*

$$a_{k_1,\dots,k_m} \to 0 \quad \text{as } k_1 + \cdots k_m \to \infty \tag{12}$$

and there exist numbers $C_{k_1,\dots,k_m}$ *such that*

$$C_{k_1,\dots,k_m} \to 0 \quad \text{as } k_1 + \cdots k_m \to \infty, \tag{13}$$

$$C_{k_1,\dots,k_n} \geqslant \left|\bar{\Delta} b_{k_1,\dots,k_m}\right| \tag{14}$$

for all $k_1,\dots,k_m$ *and the following series converges*:

$$\sum_{k=0}^{\infty} k_1 \cdots k_\nu (k_1+1)\cdots(k_m+1)\left|\bar{\Delta} c_{k_1,\dots,k_m}\right|. \tag{15}$$

Then the series (1) *converges uniformly in each cube* $[\varepsilon, \pi]^m$, $\varepsilon > 0$; *the function* $f(x_1,\dots,x_m)$ *which is the sum of* (1) *is integrable over* $[0, \pi]^m$, *and*

$$\int_0^\pi \cdots \int_0^\pi |f(x_1,\dots,x_m)|\, dx_1 \cdots dx_m$$

$$\leqslant 2^m \sum_{k=0}^{\infty} k_1 \cdots k_\nu (k_1+1)\cdots(k_m+1)\left|\bar{\Delta} c_{k_1,\dots,k_m}\right|. \tag{16}$$

PROOF. We remark that the conditions imposed on the numbers $C_{k_1,\dots,k_m}$ imply that they are nonnegative, and that (by Lemma 1) the following series converges:

$$\sum_{k=0}^{\infty} k_1 \cdots k_\nu C_{k_1,\dots,k_m}. \tag{17}$$

Write (1) as an m-variable cosine series differentiated termwise with respect to the first ν variables

$$\sum_{k=0}^{\infty}{}' (-1)^\nu b_{k_1,\dots,k_m} \frac{\partial^\nu}{x_1 \cdots \partial x_\nu} \cos k_1 x_1 \cdots \cos k_m x_m, \tag{18}$$

where the prime in the summation has the same meaning as in (1).

Consider a rectangular partial sum of (18):

$$s_n(x) = \sum_{k_1=0}^{n_1} \cdots \sum_{k_m=0}^{n_m}{}' (-1)^\nu b_{k_1,\dots,k_m} \frac{\partial^\nu}{\partial x_1,\dots,\partial x_\nu} \cos k_1 x_1 \cdots \cos k_m x_m$$

and perform Abel's transformation in it on each summation index.

For simplicity of notation we shall carry out the detailed exposition for $m = 2$, $\nu = 1$. We have

$$s_n(x) = -\sum_{k_1=0}^{n_1-1}\sum_{k_2=0}^{n_2-1}\frac{\partial}{\partial x_1}D_{k_1}(x_1)D_{k_2}(x_2)\bar{\Delta}b_{k_1,k_2}$$
$$-\sum_{k_1=0}^{n_1-1}\frac{\partial}{\partial x_1}D_{k_1}(x_1)D_{n_2}(x_2)\Delta_1 b_{k_1,n_2}$$
$$-\sum_{k_2=0}^{n_2-1}\frac{\partial}{\partial x_1}D_{n_1}(x_1)D_{k_2}(x_2)\Delta_2 b_{n_1,k_2} - \frac{\partial}{\partial x_1}D_{n_1}(x_1)D_{n_2}(x_2)b_{n_1,n_2}. \quad (19)$$

But, uniformly on all parameters,

$$\frac{\partial}{\partial x_1}D_{k_1}(x)D_{n_2}(x_2) = O\left(\frac{k_1}{x_1^2 x_2}\right)$$

and by (12) and (14)

$$\sum_{k_1=0}^{n_1-1}k_1\left|\Delta_1 b_{k_1,n_2}\right| = \sum_{k_1=0}^{n_1-1}k_1\left|\sum_{k_2=n_2}^{\infty}\bar{\Delta}b_{k_1,k_2}\right|$$
$$\leqslant \sum_{k_1=0}^{\infty}\sum_{k_2=n_2}^{\infty}k_1\left|\bar{\Delta}b_{k_1,k_2}\right| \leqslant \sum_{k_1=0}^{\infty}\sum_{k_2=n_2}^{\infty}k_1 C_{k_1,k_2},$$

and, as we have noted, $\sum_{k,k_2=0}^{\infty}k_1 C_{k_1,k_2}$ converges.

For the quantity obtained in the estimate of the third sum on the right-hand side of (19) we have, in exactly the same manner,

$$\sum_{k_2=0}^{n_2-1}n_1\left|\Delta_2 b_{n_1,k_2}\right| = \sum_{k_2=0}^{n_2-1}n_1\left|\sum_{k_1=n_1}^{\infty}\bar{\Delta}b_{k_1,k_2}\right| \leqslant \sum_{k_1=n_1}^{\infty}\sum_{k_2=0}^{\infty}k_1 C_{k_1,k_2}.$$

Finally,

$$n_1\left|b_{n_1,n_2}\right| \leqslant \sum_{k_1=n_1}^{\infty}\sum_{k_2=n_2}^{\infty}n_1\left|\bar{\Delta}b_{k_1,k_2}\right| \leqslant \sum_{k_1=n_1}^{\infty}\sum_{k_2=n_2}^{\infty}k_1 C_{k_1,k_2}.$$

For arbitrary m and ν the estimates are analogous. They show that as $\min(n_1,\dots,n_m)\to\infty$

$$\lim\left(s_n(x) - \sum_{k_1=0}^{n_1-1}\cdots\sum_{k_m=0}^{n_m-1}(-1)^{\nu}\frac{\partial^{\nu}}{\partial x_1\cdots\partial x_{\nu}}D_{k_1}(x_1)\cdots D_{k_m}(x_m)\bar{\Delta}b_{k_1,\dots,k_m}\right) = 0. \quad (20)$$

uniformly in each cube $[\varepsilon,\pi]^m$, $\varepsilon > 0$.

Using the estimate

$$\frac{\partial^{\nu}}{\partial x_1,\dots,\partial x_{\nu}}D_{k_1}(x_1)\cdots D_{k_m}(x_m)\bar{\Delta}b_{k_1,\dots,k_m} = O\left(\frac{k_1\cdots k_{\nu}}{x_1^2\cdots x_{\nu}^2 x_{\nu+1}\cdots x_m}C_{k_1,\dots,k_m}\right)$$

we find that the series

$$\sum_{k=0}^{\infty}(-1)^{\nu}\frac{\partial^{\nu}}{\partial x_1\cdots\partial x_{\nu}}D_{k_1}(x_1)\cdots D_{k_m}(x_m)\bar{\Delta}b_{k_1,\dots,k_m}$$

is majorized by the convergent series (17) with nonnegative terms. It therefore follows from (20) that in each such cube the series (1) converges uniformly, and if the sum of this series is denoted by $f(x) = f(x_1,\dots,x_m)$, then for all x_i with $0 \leqslant x_i \leqslant \pi$, $i = 1,\dots,m$,

$$f(x) = \sum_{k=0}^{\infty} (-1)^{\nu} \frac{\partial^{\nu}}{\partial x_1 \cdots \partial x_{\nu}} D_{k_1}(x_1) \cdots D_{k_m}(x_m) \bar{\Delta} b_{k_1,\dots,k_m}. \tag{21}$$

We shall use the notation

$$\alpha_k = \begin{cases} (-1)^{\nu} \dfrac{\bar{\Delta} b_{k_1,\dots,k_m}}{C_{k_1,\dots,k_m}}, & \text{if } C_{k_1,\dots,k_m} > 0, \\ 0, & \text{if } C_{k_1,\dots,k_m} = 0, \end{cases}$$

and

$$c_{k_1,\dots,k_m}(x) = \alpha_k \frac{\partial^{\nu}}{\partial x_1 \cdots \partial x_{\nu}} D_{k_1}(x_1) \cdots D_{k_1 m}(x_m).$$

We note that $|\alpha_k| \leqslant 1$ and, uniformly in each cube $[\varepsilon, \pi]^m$, $\varepsilon > 0$,

$$c_{k_1,\dots,k_m}(x) = O(k_1 \cdots k_{\nu}). \tag{22}$$

With this notation (21) can be written as follows:

$$f(x) = \sum_{k=0}^{\infty} C_{k_1,\dots,k_m} c_{k_1,\dots,k_m}(x). \tag{23}$$

Consider the rectangular partial sum of the series (23)

$$S_n(x) = \sum_{k_1=0}^{n_1} \cdots \sum_{k_m=0}^{n_m} C_{k_1,\dots,k_m} c_{k_1,\dots,k_m}(x)$$

and perform Abel's transformation on each index of summation. We use the notation

$$d_{k_1,\dots,k_m} = d_{k_1,\dots,k_m}(x) = \sum_{i_1=0}^{k_1} \cdots \sum_{i_m=0}^{k_m} c_{i_1,\dots,i_m}(x).$$

It follows from (22) that, uniformly in each cube $[\varepsilon, \pi]^m$, $\varepsilon > 0$,

$$d_{k_1,\dots,k_m} = O(k_1 \cdots k_{\nu}(k_1 + 1) \cdots (k_m + 1)). \tag{24}$$

As above, we carry out the detailed exposition only for $m = 2$ and $\nu = 1$. We have

$$S_n(x) = \sum_{k_1=0}^{n_1-1} \sum_{k_2=0}^{n_2-1} d_{k_1,k_2} \bar{\Delta} C_{k_1,k_2} + \sum_{k_1=0}^{n_1-1} d_{k_1,k_2} \Delta_1 C_{k_1,n_2}$$
$$+ \sum_{k_2=0}^{n_2-1} d_{n_1,k_2} \Delta_2 C_{n_1,k_2} + d_{n_1,n_2} C_{n_1,n_2} = I_1 + I_2 + I_3 + I_4.$$

Using (24) and (4), we obtain

$$I_4 = O\left(\sum_{k_1=n_1}^{\infty} \sum_{k_2=n_2}^{\infty} n_1^2(n_2 + 1) \left| \bar{\Delta} C_{k_1,k_2} \right| \right).$$

Hence, by the convergence of (15), $I_4 \to 0$ as $\min(n_1, n_2) \to \infty$.

Applying (24) to I_3, we obtain the sum $\sum_{k_2=0}^{n_2-1} n_1^2(k_2+1)|\Delta_2 C_{n_2,k_2}|$, for which we find, using (3),

$$\sum_{k_2=0}^{n_2-1} n_1^2(k_2+1)\left|\Delta_2 C_{n_1,k_2}\right| \leqslant \sum_{k_2=0}^{n_2-1} \sum_{k_1=n_1}^{\infty} n_1^2(k_2+1)\left|\overline{\Delta} C_{k_1,k_2}\right|$$
$$\leqslant \sum_{k_1=n_1}^{\infty} \sum_{k_2=0}^{\infty} k_1^2(k_2+1)\left|\overline{\Delta} C_{k_1,k_2}\right|,$$

and since (15) converges, $I_3 \to 0$ as $\min(n_1, n_2) \to \infty$.

To estimate I_2 we proceed in exactly the same way with the sum

$$\sum_{k_1=0}^{n_1-1} k_1^2(n_2+1)\left|\Delta_1 C_{k_1,k_2}\right| \leqslant \sum_{k_1=0}^{\infty} \sum_{k_2=n_2}^{\infty} k_1^2(k_2+1)\left|\overline{\Delta} C_{k_1,k_2}\right|.$$

From this we conclude that $I_2 \to 0$ as $\min(n_1, n_2) \to \infty$.

For arbitrary m and ν the estimates are analogous. Relying on these estimates and the convergence of (23), we see that

$$f(x) = \sum_{k=0}^{\infty} d_{k_1,\ldots,k_m} \overline{\Delta} C_{k_1,\ldots,k_m}, \tag{25}$$

and the series on the right converges uniformly in each cube $[\varepsilon, \pi]^m$, $\varepsilon > 0$, because of (24) and the convergence of (15).

We shall estimate the integral of the function f using (25). We have

$$\int_0^\pi \cdots \int_0^\pi |f(x)|\, dx_1 \cdots dx_m \leqslant \sum_{k=0}^{\infty} \left|\Delta C_{k_1,\ldots,k_m}\right| \int_0^\pi \cdots \int_0^\pi \left|d_{k_1,\ldots,k_m}\right| dx_1 \cdots dx_m. \tag{26}$$

Since $d_{k_1,\ldots,k_m}(x)$ is a trigonometric polynomial of degree k_1 in $x_1, \ldots, k_m$, in x_m, it follows from Bernstein's inequality for derivatives of a trigonometric polynomial (cf., for example, [9], §2.7) that

$$\int_0^\pi \cdots \int_0^\pi \left|d_{k_1,\ldots,k_m}\right| dx_1 \cdots dx_m$$
$$\leqslant k_1 \cdots k_\nu \int_0^\pi \cdots \int_0^\pi \left| \sum_{i_1=0}^{k_1} \cdots \sum_{i_m=0}^{k_m} \alpha_k D_{k_1}(x_1) \cdots D_{k_m}(x_m) \right| dx_1 \cdots dx_m. \tag{27}$$

But $|\alpha_k| \leqslant 1$; therefore by Lemma 2

$$\int_0^\pi \cdots \int_0^\pi \left|d_{k_1,\ldots,k_m}\right| dx_1 \cdots dx_m \leqslant 2^m k_1 \cdots k_\nu (k_1+1) \cdots (k_m+1),$$

and, returning to (26), we obtain (16).

The theorem is proved.

If in Theorem 1 the numbers $C_{k_1,\ldots,k_m}$ are such that

$$\overline{\Delta} C_{k_1,\ldots,k_m} \geqslant 0 \quad \text{for all } k_1, \ldots, k_m \tag{28}$$

then by Lemma 1 the estimate (16) is equivalent to

$$\int_0^\pi \cdots \int_0^\pi |f(x)|\, dx_1 \cdots dx_m \leqslant 2^{m+\nu} \sum_{k=0}^{\infty} k_1 \cdots k_\nu C_{k_1,\ldots,k_m}. \tag{29}$$

For double cosine series, i.e. for $m = 2$ and $\nu = 0$, this assertion, without the estimate of the numerical coefficient in (29) was proved by Nosenko [5].([1])

We now present the integrability conditions for the series (1) in terms of the second differences of the sequence $b_{k_1,\dots,k_m}$.

THEOREM 2. *Suppose the coefficients of the series* (1) *satisfy condition* (12) *and that the following series converges*:

$$\sum_{k=0}^{\infty} k_1 \cdots k_\nu (k_1+1) \cdots (k_m+1) \left|\bar{\Delta}^2 b_{k_1,\dots,k_m}\right|. \tag{30}$$

Then the series (1) *converges uniformly in each cube* $[\varepsilon, \pi]^m$, $\varepsilon > 0$; *the function* $f(x_1,\dots,x_m)$ *which is the sum of this series is integrable over* $[0, \pi]^m$, *and*

$$\int_0^\pi \cdots \int_0^\pi |f(x_1,\dots,x_m)|\, dx_1 \cdots dx_m$$
$$\leqslant 2^m \sum_{k=0}^{\infty} k_1 \cdots k_\nu (k_1+1) \cdots (k_m+1) \left|\bar{\Delta}^2 b_{k_1,\dots,k_m}\right|. \tag{31}$$

This theorem is a simple corollary of Theorem 1. Indeed, by the convergence of (30), the first differences of the numbers $b_{k_1,\dots,k_m}$ can be represented by the second using (4):

$$\bar{\Delta} b_{k_1,\dots,k_m} = \sum_{\substack{i_1=k_1\\ \cdots\\ i_m=k_m}}^{\infty} \bar{\Delta}^2 b_{i_1,\dots,i_m}.$$

Therefore if we set

$$C_{k_1,\dots,k_m} = \sum_{\substack{i_1=k_1\\ \cdots\\ i_m=k_m}}^{\infty} \left|\bar{\Delta}^2 b_{i_1,\dots,i_m}\right|,$$

it is obvious that conditions (13) and (14) are satisfied. But since in this case

$$\bar{\Delta} C_{k_1,\dots,k_m} = \left|\bar{\Delta}^2 b_{k_1,\dots,k_m}\right|,$$

the series (31) coincides with (15).

We shall show that Theorem 2 is equivalent to the conditions of integrability of multiple trigonometric series obtained in [3].

In [3] we introduced two families of operators $A_i(c)$ and $A_i(s)$, $i = 1,\dots,m$, acting on the sequence $\{a_{k_1,\dots,k_m}\}$. The operator $A_i(c)$ transforms $\{a_{k_1,\dots,k_m}\}$ into the sequence

$$\left\{k_i \Delta_i^2 a_{k_1,\dots,k_{i-1},k_i-1,\,k_{i+1},\dots,k_m}\right\},$$

while $A_i(s)$ transforms it into $\{(1/k_i)a_{k_1,\dots,k_m}\}$, $k_i = 1,2,\dots$. For $i \neq j$ the operators A_i and A_j commute. The elements of the sequence obtained after transformation of $\{a_{k_1,\dots,k_m}\}$ by operators $A_1(\theta_1),\dots,A_m(\theta_m)$, where c or s plays the role of θ_i, are denoted by $A_i(\theta_i) \cdots A_m(\theta_m) a_{k_1,\dots,k_m}$.

([1]) Condition (28) does not occur in [5], but, as Nosenko pointed out to the author, it is necessary to add it.

It was proved in [3] that if the coefficients of (1) satisfy condition (12) and for all $\theta_1,\dots,\theta_\nu$ assuming the values c and s the following series converge:

$$J(\theta_1,\dots,\theta_\nu)=\sum_{k=1}^{\infty}\left|A_1(\theta_1)\cdots A_\nu(\theta_\nu)A_{\nu+1}(c)\cdots A_m(c)a_{k_1,\dots,k_m}\right|, \tag{32}$$

then the following estimate holds for the sum of (1):

$$\int_0^\pi\cdots\int_0^\pi|f|\,dx_1\cdots dx_m\leqslant M\sum_{\theta_1,\dots,\theta_\nu=c,s}J(\theta_1,\dots,\theta_\nu), \tag{33}$$

where M depends only on m.

For $\nu=0$, i.e. in the case of cosine series, the series (30) and (32) coincide.

To compare the integrability conditions just given for $\nu\geqslant 1$, we shall need the following elementary estimates.

LEMMA 3. *Suppose* $\alpha_p\to 0$ *and* $\beta_p=\alpha_p/p$, $p=1,2,\dots$. *Then*

$$\sum_{p=1}^{\infty}\frac{|\alpha_p|}{p}\leqslant\frac{1}{2}\sum_{p=1}^{\infty}p(p+1)\left|\Delta^2\beta_p\right|, \tag{34}$$

$$\sum_{p=0}^{\infty}(p+1)\left|\Delta^2\alpha_p\right|\leqslant 2\sum_{p=1}^{\infty}p(p+1)\left|\Delta^2\beta_p\right| \tag{35}$$

and

$$\sum_{p=0}^{\infty}p(p+1)\left|\Delta^2\beta_p\right|\leqslant 3\sum_{p=1}^{\infty}(p+1)\left|\Delta^2\alpha_p\right|+2\sum_{p=1}^{\infty}\frac{|\alpha_p|}{p}. \tag{36}$$

PROOF. The inequalities

$$\sum_{p=1}^{\infty}\frac{|\alpha_p|}{p}=\sum_{p=1}^{\infty}|\beta_p|=\sum_{p=1}^{\infty}\left|\sum_{q=p}^{\infty}\Delta\beta_q\right|\leqslant\sum_{q=1}^{\infty}q|\Delta\beta_q|$$

and

$$\sum_{q=1}^{\infty}q|\Delta\beta_q|=\sum_{q=1}^{\infty}q\left|\sum_{p=1}^{\infty}\Delta^2\beta_p\right|\leqslant\frac{1}{2}\sum_{p=1}^{\infty}p(p+1)\left|\Delta^2\beta_q\right|. \tag{37}$$

lead to (34). Furthermore, $\Delta^2\alpha_p=p\Delta^2\beta_p+2\Delta\beta_{p+1}$, and so

$$\sum_{p=1}^{\infty}(p+1)\left|\Delta^2\alpha_p\right|\leqslant\sum_{p=1}^{\infty}p(p+1)\left|\Delta^2\beta_p\right|+2\sum_{p=1}^{\infty}(p+1)\left|\Delta\beta_{p+1}\right|$$

and, using (37), we arrive at (35).

Finally, (36) follows in an obvious manner from the representation

$$p\Delta^2\beta_p=\Delta^2\alpha_p+\frac{2}{p+1}\Delta\alpha_{p+1}+\frac{2}{(p+1)(p+2)}\alpha_{p+2}.$$

The lemma is proved.

Applying (36) to the series (30) sequentially for each of the first ν indices of summation, we see that this series can be estimated by the sum on the right-hand side of (33).

Conversely, by (34) and (35) neither of the series (32) exceeds series (30).

Thus Theorem 2 is equivalent to the indicated conditions of integrability from [3]. At the same time it is more convenient to use Theorem 2, since in this case it is necessary to examine one series instead of 2^ν series.

We remark that in [3] it was not clarified how the coefficient M depends on m in (33). Here we have given estimates of the corresponding coefficients, since in certain applications (cf., for example, [10]) the nature of this dependence has turned out to be essential.

BIBLIOGRAPHY

1. S. A. Telyakovskiĭ, *On a sufficient condition of Sidon for the integrability of trigonometric series*, Mat. Zametki **14** (1973), 317–328; English transl. in Math. Notes **14** (1973).

2. ______, *On the integrability of sine series*, Trudy Mat. Inst. Steklov. **103** (1984) (to appear); English transl., to appear in Proc. Steklov Inst. Math.

3. ______, *Some estimates for trigonometric series with quasiconvex coefficients*, Mat. Sb. **63(105)** (1964), 426–444; English transl. in Amer. Math. Soc. Transl. (2) **86** (1970).

4. Takeshi Kano, *Coefficients of some trigonometric series*, J. Fac. Sci. Shinshu Univ. **3** (1968), 153–162.

5. Yu. L. Nosenko, *Conditions of Sidon type for integrability of double trigonometric series*, Theory of Functions and Mappings (G. D. Suvorov, editor), "Naukova Dumka", Kiev, 1979, pp. 132–149. (Russian)

6. S. Sidon, *Hinreichende Bedingungen für den Fourier-charakter einer trigonometrischen reihe*, J. London Math. Soc. **14** (1939), 158–160.

7. G. A. Fomin, *On linear methods of summing Fourier series*, Mat. Sb. **65(107)** (1964), 144–152. (Russian)

8. A. A. Fomin, *Linear methods of summing double Fourier series*, Izv. Vyssh, Uchebn. Zaved Mat. **1969**, no. 2 (81), 106–116. (Russian)

9. S. M. Nikol′skiĭ, *Approximation of functions of several variables and embedding theorems*, 2nd rev. aug. ed., "Nauka", Moscow, 1977; English transl. of 1st ed., Springer-Verlag, 1974.

10. Yu. N. Shakhov, *On the computation of integrals of increasing multiplicity*, Zh. Vychisl. Mat. in Mat. Fiz. **5** (1965), 911–916; English transl. in USSR Comput. Math. and Math. Phys. **5** (1965).

Translated by R. L. COOKE

Труды Матем. Инст. Стеклов.
164(1983)

Proceedings of the Steklov
Institute of Mathematics
1985, Issue 2

ON RELATIONS BETWEEN BEST APPROXIMATIONS OF FUNCTIONS ANALYTIC IN THE BIDISK

UDC 517.5

V. N. TEMLYAKOV

ABSTRACT. The author studies, in uniform and integral metrics, full and partial best approximations of functions, that is, approximations by polynomials in both variables and by polynomials in one variable with coefficients depending on the other variable. An estimate is obtained for the full best approximation in terms of the sum of the partial best approximations, and with an asymptotically sharp constant. The proof of this estimate is based on some properties of trigonometric polynomials with disjoint spectra.

Bibliography: 9 titles.

The principal goal of this paper is to obtain relations between full and partial best approximations of functions of several variables. Such questions were first considered by S. N. Bernstein [1].

It was proved in [1] that

$$E_{n,m}(f)_C \leqslant \left[\left(\frac{2}{\pi} + o(1)\right) \log(\min(n, m) + 2)\left(E_{n,\infty}(f)_C + E_{\infty,m}(f)_C\right),\right. \quad (1)$$

where $E_{n,m}(f)$ is the best approximation to the periodic function $f(x, y)$ in the uniform metric by trigonometric polynomials of order n in x and order m in y; similarly $E_{n,\infty}(f)_C$ is the best approximation to f by trigonometric polynomials of order n in x with coefficients depending on y; and $E_{\infty,m}(f)$ is defined analogously.

In the author's paper [2] (where references are given to other papers related to this question) it is proved that inequality (1), as well as an analogous inequality in an integral metric, is best possible.

In this paper we consider approximations of analytic functions $f(z)$ of two variables in the bidisk $U^2 = \{z = (z_1, z_2): |z_1|, |z_2| < 1\}$.

Let Y denote either the space $A(U^1)$ of complex-valued functions analytic in U^2 and continuous in the closure $\overline{U}^2$ or the Hardy space $H^1(U^2)$, and let

$$\|f\|_Y = \begin{cases} \sup\limits_{z \in U^2} |f(z)| = \sup\limits_{z \in \overline{U}^2} |f(z)|, & Y = A(U^2), \\ \sup\limits_{0 \leqslant r < 1} \dfrac{1}{(2\pi)^2} \displaystyle\int_0^{2\pi}\int_0^{2\pi} |f(re^{i\theta_1}, re^{i\theta_2})|\, d\theta_1\, d\theta_2 = \lim_{r \to 1} \dfrac{1}{(2\pi)^2} \int_{\pi_2} |f(re^{i\theta})|\, d\theta, & \\ & Y = H^1(U^2) \end{cases}$$

(cf. for instance [3]).

1980 *Mathematics Subject Classification*. Primary 42A10; Secondary 41A50, 41A63.

We shall study relations between the following quantities:

$$e_{n,m}(f)_Y = \inf_{c_{k_1,k_2}} \left\| f(z) - \sum_{\substack{0 \leqslant k_1 \leqslant n \\ 0 \leqslant k_2 \leqslant m}} c_{k_1,k_2} z_1^{k_1} z_c^{k_2} 2 \right\|_Y,$$

$$e_{n,\infty}(f)_Y = \inf_{\varphi_k(z_j 2)} \left\| f(z) - \sum_{k=0}^{n} \varphi_k(z_2) z_1^k \right\|_Y,$$

where the $\varphi_k(z_2)$ are functions analytic in the disk $|z_2| < 1$, and $e_{\infty,m}(f)_Y$ is defined analogously.

We shall prove the following assertion:

THEOREM 2.1. *Let Y denote either the space $A(U^2)$ or the space $H^1(U^2)$. Then*

$$\sup_{f \in Y} e_{n,m}(f)_Y / (e_{n,\infty}(f)_Y + e_{\infty,m}(f)_Y) = \frac{1}{\pi} \log(\min(n, m) + 2) + O(1).$$

The proof of this theorem is based on some lemmas about the properties of trigonometric polynomials with disjoint spectra (§1). The proof of these lemmas depends on simple and well-known methods. At the same time they lead to some new results for functions of one variable (see Theorems 1.2 and 1.3) which in turn enable us to solve a question in the intgral metric (see Theorem 1.4) that was answered in the uniform metric by A. L. Garkavi [4].

§1. The distance between trigonometric polynomials having no common harmonics

Let $x = (x_1, \ldots, x_d)$, $\pi_d = \prod_{j=1}^{d}[0, 2\pi]$, let $f(x)$ be a function of period 2π in each variable, and let

$$\|f\|_p = \left\{ \frac{1}{(2\pi)^d} \int_{\pi_d} |f(x)|^p \, dx \right\}^{1/p}, \qquad 1 \leqslant p < \infty,$$

$$\|f\|_\infty = \operatorname*{ess\,sup}_{x \in \pi_d} |f(x)|, \qquad p = \infty.$$

We denote by $\tilde{G}_1$ and $\tilde{G}_2$ two disjoint sets of integer lattice points in R^d and by G_i^p, $i = 1, 2$, the set of functions in L_p whose Fourier harmonics lie in the set $\tilde{G}_i$. We set

$$M(G_1, G_2)_p = \sup_{f_1 \in G_1^p, f_2 \in G_2^p} \frac{\|f_1\|_p}{\|f_1 + f_2\|_p}, \tag{1.1}$$

for $1 \leqslant p \leqslant \infty$.

Note that always $M(G_1, G_2)_p \geqslant 1$.

THEOREM 1.1. *For all disjoint sets $\tilde{G}_1$ and $\tilde{G}_2$*

$$M(G_1, G_2)_1 \leqslant M(G_1, G_2)_\infty.$$

PROOF. Let $M(G_1, G_2)_1 = M < \infty$. Given $\varepsilon > 0$, choose $f_i^\varepsilon \in G_i^1$, $i = 1, 2$, such that

$$\|f_1^\varepsilon\|_1 \geqslant (M - \varepsilon)\|f_1^\varepsilon + f_2^\varepsilon\|_1.$$

Let

$$g_i^\varepsilon(x) = \frac{1}{(2\pi)^d} \int_{\pi_d} f_i^\varepsilon(x+t) \operatorname{sgn} f_1^\varepsilon(t)\, dt, \qquad i = 1,2.$$

Then $g_i^\varepsilon \in G_i^\infty$ and

$$g_1^\varepsilon(0) = \|f_1^\varepsilon\|_1 \geqslant (M - \varepsilon)\|f_1^\varepsilon + f_1^\varepsilon\|_1, \tag{1.2}$$

$$\|g_1^\varepsilon + g_2^\varepsilon\|_\infty \leqslant \|f_1^\varepsilon + f_1^\varepsilon\|_1. \tag{1.3}$$

Combining (1.2) and (1.3), we find that for arbitrary $\varepsilon > 0$

$$M(G_1, G_2)_\infty \geqslant M(G_1, G_2)_1 - \varepsilon.$$

Consequently $M(G_1, G_2)_\infty \geqslant M(G_1, G_2)_1$.

$$M(G_1, G_2)_\infty \geqslant M(G_1, G_2)_1.$$

The proof in the case $M(G_1, G_2)_1 = \infty$ is analogous.

We now compute the quantities $M(G_1, G_2)$ for some examples. If $\tilde{G}_1 = [-n, n]$ and $\tilde{G}_2 = (-\infty, -n) \cup (n, +\infty)$ then $M(G_1, G_2)_\infty$ coincides with $M(G_1, G_2)_1$ and equals the Lebesgue constant L_n. By a theorem of Fejér (see [5], Vol. I, p. 67).

$$L_n = \frac{4}{\pi^2} \log n + O(1).$$

Let $\tilde{F}_1 = [0, n]$ and $\tilde{F}_2 = [-n - m - 1, -m - 1]$.

LEMMA 1.1. *For all nonnegative integers n and m*

$$M(F_1, F_2)_\infty \leqslant \frac{1}{\pi} \log\left(1 + \frac{n}{m+1}\right) + O(1).$$

PROOF. Let $f_i \in F_i^\infty$, $i = 1, 2$, and

$$g_{n-1}(x) = \sum_{k=0}^{n-1} \left(1 - \frac{k}{n}\right) e^{ikx}, \qquad \psi_{n,m}(x) = g_{n-1}(x) + g_m(-x) - 1. \tag{1.4}$$

It is easy to check that

$$\|\psi_{n,m}\|_1 = \frac{1}{\pi} \log\left(1 + \frac{n}{m+1}\right) + O(1).$$

Set

$$\theta_{n,m}(x) = \psi_{n,m}(x) + 2e^{inx} K_n(x)$$

where $K_n(x)$ is the Fejér kernel. It is not hard to see that

$$f_1(x) = \frac{1}{2\pi} \int_0^{2\pi} (f_1(t) + f_2(t))\theta_{n,m}(x - t)\, dt.$$

Consequently

$$|f_1(x)| \leqslant \|f_1 + f_2\|_\infty \|\theta_{n,m}\|_1,$$

and

$$M(F_1, F_2)_\infty \leqslant \|\theta_{n,m}\|_1 = \frac{1}{\pi} \log\left(1 + \frac{n}{m+1}\right) + O(1),$$

which proves the lemma.

LEMMA 1.2. *For all nonnegative integers n and m*

$$M(F_1, f_2)_1 \geqslant \frac{1}{\pi}\log\left(1 + \frac{n}{m+1}\right) + O(1).$$

PROOF. Assuming $n \geqslant m + 1$, we set

$$f_1(x) = g_{[n/(m+1)]-1}((m+1)x),$$
$$f_2(x) = g_{[n/(m+1)]-1}(-(m+1)x) - 1,$$

where $g_{l-1}(x)$ is defined by (1.4). Then

$$\|f_1 + f_2\|_1 = \left\|2K_{[m/(m+1)]}((m+1)x)\right\|_1 = 1. \tag{1.5}$$

From the estimate for $\|\psi_{n,0}\|_1$ it follows that

$$\|f_1\|_1 = \frac{1}{\pi}\log\left(1 + \frac{n}{m+1}\right) + O(1). \tag{1.6}$$

Combining (1.5) and (1.6) gives the assertion of the lemma.

Together with Lemmas 1.1 and 1.2, Theorem 1.1 implies

THEOREM 1.2. *Let* $\tilde{F}_1 = [0, n]$ *and* $\tilde{F}_2 = [-n - m - 1, -m - 1]$, *and let* F_1 *and* F_2 *be the sets of functions whose Fourier harmonics lie in* $\tilde{F}_1$ *and* $\tilde{F}_2$ *respectively. Then*

$$M(F_1, F_2)_p = \frac{1}{\pi}\log\left(1 + \frac{n}{m+1}\right) + O(1), \qquad p = 1, \infty.$$

The following theorem (cf. [4]) gives $M(G_1, G_2)_\infty$ for $\tilde{G}_1 = [-n, n]$ and $\tilde{G}_2 = (-\infty, -n - m - 1] \cup [n + m + 1, +\infty)$.

THEOREM (S. B. STECHKIN). *If* $U_{n,m}$ *is the set of functions such that* $\int_0^{2\pi} f(x)e^{ikx}\,dx = 0$, $n < |k| \leqslant n + m$, *then*

$$\sup_{f \in U_{n,m}} \|S_n(f)\|_\infty / \|f\|_\infty = \frac{4}{\pi^2}\log\left(1 + \frac{n}{m+1}\right) + O(1).$$

We now prove a similar proposition for the space L_1.

THEOREM 1.3. *For all nonnegative integers n and m*

$$\alpha_{n,m} = \sup_{f \in U_{n,m}} \|S_n(f)\|_1 / \|f\|_1 = \frac{4}{\pi^2}\log\left(1 + \frac{n}{m+1}\right) + O(1).$$

PROOF. Finding the upper bound for $\alpha_{n,m}$ is no problem. For example, it can be obtained from Stechkin's theorem and Theorem 1.1.

We derive the lower bound. Let $b = [n/2(m+1)] + 1$. We set

$$V_{b,s}(t) = 2\left(1 + \frac{b}{s-b}K_s(t) - \frac{2b}{s-b}\right)K_b(t), \qquad b < s,$$

where $K_s(t)$ is the Fejér kernel, and

$$f(x) = V_{b,s}(2(m+1)x)e^{iqx},$$

where q is determined in the following way. If $2b(m+1) > n + m$ then $q = 0$, and in the contrary case $q = m$. Then $f(x) \in U_{n,m}$ $\|f\|_1 \leqslant 1 + 2b/(s-b)$; on the other hand,

$$|S_n(f)| = 2\left|D_{[n/2(m+1)]}(2(m+1)x)\right| + O(1),$$

where $D_l(t)$ is the Dirichlet kernel. Therefore for every $s > b$

$$\alpha_{n,m} \geqslant \left(1 - \frac{2b}{s-b}\right)\frac{4}{\pi^2}\log\left(1 + \frac{n}{m+1}\right) + O(1),$$

and so the required lower bound for $\alpha_{n,m}$ has been found.

The theorem is proved.

Using Stechkin's theorem, Garkavi [4] proved the following result.

THEOREM (A. L. GARKAVI). *Let*

$$C_{n,r}^{\infty} = \sup_{f \in W^{(r)}} \inf_{T_n} \max_{0 \leqslant s \leqslant r} \frac{\left\|f^{(s)} - T_n^{(s)}\right\|_{\infty}}{E_n\left(f^{(s)}\right)_{\infty}},$$

where T_n is a trigonometric polynomial of order n, $E_n(\varphi)_\infty$ is the best approximation to φ in the L_∞ metric by trigonometric polynomials of order n, and $p = \min(n, r)$. Then for all natural numbers n and r

$$C_{n,r}^{\infty} = \frac{4}{\pi^2}\log(p+1) + O(\log\log\log(p+20)).$$

By examining the proof of this theorem it is easy to see that the same arguments, using Theorem 1.3 in place of Stechkin's result, determine the value of $C_{n,r}^1$.

THEOREM 1.4. *Let $p = \min(n, r)$. For all natural numbers n and r*

$$C_{n,r}^{1} = \frac{4}{\pi^2}\log(p+1) + O(\log\log\log(p+20)).$$

We are interested in the following question: what is the rate of growth in N of the quantity

$$r_{N,p} = \sup_{G_1 : |\tilde{G}_1| \leqslant N} \sup_{G_2} M(G_1, G_2)_p ?$$

THEOREM 1.5. *The estimate*

$$CN^{|(p-2)/2p|} \leqslant r_{N,p} \leqslant N^{|(p-2)/2p|}, \qquad 1 \leqslant p \leqslant \infty,$$

holds with an absolute constant C.

PROOF. First we prove the upper bound. Let $2 \leqslant p \leqslant \infty$. For arbitrary disjoint sets $\tilde{G}_1$ and $\tilde{G}_2$ and arbitrary functions $f \in G_1^p$ and $g \in G_2^p$ we have

$$\|f + g\|_p \geqslant \|f + g\|_2 \geqslant \|f\|_2 = \left(\sum_{k \in \tilde{G}_1} |\hat{f}(k)|^2\right)^{1/2}. \tag{1.8}$$

On the other hand, by the theorem of Hausdorff-Young-Riesz ([5], Vol. II, p. 101)

$$\|f\|_p \leqslant \left(\sum_{k \in \tilde{G}_1} |\hat{f}(k)|^q\right)^{1/q}, \qquad 1/p + 1/q = 1. \tag{1.9}$$

Applying Hölder's inequality with exponents $2/q$ and $2/(2-q)$, we obtain

$$\sum_{k \in \tilde{G}_1} |\hat{f}(k)|^q \leqslant \left(\sum_{k \in \tilde{G}_1} |\hat{f}(k)|^2\right)^{q/2} \left(\sum_{k \in \tilde{G}_2} 1\right)^{1-q/2}. \tag{1.10}$$

It follows from (1.8)–(1.10) that

$$\|f\|_p \leqslant \|f + g\|_p N^{(2-q)/2q}.$$

We complete the proof of the upper estimate for $2 \leqslant p \leqslant \infty$ by noting that $(2-q)/2q = (p-2)/2p$.

From the estimate $r_{N,\infty} < \sqrt{N}$ already proved it follows by Theorem 1.1 that $r_{1,\infty} \leqslant \sqrt{N}$. It remains to consider the case $1 < p < 2$.

For an arbitrary set $\tilde{G}_2$ disjoint from $\tilde{G}_1$ and arbitrary functions $f \in G_1^p$ and $g \in G_2^p$ we have by the Hausdorff-Young-Riesz theorem

$$\|f + g\|_p \geqslant \left(\sum_{k \in \tilde{G}_1 \cup \tilde{G}_2} |\hat{f}(k) + \hat{g}(k)|^q \right)^{1/q} \geqslant \left(\sum_{k \in \tilde{G}_1} |\hat{f}(k)|^q \right)^{1/q}.$$

Furthermore

$$\|f\|_p \leqslant \|f\|_2 = \left(\sum_{k \in \tilde{G}_1} |\hat{f}(k)|^2 \right)^{1/2}.$$

Applying Hölder's inequality, we obtain

$$\sum_{k \in \tilde{G}_1} |\hat{f}(k)|^2 \leqslant \left(\sum_{k \in \tilde{G}_1} |\hat{f}(k)|^q \right)^{1/q} \left(\sum_{k \in \tilde{G}_1} 1 \right)^{1-2/q},$$

and consequently

$$\|f\|_p \leqslant \|f + g\|_p N^{(2-p)/2p}.$$

This completes the proof of the upper bound in the theorem.

We obtain the lower bound in Theorem 1.5 from the following lemma.

LEMMA 1.3. *The set of integers in* $[-2N, 2n]$ *can be divided into two disjoint sets* $\tilde{G}_1$ *and* $\tilde{G}_2$ *such that*

$$M(G_1, G_2)_p \geqslant CN^{|(p-2)/2p|}, \qquad 1 \leqslant p \leqslant \infty,$$

with an absolute constant C.

PROOF. By a theorem of Rudin and Shapiro (see [6], p. 35) there are $\varepsilon_j = \pm 1$ such that

$$\left\| \sum_{j=1}^{N} \varepsilon_j e^{ijx} \right\|_\infty \leqslant C_1 \sqrt{N}. \tag{1.11}$$

Let $\tilde{G}_1$ be the set of integers j in $[1, N]$ for which $\varepsilon_j = 1$, and let $\tilde{G}_2$ be the complementary set of integers in $[-2N, 2N]$.

First suppose that $2 \leqslant p \leqslant \infty$. We set

$$f_1(x) = \sum_{k \in \tilde{G}_1} e^{ikx}, \qquad f_2(x) = -\sum_{k \in \tilde{G}_2 \cap [1, N]} e^{ikx}.$$

Then, using (1.11), we find that

$$\|f_1 + f_2\|_p \leqslant \|f_1 + f_2\|_\infty \leqslant C_1 \sqrt{N}. \tag{1.12}$$

On the other hand, in view of Jackson's inequality (see [7], §4.9.2)

$$\|f_1\|_p \geqslant C_2 N^{-1/p} \|f_1\|_\infty \geqslant C_2 N^{-1/p} |\tilde{G}_1|. \tag{1.13}$$

From (1.11) it follows that $|\tilde{G}_1| > N/2 - C_1\sqrt{N}$, so we find from (1.13) that for sufficiently large N

$$\|f_1\|_p \geqslant C_3 N^{1-1/p}. \tag{1.14}$$

Combining (1.12) and (1.14), we obtain the required lower bound.

Now suppose that $1 \leqslant p \leqslant 2$. Set $f_3(x) = 2V_N(x) - f_1(x)$, where $V_N(x)$ is the Vallée-Poussin kernel of order $2N - 1$. Then $f_3 \in G_2^p$ and

$$\|f_1 + f_3\|_1 \leqslant C_4, \tag{1.15}$$

and in view of Nikol'skiĭ's inequality (see [7], §4.9.2)

$$\|f_1 + f_3\|_p \leqslant C_5 N^{1-1/p}\|f_1 + f_3\|_1 \leqslant C_6 N^{1-1/p}. \tag{1.16}$$

On the other hand,

$$\|f_1\|_1 \geqslant C_7\sqrt{N}. \tag{1.17}$$

Indeed,

$$|\tilde{G}_1| = (f_1, f_1 + f_2) \leqslant \|f_1\|_1\|f_1 + f_2\|_\infty,$$

whence (1.17) follows. Combining (1.16) and (1.17), we obtain the required lower bound. The lemma is proved.

§2. Proof of Theorem 2.1

First we prove the upper bound. Let $\varepsilon > 0$ be given, and let

$$P_1(z_1, z_2) = \sum_{k=0}^{n} \varphi_k^1(z_2) z_1^k, \tag{2.1}$$

$$P_2(z_1, z_2) = \sum_{k=0}^{m} \varphi_k^2(z_1) z_2^k, \tag{2.2}$$

be such that

$$\|f - P_1\|_Y \leqslant e_{n,\infty}(f)_Y + \varepsilon, \tag{2.3}$$

$$\|f - P_2\|_Y \leqslant e_{\infty,m}(f)_Y + \varepsilon. \tag{2.4}$$

For definiteness we suppose that $n \leqslant m$. From (2.3) and (2.4) it follows that

$$\|P_2 - P_1\|_Y \leqslant e_{n,\infty}(f)_Y + e_{\infty,m}(f)_Y + 2\varepsilon. \tag{2.5}$$

Let $S_n(\varphi, z)$ denote the partial sum of order n of the Taylor series of the analytic function $\varphi(z)$. Set

$$P(z_1, z_2) = \sum_{k=0}^{m} S_n(\varphi_k^2, z_1) z_2^k = S_n(P_2(\cdot, z_2), z_1).$$

We estimate the difference $f - P$. We have

$$\|f - P\|_Y \leqslant \|f - P_2\|_Y + \|P_2 - P_1\|_Y. \tag{2.6}$$

To estimate the second term we use the following identity:

$$P_2(z_1, z_2) - P(z_1, z_2) \equiv P_2(z_1, z_2) - P_1(z_1, z_2) - S_n((P_2 - P_1)(\cdot, z_2), z_1),$$

from which, by (2.5) and the theorem of Landau on estimating norms of partial sums of Taylor series [8], we obtain

$$\|P_2 - P\|_Y \leqslant (e_{n,\infty}(f)_Y + e_{\infty,m}(f)_Y + 2\varepsilon)\Big(\frac{1}{\pi}\log(n+2) + O(1)\Big). \tag{2.7}$$

The required upper bound follows from (2.4), (2.6), and (2.7).

Now we prove the lower bound. Let $N = n + m + 1$, $\tilde{G}_1 = \{k = (k_1, k_2)\colon k_1 + k_2 = N\}$, $z_1 = e^{ix}$, $z_2 = e^{iy}$ and $n \leqslant m$. It is not hard to see (see, for example, [9], Lemma 1) that for arbitrary $\tilde{G}_2$ disjoint from $\tilde{G}_1$ we have

$$M(G_1, G_2)_p = 1. \tag{2.8}$$

Let $\tilde{\mathscr{G}}_1 = \{k\colon k_1 + k_2 = N, 0 \leqslant k_1 \leqslant n\}$ and $\tilde{\mathscr{G}}_2 = \{k\colon k_1 + k_2 = N, n < k_1 \leqslant 2n + 1\}$. It follows from Theorem 1.2 that

$$M(\mathscr{G}_1, \mathscr{G}_2) = \frac{1}{\pi}\log(n+1) + O(1), \qquad p = 1, \infty. \tag{2.9}$$

Let $f_1 \in \mathscr{G}_1$ and $f_2 \in \mathscr{G}_2$ be such that $\|f_1 + f_2\|_p = 1$ and

$$\|f_1\|_p \geqslant M(\mathscr{G}_1, \mathscr{G}_2)_p. \tag{2.10}$$

Then, because of (2.8),

$$e_{n,m}(f_1)_Y \geqslant \|f_1\|_p, \qquad Y = H^1(U^2), A(U^2), \qquad p = 1, \infty. \tag{2.11}$$

On the other hand,

$$\begin{aligned} &e_{n,\infty}(f_1)_Y = 0, \\ &e_{\infty,m}(f_1)_Y \leqslant \|f_1 + f_2\|_p = 1. \end{aligned} \tag{2.12}$$

Comparing (2.11), (2.12), and (2.10), we obtain

$$e_{n,m}(f_1)_Y / (e_{n,\infty}(f_1)_Y + e_{\infty,m}(f_1)_Y) \geqslant M(\mathscr{G}_1, \mathscr{G}_2)_p,$$

and to conclude the proof it suffices to apply (2.9).

Bibliography

1. S. N. Bernshteĭn [Bernstein], *On the best approximation of functions of several variables by means of polynomials or trigonometric sums*, Trudy Mat. Inst. Steklov. **38** (1951), 24–29. (Russian)

2. V. N. Temlyakov, *Best approximations for functions of two variables*, Dokl. Akad. Nauk SSSR **223** (1975), 1079–1082; English transl. in Soviet Math. Dokl. **16** (1975).

3. Walter Rudin, *Function theory in polydiscs*, Benjamin, 1969.

4. A. L. Garkavi, *Simultaneous approximation to a periodic function and its derivatives by trigonometric polynomials*, Izv. Akad. Nauk SSSR Ser. Mat. **24** (1960), 103–128. (Russian)

5. A. Zygmund, *Trigonometric series*, 2nd rev. ed., Vols. I, II, corrected reprint, Cambridge Univ. Press, 1977.

6. Jean-Pierre Kahane, *Séries de Fourier absolument convergentes*, Springer-Verlag, 1970.

7. A. F. Timan, *Theory of approximation of functions of a real variable*, Fizmatgiz, Moscow, 1960; English transl., Pergamon Press, Oxford, and Macmillan, New York, 1963.

8. Edmund Landau, *Abschätzung der Koeffizientensumme einer Potenzreihe*, Arch. Math. Phys. (3) **21** (1913), 42–50.

9. V. N. Temlyakov, *Relations between best approximations of functions of two variables*, Mat. Zametki **29** (1981), 95–106; English transl. in Math. Notes **29** (1981).

Translated by H. P. BOAS

Труды Матем. Инст. Стеклов.
164(1983)

Proceedings of the Steklov
Institute of Mathematics
1985, Issue 2

BASES AND K-FUNCTIONALS FOR SOBOLEV SPACES ON COMPACT MANIFOLDS OF CLASS C^∞

UDC 517.5

Z. CHESEL'SKIĬ [CIESIELSKI]

ABSTRACT. Some new properties are established for the bases constructed by the author and Fiegiel in the joint paper [5] for Sobolev spaces on smooth compact manifolds. For given integers $l < k$ it is shown that the multiplier $(n^{(k-l)/d}, n = 1, 2, \dots)$ corresponding to the basis under consideration induces a linear isomorphism between W_p^k and W_p^l $(1 < p < \infty)$. Further, new asymptotic formulas are established for the Peetre functionals (K-functionals) for the pair $[W_p^l, W_p^k]$ of spaces.

Bibliography: 5 titles.

§1. Introduction

Following joint work of the author and T. Figiel (see [3]–[5]), we continue here our study of bases in Sobolev spaces on complex C^∞-manifolds. Moreover, we adhere to the same system as in [5] for presenting results. Namely, let M be a compact d-dimensional C^∞-manifold, with or without boundary. Assume also that μ is a smooth measure (henceforth fixed) on M that is locally equivalent to Lebesgue measure on R^d.

One possible way of defining Sobolev spaces on R^d for $1 < p < \infty$ uses the techniques of the Fourier transformation and Fourier multipliers. Theorem 2.1 is an analogue of this approach to Sobolev spaces on smooth compact manifolds. A similar result for Besov spaces follows from Theorem B in [5]. Since inequalities of the Bernstein and Jackson type are satisfied for the bases under consideration ([5], Corollary 11.2), it is possible to derive an equivalent K-functional representation for pairs $[W_p^l, W_p^k]$, $l < k$, $1 < p < \infty$, of Sobolev spaces. In the particular case $1 < p < \infty$ we have a new formula for a K-functional (a modulus of smoothness), as represented in Theorem 3.2. The analogue of Theorem 3.2 for the trigonometric system—Theorem 4.1—is apparently new.

§2. Isomorphisms between Sobolev spaces

The basis whose construction is contained in Theorem 2.1 coincides with the basis in Theorem A of [5] to within a reordering. We use another notation for this basis in order to underscore the difference in the enumeration. The definitions of the various Sobolev spaces on M can be found in [4]. In this section we assume that $1 < p < \infty$, and in this case the definitions are simpler.

1980 *Mathematics Subject Classification*. Primary 46B15, 46E35, 58C05; Secondary 42B15.

THEOREM 2.1. *For a particular positive integer m there are two systems $(F_n, n = 1,2,\dots)$ and $(\mathring{F}_n, n = 1,2,\dots)$ of functions on M such that $F_n \in C^m(M)$, $\mathring{F}_n \in \mathring{C}^m(M)$ and*

$$(F_i, \mathring{F}_j) = \int_M F_i \mathring{F}_j \, d\mu = \delta_{i,j}, \qquad i,j = 1,2,\dots.$$

For $1 < p < \infty$ and each integer l with $|l| \leqslant m$ the systems (F_n) and $(\mathring{F}_n)$ are unconditional bases in $W_p^l(M)$ and $\mathring{W}_p^l(M)$, respectively. Further, if for $-m \leqslant l < k \leqslant m$

$$T_{k,l}f = \sum_{n=1}^{\infty} n^{(k-l)/d}(f, \mathring{F}_n)F_n, \qquad \mathring{T}_{k,l}g = \sum_{n=1}^{\infty} n^{(k-l)/d}(g, F_n)\mathring{F}_n,$$

then

$$T_{k,l}\colon W_p^{(k)}(M) \to W_p^l(M) \quad \text{and} \quad \mathring{T}_{k,l}\colon \mathring{W}_p^k(M) \to \mathring{W}_p^l(M)$$

are linear isomorphisms between the corresponding spaces.

We prove the theorem in three stages.

The first stage consists in reducing the problem to the case $0 \leqslant l < k \leqslant m$. Consequently, we assume that the theorem is true for all $0 \leqslant l < k \leqslant m$ and $1 < p < \infty$. By Banach's theorem,

$$\left\|(T_{k,l})^*\colon (W_p^l)^* \to (W_p^k)\right\| = \left\|T_{k,l}\colon W_p^k \to W_p^l\right\|.$$

Since $1 < p < \infty$, it is possible to establish the following identities:

$$(W_p^l)^* = \mathring{W}_{p'}^{-l}, \quad (W_p^k)^* = \mathring{W}_{p'}^{-k}, \qquad p' = p/(p-1),$$

where the spaces on the right-hand sides of these equalities are understood as the closure of $\mathring{C}^\infty(M)$ in the corresponding norms (see, for example, [4]). Therefore, the operator $(T_{k,l})^*$ can be identified with $\mathring{T}_{-l,-k}$. Similar arguments are applicable also to $\mathring{T}_{k,l}$. From this, the theorem is valid if $0 \leqslant l < k \leqslant m$ or $-m \leqslant l < k \leqslant 0$. But for $-m \leqslant l < 0 < k < m$ we have

$$T_{k,l} = T_{l,0} \circ T_{0,k} \quad \text{and} \quad \mathring{T}_{k,l} = \mathring{T}_{l,0} \circ \mathring{T}_{0,k},$$

which concludes the first step of the proof.

The second step consists in proving an analogue of Theorem 2.1 for the standard spaces $W_p^k(Q)\mathbf{z}$ and $W_p^l(Q)\mathbf{z}$, where $0 \leqslant l \leqslant m$, $Q = I^d = I_1 \times \cdots \times I_d$, $I_j = \langle 0,1 \rangle$, $Z = Q \setminus (I_1 \setminus Z_1) \times \cdots \times (I_d \setminus Z_d)$, $Z_j \subset \partial I_j = \{0,1\}$, $\mathbf{Z}' = Q \setminus (I_1 \setminus Z_1') \times \cdots \times (I_d \setminus Z_d')$ and $Z_j' = \{0,1\} \setminus Z_j$. The space $W_p^k(Q)_{\mathbf{Z}}$ is the closure in $W_p^k(Q)$ of the set of functions in $C^\infty(Q)$ equal to zero in a neighborhood of $\mathbf{Z}$. More detailed information about these spaces and bases in them is contained in [5]. The definitions and properties of the systems $(F_n^{(m,k)}, \mathbf{n} \geqslant \mathbf{n}(\mathbf{Z},\mathbf{k}))$, are in §9 of [5]. Here we only recall that $\mathbf{n} = (n_1,\dots,n_d)$, $\mathbf{n}(\mathbf{Z},\mathbf{k}) = (n(Z_1,k_1),\dots,n(Z_d,k_d))$, $|k_j| \leqslant m$, $|k_j| - m \leqslant n(Z_j,k_j) \leqslant 1$ and $\mathbf{k} \leqslant \mathbf{n}$ means that $k_j \leqslant n_j$ for $j = 1,\dots,d$; $\mathbf{k} < \mathbf{n}$ has a similar meaning.

LEMMA 2.2. *Suppose that $m \geqslant 0$, $\tilde{m} \geqslant 0$, $d \geqslant 1$, and $1 < p < \infty$. Let $\mathbf{k}$ and $\tilde{\mathbf{k}}$ be such that $|k_j| \leqslant m$ and $|\tilde{k}_j| \leqslant \tilde{m}$ for $j = 1,\dots,d$. Then there exists a $C = C(m, \tilde{m}, p, d)$ such that*

$$\left\| \sum_{\mathbf{n} \in E} \varepsilon_{\mathbf{n}}\left(f, \tilde{F}_{\mathbf{n}}^{(\tilde{m},-\tilde{k})}(\cdot\,; \tilde{\mathbf{Z}})\right) \tilde{F}_{\mathbf{n}}^{(m,\mathbf{k})}(\cdot\,; \mathbf{Z}) \right\|_{L^p(Q)} \leqslant C\|f\|_{L^p(Q)},$$

where

$$E = \{\mathbf{n}: n_j \geq \max(n(Z_j, |k_j|), n(\tilde{Z}_j, |\tilde{k}_j|)), j = 1, \ldots, d\},$$

and $\tilde{F}_n^{(m,k)}$ *is the function* $F_n^{(m,k)}$, *normalized in* $L^2(Q)$.

PROOF. *The case* $\varepsilon_{\mathbf{n}} = 1$ *for all* n. In the one-dimensional case the proof of Lemma 2.2 goes as follows. By Lemma 9.9 in [5], the system $(F_n^{(m,k)}(\cdot; Z_0))$ is a Riesz basis (unconditional) in $L^2(I)$. Further, by (8.12), (8.28), and Definitions 9.1 and 9.4 in [5],

$$|F_n^{(m,k)}(t; Z_0)| \leqslant Cn^{1/2+k+m(Z_0)}q^{n|t-t_n|}, \qquad 0 < q < 1,$$

where $m(Z_0) = m + 2$ for $Z_0 = \varnothing$ or $\{0\}$, $m(Z_0) = -(m+2)$ for $Z_0 = \{1\}$ or $\{0,1\}$, and $t_n = (2i-1)/2^{(j+1)}$ for $n = 2^j + i$ and $1 \leqslant i \leqslant 2^j$. We can now repeat the same arguments used in [2] to prove Theorem 3.2. Passage to the d-dimensional case can then be realized by using a well-known lemma about products of operators acting in the spaces L^p. The extension to the case of an arbitrary sequence $\varepsilon_n = \pm 1$ now follows from the fact that the basis $(F_{\mathbf{n}}^{(m,\mathbf{k})})(\cdot; \mathbf{Z})$ in $L^p(Q)$ is unconditional (for comparison we refer to Lemma 9.1 in [5]).

COROLLARY 2.3. *For* $m \geqslant 0$ *and* $1 < p < \infty$ *such that* $|k_j| \leqslant m$ *and* $|\tilde{k}_j| \leqslant m$ *there is a constant* $C = C(p, m, d)$ *for which*

$$\Big\| \sum_{\mathbf{n} \geqslant \mathbf{n}(\mathbf{Z},\mathbf{k}) \vee \mathbf{n}(\mathbf{Z},\tilde{\mathbf{k}})} a_{\mathbf{n}} F_{\mathbf{n}}^{(m,\mathbf{k})}(\cdot; \mathbf{Z}) \Big\|_{L^p(Q)} \leqslant C \Big\| \sum_{\mathbf{n} \geqslant \mathbf{n}(\mathbf{Z};\,\mathbf{k})} \lambda_{\mathbf{n}} a_{\mathbf{n}} F_{\mathbf{n}}^{(m,\tilde{\mathbf{k}})}(\cdot; Z) \Big\|_{L^p(Q)},$$

where $\lambda_n = (\mathbf{n} + \mathbf{m} + \mathbf{1})^{\mathbf{k}-\tilde{\mathbf{k}}}$, $\mathbf{m} = (m, \ldots, m)$, $\mathbf{1} = (1, \ldots, 1)$, *and*

$$\mathbf{r} \vee s = (r_1 \vee s_1, \ldots, r_d \vee s_d), \quad \textit{with } r \vee s = \max(r, s).$$

This inequality follows from Lemma 2.2 if we set $\tilde{m} = m$, $\mathbf{Z} = \mathbf{Z}'$, $\varepsilon_{\mathbf{n}} = 1$ for all $\mathbf{n}$, and

$$f = \sum_{\mathbf{n} \geqslant \mathbf{n}(\mathbf{Z},\mathbf{k})} \lambda_{\mathbf{n}} a_{\mathbf{n}} F_{\mathbf{n}}^{(m,\tilde{\mathbf{k}})}(\cdot; \mathbf{Z}),$$

and make use of the fact that $(F_{\mathbf{n}}^{(m,\mathbf{k})}(\cdot; \mathbf{Z}))$ is unconditional in $L^p(Q)$, along with the asymptotic formula (see Definition 9.1 in [5])

$$\|F_{\mathbf{n}}^{(m,-\tilde{\mathbf{k}})}(\cdot; \mathbf{Z}')\|_{L^2(Q)} \cdot \|F_{\mathbf{n}}^{(m,\mathbf{k})}(\cdot, \mathbf{Z})\|_{L^2(Q)} \sim (\mathbf{n} + \mathbf{n} + 1)^{\mathbf{k}-\tilde{\mathbf{k}}}.$$

We can now prove the following assertion.

LEMMA 2.4. *Let* $1 < p < \infty$ *and* $0 \leqslant l < k \leqslant m$. *Then the operator*

$$T_{k,l}(\mathbf{Z}): W_p^k(Q)\mathbf{z} \to W_p^l(Q)\mathbf{z}$$

defined by

$$T_{k,l}(\mathbf{Z})(f) = \sum_{\mathbf{n} \geqslant \mathbf{n}(\mathbf{Z})} |\mathbf{n} + \mathbf{m} + \mathbf{1}|^{k-l} (f, F_{\mathbf{n}}^{(m)}(\cdot; \mathbf{Z}'))\, F_{\mathbf{n}}^{(m)}(\cdot; \mathbf{Z}),$$

is a linear isomorphism.

PROOF. For convenience we drop the boundary condition index $\mathbf{Z}$ and denote by $\| \ \|_p^{(k)}$ the norm in $W_p^k(Q)_{\mathbf{Z}}$. From the definition of the norm and Corollary 2.3,

$$\Big\|\sum_{\mathbf{n}} a_{\mathbf{n}} F_{\mathbf{n}}^{(m)}\Big\|_p^{(k)} \leqslant \Big\|\sum_{\mathbf{n}} a_{\mathbf{n}} F_{\mathbf{n}}^{(m)}\Big\|_p^{(l)} + \sum_{0<|s|\leqslant k-l} \sum_{|\mathbf{u}|=l} \Big\| D^{\mathbf{u}} \sum_{\mathbf{n}} a_{\mathbf{n}} F_{\mathbf{n}}^{(m,s)}\Big\|_p$$

$$\leqslant \sum_{0\leqslant|s|<k-l} \Big\|\sum_{\mathbf{n}} a_{\mathbf{n}} G_{\mathbf{n}}^{(m,s)}\Big\|_p^{(l)} \leqslant C \sum_{0\leqslant|s|\leqslant k-l} \Big\|\sum_{\mathbf{n}} (\mathbf{n}+\mathbf{m}+1)^{\mathbf{s}} a_{\mathbf{n}} F_{\mathbf{n}}^{(m)}\Big\|_p^{(l)}$$

$$\leqslant C\Big\|\sum_{\mathbf{n}} |\mathbf{n}+\mathbf{m}+\mathbf{1}|^{k-l} a_{\mathbf{n}} F_{\mathbf{n}}^{(m)}\Big\|_p^{(l)}.$$

To prove the converse inequality, we argue as follows:

$$\Big\|\sum_{\mathbf{n}\geqslant\mathbf{n}(\mathbf{Z})} |\mathbf{n}+\mathbf{m}+\mathbf{1}|^{k-l} a_{\mathbf{n}} F_{\mathbf{n}}^{(m)}\Big\|_p^{(l)} = \sum_{|\mathbf{s}|\leqslant l} \Big\|\sum_{\mathbf{n}\geqslant\mathbf{n}(\mathbf{Z},s)} |\mathbf{n}+\mathbf{m}+\mathbf{1}|^{k-l} a_{\mathbf{n}} F_{\mathbf{n}}^{(m,s)}\Big\|_p$$

$$\leqslant C\sum_{|\mathbf{s}|\leqslant l} \sum_{|\mathbf{u}|=k-l} \Big\|\sum_{\mathbf{n}\geqslant\mathbf{n}(\mathbf{Z},s)} (\mathbf{n}+\mathbf{m}+1)^{\mathbf{u}} a_{\mathbf{n}} F_{\mathbf{n}}^{(m,s)}\Big\|_p.$$

Further, for fixed s and $\mathbf{u}$,

$$\{\mathbf{n}\colon \mathbf{n}\geqslant\mathbf{n}(\mathbf{Z},s)\} = \bigcup_{e\subset\{1,2,\dots,d\}} N(e),$$

where

$$N(e) = \big\{\mathbf{n}\colon \mathbf{n}\geqslant\mathbf{n}(\mathbf{Z},s),\ n_j < n\big(Z_j, s_j+u_j\big) \text{ for } j\in e \text{ and } n_j \geqslant n\big(Z_j, s_j+u_j\big) \text{ for } j\notin e\big\}.$$

Let

$$u_j(e) = \begin{cases} u_j & \text{for } j\notin e,\\ 0 & \text{for } j\in e.\end{cases}$$

We see that $N(e)\subset\{\mathbf{n}\colon \mathbf{n}\geqslant\mathbf{n}(\mathbf{Z},\mathbf{u}(\mathrm{e})+\mathrm{s})\}$, and thus, by Corollary 2.3,

$$\Big\|\sum_{\mathbf{n}\geqslant\mathbf{n}(\mathbf{Z},\mathbf{s})} (\mathbf{n}+\mathbf{m}+1)^{\mathbf{u}} a_{\mathbf{n}} F_{\mathbf{n}}^{(m,s)}\Big\|_p$$

$$\leqslant C\sum_{e\subset\{1,\dots,d\}} \Big\|\sum_{\mathbf{n}>\mathbf{n}(\mathbf{Z},\mathbf{u}(e)+s)} (\mathbf{n}+\mathbf{m}+\mathbf{1})^{\mathbf{u}(e)} a_{\mathbf{n}} F_{\mathbf{n}}^{(m,s)}\Big\|_p$$

$$\leqslant C\sum_{e\subset\{1,\dots,d\}} \Big\|\sum_{\mathbf{n}>\mathbf{n}(\mathbf{Z},\mathbf{u}(e)+s)} a_{\mathbf{n}} F_{\mathbf{n}}^{(m,\mathbf{s}+\mathbf{u}(e))}\Big\|_p$$

$$\leqslant C\sum_{e\subset\{1,\dots,d\}} \Big\|D^{s+\mathbf{u}(e)} \sum_{\mathbf{n}\geqslant\mathbf{n}(\mathbf{Z})} a_{\mathbf{n}} F_{\mathbf{n}}^{(m)}\Big\|_p.$$

This proves the required inequality.

The third step concludes the proof of Theorem 2.1. For $|l|\leqslant m$ there exists a linear isomorphism (see Theorem 4.9 in [5])

$$T\colon W_p^{(l)}(M) \to \sum_{j=1}^{N} \oplus\, W_p^l(Q)_{Z_j}.$$

In each of the standard spaces $W_p^l(Q)_{\mathbf{Z}}$, regarded as a subspace of the direct sum given above, we choose the system $(F_n^{(m)}(\cdot;\, \mathbf{Z}_j),\ \mathbf{n} \geqslant \mathbf{n}(\mathbf{Z}_i))$ as a basis for it. With the notation $\|n\| = \max(n_i\colon i = 1,\ldots,d)$, consider for a particular $j = 1,\ldots,N$ the following decomposition of the whole set of indices into blocks:

$$\{n \geqslant n(Z_j)\} = \{n \geqslant n(Z_j)\colon \|n\| \leqslant 1\} \cup \bigcup_{h=0}^{\infty} \{n \geqslant n(Z_j)\colon 2^h < \|n\| \leqslant 2^{h+1}\}.$$

For a given j the enumeration of the system is such that the functions corresponding to the block with index h precede the functions corresponding to the block with index $h+1$. Let $(F_{n,j},\ n = 1,2,\ldots)$ denote the basis ordered in this way. Finally, for $j = 1,\ldots,N$ and $n = 1,\ldots$ let

$$F_{(j-1)+n} = T^{-1}(F_{n,j}).$$

The system $(\mathring{F}_n,\ n = 1,2,\ldots)$ is defined similarly. These systems have the desired properties, and this concludes the proof.

§3. Peetre functionals

For $p = \infty$ the spaces $W_p^k(M)$ and $\mathring{W}_p^k(M)$ in this section are understood as $C^k(M)$ and $\mathring{C}^k(M)$, respectively. To formulate the next result we need the following notation:

$$m(h) = \operatorname{card} \bigcup_{j=1}^{N} \{\mathbf{n}\colon \mathbf{n}(\mathbf{Z}_j) \leqslant \mathbf{n} \leqslant (2^h,\ldots,2^h)\}, \qquad h = 0,1,\ldots,$$

$$Q_h f = \sum_{n \leqslant m(h)} (f, \mathring{F}_n) F_n,\ \mathring{Q}_h f = \sum_{n \leqslant m(h)} (f, F_n) \mathring{F}_n.$$

See [1] for the definitions and properties of the Peetre functionals (K-functionals) (see also [4]).

THEOREM 3.1. *Suppose that $1 \leqslant p \leqslant \infty$ and $-m \leqslant l < k \leqslant m$. If $f \in W_p^l(M)$ and $1/2 \leqslant t \cdot 2^h \leqslant 1$, $h = 0,1,\ldots,$ then*

$$K(t^{k-l}, f;\, W_p^l(M), W_p^k(M)) \sim \|f - Q_h f\|_p^{(l)} + 2^{-h(k-l)} \|Q_h f\|_p^{(k)},$$

where the constants in these inequalities depend only on m and M.

PROOF. On the one hand, by the definition of a K-functional,

$$K(t^{k-l}, f;\, W_p^l, W_p^k) = \inf\{g \in W_p^k\colon \|f - g\|_p^{(l)} + t^{k-l}\|g\|_p^{(k)}\}$$
$$\leqslant \|f - Q_h f\|_p^{(l)} + 2^{-h(k-l)} \|Q_h f\|_p^{(k)}.$$

To establish the converse inequality we use the following estimates of Bernstein and Jackson type (see Theorem 9.20 and Corollary 11.2 in [5] for proofs):

$$\|Q_h f\|_p^{(k)} \leqslant C 2^{h(k-l)} \|Q_h f\|_p^{(l)},$$

and

$$\|f - Q_h f\|_p^{(l)} \leqslant C \cdot 2^{-h(k-l)} \|f - Q_h f\|_p^{(k)}.$$

Further, for $f \in W_p^l(M)$ and $g \in W_p^k(M)$

$$\begin{aligned}\|f - Q_h f\|_p^{(l)} + 2^{-h(k-l)}\|Q_h f\|_p^{(k)} &\leqslant \|f-g\|_p^{(l)} + \|Q_h(g-f)\|_p^{(l)} + \|g - Q_h g\|_p^{(l)} \\ &\quad + 2^{-h(k-l)}\|Q_h(f-g)\|_p^{(k)} + 2^{-h(k-l)}\|Q_h g\|_p^{(k)} \\ &\leqslant C\|f-g\|_p^{(l)} + 2^{-h(k-l)}\|g - Q_h g\|_p^{(k)} + 2^{-h(k-l)}\|Q_h g\|_p^{(k)} \\ &\leqslant CK\big(t^{k-l}, f; W_p^l(M), W_p^k(M)\big).\end{aligned}$$

REMARK. Theorem 3.1 remains true if W and Q are replaced by $\mathring{W}$ and $\mathring{Q}$, respectively.

THEOREM 3.2. *Let $1 < p < \infty$. Then for $1/2 \leqslant t \cdot 2^h \leqslant 1$ and $f \in W_p^l(M)$*

$$K\big(t^{k-1}, f; W_p^l(M), W_p^k(M)\big) \sim \|I_{k,l,t} f\|_p^{(l)}(M),$$

where the constants depend only on M, m, and p, and

$$I_{k,l,t} f = \sum_1^\infty \big(\min(1, tn^{1/d})\big)^{k-l} (f, \mathring{F}_n) F_n.$$

PROOF. By Theorem 3.1,

$$K\big(t^{k-l}, f; W_p^l, W_p^k\big) \sim \|f - Q_h f\|_p^{(l)} + 2^{-h(k-l)}\|Q_h f\|_p^{(k)}.$$

Further, the fact that the basis (F_n) is unconditional and Theorem 2.1 give us that

$$\begin{aligned}\|f - Q_h f\|_p^{(l)} &= \left\| \sum_{m(h)+1}^\infty (f, \mathring{F}_n) F_n \right\|_p^{(l)} \leqslant C \left\| \sum_1^\infty \big(\min(1, tn^{1/d})\big)^{k-l} (f, \mathring{F}_n) F_n \right\|_p^{(l)} \\ &\leqslant C\|I_{k,l,t} f\|_p^{(l)},\end{aligned}$$

$$2^{-h(k-l)}\|Q_h f\|_p^{(k)} \sim \left\| \sum_{n=1}^{m(h)} (tn^{1/d})^{k-1} (f, \mathring{F}_n) F_n \right\|_p^{(l)} \leqslant C\|I_{k,l,t} f\|_p^{(l)},$$

$$\begin{aligned}\|f - Q_h f\|_p^{(l)} + 2^{-h(k-l)}\|Q_h f\|_p^{(k)} &\geqslant C \left\{ \left\| \sum_{m(h)+1}^\infty (f, \mathring{F}_n) F_n \right\|_p^{(l)} \right. \\ &\quad \left. + \left\| \sum_1^{m(h)} (tn^{1/d})^{k-l} (f, \mathring{F}_n) F_n \right\|_p^{(l)} \right\} \\ &\geqslant C\|I_{k,l,t} f\|_p^{(l)}.\end{aligned}$$

REMARK. Theorem 3.2 remains true if W is replaced by $\mathring{W}$ and F by $\mathring{F}$.

§4. The trigonometric case on the one-dimensional torus

Suppose for simplicity that $T = \langle 0, 1\rangle$ and $e_n(t) = e^{2\pi i n t}$, and let $e_n(f) = \int_T f\,\bar{e}_n$.

THEOREM 4.1. *Suppose that $f \in L^p(T)$, $1 < p < \infty$. Then the modulus of smoothness of order k satisfies the estimate*

$$\omega_k(f; \delta)_{L^p(T)} \sim \left\| \sum_{|n|>0} \big(\min(1, \delta|n|)\big)^k \hat{e}_n(f) e_n \right\|_{L^p(T)},$$

where the constants depend only on p and k.

The idea of the proof is the same as in the case of a manifold. There is one essential difference: the system (e_n) is not an unconditional basis in $L^p(T)$. However, to prove this result it suffices to have the Riesz theorem that (e_n) is a basis in $L^p(T)$ and the Marcinkiewicz theorem on multipliers of Fourier series. We do not dwell here on the details.

REMARK. Is it not completely clear how to prove such a result for the case $L^p(T^d)$ with $d > 1$ and $1 < p < \infty$?

BIBLIOGRAPHY

[1] Jöran Bergh and Jörgen Löfström, *Interpolation spaces: An introduction*, Springer-Verlag, 1976.

[2] Z. Ciesielski, *Equivalence, unconditionality and convergence a.e. of the spline bases in L_p spaces*, Approximation Theory, Banach Center Publ. No. 4, PWN, Warsaw, 1979, pp. 55–68.

[3] Z. Ciesielski and T. Figiel, *Construction of Schauder bases in function spaces on smooth compact manifolds*, Approximation and Function Spaces (Gdańsk, 1979; Z. Ciesielski, editor), PWN, Warsaw and North-Holland, Amsterdam, 1981, pp. 217–232.

[4] ______, *Spline approximation and Besov spaces on compact manifolds*, Studia Math. **75** (1982), 13–36.

[5] ______, *Spline bases in classical function spaces on compact C^∞ manifolds*. I, II, Studia Math. **76** (1983), 1–58, 95–136.

Translated by H. H. McFADEN

Труды Матем. Инст. Стеклов.
164(1983)

Proceedings of the Steklov
Institute of Mathematics
1985, Issue 2

SOME PROBLEMS OF EXTREMAL INTERPOLATION IN THE MEAN FOR LINEAR DIFFERENTIAL OPERATORS

UDC 517:518.12

V. T. SHEVALDIN

ABSTRACT. Let $\mathscr{L}_n = \mathscr{L}_n(D)$ be an arbitrary nth-order linear differential operator with constant real coefficients, let $\Delta^{(h)}_{\mathscr{L}_n}$ be the corresponding generalized difference with step $h > 0$, and let $1 \leqslant p \leqslant \infty$ and $h_1 \geqslant 0$. For the sequence class

$$Y_{h,p} = \left\{ y = \{y_m\}_{m \supset -\infty}^{\infty} : \left\|\Delta^h_{\mathscr{L}_n} y\right\|_{l_p} \leqslant 1 \right\}$$

and the function class

$$F_{h,h_1,p}(y) = \left\{ f : f(n-1) \in AC; \mathscr{L}_n(D) f \in L_p(\mathbf{R}); \right.$$
$$\left. \frac{1}{h_1}\int_{-h_1/2}^{h_1/2} f(mh+t)\,dt = y_m,\ m = 0, \pm 1, \pm 2, \ldots \right\},$$

with $0 \leqslant h_1 \leqslant h$, $0 < h < h_0$ (h_0 some number), and $\omega = \omega(\delta)$ a convex modulus of continuity, the following quantities are computed:

$$A(\mathscr{L}_n, p, h, h_1) = \sup_{y \in Y_{h,p}} \inf_{f \in F_{h,h_1,p}(y)} \|\mathscr{L}_n(D) f\|_{L_p(\mathbf{R})},$$

$$B(\mathscr{L}_n, \omega, h) = \sup_{y \in Y_{h,\infty}} \inf_{f \in F_{h,0,\infty}(y)} \sup_{\delta > 0} \frac{\omega(\delta, \mathscr{L}_n(D) f)}{\omega(\delta)}.$$

Bibliography: 23 titles.

§1. Introduction. Statement of the problems

Let n be a particular positive integer, and $\mathscr{L}_n = \mathscr{L}_n(D)$ (D the differential symbol) an arbitrary nth-order linear differential operator with constant real coefficients that can be written in the form

$$\mathscr{L}_n(D) = \prod_{s=1}^{k} \left(D^2 - 2\gamma_s D + \gamma_s^2 + \alpha_s^2\right) \prod_{j=1}^{n-2k} \left(D - \beta_j\right), \tag{1.1}$$

where α_s, β_j, and γ_s are real numbers, and it can be assumed that $\alpha_s > 0$. For $h > 0$ and an arbitrary sequence $y = \{y_m\}_{-\infty}^{\infty}$ of real numbers we define the generalized difference $\Delta^h_{\mathscr{L}_n} y_m$ corresponding to $\mathscr{L}_n$:

$$\Delta^h_{\mathscr{L}_n y_m} = \prod_{s=1}^{k} \left(T^2 - 2Te^{\gamma_s h}\cos\alpha_s h + e^{2\gamma_s h}E\right) \prod_{j=1}^{n-2k} \left(T - e^{\beta_j h}E\right) y_m, \tag{1.2}$$

1980 *Mathematics Subject Classification*. Primary 47E05; Secondary 30E05, 39A10, 41A15, 41A46, 46E30, 49D25.

0081-5438/85 $1.00 + $.25 per page

where $Ty_m = y_{m+1}$ and E is the identity operator. For any function f on $\mathbf{R} = (-\infty, +\infty)$ let

$$\left(\Delta^h_{\mathscr{L}_n} f\right)(x + mh) = \Delta^h_{\mathscr{L}_n} y_m,$$

where $y_m = y_m(x) = f(x + mh)$. We remark that for any solution φ of the homogeneous linear differential equation $\mathscr{L}_n(D)f = 0$

$$\left(\Delta^h_{\mathscr{L}_n}\varphi\right)(x + mh) = 0.$$

It is easy to put $\Delta^h_{\mathscr{L}_n} y_m$ in the form

$$\Delta^h_{\mathscr{L}_n} y_m = \sum_{s=0}^{n} (-1)^{n-s} \mu_s^{(n)} y_{m+s}, \tag{1.3}$$

where $\mu_s^{(n)} = \mu_s(\mathscr{L}_n, h)$ and does not depend on y_m. Let l_p, $1 \leqslant p \leqslant \infty$, denote the sequence space with the norm

$$\|y\|_{l_p} = \begin{cases} \left(\sum\limits_{m=-\infty}^{\infty} |y_m|^p\right)^{1/p}, & 1 \leqslant p < \infty, \\ \sup\limits_m |y_m|, & p = \infty. \end{cases}$$

Let AC be the class of locally absolutely continuous functions, and $L_p(\mathbf{R})$ $(1 \leqslant p \leqslant \infty)$ the space of functions on $\mathbf{R}$ with norm

$$\|f\|_{L_p(\mathbf{R})} = \begin{cases} \left(\int_{\mathbf{R}} |f(t)|^p \, dt\right)^{1/p}, & 1 \leqslant p < \infty, \\ \operatorname{ess\,sup}\limits_{t \in \mathbf{R}} |f(t)|, & p = \infty. \end{cases}$$

For $h > 0$, $h_1 \geqslant 0$ and $1 \leqslant p \leqslant \infty$ we consider the sequences class

$$Y_{h,p} = \left\{ y = \{y_m\}_{m=-\infty}^{\infty} : \left\|\Delta^h_{\mathscr{L}_n} y\right\|_{l_p} \leqslant 1 \right\} \tag{1.4}$$

and the function class

$$F_{h,h_1,p}(y) = \Bigg\{ f : f^{(n-1)} \in AC; \mathscr{L}_n(D)f \in L_p(\mathbf{R}); \\ \frac{1}{h_1} \int_{-h_1/2}^{h_1/2} f(mh + t) \, dt = y_m, \; m = 0, \pm 1, \pm 2, \dots \Bigg\} \tag{1.5}$$

(here we set $f(mh) = y_m$ when $h_1 = 0$).

Now we can formulate the problems to be considered below.

1. What conditions must the numbers p, h, and h_1 satisfy in order that there exist a function $f \in F_{h,h_1,p}(y)$ for any sequence $y \in V_{h,p}$?

2. For the values of p, h, and h_1 found, compute the quantities

$$A(\mathscr{L}_n, p, h, h_1) = \sup_{y \in Y_{h,p}} \inf_{f \in F_{h,h_1,p}(y)} \|\mathscr{L}_n(D) f\|_{L_p(\mathbf{R})}, \tag{1.6}$$

$$B(\mathscr{L}_n, \omega, h) = \sup_{y \in Y_{h,\infty}} \inf_{f \in F_{h,0,\infty}(y)} \sup_{\delta > 0} \frac{\omega(\delta, \mathscr{L}_n(D)f)}{\omega(\delta)}, \tag{1.7}$$

where $\omega(\delta, \mathscr{L}_n(D)f)$ is the modulus of continuity of the function $\mathscr{L}_p(D)f$, and $\omega = \omega(\delta)$ is a given modulus of continuity.

A number of publications have dealt with problem 1 and the computation of the quantities $A(\mathscr{L}_n, p, h, h_1)$ and $B(\mathscr{L}_n, \omega, h)$. We mention first those which involve the operator $\mathscr{L}_n(D) = D^n$. Ryaben'kiĭ [1] proved that $A(\mathscr{L}_n, p, h, h_1)$ is bounded for this operator when $p = \infty$ and $h_1 = 0$ (see also Ryaben'kiĭ and Filippov [2]). Sobolev [3] showed that the Ryaben'kiĭ process gives us that $A(\mathscr{L}_n, p, h, 0)$ is bounded also when $1 \leqslant p < \infty$. The precise values of $A(\mathscr{L}_n, p, h, h_1)$ and $B(\mathscr{L}_n, \omega, h)$ for the operator $\mathscr{L}_n(D) = D^n$ when $1 \leqslant p \leqslant \infty$, $0 \leqslant h_1 \leqslant h$, and $\omega = \omega(\delta)$ is a convex modulus of continuity were computed by Subbotin in [4]–[7]. In [8] the same author solved the analogous problems in certain more general cases, replacing the class $Y_{h,p}$ by the class $Y_{h,q}$ $(1 \leqslant q \leqslant \infty)$. The precise value of $A(\mathscr{L}_n, p, h, 0)$ for $\mathscr{L}_n(D) = D^n$ when $p = 1, 2$, or ∞ was also found by Schoenberg [9], and for $p = 2$ he determined a best extension for an individual sequence y. In [10] Sharma and Tzimbalario used Subbotin's method from [4]–[7] to compute $A(\mathscr{L}_n, p, h, h_1)$ for the operator $\mathscr{L}_n(D) = \prod_{j=1}^{n}(D - \beta_j)$ under the condition that it is formally selfadjoint, $h_1 = 0$, and $p = \infty$. The author later (see [11] and [12]) computed $A(\mathscr{L}_n, p, h, h_1)$ in a similar way for an arbitrary linear differential operator $\mathscr{L}_n = \mathscr{L}_n(D)$ with constant real coefficients when $0 < h < h_0 = \pi/\max_{1\leqslant s\leqslant k}\sigma_s$, $h_1 = 0$ and $p = \infty$.

In the present article we give some sufficient conditions for a solution of problem 1 and compute $A(\mathscr{L}_n, p, h, h_1)$ and $B(\mathscr{L}_n, \omega, h)$ for an arbitrary linear differential operator $\mathscr{L}_n = \mathscr{L}_n(D)$ with constant real coefficients when $1 \leqslant p \leqslant \infty$, $0 < h < h_0 = \pi/\max_{1\leqslant s\leqslant k}\alpha_s$, $0 \leqslant h_1 \leqslant h$, and $\omega = \omega(\delta)$ is a convex modulus of continuity. The scheme for obtaining lower and upper estimates of $A(\mathscr{L}_n, p, h, h_1)$ and $B(\mathscr{L}_n, \omega, h)$ is completely analogous to the scheme in [4]–[12] and goes as follows.

To obtain lower estimates of $A(\mathscr{L}_n, p, h, h_1)$ and $B(\mathscr{L}_n, \omega, h)$ we construct a sequence $y^* \in y_{h,p}$ such that for any $f \in F_{h,h_1,p}(y^*)$ the quantities $\|\mathscr{L}_n(D)f\|_{L_p(\mathbf{R})}$ and $\sup_{\delta>0}\omega(\delta, \mathscr{L}_n(D)f)/\omega(\delta)$ are bounded below by certain constants depending only on $\mathscr{L}_n$, p, h, h_1, and $\omega = \omega(\delta)$. To obtain upper estimates and to solve Problem 1 we use functions to be called generalized $\mathscr{L}$-splines with uniform nodes.

DEFINITION 1. A function $s_n(x) \in C^{(n-1)}(\mathbf{R})$ is called a *generalized $\mathscr{L}$-spline with uniform nodes* $(l + \theta)h$ $(h > 0; 0 \leqslant \theta < 1; l = 0, \pm 1, \pm 2, \dots)$ corresponding to an operator $\mathscr{L}_n = \mathscr{L}_n(D)$, a parameter θ, and a measurable Lebesgue-integrable function $g(x)$ on the interval $[\theta h, (\theta + 1)h]$ if for any $l = 0, \pm 1, \pm 2, \dots$

$$\mathscr{L}_n(D)s_n((l-1)h + x) = Z_l g(x), \qquad \theta h \leqslant x < (\theta + 1)h,$$

where the Z_l are constants.

In deriving upper estimates for the numbers Z_l we obtain difference equations whose characteristic polynomials $U_r(x)$ satisfy the conditions in the following theorem, which was proved by M. G. Kreĭn and Subbotin ([7], Theorem B, Lemma 1.1, Remarks 1.1 and 1.2):

THEOREM A. *If all the zeros of the polynomial $U_r(x) = \sum_0^r c_l x^l$ are negative and simple and $U_r(-1) \neq 0$, then the difference equation $\sum_0^r c_l Z_{m+1} = K_m$ $(m = 0, \pm 1, \pm 2, \dots)$, where $\{K_m\}_{-\infty}^{\infty} \in l_p$ $(1 \leqslant p \leqslant \infty)$, has a unique solution $\{Z_m^0\}_{-\infty}^{\infty} \in l_p$, which can be expressed by the formula*

$$Z_m^0 = \sum_{s=-\infty}^{\infty} a_{-s-m} K_s,$$

where $\sum_{-\infty}^{\infty} a_s x^s = 1/U_r(x)$, *and it satisfies the estimate*

$$\|Z^0\|_{l_p} \leqslant \|K\|_{l_p}/|U_r(-1)|.$$

The problems of computing the Kolmogorov widths of certain classes of 2π-periodic functions are important applications of problems 1 and 2.

DEFINITION 2. The *m-dimensional Kolmogorov width* of a subset F of a Banach space X is defined to be

$$d_m(F, X) = \inf_{R_m \subset X} \sup_{x \in F} \inf_{y \in R_m} \|x - y\|_X,$$

where R_m is an arbitrary m-dimensional subspace of X.

The widths were first considered by Kolmogorov in [13]. A survey of investigations on the question of widths is contained, for example, in Tikhomirov's monograph [14]. Denote by $C_{2\pi}^{(n)}$ $(C^{(0)})_{2\pi} = C_{2\pi}$ the space of 2π-periodic n times continuously differentiable functions, and let

$$W^{\mathscr{L}_n}H^\omega = \{f \in C_{2\pi}^{(n)} \colon \omega(\delta, \mathscr{L}_n(D)f) \leqslant \omega(\delta)\}, \tag{1.8}$$

where $\omega = \omega(\delta)$ is a given convex modulus of continuity. In this article we use the solutions of problems 1 and 2 to establish in a way analogous to that in [6] the lower estimate

$$d_{2m-1}(W^{\mathscr{L}_n}H^\omega, C) \geqslant \left(B\left(\mathscr{L}_n, \omega, \frac{\pi}{m}\right) \sum_{s=0}^{n} \mu_s\left(\mathscr{L}_n, \frac{\pi}{m}\right)\right)^{-1} \tag{1.9}$$

for the width $d_{2m-1}(W^{\mathscr{L}_n}H^\omega, C)$ when $m \geqslant 2\max_{1 \leqslant s \leqslant k} \sigma_s$. We remark that Lorentz [15] and Subbotin [6] obtained the lower estimate (1.9) for the width $d_{2m-1}(W^{\mathscr{L}_n}H^\omega, C)$ in the case of the operator $\mathscr{L}_n(D) = D^n$ under the condition that $\omega = \omega(\delta)$ is convex. Korneĭchuk [16], [17] showed that this estimate is sharp, namely, that it coincides with the value of the best approximation of the class $W^{\mathscr{L}_n}H^\omega$ by the space T_{m-1} of trigonometric polynomials of order $m - 1$. The even widths were computed for $\mathscr{L}_n(D) = D^n$ by Ruban [18].

To solve problems 1 and 2 we need some auxiliary results which we formulate and prove in the next section.

§2. Properties of the auxiliary functions

With an operator $\mathscr{L}_n = \mathscr{L}_n(D)$ of the form (1.1) we associate a sequence of differential operators $\mathscr{L}_r = \mathscr{L}_r(D)$ $(r = 2, 4, \ldots, 2k, 2k+1, \ldots, n)$:

$$\begin{aligned} \mathscr{L}_2(D) &= D^2 - 2\gamma_1 D + \gamma_1^2 + \alpha_1^2, \\ \mathscr{L}_4(D) &= (D^2 - 2\gamma_2 D + \gamma_2^2 + \alpha_2^2)\mathscr{L}_2(D), \ldots, \mathscr{L}_{2k}(D) \\ &= (D^2 - 2\gamma_k D + \gamma_k^2 + \alpha_k^2)\mathscr{L}_{2k-2}(D), \\ \mathscr{L}_{2k+1}(D) &= (D - \beta_1)\mathscr{L}_n(D) = \cdots = (D - \beta_{n-2k})\mathscr{L}_{n-1}(D). \end{aligned} \tag{2.1}$$

With each operator $\mathscr{L}_r = \mathscr{L}_r(D)$ in this sequence we associate, in turn, the operator

$$\mathscr{L}_{r+1}^0 = \mathscr{L}_{r+1}^0(D) = D\mathscr{L}_r(D). \tag{2.2}$$

For each operator $\mathscr{L}_r$ (respectively, for $\mathscr{L}_{r+1}^0$) $(r = 2, 4, \ldots, 2k, 2k+1, \ldots, n)$ we determine a function φ_r (φ_{r+1}^0) such that

$$\begin{aligned} &\mathscr{L}_r(D)\varphi_r(t) = 0, \quad \varphi_r^{(j)}(0) = \delta_{j,r-1} \quad (j = 0, 1, \ldots, r-1), \\ &(\mathscr{L}_{r+1}^0(D)\varphi_{r+1}^0(t) = 0, \quad \varphi_{r+1}^{0(j)}(0) = \delta_{j,r} \quad (j = 0, 1, \ldots, r)), \end{aligned} \tag{2.3}$$

where $\delta_{j,r-1}$ ($\delta_{j,r}$) is the Kronecker symbol. For an arbitrary sequence $y = \{y_m\}_{-\infty}^{\infty}$ of real numbers and $h > 0$ we define as in (1.2) the generalized difference $\Delta^h_{\mathscr{L}_r} y_m$ ($\Delta^h_{\mathscr{L}^0_{r+1}} y_m$) with step h corresponding to the operator $\mathscr{L}_r$ ($\mathscr{L}^0_{r+1}$) ($r = 2, 4, \ldots, 2k, 2k + 1, \ldots, n$):

$$\Delta^h_{\mathscr{L}_r} y_m = \begin{cases} \prod_{s=1}^{r} \left(T^2 - 2Te^{\gamma_s h} \cos \alpha_s h + e^{2\gamma_s^h} E\right) y_m, & r \leqslant 2k, \\ \prod_{s=1}^{k} \left(T^2 - 2Te^{\gamma_s^h} \cos \alpha_s h + e^{2\gamma_s^h} E\right) \prod_{j=1}^{r-2k} \left(T - e^{\beta_j h} E\right) y_m, & r > 2k, \end{cases} \tag{2.4}$$

$$\Delta^h_{\mathscr{L}^0_{r+1}} y_m = (T - E) \Delta^h_{\mathscr{L}_r} y_m. \tag{2.5}$$

For any function f on $\mathbf{R}$, let

$$\left(\Delta^h_{\mathscr{L}_r} f\right)(x + mh) = \Delta^h_{\mathscr{L}_r} y_m, \tag{2.6}$$

where $y_m = y_m(x) = f(x + mh)$. Note that for any function φ that solves the homogeneous linear differential equation $\mathscr{L}_r(D) f = 0$

$$\left(\Delta^h_{\mathscr{L}} \varphi\right)(x + mh) = 0. \tag{2.7}$$

It is easy to put $d_{2m-1}(W^{\mathscr{L}_n} H^\omega, C)$ in the form $\Delta^h_{\mathscr{L}_r} y_m$ ($\Delta^h_{\mathscr{L}^0_{r+1}} y_m$) in the form

$$\Delta^h_{\mathscr{L}_r} y_m = \sum_{s=0}^{r} (-1)^{r-s} \mu_s^{(r)} y_{m+s},$$
$$\left(\Delta^h_{\mathscr{L}^0_{r+1}} y_m = \sum_{s=0}^{r+1} (-1)^{r+1-s} \mu_s^{0(r+1)} y_{m+s}\right), \tag{2.8}$$

where $\mu_s^{(r)} = \mu_s(\mathscr{L}_r, h)$ and $\mu_s^{0(r+1)} = \mu_s(\mathscr{L}^0_{r+1}, h)$, and they do not depend on y_m. By using (2.4) and (2.5) it is not hard to verify that

$$\mu_s^{(r)} = e^{\beta_{r-2k} h} \mu_s^{(r-1)} + \mu_{s-1}^{(r-1)} \qquad (s = 0, 1, \ldots, r;\ r = 2k + 1, \ldots, n) \tag{2.9}$$

(here $\mu_{-1}^{(r-1)}$ and $\mu_r^{(r-1)}$ are set equal to zero),

$$\mu_s^{(r)} = e^{2\gamma_{r/2} h} \mu_s^{(r-2)} + 2e^{\gamma r/2^h} \cos \alpha_{r/2} h \mu_{s-1}^{(r-2)} + \mu_{s-2}^{(r-2)}$$
$$(s = 0, 1, \ldots, r;\ r = 2, 4, \ldots, 2k) \tag{2.10}$$

(here $\mu_{-2}^{(r-2)}$, $\mu_{-1}^{(r-2)}$, $\mu_{r-1}^{(r-2)}$, and $\mu_r^{(r-2)}$ are set equal to zero), and

$$\mu_s^{0(r+1)} = \mu_s^{(r)} + \mu_{s-1}^{(r)}$$
$$(s = 0, 1, \ldots, r + 1;\ r = 2, 4, \ldots, 2k, 2k + 1, \ldots, n) \tag{2.11}$$

(here $\mu_{-1}^{(r)}$ and $\mu_{r+1}^{(r)}$ are set equal to zero).

It follows from the definition of φ_r (φ^0_{r+1}) that

$$\begin{gathered} \varphi_r' - \beta_{r-2k} \varphi_r = \varphi_{r-1} \qquad (r = 2k + 1, \ldots, n), \\ \varphi_r'' - 2\gamma_{r/2} \varphi_r' + \left(\gamma_{r/2}^2 + \alpha_{r/2}^2\right) \varphi_r = \varphi_{r-2} \qquad (r = 4, 6, \ldots, 2k), \\ \varphi^{0'}_{r+1} = \varphi_r \qquad (r = 2, 4, \ldots, 2k, 2k + 1, \ldots, n). \end{gathered} \tag{2.12}$$

For each operator $\mathscr{L}_r$ ($\mathscr{L}^0_{r+1}$) we define for $0 \leqslant t \leqslant 1$ and $-\infty < x < \infty$ the function

$$\Pi_r(x, t) = \sum_{l=-\infty}^{\infty} x^l \Delta^h_{\mathscr{L}_r} \varphi_r((l - r + 1 - t) h_+)$$
$$\left(\Pi^0_{r+1}(x, t) = \sum_{l=-\infty}^{\infty} x^l \Delta^h_{\mathscr{L}^0_{r+1}} \varphi^0_{r+1}((l - r - t) h_+)\right), \tag{2.13}$$

where

$$\varphi_r(u_+) = \begin{cases} \varphi_r(u), & u \geqslant 0, \\ 0, & u < 0, \end{cases} \qquad \varphi_{r+1}^0(u_+) = \begin{cases} \varphi_{r+1}^0(u), & u \geqslant 0, \\ 0, & u < 0. \end{cases}$$

Let us transform the function $\Pi_r(x, t)$ ($\Pi_{r+1}^0(x, t)$) for $0 \leqslant t \leqslant 1$. Using (2.7), we have

$$\begin{aligned} \Pi_r(x, t) &= \sum_{l=-\infty}^{\infty} x^l \Delta_{\mathscr{L}_r}^h \varphi_r((l - r + 1 - t)h_+) \\ &= \sum_{l=0}^{r-1} x^l \sum_{s=r-l}^{r} (-1)^{r-s} \mu_s^{(r)} \varphi_r((s + l - r + 1 - t)h) \\ &= \sum_{l=0}^{r-1} x^l \sum_{s=r-l}^{r} (-1)^{r-s} u_s^{(r)} \varphi_r((s + l - r + 1 - t)h). \end{aligned} \tag{2.14}$$

Similarly,

$$\Pi_{r+1}^0(x, t) = \sum_{l=0}^{r} x^l \sum_{s=r+1-l}^{r+1} (-1)^{r+1-s} \mu_s^{0(r+1)} \varphi_{r+1}^0((s + l - r - t)h). \tag{2.15}$$

It follows from (2.14) and (2.15) that for any $r = 2, 4, \ldots, 2k, 2k + 1, \ldots, n$ the function $\Pi_r(x, t)$ ($\Pi_{r+1}^0(x, t)$) is a polynomial of degree $r - 1$ (r) in x which $0 < t \leqslant 1$, and degree $r - 2$ ($r - 1$) when $t = 0$. It also follows from (2.14) and (2.15) that

$$\Pi_r(x, 1) = x\Pi_r(x, 0) \qquad \left(\Pi_{r+1}^0(x, 1) = x\Pi_{r+1}^0(x, 0)\right). \tag{2.16}$$

Using (2.16), we extend the function $\Pi_r(x, t)$ ($\Pi_{r+1}^0(x, t)$) for each particular x as a function of t to the whole number line by means of the equality

$$\Pi_r(x, t + 1) = x\Pi_r(x, t) \qquad \left(\Pi_{r+1}^0(x, t + 1) = x\Pi_{r+1}^0(x, t)\right). \tag{2.17}$$

Differentiating (2.14) and (2.15) with respect to t ($D = d/dt$) for $0 < t < 1$ and using (2.9)–(2.12), we get the following equalities:

$$(D + \beta_{r-2k}h)\Pi_r(x, t) = -h(1 - xe^{\beta_{r-2k}h})\Pi_{r-1}(x, t) \qquad (r = 2k + 1, \ldots, n), \tag{2.18}$$

$$\begin{aligned} &\left(D^2 + 2\gamma_{r/2}hD + \left(\gamma_{r/2}^2 + \alpha_{r/2}^2\right)h^2\right)\Pi_r(x, t) \\ &\quad = h^2\left(x^2 e^{2\gamma_{r/2}h} - 2xe^{\gamma_{r/2}h} \cos \alpha_{r/2}h + 1\right)\Pi_{r-2}(x, t) \qquad (r = 4, 6, \ldots, 2k), \end{aligned} \tag{2.19}$$

$$D\Pi_{r+1}^0(x, t) = -h(1 - x)\Pi_r(x, t) \qquad (r = 2, 4, \ldots, 2k, 2k + 1, \ldots, n). \tag{2.20}$$

The equalities (2.17)–(2.20) show that for each particular x the function $\Pi_r(x, t)$ ($\Pi_{r+1}^0(x, t)$), as a function of t, is in $C^{(r-2)}(\mathbf{R})$ ($C^{(r-1)}(\mathbf{R})$).

LEMMA 1. *Let* $0 < h < h_0 = \pi/\max_{1 \leqslant s \leqslant k} \sigma_s$. *For any* $r = 2, 4, \ldots, 2k, 2k + 1, \ldots, n$ *the function* $\Pi_r(x, t)$ ($\Pi_r^0(x, t)$) *has for each particular* $x < 0$ *a unique zero on the half-open interval* $[0, 1)$, *and this zero is simple.*

PROOF. The scheme of the proof of Lemma 1 is given in the author's papers [11] and [12]. In the present article we give a complete proof of Lemma 1 needed in what follows. We note that for $k = 0$ Lemma 1 was proved by Micchelli ([19], Corollary 2.2) (another proof was given by Schoenberg in [20], Theorem 2). Therefore, we assume at once that $k \neq 0$. Note that for any $r = 2, 4, \dots, 2k, 2k+1, \dots, n$ the function $\Pi_r(x, t)$ has for each particular $x < 0$ at least one zero in $[0, 2)$, since it follows from (2.16) that $\Pi_r(x, t)$ has different signs at the points $t = 0$ and $t = 1$. Lemma 1 will be proved by contradiction.

Suppose that for some r ($r = 2, 4, \dots, 2k, 2k+1, \dots, n$) and $x_0 < 0$ the function $\Pi(t) = \Pi_r(x_0, t)$ has at least two zeros on $[0, 1)$, counting multiplicity. It follows from (2.16) with $x = x_0 < 0$ that $\Pi(t)$ has an odd number of zeros on $[0, 1)$ (counting multiplicity); consequently, there are at least three. For any differentiable function $\varphi(t)$ it is easy to verify the identities

$$(D + \beta)\varphi(t) = e^{-\beta t} D\left(e^{\beta t}\varphi(t)\right), \tag{2.21}$$

$$\begin{aligned}(D^2 - 2\gamma D + \gamma^2 + \alpha^2)\varphi(t) \\ = \frac{e^{\gamma(t-t_0)}}{\sin\alpha(t - t_0)} D\left(\sin^2\alpha(t - t_0) D\left(\frac{e^{-\gamma(t-t_0)}}{\sin\alpha(t - t_0)}\varphi(t)\right)\right).\end{aligned} \tag{2.22}$$

Without loss of generality we assume that $r > 2k$. Setting $\beta = \beta_{r-2k}h$ and $\varphi(t) = \Pi(t) = \Pi_r(x_0, t)$ in (2.21), we show that $\Pi_{r-1}(x_0, t)$ has at least two zeros (counting multiplicity) on $[0, 1)$ (the proof of this assertion given here is due to Schoenberg [20], Theorem 2). We consider separately three cases.

1. Let $\Pi(0) \neq 0$; without loss of generality we assume that $\Pi(0) > 0$. We then get from (2.16) with $x_0 < 0$ that $\Pi(1) < 0$. Consequently, $\Pi(t)$ has at least three zeros (counting multiplicity) on $(0, 1)$; applying Rolle's theorem, we get from (2.18) and (2.21) that $\Pi_{r-1}(x_0, t)$ has at least two zeros (counting multiplicity) on $[0, 1)$.

2. Suppose that $\Pi(0) = 0$ and $\Pi'(0) \neq 0$; without loss of generality we assume that $\Pi'(0) > 0$. From (2.17) with $x_0 < 0$ and the fact that $\Pi(t) \in C^{(r-2)}(\mathbf{R})$ we then get $\Pi(1) = 0$ and $\Pi'(1) < 0$. Therefore, $\Pi(t)$ has at least three zeros on $[0, 1]$, and $t = 1$ is a simple zero. Again we get from (2.18), (2.21), and Rolle's theorem that $\Pi_{r-1}(x_0, t)$ has at least two zeros (counting multiplicity) on $[0, 1)$.

3. Suppose that $\Pi(0) = \Pi'(0) = 0$. Then $\Pi(1) = \Pi'(1) = 0$, and again from (2.18), (2.21), and Rolle's theorem we get that $\Pi_{r-1}(x_0, t)$ has at least two zeros (counting multiplicity) on $[0, 1)$.

Applying the same arguments to the functions $\Pi_{r-1}(x_0, t)$ $\Pi_{r-2}(x_0, t), \dots,$ we get that $\Pi_{2k}(x_0, t)$ has at least two zeros (counting multiplicity) on $[0, 1)$. We now prove that $\Pi_{2k-2}(x_0, t)$ has the same property for $0 < h < h_0 = \pi/\max_{1 \leqslant s \leqslant k} \alpha_s$ and $k \geqslant 4$. Here three cases are possible.

1. Let $\Pi_{2k}(x_0, 0) = 0$ and $\Pi'_{2k}(x_0, 0) \neq 0$; without loss of generality we assume that $\Pi'_{2k}(x_0, 0) > 0$. Then from (2.17) and the fact that $\Pi_{2k}(x_0, t) \in C^{(2k-2)}(\mathbf{R})$ we get that $\Pi_{2k}(x_0, 1) = 0$ and $\Pi'_{2k}(x_0, 1) < 0$. Consequently, $t = 0$ and $t = 1$ are simple roots of $\Pi_{2k}(x_0, t)$, and $\Pi_{2k}(x_0, t)$ has in all at least four zeros on $[0, 1]$. From this and (2.22) with $t_0 = 0$, $\gamma = \gamma_k h$, $\alpha = \alpha_k h$, $0 < h < h_0$, and $\varphi(t) = \Pi_{2k}(x_0, t)$ and from (2.19) and Rolle's theorem it follows that $\Pi_{2k-2}(x_0, t)$ has at least two zeros (counting multiplicity) in $[0, 1)$.

2. Suppose that $\Pi_{2k}(x_0, 0) = \Pi'_{2k}(x, 0) = 0$. Then $\Pi_{2k}(x_0, 1) = \Pi'_{2k}(x_0, 1) = 0$. From (2.22) with $t_0 = 0$, $\gamma = \gamma_k h$, $0 < h < h_0$ and $\varphi(t) = \Pi_{2k}(x_0, t)$ and from (2.19) and Rolle's theorem we get that $\Pi_{2k-2}(x_0, t)$ has at least two zeros (counting multiplicity) on $[0, 1)$.

3. Let $\Pi_{2k}(x_0, 0) \neq 0$; without loss of generality we assume that $\Pi_{2k}(x_0, 0) > 0$. From (2.16) with $r = 2k$ we then get that $\Pi_{2k}(x_0, 1) < 0$. Consequently, all the zeros (and there are at least three) of $\Pi_{2k}(x_0, t)$ are in $(0, 1)$. Let $t = t_0$ $(0 < t_0 < 1)$ be the zero of $\Pi_{2k}(x_0, t)$ in $(0, 1)$ closest to 0. We get from (2.17) that $\Pi_{2k}(x_0, t)$ has at least three zeros (counting multiplicity) on $[t_0, t_0 + 1)$. As in the preceding cases, we apply (2.19), (2.22), and Rolle's theorem with $0 < h < h_0$ and get that $\Pi_{2k-2}(x_0, t)$ has at least two zeros (counting multiplicity) on $[t_0, t_0 + 1)$. Using (2.17), we then conclude that $\Pi_{2k-2}(x_0, t)$ has the same property on $[0, 1)$.

Thus, for $0 < h < h_0$ has $k \geqslant 4$ the function $\Pi_{2k-2}(x_0, t)$ always has at least two zeros (counting multiplicity) on $[0, 1)$. By induction we establish that $\Pi_2(x_0, t)$ also has the analogous property. On the other hand, computing $\Pi_2(x_0, t)$ directly, we get

$$\Pi_2(x_0, t) = (1/\alpha_1) e^{\gamma 1(1-t)h} \sin \alpha_1 (1 - t) h + x_0 (1/\alpha_1) e^{\gamma 1(2-t)h} \sin \alpha_1 t h. \quad (2.23)$$

But for $0 < h < h_0$ and $x_0 < 0$ the function $\Pi_2(x_0, t)$ has exactly one zero on $[0, 1)$, and this zero is simple. Contradiction. Thus, Lemma 1 is proved for $\Pi_r(x_0, t)$. The proof is carried out similarly for $\Pi^0_{r+1}(x_0, t)$.

COROLLARY 1. *Let h $(0 < h < h_0 = \pi/\max_{1 \leqslant s \leqslant k} \alpha_s)$ and x $(x < 0)$ be fixed, and let α be the unique zero of the function $\Pi_r(x, t)$ $(r = 2, 4, \ldots, 2k, 2k+1, \ldots, n)$ with respect to the variable t on $[0, 1)$. Then*

$$\Pi^0_{r+1}(x, \alpha) \neq 0.$$

The proof follows from Lemma 1 and (2.20).

LEMMA 2. *Let t $(0 \leqslant t < 1)$ and h $(0 < h_0 = \pi/\max_{1 \leqslant s \leqslant k} \alpha_s)$ be fixed. For each $r = 2, 4, \ldots, 2k, 2k+1, \ldots, n$ all the zeros of the function $\Pi_r(x, t)$ $(\Pi^0_{r+1}(x, t))$ with respect to the variable x are negative and simple.*

Lemma 2 was proved by the author in [12] for the function $\Pi^0_{r+1}(x, t)$ (see also [11], [19] and [20]). This proof can be kept without change for the function $\Pi_r(x, t)$; therefore, we do not give it here.

Consider the following functions for $0 \leqslant t \leqslant 1$:

$$R_{n-1}(x, t) = x^{n-1} \Pi_n\left(\frac{1}{x}, t\right), \qquad R^0_n(x, t) = x^n \Pi^0_{n+1}\left(\frac{1}{x}, t\right). \quad (2.24)$$

It follows from (2.14) and (2.15) with $r = n$ and $0 \leqslant t \leqslant 1$ that the functions $R_{n-1}(x, t)$ and $R^0_n(x, t)$ are polynomials in the variable x, and

$$\begin{aligned} R_{n-1}(x, t) &= \sum_{l=0}^{n-1} x^l \sum_{s=0}^{l} (-1)^{n-1-s} \mu_s^{(n)} \varphi_n((s - l - t)h), \\ R^0_n(x, t) &= \sum_{l=0}^{n} x^l \sum_{s=0}^{l} (-1)^{n-s} \mu_s^{0(n+1)} \varphi^0_{n+1}((s - l - t)h). \end{aligned} \quad (2.25)$$

LEMMA 3. *Let* t $(0 \leqslant t < 1)$ *and* h $(0 < h < h_0 = \pi/\max_{1 \leqslant s \leqslant k} \alpha_s)$ *be fixed, and let* $\eta_j(t)$ $(j = 1,\dots,n-1)$ *be the zeros of the polynomial* $R_{n-1}(x,t)$ *for* $0 < t < 1$, *and* η_j $(j = 1,\dots,n-1)$ *the zeros of* $R_{n-1}(x,0)$, *arranged in decreasing order. For any* $j = 1,\dots,n-1$ *the function* $\eta_j(t)$ *is a strictly decreasing function on* $[0,1)$, *and, moreover, for* $0 < t < 1$

$$\eta_2 < \eta_1(t) < \eta_1 = 0, \qquad \eta_j < \eta_{j-1}(t) < \eta_{j-1} \quad (j = 3,4,\dots,n-1),$$
$$-\infty < \eta_{n-1}(t) < \eta_{n-1}.$$

PROOF. Using (2.14) with $r = n$ and Lemma 2, we number all the zeros of $\Pi_n(x,t)$ with respect to the variable x with $0 \leqslant t < 1$ and $0 < h < h_0$ as follows: let

$$\lambda_{n-1}^{(n)}(t) < \lambda_{n-2}^{(n)}(t) < \cdots < \lambda_1^{(n)}(t) < 0 \tag{2.26}$$

be the zeros of $\Pi_n(x,t)$ for $0 < t < 1$, and let

$$\lambda_{n-2}^{(n)} < \lambda_{n-3}^{(n)} < \cdots < \lambda_1^{(n)} < 0 \tag{2.27}$$

be the zeros of $\Pi_n(x,0)$. Since $\Pi_n(x,1) = x\Pi_n(x,0)$, it is natural to define the functions $\lambda_j^{(n)}(t)$ $(j = 1,\dots,n-1)$ at the points $t = 0$ and $t = 1$ in the following way:

$$\begin{aligned} &\lambda_j^{(n)}(0) = \lambda_j^{(n)} \quad (j = 1,2,\dots,n-2), \qquad \lambda_{n-1}^{(n)}(0) = -\infty, \\ &\lambda_j^{(n)}(1) = \lambda_{j-1}^{(n)} \quad (j = 2,3,\dots,n-1), \qquad \lambda_1^{(n)}(1) = 0. \end{aligned} \tag{2.28}$$

Then for any $j = 1,\dots,n-1$ the function $\lambda_j^{(n)}(t)$ is continuous on $[0,1]$ except for the point $t = 0$ for the function $\lambda_{n-1(t)}^{(n)}$, where $\lim_{t\to+0} \lambda_{n-1}^{(n)}(t) = -\infty$. Suppose that $0 \leqslant t_1 < t_2 < 1$ or $0 < t_1 < t_2 \leqslant 1$. We prove that for any $i,j = 1,\dots,n-1$

$$\lambda_i^{(n)}(t_1) \neq \lambda_j^{(n)}(t_2). \tag{2.29}$$

Assume that for some i, j, t_1, and t_2

$$\lambda_i^{(n)}(t_1) = \lambda_j^{(n)}(t_2) = \lambda < 0,$$

But then $\Pi_n(\lambda, t_1) = \Pi_n(\lambda, t_2) = 0$, which contradicts Lemma 1. By (2.29), the functions $\lambda_j^{(n)}(t)$ $(j = 1,\dots,n-1)$ are strictly monotone on $[0,1]$. From this (2.26)–(2.28) we get that for any $j = 1,\dots,n-1$ the function $\lambda_j^{(n)}(t)$ is a strictly increasing function on $[0,1]$, and for $0 < t < 1$

$$\begin{aligned} &\lambda_1^{(n)} < \lambda_1^{(n)}(t) < 0, \qquad \lambda_j^{(n)} < \lambda_j^{(n)}(t) < \lambda_{j-1} \quad (j = 2,3,\dots,n-2), \\ &\qquad -\infty < \lambda_{n-1}^{(n)}(t) < \lambda_{n-2}^{(n)}. \end{aligned} \tag{2.30}$$

The use of (2.24) completes the proof of Lemma 3.

LEMMA 4. *For* $0 < t \leqslant 1$ *and* $0 < h < h_0 = \pi/\max_{1 \leqslant s \leqslant k} \alpha_s$

PROOF. From (2.25) with $0 < t \leqslant 1$,

$$R_{n-1}(0,t) = (-1)^{n-1}\mu_0^{(n)}\varphi n(-th). \tag{2.31}$$

From (2.4) and (2.8) with $r = n$,

$$\mu_0^{(n)} = e^{\tau}, \tag{2.32}$$

where $\tau = (2\Sigma_1^k \gamma_s + \Sigma_1^{n-2k}\beta_j)h$.

Using the definition of φ_r and (2.12), we get the recursion relations

$$\varphi_r(t) = \int_0^t e^{\beta_{r-k}(t-x)}\varphi_{r-1}(x)\,dx \qquad (r = 2k+1,\dots,n),$$
$$\varphi_r(t) = \int_0^t \frac{\sin \alpha_{r/2}(t-x)}{\alpha_{r/2}} e^{\gamma_{r/2}(t-x)}\varphi_{r-2}(x)\,dx \qquad (r = 4,6,\dots,2k). \tag{2.33}$$

From this we get by induction on r that for $0 < t \leqslant 1$ and $0 < h < h_0$

$$\operatorname{sgn} \varphi_n(-th) = (-1)^{n-1}. \tag{2.34}$$

Then it follows from (2.31), (2.32), and (2.34) that for $0 < t \leqslant 1$ and $0 < h < h_0$

$$R_{n-1}(0,t) > 0.$$

The lemma is proved.

LEMMA 5. *Suppose that* $0 < t < 1$, $0 < h < h_0 = \pi/\max_{1 \leqslant s \leqslant k} \alpha_s$, $\eta_1(t),\dots,\eta_{n-1}(t)$ *are the zeros of the polynomial* $R_{n-1}(x,t)$, *and* $\eta_1,\dots,\eta_{n-1}$ *are the zeros of* $R_{n-1}(x,0)$, *arranged in decreasing order. Then*

$$\operatorname{sgn} R_{n-1}(\eta_j(t),u) = \begin{cases} (-1)^j, & 0 \leqslant u < t, \\ (-1)^{j+1}, & t < u < 1, \end{cases}$$
$$\operatorname{sgn} R_{n-1}(\eta_j,t) = (-1)^{j+1}, \qquad 0 < t < 1 \ (j = 1,2,\dots,n-1).$$

PROOF. From the relations

$$R_{n-1}(0,u) > 0, \quad R_{n-1}(\eta_1(u),u) = 0 \qquad (0 < u < 1),$$

and the fact that $x = \eta_1(t)$ is a simple zero of $R_{n-1}(x,t)$ for $0 < t < 1$, it follows that

$$\operatorname{sgn} R_{n-1}(\eta_1(t),u) = \begin{cases} 1, & t < u, \\ -1, & u < t. \end{cases}$$

Further, for $t = 1$

$$R_{n-1}(\eta_1(1),u) < 0, \qquad 0 < u < 1,$$

and since $\eta_1(1) = \eta_2(0)$, it follows that

$$R_{n-1}(\eta_2(0),u) < 0, \qquad 0 < u < 1.$$

From this and the fact that $x = \eta_2(t)$ is a simple zero of $R_{n-1}(x,t)$ for $0 < t < 1$ we get that

$$\operatorname{sgn} R_{n-1}(\eta_2(t),u) = \begin{cases} -1, & t < u, \\ 1, & u < t. \end{cases}$$

Applying the analogous arguments to $\eta_2(t),\dots,\eta_{n-1}(t)$, we get that

$$\operatorname{sgn} R_{n-1}(\eta_j(t),u) = \begin{cases} (-1)^j, & 0 \leqslant u < t, \\ (-1)^{j+1}, & t < u < 1, \end{cases}$$
$$\operatorname{sgn} R_{n-1}(\eta_j,t) = (-1)^{j+1}, \qquad 0 < t < 1 \ (j = 1,2,\dots,n-1).$$

Lemma 5 is proved.

LEMMA 6. *Let* γ $(0 \leqslant \gamma < 1)$ *and* h $(0 < h < h_0 = \pi/\max_{1 \leqslant s \leqslant k} \alpha_s)$ *be fixed, and suppose that the function* $A(t) \geqslant 0$ *is integrable on* $[0,1]$, *is nonzero on sets of positive measure in* $(0,\gamma)$ *and* $(\gamma,1)$ *if* $\gamma \neq 0$, *and is nonzero on a set of positive measure in*

$(0,1)$ *if* $\gamma = 0$. *Then the polynomial of degree n in x given by*

$$Q_n(x,\gamma) = \int_0^{\gamma} A(t)R_{n-1}(x,t)\,dt + \int_{\gamma}^{1} xA(t)R_{n-1}(x,t)\,dt \tag{2.35}$$

has exactly n real simple roots. For $0 < \gamma < 1$ *they are all negative, while* $Q_n(0,0) = 0$ *and* $Q_n(x,0)$ *has* $n-1$ *negative roots in the case* $\gamma = 0$.

PROOF. Consider first the case $0 < \gamma < 1$. Let $\eta_j(\gamma)$ $(j = 1,2,\ldots,n-1)$ be the zeros of $R_{n-1}(x,\gamma)$, in decreasing order. It follows from Lemmas 3 and 5 and (2.35) that

$$\operatorname{sgn} Q_n(\eta_j(\gamma),\gamma) = (-1)^j \qquad (j = 1,\ldots,n-1). \tag{2.36}$$

By (2.36), the polynomial $Q_n(x,\gamma)$ has at least $n-2$ negative zeros on the interval $(\eta_{n-1}(\gamma), \eta_1(\gamma))$. It follows from (2.25) that $R_{n-1}(x,t)$ can be written for $0 \leqslant t < 1$ in the form

$$R_{n-1}(x,t) = \sum_{l=0}^{n-1} b_l(t)x^l,$$

where

$$b_l(t) = \sum_{s=0}^{l} (-1)^{n-1-s}\mu_s^{(n)}\varphi_n((s-l-t)h) \qquad (l = 0,1,\ldots,n-1),$$

and, in particular,

$$b_0(t) = (-1)^{n-1}\mu_0^{(n)}\varphi_n(-th), \qquad b_{n-1}(t) = \mu_n^{(n)}\varphi_n((1-t)h).$$

By using (2.33) and the fact that $\mu_n^{(n)} = 1$ it is easy to get that $b_{n-1}(t) > 0$ for $0 \leqslant t < 1$. From this, Lemma 4, (2.35), and (2.36) with $j = 1$ and $j = n-1$ we get that $Q_n(x,\gamma)$ has at least two more roots, one on $(-\infty, \eta_{n-1}(\gamma))$ and the other on $(\eta_1(\gamma), 0)$, and Lemma 6 is proved for $0 < \gamma < 1$.

Let $\gamma = 0$. Let η_j $(j = 1,\ldots,n-1)$ be the zeros of $R_{n-1}(x,0)$, arranged in decreasing order. We get from Lemmas 3 and 5 and from (2.35) that

$$\operatorname{sgn} Q_n(\eta_j, 0) = (-1)^j \qquad (j = 1,\ldots,n-1), \tag{2.37}$$

and, consequently, $Q_n(x,0)$ has at least $n-2$ negative zeros on (η_{n-1}, η_1). On the other hand, since $b_{n-1}(0) > 0$, we get from (2.37) with $j = n-1$ that $Q_n(x,0)$ has at least one more root on $(-\infty, \eta_{n-1})$. Since, moreover, $Q_n(0,0) = 0$, the assertion of Lemma 6 is completely proved.

REMARK 1. Results analogous to Lemmas 3–6 can be established similarly for $R_n^0(x,t)$.

For each $r = 2,4,\ldots,2k,2k+1,\ldots,n$ let

$$P_r(t) = (-1)^r\Pi_r(-1,t), \qquad P_{r+1}^0(t) = (-1)^{r+1}\Pi_{r+1}^0(-1,t), \tag{2.38}$$

where $\Pi_r(x,t)$ and $\Pi_{r+1}^0(x,t)$ are defined by (2.13) and (2.17). Using (2.14) and (2.15), we get that for $0 \leqslant t \leqslant 1$

$$P_r(t) = \sum_{j=0}^{r}(-1)^j\sum_{s=j}^{r}(-1)^{r-s}\mu_s^{(r)}\varphi_r((s-j+1-t)h),$$

$$P_{r+1}^0(t) = \sum_{j=0}^{r+1}(-1)^j\sum_{s=j}^{r+1}(-1)^{r+1-s}\mu_s^{0(r+1)}\varphi_{r+1}^0((s-j+1-t)h). \tag{2.39}$$

For $0 \leqslant t \leqslant 1$ let

$$a_{j,r}(t,h,h_1) = \begin{cases} \dfrac{h^2}{h_1}\displaystyle\int_{-h_1/2h}^{h_1/2h} \sum_{l=0}^{r} (-1)^{r-l}\mu_l^{(r)}\varphi_r((l+x+1-j-t)h_+)\,dx, & h_1 = 0 \\ h\displaystyle\sum_{l=0}^{r}(-1)^{r-l}\mu_l^{(r)}\varphi_r((l+l-j-t)h_+), & h_1 = 0 \end{cases} \tag{2.40}$$

$$(j = 0,1,\dots,r+1;\ r = 2,4,\dots,2k,2k+1,\dots,n),$$

$$S_r(t) = \sum_{j=0}^{r+1}(-1)^j a_{j,r}(t,h,h_1). \tag{2.41}$$

Observe that for $h_1 = 0$ and $0 \leqslant t \leqslant 1$

$$S_r(t) = hP_r(t). \tag{2.42}$$

Using (2.20), (2.38), and the fact that $\Pi_r(-1,t) \in C^{(r-2)}(\mathbf{R})$ and $\Pi_{r+1}^0(-1,t) \in C^{(r-1)}(\mathbf{R})$, we get that for any $t \in \mathbf{R}$

$$P_{r+1}^{0\prime}(t) = 2hP_r(t). \tag{2.43}$$

It follows from (2.17) and (2.38) that

$$P_r(t+1) = -P_r(t), \qquad P_{r+1}^0(t+1) = -P_{r+1}^0(t). \tag{2.44}$$

We transform the function $S_r(t)$ $(r = 2,4,\dots,2k,2k+1,\dots,n)$ for $0 \leqslant t \leqslant 1$ and $0 < h_1 \leqslant h$. By (2.43) and (2.44),

$$S_r(t) = \frac{h^2}{h_1}\int_{-h_1/2h}^{h_1/2h}\sum_{j=0}^{r+1}(-1)^j\sum_{l=0}^{r}(-1)^{r-l}\mu_l^{(ir)}\varphi_r((l+x+1-j-t)h_+)\,dx$$

$$= \begin{cases} \dfrac{h^2}{h_1}\displaystyle\int_{-h_1/2h}^{t} P_r(t-x)\,dx - \dfrac{h^2}{h_1}\int_t^{h_1/2h} P_r(1-x+t)\,dx, & 0 \leqslant t \leqslant h_1/2h, \\ \dfrac{h^2}{h_1}\displaystyle\int_{-h_1/2h}^{h_1/2h} P_r(t-x)\,dx, & h_1/2h \leqslant t \leqslant 1-h_1/2h, \\ \dfrac{h^2}{h_1}\displaystyle\int_{-h_1/2h}^{t-1} P_r(t-1-x)\,dx + \dfrac{h^2}{h_1}\int_{t-1}^{h_1/2h} P_r(t-x)\,dx, & 1-h_1/2h \leqslant t \leqslant 1, \end{cases}$$

$$= \begin{cases} \dfrac{h}{2h_1}\left[P_{r+1}^0\left(t+\dfrac{h_1}{2h}\right) + P_{r+1}^0\left(1-\dfrac{h_1}{2h}+t\right)\right], & 0 \leqslant t \leqslant h_1/2h, \\ \dfrac{h}{2h_1}\left[-P_{r+1}^0\left(t-\dfrac{h_1}{2h}\right) + P_{r+1}^0\left(t+\dfrac{h_1}{2h}\right)\right], & h_1/2h \leqslant t \leqslant 1-h_1/2h, \\ \dfrac{h}{2h_1}\left[-P_{r+1}^0\left(t-1+\dfrac{h_1}{2h}\right) - P_{r+1}^0\left(t-\dfrac{h_1}{2h}\right)\right], & 1-h_1/2h \leqslant t \leqslant 1, \end{cases}$$

$$= \frac{h}{2h_1}\left[P_{r+1}^0\left(t+\frac{h_1}{2h}\right) - P_{r+1}^0\left(t-\frac{h_1}{2h}\right)\right]. \tag{2.45}$$

It follows from (2.42), (2.44), and (2.45) that for $0 \leqslant h_1 \leqslant h$

$$S_r(1) = -S_r(0). \tag{2.46}$$

The equality (2.46) enables us to extend $S_r(t)$ for $0 \leqslant h_1 \leqslant h$ from $[0, 1]$ to the whole number line by the equality

$$S_r(t+1) = -S_r(t). \tag{2.47}$$

Moreover, since $\Pi_r(-1, t) \in C^{(r-2)}(\mathbf{R})$ and $\Pi^0_{r+1}(-1, t) \in C^{(r-1)}(\mathbf{R})$, we get from (2.38), (2.42), and (2.45) that $S_r(t) \in C^{(r-1)}(\mathbf{R})$ for $h_1 > 0$, while $S_r(t) \in C^{(r-2)}(\mathbf{R})$ for $h_1 = 0$. Considering (2.44) and (2.47), we can assume that (2.42) and (2.45) hold for any $t \in \mathbf{R}$, and, thus,

$$S_r(t) = \begin{cases} \dfrac{h}{2h_1}\left[P^0_{r+1}\left(t + \dfrac{h_1}{2h}\right) - P^0_{r+1}\left(t - \dfrac{h_1}{2h}\right)\right], & h_1 > 0, \\ hPr(t), & h_1 = 0 \end{cases} \tag{2.48}$$

$$(r = 2, 4, \dots, 2k, 2k+1, \dots, n).$$

LEMMA 7. *Let $0 < h < h_0 = \pi/\max_{1 \leqslant s \leqslant k} \alpha_s$ and $0 \leqslant h_1 \leqslant h$. For each $r = 2, 4, \dots, 2k, 2k+1, \dots, n$ the function $S_r(t)$ has a unique zero on $[0, 1)$, and this zero is simple.*

PROOF. For $h_1 = 0$ Lemma 7 follows from (2.38), (2.48), and Lemma 1. Suppose that $0 < h_1 \leqslant h < h_0 = \pi/\max_{1 \leqslant s \leqslant k} \alpha_s$. Using (2.18)–(2.20), (2.38), and (2.48), we get that for any $t \in \mathbf{R}$ $(D = d/dt)$

$$\begin{aligned} &(D + \beta_{r-2k}h)S_r(t) = -h(1 + e^{\beta_{r-2k}h})S_{r-1}(t) \qquad (r = 2k+1, \dots, n), \\ &\left(D^2 + 2\gamma_{r/2}hD + \left(\gamma^2_{r/2} + \alpha^2_{r/2}\right)h^2\right)S_r(t) \\ &\quad = h^2\left(e^{2\gamma_{r/2}h} + 2e^{\gamma_{r/2}h}\cos\alpha_{r/2}h + 1\right)S_{r-2}(t) \qquad (r = 4, 6, \dots, 2k). \end{aligned} \tag{2.49}$$

The proof of Lemma 7 for $0 < h_1 \leqslant h < h_0$ is analogous to that of Lemma 1, namely, by induction on r. We first verify Lemma 7 for the functions $S_1(t) = S_1^{(1)}(t)$ and $S_1(t) = S_1^{(2)}(t)$ corresponding to the operators $\mathscr{L}_1^{(1)}(D) = D$ and $\mathscr{L}_1^{(2)}(D) = D - \beta_1$, where $\beta_1 \neq 0$. In the case of operator $\mathscr{L}_1^{(1)}(D)$ the function $S_1^{(1)}(t)$ has the form

$$S_1^{(1)}(t) = \begin{cases} -2th^2/h_1, & 0 \leqslant t \leqslant h_1/2h, \\ -h, & h_1/2h \leqslant t \leqslant 1 - h_2/2h, \\ 2h^2(t-1)/h_1, & 1 - h_1/2h \leqslant t \leqslant 1, \end{cases}$$

and Lemma 7 follows at once from this. In the case of the operator $\mathscr{L}_1^{(2)}(D) = D - \beta_1$, with $\beta_1 \neq 0$, the function $S_1^{(2)}(t)$ corresponding to this operator has for $0 \leqslant t \leqslant 1$ the form

$$S_1^{(2)}(t) = \begin{cases} \dfrac{h}{\beta_1 h_1}\left[e^{-\beta_1 th}\left(e^{\beta_1 h - \beta_1 h_1/2} + e^{\beta_1 h_1/2}\right) - \left(e^{\beta_1 h} + 1\right)\right], & 0 \leqslant t \leqslant \dfrac{h_1}{2h}, \\ \dfrac{h}{\beta_1 h_1} e^{\beta_1 h - \beta_1 th}\left(e^{-\beta_1 h_1/2} - e^{\beta_1 h_1/2}\right), & \dfrac{h_1}{2h} \leqslant t \leqslant 1 - \dfrac{h_1}{2h}, \\ \dfrac{h}{\beta_1 h_1}\left[e^{\beta_1 h - \beta_1 th}\left(-e^{\beta_1 h - \beta_1 h_1/2} - e^{\beta_1 h_1/2}\right) + e^{\beta_1 h_1} + 1\right], & 1 - \dfrac{h_1}{2h} \leqslant t \leqslant 1. \end{cases} \tag{2.50}$$

An analysis of (2.50) shows that $S_1^{(2)}(t)$ has for $0 < h_1 \leqslant h$ precisely one zero on $[0, 1)$, and this zero is simple. The verification of the induction base for the function $S_2(t)$ corresponding to the operator $\mathscr{L}_2(D) = D^2 - 2\gamma_1 D + \gamma_1^2 + \alpha_1^2$ involves cumbersome formulas which express $S_2(t)$ in terms of the coefficients of $\mathscr{L}_2(D)$. But it turns out that to prove Lemma 7 it suffices to verify that the function

$$S_1^{(3)}(t) = \frac{h}{2h_1}\left[P_2\left(t + \frac{h_1}{2h}\right) - P_2\left(t - \frac{h_1}{2h}\right)\right], \tag{2.51}$$

where the function $P_2(t)$ corresponds to the operator $\mathscr{L}_2(D)$, has exactly one zero in $[0, 1)$, and this zero is simple. Indeed, $S_1^{(3)}(t)$ has the form

$$S_1^{(3)}(t) = \begin{cases} \dfrac{h}{2h_1\alpha_1} S_1^{(4)}(t), & 0 \leqslant t \leqslant \dfrac{h_1}{2h}, \\[2ex] \dfrac{h}{2h_1\alpha_1} S_1^{(5)}(t), & \dfrac{h_1}{2h} \leqslant t \leqslant 1 - \dfrac{h_1}{2h}, \\[2ex] \dfrac{h}{2h_1\alpha_1} S_1^{(6)}(t), & 1 - \dfrac{h_1}{2h} \leqslant t \leqslant 1, \end{cases} \tag{2.52}$$

where

$$\begin{aligned} S_1^{(4)}(t) = e^{\gamma_1(h - th - h_1/2)}\Big[&-e^{\gamma_1 h} \sin\alpha_1\left(th + \frac{h_1}{2}\right) + \sin\alpha_1\left(h - th - \frac{h_1}{2}\right) \\ &- e^{\gamma_1 h_1} \sin\alpha_1\left(h + th - \frac{h_1}{2}\right) + e^{-\gamma_1 h - + \gamma_1 h_1} \sin\alpha_1\left(-th + \frac{h_1}{2}\right)\Big], \\ S_1^{(5)}(t) = e^{\gamma_1(h - th - h_1/2)}\Big[&e^{\gamma_1 h + \gamma_1 h_1} \sin\alpha_1\left(th - \frac{h_1}{2}\right) - e^{\gamma_1 h_1} \sin\alpha_1\left(h - th - \frac{h_1}{2}\right) \\ &- e^{\gamma_1 h} \sin\alpha_1\left(th + \frac{h_1}{2}\right) + \sin\alpha_1\left(h - th - \frac{h_1}{2}\right)\Big], \\ S_1^{(6)}(t) = e^{\gamma_1(h - th - h_1/2)}\Big[&e^{2\gamma_1 h} \sin\alpha_1\left(th - h + \frac{h_1}{2}\right) - e^{\gamma_1 h} \sin\alpha_1\left(2h - th - \frac{h_1}{2}\right) \\ &+ e^{\gamma_1 h + \gamma_1 h_1} \sin\alpha_1\left(th - \frac{h_1}{2}\right) - e^{\gamma_1 h_1} \sin\alpha_1\left(h - th + \frac{h_1}{2}\right)\Big]. \end{aligned} \tag{2.53}$$

It is not hard to verify that $S_1^{(3)}(t)$ is a continuous function on $[0, 1]$, and also that the functions $S_1^{(4)}(t)$, $S_1^{(5)}(t)$, and $S_1^{(6)}(t)$ can have at most one zero, and that one simple, on the closed intervals where they are defined. On the other hand, an analysis of the signs of $S_1^{(3)}(t)$ at the points $t = 0$, $t = h_1/2h$, $t = 1 - h_1/2h$, and $t = 1$ shows that $S_1^{(3)}(t)$ has exactly one sign change in the indicated sequence of points. Consequently, $S_1^{(3)}(t)$ has exactly one zero on $[0, 1)$, and this zero is simple. Thus, we have completely verified the induction base. The induction step can be proved just as in Lemma 1 by using (2.21), (2.22), (2.49), and Rolle's theorem.

Lemma 7 is proved.

Note that the proof of Lemma 7 actually contains the following assertion.

COROLLARY 2. *Suppose that* $0 < h < h_0$ *and* $0 < h_1 \leqslant h$. *For each* $r = 2, 4, \dots, 2k, 2k+1, \dots, n$ *the function*

$$\tilde{S}_r(t) = P_r\left(t + \frac{h_1}{2h}\right) - P_r\left(t - \frac{h_1}{2h}\right)$$

has a unique zero on $[0, 1)$, *and this zero is simple.*

LEMMA 7, Corollary 2, and the equalities (2.43) and (2.47) combine to give us

COROLLARY 3. *Let* $0 < h < h_0$ *and* $0 < h_1 \leqslant h$, *and let* θ *be the unique zero of* $S_n(t)$ *on* $[0, 1)$. *Then* $S_n(t)$ *has exactly one extremum point on the interval* $(\theta, \theta + 1)$.

LEMMA 8. *Let* $0 < h < h_0$, $0 < h_1 \leqslant h$ *and* $1 < q < \infty$, *and let* θ *be the unique zero of* $S_n(t)$ *on* $[0, 1)$. *The function*

$$\Phi_n(t) = \left|S_n\left(t + \frac{h_1}{2h}\right)\right|^{q-1} - \left|S_n\left(t - \frac{h_1}{2h}\right)\right|^{q-1} \tag{2.54}$$

has a unique zero on $[\theta + h_1/2h, 1 + \theta - h_1/2h]$, *and it changes sign on passing through this zero.*

PROOF. Observe that

$$\Phi_n\left(\theta + \frac{h_1}{2h}\right) = \left|S_n\left(\phi + \frac{h_1}{h}\right)\right|^{q-1}, \quad \Phi_n\left(1 + \theta - \frac{h_1}{2h}\right) = -\left|S_n\left(\theta - \frac{h_1}{h}\right)\right|^{q-1}, \tag{2.55}$$

and, consequently, $\Phi_n(t)$ has at least one zero on $[\theta + h_1/2h, \Theta + 1 - h_1/2h]$ and in all it has an odd number of zeros on this interval when all multiplicities are counted. Our problem is to show that $\Phi_n(t)$ does not have more zeros on the interval. Note that $S_n(t)$ does not change sign on $(\theta, \theta + 1)$, and thus to prove Lemma 8 it suffices to verify that the function

$$\tilde{\Phi}_n(t) = S_n(t + h_1/2h) - S_n(t - h_1/2h) \tag{2.56}$$

has exactly one zero on $[\Theta + h_1/2h, 1 + \theta - h_1/2h]$. From Lemma 7, Corollary 3, and (2.47) we conclude that $S_n(t)$ has the following properties on $[\theta, \theta + 1]$:

1) $S_n(\theta) = S_n(\theta + 1) = 0$.
2) $S_n(t)$ is of constant sign on $(\theta, \theta + 1)$.
3) $S_n(t)$ has exactly one extremum point on $(\theta, \theta + 1)$.

Without loss of generality it will be assumed that $S_n(t) > 0$ on $(\pi, \theta + 1)$; let ε be the maximum point of $S_n(t)$ on $(\theta, \theta + 1)$. By properties 1)– 3), $S_n(t)$ is monotonically increasing on $[\theta, \varepsilon]$ and monotonically decreasing on $[\varepsilon, \theta + 1]$. Without loss of generality we assume that $h_1 < h/2$. We consider separately three cases.

1. $\varepsilon \in (\theta, \theta + h_1/h)$. In this case $S_n(t + h_1/2h)$ is monotonically decreasing on $[\theta + h_1/2h, \theta + 1 - h_1/2h]$, and $S_n(\varepsilon) > S_n(h_1/h) > S_n(\theta + 1) = 0$. The function $-S_n(t - h_1/2h)$ is also monotonically decreasing, from 0 to $-S_n(\varepsilon)$, on $[\theta + h_1/2h, \varepsilon + h_1/2h]$. Consequently, $\tilde{\Phi}_n(t)$ is also monotonically decreasing on $[\theta + h_1/2h, \varepsilon + h_1/2h]$ and, moreover, has different signs at the endpoint of this interval. Therefore, $\tilde{\Phi}_n(t)$ has exact one zero on $(\theta + h_1/2h, \varepsilon + h_1/2h)$. On the other hand, $-S_n(t - h_1/2h)$ is monotonically increasing from $-S_n(\varepsilon)$ to 0 on

$[\varepsilon + h_1/2h, \theta + 1 - h_1/2h]$. Furthermore, on this interval

$$S_n(t + h_1/2h) - S_n(t - h_1/2h) < 0.$$

Consequently, $\tilde{\Phi}_n(t)$ has exactly one zero on $[\theta + h_1/2h, 1 + \theta - h_1/2h]$ in case 1.

2. $\varepsilon \in (\theta + 1 - h_1/2h, \theta + 1)$. In this case $-S_n(t - h_1/2h)$ is monotonically decreasing on $[\theta + h_1/2h, \theta + 1 - h_1/2h]$, and $-S_n(\theta) > -S_n(\theta + 1 - h_1/2h) > -S_n(\varepsilon)$. The function $S_n(t + h_1/2h)$ is monotonically increasing from $S_n(h_1/h)$ to $S_n(\varepsilon)$ on $[\theta + h_1/2h, \varepsilon - h_1/2h)$, and on this interval

$$S_n(t + h_1/2h) - S_n(t - h_1/2h) > 0;$$

consequently, $\tilde{\Phi}_n(t)$ does not have zeros on $[\theta + h_1/2h, \varepsilon - h_1/2h]$. On $[\varepsilon - h_1/2h, \theta + 1 - h_1/2h]$ the function $S_n(t + h_1/2h)$ is monotonically decreasing, and, consequently, $\tilde{\Phi}_n(t)$ is also monotonically decreasing on this interval. Since, moreover, $\tilde{\Phi}_n(t)$ has different signs at the endpoints of $[\varepsilon - h_1/2h, \theta + 1 - h_1/2h]$, we get as in case 1 that $\tilde{\Phi}_n(t)$ has exactly one zero on $[\theta + h_1/2h, 1 + \theta - h_1/2h]$

3. $\varepsilon \in (\theta + h_1/h, \theta + 1 - h_1/h)$. We partition $[\theta + h_1/2h, \theta + 1 - h_1/2h]$ by the points $\varepsilon - h_1/2h$ and $\varepsilon + h_1/2h$ into three parts and investigate the behavior of $\tilde{\Phi}_n(t)$ on each of the intervals

$$[\theta + h_1/2h, \varepsilon - h_1/2h], \quad [\varepsilon - h_1/2h, \varepsilon + h_1/2h], \quad [\varepsilon + h_1/2h, \theta + 1 - h_1/2h]$$

On $[\theta + h_1/2h, \varepsilon - h_1/2h]$ the function $S_n(t + h_1/2h)$ is monotonically increasing from $S_n(\theta + h_1/h)$ to $S_n(\varepsilon)$, while $-S_n(t - h_1/2h)$ is monotonically decreasing from 0 to $-S_n(\varepsilon - h_1/h)$, and

$$S_n(t + h_1/2h) - S_n(t - h_1/2h) > 0,$$

consequently, $\tilde{\Phi}_n(t)$ does not have zeros on $[\theta + h_1/2h, \varepsilon - h_1/2h]$. Both the functions $S_n(t + h_1/2h)$ and $-S_n(t - h_1/2h)$ are monotonically decreasing on $[\varepsilon - h_1/2h, \varepsilon + h_1/2h]$. Therefore, $\tilde{\Phi}_n(t)$ is also monotonically decreasing, and since it also has different signs at the endpoints of this interval, we get from this that $\tilde{\Phi}_n(t)$ has exactly one zero on $[\varepsilon - h_1/2h, \varepsilon + h_1/2h]$. As before, it is not hard to verify that on $[\varepsilon + h_1/2h, \theta + 1 - h_1 2h]$

$$S_n(t + h_1/2h) - S_n(t - h_1/2h) < 0.$$

Thus, $\tilde{\Phi}_n(t)$ has exactly one zero on $[\theta + h_1/2h, 1 + \theta - h_1/2h]$ also in case 3.

The cases $h_1 \geqslant h/2$, $\varepsilon = \theta + h_1/h$ and $\varepsilon = \theta + 1 - h_1/h$ can be investigated similarly. Thus, the proof of Lemma 8 is complete.

Suppose again that $0 < h < h_0 = \pi/\max_{1 \leqslant s \leqslant k} \alpha_s$, $0 < h_1 \leqslant h$, $1 < q < \infty$, and θ is the unique zero of $S_n(t)$ on $[0, 1)$. We define the function

$$T_{n+2}(x) = \begin{cases} \dfrac{1}{h(1-x)} T_{n+3}^{(1)}(x), & 0 \leqslant \theta < h_1/2h, \\ \dfrac{1}{h(1-x)} T_{n+3}^{(2)}(x), & h_1/2h \leqslant \theta \leqslant 1 - h_1/2h, \\ \dfrac{1}{h(1-x)} T_{n+3}^{(3)}(x), & 1 - h_1/2h < \theta < 1, \end{cases} \tag{2.57}$$

where

$$T_{n+3}^{(1)}(x) = \int_0^{\theta+h_1/2h}\left[x^2\left|S_n\left(t+\frac{h_1}{2h}\right)\right|^{q-1} - x\left|S_n\left(t-\frac{h_1}{2h}\right)\right|^{q-1}\right]R_n^0(x,t)\,dt$$
$$+\int_{\theta+h_1/2h}^{1+\theta-h_1/2h}\left[\left|S_n\left(t+\frac{h_1}{2h}\right)\right|^{q-1} - \left|S_n\left(t-\frac{h_1}{2h}\right)\right|^{q-1}\right]x^2R_n^0(x,t)\,dt$$
$$+\int_{1+\theta-h_1/2h}^{1}\left[x^3\left|S_n\left(t+\frac{h_1}{2h}\right)\right|^{q-1} - x^2\left|S_n\left(t-\frac{h_1}{2h}\right)\right|^{q-1}\right]R_n^0(x,t)\,dt, \tag{2.58}$$

$$T_{n+3}^{(2)}(x) = \int_0^{\theta-h_1/2h}\left[\left|S_n\left(t+\frac{h_1}{2h}\right)\right|^{q-1} - \left|S_n\left(t-\frac{h_1}{2h}\right)\right|^{q-1}\right]xR_n^0(x,t)\,dt$$
$$+\int_{\theta-h_1/2h}^{\theta+h_1/2h}\left[\left|S_n\left(t+\frac{h_1}{2h}\right)\right|^{q-1}x^2 - \left|S_n\left(t-\frac{h_1}{2h}\right)\right|^{q-1}x\right]R_n^0(x,t)\,dt$$
$$+\int_{\theta+h_1/2h}^{1}\left[\left|S_n\left(t+\frac{h_1}{2h}\right)\right|^{q-1} - \left|S_n\left(t-\frac{h_1}{2h}\right)\right|^{q-1}\right]x^2R_n^0(x,t)\,dt, \tag{2.59}$$

$$T_{n+3}^{(3)}(x) = \int_0^{\theta+h_1/2h-1}\left[\left|S_n\left(t+\frac{h_1}{2h}\right)\right|^{q-1}x - \left|S_n\left(t-\frac{h_1}{2h}\right)\right|^{q-1}\right]R_n^0(x,t)\,dt$$
$$+\int_{\theta+h_1/2h-1}^{\theta-h_1/2h}\left[\left|S_n\left(t+\frac{h_1}{2h}\right)\right|^{q-1} - \left|S_n\left(t-\frac{h_1}{2h}\right)\right|^{q-1}\right]xR_n^0(x,t)\,dt$$
$$\times\int_{\theta-h_1/2h}^{1}\left[\left|S_n\left(t+\frac{h_1}{2h}\right)\right|^{q-1}x^2 - \left|S_n\left(t-\frac{h_1}{2h}\right)\right|^{q-1}x\right]R_n^0(x,t)\,dt \tag{2.60}$$

and $R_n^0(x,t)$ is defined in (2.24). Observe that $T_{n+2}(x)$, as a function of θ, is continuous on $[0,1)$.

LEMMA 9. *$T_{n+2}(x)$ is a polynomial of degree at most $n+2$.*

PROOF. By (2.25) and (2.57)–(2.60), to prove Lemma 9 it suffices to verify that $x=1$ is a root of each of the polynomials $T_{n+3}^{(i)}(x)$ $(i=1,2,3)$. Indeed, by (2.54),

$$T_{n+3}^{(i)}(1) = \int_0^1\left[\left|S_n\left(t+\frac{h_1}{2h}\right)\right|^{q-1} - \left|S_n\left(t-\frac{h_1}{2h}\right)\right|^{q-1}\right]R_n^0(1,t)\,dt$$
$$= \int_0^1\Phi_n(t)R_n^0(1,t)\,dt \qquad (i=1,2,3). \tag{2.61}$$

Integrating (2.61) by parts and using (2.20) and (2.24), we get that

$$T_{n+3}^{(i)}(1) = \int_0^1\Phi_n(t)R_n^0(1,t)\,dt = \left(R_n^0(1,t)\int_0^t\Phi_n(u)\,du\right)_0^1 = R_n^0(1,1)\int_0^1\Phi_n(u)\,du$$
$$= R_n^0(1,1)\left(\int_0^1|S_n(t)|^{q-1}dt - \int_0^1|S_n(t)|^{q-1}dt\right) = 0 \qquad (i=1,2,3).$$

Lemma 9 is proved.

LEMMA 10. 1. *$T_{n+2}(x)$ has exactly $n+1$ negative simple roots for $0 \leqslant \theta < h_1/2h$, and $T_{n+2}(0) = 0$.*

2. *$T_{n+2}(x)$ is a polynomial of degree $n+1$ with exactly n negative simple roots and with $T_{n+2}(0) = 0$ for $h_1/2h \leqslant \theta \leqslant 1 - h_1/2h$.*

3. *$T_{n+2}(x)$ is a polynomial of degree $n+1$ with all roots negative and simple for $1 - h_1/2h < \theta < 1$.*

PROOF. 1. For $0 \leqslant \theta < h_1/2h$ we have from (2.57) that

$$T_{n+2}(x) = \frac{1}{h(1-x)} T_{n+3}^{(1)}(x), \tag{2.62}$$

where $T_{n+3}^{(1)}(x)$ is defined by (2.58) and can be represented in the form

$$T_{n+3}^{(1)}(x) = \sum_{l=0}^{n+3} b_l^{(1)} x^l. \tag{2.63}$$

By using (2.58) and (2.25) it is not hard to verify that

$$b_0^{(1)} = 0, \qquad b_1^{(1)} = (-1)^{n+1} \mu_0^{0(n+1)} \int_0^{\theta + h_1/2h} \varphi_{n+1}^0(-th) \left| S_n\left(t - \frac{h_1}{2h}\right)\right|^{q-1} dt,$$

$$b_{n+3}^{(1)} = \mu_{n+1}^{0(n+1)} \int_{1+\theta-h_1/2h}^{1} \varphi_{n+1}^0((1-t)h) \left| S_n\left(t + \frac{h_1}{2h}\right)\right|^{q-1} dt. \tag{2.64}$$

From (2.5) and (2.8) with $r = n$ it follows that

$$\mu_0^{0(n+1)} > 0, \qquad \mu_{n+1}^{0(n+1)} = 1. \tag{2.65}$$

On the other hand, it follows from (2.34) and the equality

$$\varphi_{n+1}^0(t) = \int_0^t \varphi_n(u)\, du \tag{2.66}$$

that for $0 < t \leqslant 1$

$$\operatorname{sgn} \varphi_{n+1}^0(-th) = (-1)^n, \tag{2.67}$$

and for $0 \leqslant t < 1$

$$\varphi_{n+1}^0((1-t)h) > 0. \tag{2.68}$$

From (2.64), (2.65), and (2.68) we find that

$$b_0^{(1)} = 0, \qquad b_1^{(1)} < 0, \qquad b_{n+3}^{(1)} > 0. \tag{2.69}$$

Observe also that Lemma 9 gives us that

$$T_{n+3}^{(1)}(1) = 0. \tag{2.70}$$

Denote the unique zero of $\Phi_n(t)$ on $[\theta + h_1/2h, \theta + 1 - h_1/2h]$ by η ($\eta \neq 0$), and let $x_j(\eta) < 0$ ($j = 1, \ldots, n$) be the zeros of the polynomial $R_n^0(x, \eta)$, arranged in decreasing order.

Using Lemma 5 and Remark 1, we have that

$$\operatorname{sgn} R_n^0(x_j(\eta), t) = \begin{cases} (-1^j, & 0 \leqslant t < \eta, \\ (-1)^{j+1}, & \eta < t < 1 \end{cases}$$
$$(j = 1, \ldots, n). \tag{2.71}$$

From this, along with (2.58) and Lemma 8, we get

$$\operatorname{sgn} T_{n+3}^{(1)}(x_j(\eta)) = (-1)^j \qquad (j = 1,\dots,n). \tag{2.72}$$

It follows from (2.72) that $T_{n+3}^{(1)}(x)$ has at least $n-1$ negative zeros on $(x_n(\eta), x_1(\eta))$. Moreover, by (2.72) with $j=1$ and $j=n$ and by (2.69), $T_{n+3}^{(1)}(x)$ has at least two more negative roots, one on $(-\infty, x_n(\eta))$, and the other on $(x_1(\eta), 0)$. Thus, $T_{n+3}^{(1)}(x)$ has at least $n+1$ negative roots, and, moreover, $T_{n+3}^{(1)}(1) = 0$ and $T_{n+3}^{(1)}(0) = 0$, because $b_0^{(1)} = 0$. This and (2.62) complete the proof of Lemma 10 in case 1.

2. For $h_1/2h \leqslant \theta \leqslant 1 - h_1/2h$ we have

$$T_{n+2}(x) = \frac{1}{h(1-x)} T_{n+3}^{(2)}(x), \tag{2.73}$$

where the polynomial $T_{n+3}^{(2)}(x)$ can be represented in the form

$$T_{n+3}^{(2)}(x) = \sum_{l=0}^{n+3} b_l^{(2)} x^l. \tag{2.74}$$

Using (2.59) and (2.25), we get that

$$b_0^{(2)} = 0, \qquad b_1^{(2)} = (-1)^n \mu_0^{0(n+1)} \left[\int_0^{\theta - h_1/2h} \Phi_n(t) \varphi_{n+1}^0(-th)\, dt \right.$$
$$\left. - \int_{\theta - h_1/2h}^{\theta + h_1/2h} \left| S_n\left(t - \frac{h_1}{2h}\right)\right|^{q-1} \varphi_{n+1}^0(-th)\, dt \right],$$

$$b_{n+2}^{(2)} = \mu_{n+1}^{0(n+1)} \left[\int_{\theta - h_1/2h}^{\theta + h_1/2h} \left| S_n\left(t + \frac{h_1}{2h}\right)\right|^{q-1} \varphi_{n+1}^0((1-t)h)\, dt \right.$$
$$\left. + \int_{\theta + h_1/2h}^{1} \Phi_n(t) \varphi_{n+1}^0((1-t)h)\, dt \right], \qquad b_{n+3}^{(2)} = 0. \tag{2.75}$$

By Lemma 8 and (2.47) with $r = n$,

$$\Phi_n(t+1) = \Phi_n(t), \tag{2.76}$$

and $\Phi_n(t)$ has exactly one root on each of the intervals

$$[\theta + h_1/2h - 1, \theta - h_1/2h] \quad \text{and} \quad [\theta + h_1/2h, \theta + 1 - h_1/2h]$$

and changes sign in both cases in passing through the root. Consequently, $\Phi_n(t)$ has exactly one root on $[0, \theta - h_1/2h] \cup [\theta + h_1/2h, 1)$ (denote it by η) and changes sign in passing through it.

We consider two cases.

A. Let $\eta \in [0, \theta - h_1/2h]$. Then

$$\Phi_n\left(\theta - \frac{h_1}{2h}\right) = -\left| S_n\left(\theta - \frac{h_1}{h}\right)\right|^{q-1}, \tag{2.77}$$

and, consequently, $\Phi_n(0) \geqslant 0$. On the other hand,

$$\Phi_n\left(\theta + \frac{h_1}{2h}\right) = \left|S_n\left(\theta + \frac{h_1}{h}\right)\right|^{q-1}, \tag{2.78}$$

and, consequently, on $(\theta + h_1/2h, 1)$

$$\Phi_n(t) > 0. \tag{2.79}$$

By this and (2.68), (2.75), and (2.65)

$$b_{n+2}^{(2)} > 0. \tag{2.80}$$

Let $x_j(\eta)$ $(j = 1,\dots,n)$ be the zeros of $R_n^0(x, \eta)$ for $\eta \neq 0$, and let x_j $(j = 1,\dots,n)$ be the zeros of $R_n^0(x, 0)$, arranged in decreasing order. Note that the properties of $R_n^0(x, t)$ give us that $x_1 = 0$. Using Lemma 5 and Remark 1, we have

$$\operatorname{sgn} R_n^0\big(x_j(\eta), t\big) = \begin{cases} (-1)^j, & 0 \leqslant t < \eta, \\ (-1)^{j+1}, & \eta < t < 1 \end{cases} \qquad (j = 1,\dots,n) \tag{2.81}$$

If $\eta \neq 0$, and

$$\operatorname{sgn} R_n^0\big(x_j, t\big) = (-1)^{j+1}, \qquad 0 < t < 1 \; (j = 1,\dots,n). \tag{2.82}$$

By (2.59), (2.79), (2.81), (2.82), and Lemma 8,

$$\operatorname{sgn} T_{n+3}^{(3)}\big(x_j(\eta)\big) = (-1)^{j+1}, \quad \operatorname{sgn} T_{n+3}^{(2)}\big(x_j\big) = (-1)^{j+1} \qquad (j = 1, 2, \dots, n) \tag{2.83}$$

Consequently, for any $\eta \in [0, \theta - h_1/2h]$ the polynomial $T_{n+3}^{(2)}(x)$ has at least $n - 1$ negative roots on $(x_n(\eta), x_1(\eta))$ if $\eta \neq 0$, or on (x_n, x_1) if $\eta = 0$. Moreover, it follows from (2.75) and (2.80) that $T_{n+3}^{(2)}(x)$ is a polynomial of degree $n + 2$ and $T_{n+3}^{(2)}(0) = 0$, while $T_{n+3}^{(2)}(1) = 0$ by Lemma 9. Further, it follows from (2.83) with $j = n$ and (2.80) that $T_{n+3}^{(2)}(x)$ has at least one more root on $(-\infty, x_n(\eta))$ if $\eta \neq 0$, or on $(-\infty, x_n)$ if $\eta = 0$. Consequently, for any $\eta \in [0, \theta - h_1/2h]$ the polynomial $T_{n+3}^{(2)}(x)$, which is of degree $n + 2$, has exactly n negatively simple roots, $T_{n+3}^{(2)}(0) = T_{n+3}^{(2)}(1) = 0$, and Lemma 10 is proved in case A with (2.73) taken into account.

B. Let $\eta \in [\, \theta + h_1/2h, 1)$. Then

$$\Phi_n\left(\theta - \frac{h_1}{2h}\right) = -\left|S_n\left(\theta - \frac{h_1}{h}\right)\right|^{q-1},$$

and, consequently, $\Phi_n(t) < 0$ on $(0, \theta - h_1/2h)$. By this and (2.75), (2.65), and (2.67),

$$b_1^{(2)} < 0. \tag{2.84}$$

Using (2.81), we get in a way similar to that in case A that

$$\operatorname{sgn} T_{n+3}^{(2)}\big(x_j(\eta)\big) = (-1)^j \qquad (j = 1,\dots,n). \tag{2.85}$$

It follows from (2.85) that $T_{n+3}^{(2)}(x)$ has at least $n - 1$ negative roots on $(x_n(\eta), x_1(\eta))$. Moreover, (2.75) and (2.80) give us that $T_{n+3}^{(2)}(x)$ is a polynomial of degree $n + 2$ and $T_{n+3}^{(2)}(0) = 0$, and Lemma 9 gives us that $T_{n+3}^{(2)}(1) = 0$. Further, from (2.85) with $j = 1$

and from (2.84) we get that $T^{(2)}_{n+3}(x)$ has at least one more root on $(x_1(\eta), 0)$. With all this and (2.73) taken into account, Lemma 10 is also proved in case B.

3. The proof of Lemma 10 in case 3 is completely analogous to the proofs carried out in cases 1 and 2, and thus we omit it.

Lemma 10 is proved.

To conclude this section we make a remark about the lemmas.

REMARK 2. For the operator $\mathscr{L}_n(D) = D^n$ all the lemmas in this section except for Lemma 8 were proved by Subbotin [4]–[7]. For the operator $\mathscr{L}_n(D) = \prod_{j=1}^n (D - \beta_j)$, where $\beta_j \in \mathbf{R}$, Lemmas 1–3 were proved by Micchelli [19] and Schoenberg [20].

§3. Extremal interpolation in the mean in the space $L_p(\mathbf{R})$ for $1 \leqslant p \leqslant \infty$

In this section we give some sufficient conditions under which problem 1 (stated in the Introduction) has a solution for an arbitrary linear differential operator $\mathscr{L}_n = \mathscr{L}_n(D)$ of the form (1.1) with constant real coefficients, and we compute the quantity $A(\mathscr{L}_n, p, h, h_1)$ when these conditions are satisfied.

Let us rewrite the operator $\mathscr{L}_n = \mathscr{L}_n(D)$ defined by (1.1) in the form

$$\mathscr{L}_n(D) = \prod_{i=1}^{l} \left(D^2 - 2\gamma_i D + \gamma_i^2 + \alpha_i^2\right)^{k_i} \prod_{j=1}^{m} \left(D - \beta_j\right)^{s_j}, \tag{3.1}$$

where $k_1, \dots, k_l$ and $s_1, \dots, s_m$ are the corresponding multiplicities, $k_1 + \cdots + k_l = k$, $s_1 + \cdots + s_m = n - 2k$, $\beta_i \neq \beta_j$ for $i \neq j$, and $\alpha_i \neq \alpha_j$ and $\gamma_i \neq \gamma_j$ simultaneously for $i \neq j$. Fix the following fundamental system $v_1, \dots, v_n$ of solutions of the homogeneous linear differential equation $\mathscr{L}_n(D)f = 0$:

$$\begin{gathered} v_1 = e^{\beta_1 x}, \dots, v_{s_1} = x^{s_1 - 1} e^{\beta_1 x}, \dots, v_{n-2k} = x^{s_m - 1} e^{\beta_m x}, \\ v_{n-2k+1} = e^{\gamma_1 x} \sin \alpha_1 x,\ v_{n-2k+2} = e^{\gamma_1 x} \cos \alpha_1 x, \dots, v_{n-2k+2k_1-1} = x^{k_1 - 1} e^{\gamma_1 x} \sin \alpha_1 x, \\ v_{n-2k-dk_1} = x^{k_1 - 1} e^{\gamma_1 x} \cos \alpha_1 x, \dots, v_n = x^{k_l - 1} e^{\gamma_l x} \cos \alpha_1 x. \end{gathered} \tag{3.2}$$

Let $1 \leqslant p \leqslant \infty$. We consider the nonhomogeneous differential equation

$$\mathscr{L}_n(D)f = u, \tag{3.3}$$

where $u \in L_p(\mathbf{R})$. It is not hard to verify that the general solution of this equation has the form

$$f(x) = \sum_{j=1}^{n} c_j v_j(x) + \int_0^x \varphi_n(x - t) u(t)\, dt, \tag{3.4}$$

where $c_1, \dots, c_n$ are arbitrary constants, and $\varphi_n(t)$ is determined by (2.3). We replace the function u in (3.4) by $\mathscr{L}_n(D)f$. Then, using (2.6) and (2.7) with $r = n$, we get for any function f with $\mathscr{L}_n(D)f \in L_p(\mathbf{R})$ that

$$\left(\Delta^h_{\mathscr{L}_n} f\right)(x) = \left(\Delta^h_{\mathscr{L}_n} \left\{\int_0^x \varphi_n(x - t) \mathscr{L}_n(D) f(t)\, dt\right\}\right)(x). \tag{3.5}$$

Suppose that $h > 0$, $h_1 \geqslant 0$ and $y = \{y_m\}_{-\infty}^{\infty}$ is an arbitrary sequence of real numbers, and let $f \in F_{h,h_1,p}(y)$. Using (1.4), (1.5), and (3.5) with $h_1 > 0$, we have

$$\Delta^h_{\mathscr{L}_n} y_m = \frac{1}{h_1}\int_{-h_1/2}^{h_1/2} \left(\Delta^h_{\mathscr{L}_n} f\right)(mh + x)\,dx = \frac{h}{h_1}\int_{-h_1/2h}^{h_1/2h} \left(\Delta^h_{\mathscr{L}_n} f\right)((m+x)h)\,dx$$
$$= \frac{h^2}{h_1}\int_{-h_1/2h}^{h_1/2h} dx \left\{ \sum_{l=0}^{n} (-1)^{n-l}\mu_l^{(n)} \int_0^{m+l+x} \varphi_n((m+l+x-t)h)\mathscr{L}_n(D)f(th)\,dt \right\}. \tag{3.6}$$

Using (2.7) with $r = n$, we transform the expression in the curly brackets for $-h_1/2h \leqslant x \leqslant h_1/2h$:

$$\sum_{l=0}^{n} (-1)^{n-1}\mu_l^{(n)} \int_0^{m+l+x} \varphi_n((m+l+x-t)h)\mathscr{L}_n(D)f(th)\,dt$$
$$= \sum_{l=0}^{n} (-1)^{n-l}\mu_l^{(n)} \int_{m-1}^{m+l+1} \varphi_n((m+l+x-t)h_+)\mathscr{L}_n(D)f(th)\,dt$$
$$= \sum_{l=0}^{n} (-1)^{n-1}\mu_l^{(n)} \int_0^1 \sum_{j=0}^{n+1} \varphi_n((l+x+1-j-t)h_+)\mathscr{L}_n(D)f((t+m-j-1)h)\,dt. \tag{3.7}$$

From (3.6) and (3.7) with $h_1 > 0$ we get

$$\Delta^h_{\mathscr{L}_n} y_m = \frac{h^2}{h_1}\int_0^1 \sum_{j=0}^{n+1} \mathscr{L}_n(D)f((m-1+j+t)h)\,dt$$
$$\times \int_{-h_1/2h}^{h_1/2h} \sum_{l=0}^{n} (-1)^{n-l}\mu_l^{(n)}\varphi_n((l+x+1-j-t)h_+)\,dx. \tag{3.8}$$

For $h_1 = 0$ we get similarly that

$$\Delta^h_{\mathscr{L}_n} y_m = h\int_0^1 \sum_{j=0}^{n} \mathscr{L}_n(D)f((m-1+j+t)h)$$
$$\times \sum_{l=0}^{n} (-1)^{n-l}\mu_l^{(n)}\varphi_n((l+1-j-t)h_+)\,dt. \tag{3.9}$$

Suppose that $0 < h < h_0 = \pi/\max_{1\leqslant s\leqslant k}\alpha_s$, $0 \leqslant h_1 \leqslant h$, $1 \leqslant q < \infty$, and let θ be the unique zero of $S_n(t)$ (see (2.48) and Lemma 7) on $[0, 1)$. Denote by $\mathfrak{M}(\mathscr{L}_n, \theta, S_n)$ the space of functions that are generalized $\mathscr{L}$-splines with uniform nodes $(l+\theta)h$ $(l = 0, \pm 1, \pm 2, \ldots)$ corresponding to the operator $\mathscr{L}_n = \mathscr{L}_n(D)$, the parameter θ, and the function $g(t) = |S_n(t/h)|^{q-1}$ (see Definition 1).

THEOREM 1. *Suppose that* $0 < h < h_0 = \pi/\max_{1\leqslant s\leqslant k}\alpha_s$, $0 \leqslant h_1 \leqslant h$, $1 < p \leqslant \infty$ *and* $1/p + 1/q = 1$. *For any sequence* $y = \{y_m\}_{-\infty}^{\infty} \in Y_{h,p}$ *there exists a unique function* $f \in F_{h,h_1,p}(y) \cap \mathfrak{M}(\mathscr{L}_n, \theta, S_n)$.

PROOF. Any function $f \in \mathfrak{M}(\mathscr{L}_n, \theta, S_n)$ can be regarded as a solution of (3.3), where u has the form

$$u(t) = Z_m\left|S_n\left(\frac{t}{h}\right)\right|^{q-1}, \qquad (m-1+\theta)h \leqslant t < (m+\theta)h \;(m = 0, \pm 1, \pm 2, \ldots) \tag{3.10}$$

and $Z = \{Z_m\}_{-\infty}^{\infty}$ is a sequence of real numbers. We show that for any sequence $Z = \{Z_m\}_{-\infty}^{\infty} \in l_p$ a function $u(t)$ of the form (3.10) belongs to $L_p(\mathbf{R})$. Indeed, using (2.47), we get that for $1 < p < \infty$

$$\left(\int_{-\infty}^{\infty} |u(t)|^p \, dt\right)^{1/p} = \left\{h \sum_{m=-\infty}^{\infty} |Z_m|^p \int_0^1 |S_n(t)|^{(q-1)p} \, dt\right\}^{1/p}$$

$$= h^{1/p}\left\{\int_0^1 |S_n(t)|^q \, dt\right\}^{1/p} \|Z\|_{l_p}, \tag{3.11}$$

while for $p = \infty$

$$\sup_t |u(t)| = \sup_m |Z_m| = \|Z\|_{l_\infty}, \tag{3.12}$$

and, consequently, for any $1 < p \leqslant \infty$ a function $u(t)$ of the form (3.10) is in $L_p(\mathbf{R})$. Since the general solution of (3.3) has the form (3.4), it follows that, considering (3.10), to prove Theorem 1 we must choose for any $y = \{y_m\}_{-\infty}^{\infty} \in Y_{h,p}$ coefficients c_j $(j = 1, \ldots, n)$ and a sequence $Z = \{Z_m\}_{-\infty}^{\infty} \in l_p$ of real numbers such that $f \in \mathfrak{M}(\mathscr{L}_n, \theta, S_n)$ is in the class $F_{h, h_1, p}(y)$, and show that the choice can be made in a unique way.

Replacing $\mathscr{L}_n(D)f$ in (3.8) and (3.9) by a function u of the form (3.10), we get the following difference relation for the sequence $Z = \{Z_m\}_{-\infty}^{\infty}$ when $h_1 > 0$:

$$\begin{aligned}
\Delta^h_{\mathscr{L}_n} y_m = {} & \frac{h^2}{h_1} \sum_{j=0}^{n+1} Z_{m-1+j} \int_0^\theta |S_n(t)|^{q-1} \, dt \\
& \times \int_{-h_1/2h}^{h_1/2h} \sum_{l=0}^{n} (-1)^{n-l} \mu_l^{(n)} \varphi_n((l + z + 1 - j - t)h_+) \, dz \\
& + \frac{h^2}{h_1} \sum_{j=0}^{n+1} Z_{m+j} \int_0^1 |S_n(t)|^{q-1} \, dt \\
& \times \int_{-h_1/2h}^{h_1/2h} \sum_{l=0}^{n} (-1)^{n-l} \mu_l^{(n)} \varphi_n((l + z + 1 - j - t)h_+) \, dz \\
= {} & \frac{h^2}{h_1} \sum_{j=0}^{n+2} Z_{m-1+j} b_j,
\end{aligned} \tag{3.13}$$

where

$$b_0 = \int_0^\theta |S_n(t)|^{q-1} \, dt \int_{-h_1/2h}^{h_1/2h} \sum_{l=0}^{n} (-1)^{n-l} \mu_l^{(n)} \varphi_n((l + z + 1 - t)h_+) \, dz,$$

$$\begin{aligned}
b_j = {} & \int_0^\theta |S_n(t)|^{q-1} \, dt \int_{-h_1/2h}^{h_1/2h} \sum_{l=0}^{n} (-1)^{n-l} \mu_l^{(n)} \varphi_n((l + z + 1 - j - t)h_+) \, dz \\
& + \int_\theta^1 |S_n(t)|^{q-1} \, dt \int_{-h_1/2h}^{h_1/2h} \sum_{l=0}^{n} (-1)^{n-l} \mu_l^{(n)} \varphi_n((l + z + 2 - j - t)h_+) \, dz \\
& \qquad (j = 1, \ldots, n + 1),
\end{aligned}$$

$$b_{n+2} = \int_\theta^1 |S_n(t)|^{q-1} \, dt \int_{-h_1/2h}^{h_1/2h} \sum_{l=0}^{n} (-1)^{n-l} \mu_l^{(n)} \varphi_n((l + z - n - t)h_+) \, dz. \tag{3.14}$$

Similarly, for $h_1 = 0$

$$\Delta^h_{\mathscr{L}_n} y_m = h \sum_{j=0}^{n} Z_{m+j} \tilde{b}_j, \tag{3.15}$$

where

$$\tilde{b}_0 = \int_0^\theta |S_n(t)|^{q-1} \sum_{l=0}^{n} (-1)^{n-l} \mu_l^{(n)} \varphi_n((l-t)h_+)\, dt,$$

$$\tilde{b}_j = \int_0^\theta |S_n(t)|^{q-1} \sum_{l=0}^{n} (-1)^{n-l} \mu_l^{(n)} \varphi_n((l-j-t)h_+)\, dt$$

$$+ \int_0^1 |S_n(t)|^{q-1} \sum_{l=0}^{n} (-1)^{n-l} \mu_l^{(n)} \varphi_n((l+1-n-t)h_+)\, dt.$$

$$\tilde{b}_n = \int_0^1 |S_n(t)|^{q-1} \sum_{l=0}^{n} (-1)^{n-l} \mu_l^{(n)} \varphi_n(l+1-n-t)h_+) dt. \tag{3.16}$$

We investigate the characteristic polynomials

$$U_{n+2}(x, \theta) = \sum_{j=0}^{n+2} b_j x^j, \qquad \tilde{U}_n(x, \theta) = \sum_{j=0}^{n} \tilde{b}_j x^j \tag{3.17}$$

of the difference equations (3.13) and (3.15). Let

$$Q_{n+1}(x, z-t) = \begin{cases} x^2 R_{n-1}(x, -z+t+1), & 0 \leqslant z-t < 1, \\ xR_{n-1}(x, -z+t), & -1 \leqslant z-t < 0, \\ R_{n-1}(x, -z+t-1), & -2 \leqslant z-t < -1, \end{cases} \tag{3.18}$$

where $R_{n-1}(x, t)$ is defined by (2.24). Using (2.25), we get the following representation for the polynomial $U_{n+2}(x, \theta)$:

$$\begin{aligned} U_{n+2}(x, \theta) &= \int_0^\theta |S_n(t)|^{q-1} dt \int_{-h_1/2h}^{h_1/2h} Q_{n+1}(x, z-t)\, dz \\ &+ \int_\theta^1 |S_n(t)|^{q-1} dt \int_{-h_1/2h}^{h_1/2h} xQ_{n+1}(x, z-t)\, dz \\ &= \int_0^\theta |S_n(t)|^{q-1} dt \int_{-h_1/2h-t}^{h_1/2h-t} Q_{n+1}(x, z)\, dz \\ &+ \int_\theta^1 |S_n(t)|^{q-1} dt \int_{-h/2h-t}^{h_1/2h-t} xQ_{n+1}(x, z)\, dz. \end{aligned} \tag{3.19}$$

Similarly, for $\tilde{U}_n(x, \theta)$ we get

$$\tilde{U}_n(x, \theta) = \int_0^\theta |S_n(t)|^{q-1} R_{n-1}(x, t)\, dt + \int_0^1 x|S_n(t)|^{q-1} R_{n-1}(x, t)\, dt. \tag{3.20}$$

Let us now prove the equality

$$U_{n+2}(x, \theta) = T_{n+2}(x), \tag{3.21}$$

where $T_{n+2}(x)$ is the polynomial defined by (2.57). To do this we need to investigate $U_{n+2}(x, \theta)$ separately in each of three cases: 1) $0 \leqslant \theta < h_1/2h$, 2) $h_1/2h \leqslant \theta \leqslant 1 - h_1/2h$, 3) $1 - h_1/2h < \theta < 1$.

We have

1) $$U_{n+2}(x,\theta)=\int_0^\theta |S_n(t)|^{q-1}\,dt\left[\int_{-h_1/2h-t}^{0} xR_{n-1}(x,-z)\,dz + \int_0^{h_1/2h-t} x^2R_{n-1}(x,-z+1)\,dz\right]$$
$$+\int_\theta^{h_1/2h} |S_n(t)|^{q-1}\,dt\left[\int_{-h_1/2h-t}^{0} x^2R_{n-1}(x,-z)\,dz + \int_0^{h_1/2h-t} x^3R_{n-1}(x,-z+1)\,dz\right]$$
$$+\int_{h_1/2h}^{1-h_1/2h} |S_n(t)|^{q-1}\,dt\int_{-h_1/2h-t}^{h_1/2h-1} x^2R_{n-1}(x,-z)\,dz$$
$$+\int_{1-h_1/2h}^{1} |S_n(t)|^{q-1}\,dt\left[\int_{-h_1/2h-t}^{-1} xR_{n-1}(x,-z-1)\,dz + \int_{-1}^{h_1/2h-t} x^2R_{n-1}(x,-z)\,dz\right],$$

2) $$U_{n+2}(x,\theta)=\int_0^{h_1/2h} |S_n(t)|^{q-1}\,dt\left[\int_{-h_1/2h-t}^{0} xR_{n-1}(x,-z)\,dz + \int_0^{h_1/2h-t} x^2R_{n-1}(x,-z+1)\,dz\right]$$
$$+\int_{h_1/2h}^{\theta} |S_n(t)|^{q-1}\,dt\int_{-h_1/2h-t}^{h_1/2h-t} xR_{n-1}(x,-z)\,dz$$
$$+\int_\theta^{1-h_1/2h} |S_n(t)|^{q-1}\,dt\int_{-h_1/2h-t}^{h_1/2h-t} x^2R_{n-1}(x,-z)\,dz$$
$$+\int_{1-h_1/2h}^{1} |S_n(t)|^{q-1}\,dt\left[\int_{-h_1/2h-t}^{-1} xR_{n-1}(x,-z-1)\,dz + \int_{-1}^{h_1/2h-t} x^2R_{n-1}(x,-z)\,dz\right],$$

3) $$U_{n+2}(x,\theta)=\int_0^{h_1/2h} |S_n(t)|^{q-1}\,dt\left[\int_{-h_1/2h-t}^{0} xR_{n-1}(x,-z)\,dz + \int_0^{h_1/2h-t} x^2R_{n-1}(x,-z+1)\,dz\right]$$
$$+\int_{h_1/2h}^{1-h_1/2h} |S_n(t)|^{q-1}\,dt\int_{-h_1/2h-t}^{h_12h-t} xR_{n-1}(x,-z)\,dz$$
$$+\int_{1-h_1/2h}^{\theta} |S_n(t)|^{q-1}\,dt\left[\int_{-h_1/2h-t}^{-1} R_{n-1}(x,-z-1)\,dz + \int_{-1}^{h_1/2h-t} xR_{n-1}(x,-z)\,dz\right]$$
$$+\int_\theta^{1} |S_n(t)|^{q-1}\,dt\left[\int_{-h_1/2h-t}^{-1} xR_{n-1}(x,-z-1)\,dz + \int_{-1}^{h_1/2h-t} x^2R_{n-1}(x,-z)\,dz\right].$$

The transformation of $U_{n+2}(x,\theta)$ is continued by using (2.17), (2.20) with $r=n$, and (2.24). We have

1)
$$U_{n+2}(x,\theta)=\frac{1}{h(1-x)}$$
$$\times\left\{\int_0^{\theta}|S_n(t)|^{q-1}\left[-xR_n^0\left(x,t+\frac{h_1}{2h}\right)+x^2R_n^0\left(x,1+t-\frac{h_1}{2h}\right)\right]dt\right.$$
$$+\int_{\theta}^{h_1/2h}|S_n(t)|^{q-1}\left[-x^2R_n^0\left(x,t+\frac{h_1}{2h}\right)+x^3R_n^0\left(x,1+t-\frac{h_1}{2h}\right)\right]dt$$
$$+\int_{h_1/2h}^{1-h_1/2h}|S_n(t)|^{q-1}\left[x^2R_n^0\left(x,t-\frac{h_1}{2h}\right)-x^2R_n^0\left(x,t+\frac{h_1}{2h}\right)\right]dt$$
$$\left.+\int_{1-h_1/2h}^{1}|S_n(t)|^{q-1}\left[-xR_n^0\left(x,t+\frac{h_1}{2h}-1\right)+x^2R_n^0\left(x,t-\frac{h_1}{2h}\right)\right]dt\right\},$$

2)
$$U_{n+2}(x,\theta)=\frac{1}{h(1-x)}$$
$$\times\left\{\int_0^{h_1/2h}|S_n(t)|^{q-1}\left[-xR_n^0\left(x,t+\frac{h_1}{2h}\right)+x^2R_n^0\left(x,1+t-\frac{h_1}{2h}\right)\right]dt\right.$$
$$+\int_{h_1/2h}^{\theta}|S_n(t)|^{q-1}\left[xR_n^0\left(x,t-\frac{h_1}{2h}\right)-xR_n^0(x,t+h_1/2h)\right]dt$$
$$+\int_{\theta}^{1-h_1/2h}|S_n(t)|^{q-1}\left[x^2R_n^0\left(x,t-\frac{h_1}{2h}\right)-x^2R_n^0\left(x,t+\frac{h_1}{2h}\right)\right]dt$$
$$\left.+\int_{1-h_1/2h}^{1}|S_n(t)|^{q-1}\left[-xR_n^0\left(x,t+\frac{h_1}{2h}-1\right)+x^2R_n^0\left(x,t-\frac{h_1}{2h}\right)\right]dt\right\},$$

3)
$$U_{n+2}(x,\theta)=\frac{1}{h(1-x)}$$
$$\times\left\{\int_0^{h_1/2h}|S_n(t)|^{q-1}\left[-xR_n^0\left(x,t+\frac{h_1}{2h}\right)+x^2R_n^0\left(x,1+t-\frac{h_1}{2h}\right)\right]dt\right.$$
$$+\int_{1-h_1/2h}^{\theta}|S_n(t)|^{q-1}\left[-R_n^0\left(x,t+\frac{h_1}{2h}-1\right)+xR_n^0\left(x,t-\frac{h_1}{2h}\right)\right]dt$$
$$+\int_{h_1/2h}^{1-h_1/2h}|S_n(t)|^{q-1}\left[xR_n^0\left(x,t-\frac{h_1}{2h}\right)-xR_n^0\left(x,t+\frac{h_1}{2h}\right)\right]dt$$
$$\left.+\int_{\theta}^{1}|S_n(t)|^{q-1}\left[-xR_n^0\left(x,t+\frac{h_1}{2h}-1\right)+x^2R_n^0\left(x,t-\frac{h_1}{2h}\right)\right]dt\right\}.$$

Making a change of variables under the integral signs and considering (2.47) with

$r = n$, we get

$$1)\ U_{n+2}(x,\theta) = \frac{1}{h(1-x)} \times \left\{ \int_0^{\theta+h_1/2h} \left[x^2 \left| S_n\left(t+\frac{h_1}{2h}\right)\right|^{q-1} - x\left| S_n\left(t-\frac{h_1}{2h}\right)\right|^{q-1} \right] R_n^0(x,t)\,dt \right.$$
$$+ \int_{\theta+h_1/2h}^{1+\theta-h_1/2h} \left[\left| S_n\left(t+\frac{h_1}{2h}\right)\right|^{q-1} - \left| S_n\left(t-\frac{h_1}{2h}\right)\right|^{q-1} \right] x^2 R_n^0(x,t)\,dt$$
$$\left. + \int_{1+\theta-h_1/2h}^{1} \left[x^3 \left| S_n\left(t+\frac{h_1}{2h}\right)\right|^{q-1} - x^2\left| S_n\left(t-\frac{h_1}{2h}\right)\right|^{q-1} \right] R_n^0(x,t)\,dt \right\}, \tag{3.22}$$

$$2)\ U_{n+2}(x,\theta) = \frac{1}{h(1-x)} \times \left\{ \int_0^{\theta-h_1/2h} \left[\left| S_n\left(t+\frac{h_1}{2h}\right)\right|^{q-1} - \left| S_n\left(t-\frac{h_1}{2h}\right)\right|^{q-1} \right] xR_n^0(x,t)\,dt \right.$$
$$+ \int_{\theta-h_1/2h}^{\theta+h_1/2h} \left[\left| S_n\left(t+\frac{h_1}{2h}\right)\right|^{q-1} x^2 - \left| S_n\left(t-\frac{h_1}{2h}\right)\right|^{q-1} x \right] R_n^0(x,t)\,dt$$
$$\left. + \int_{\theta+h_1/2h}^{1} \left[\left| S_n\left(t+\frac{h_1}{2h}\right)\right|^{q-1} - \left| S_n\left(t-\frac{h_1}{2h}\right)\right|^{q-1} \right] x^2R_n^0(x,t)\,dt \right\}, \tag{3.23}$$

$$3)\ U_{n+2}(x,\theta) = \frac{1}{h(1-x)} \times \left\{ \int_0^{\theta+h_1/2h-1} \left[\left| S_n\left(t+\frac{h_1}{2h}\right)\right|^{q-1} x - \left| S_n\left(t-\frac{h_1}{2h}\right)\right|^{q-1} \right] R_n^0(x,t)\,dt \right.$$
$$+ \int_{\theta+h_1/2h-1}^{\theta-h_1/2h} \left[\left| S_n\left(t+\frac{h_1}{2h}\right)\right|^{q-1} - \left| S_n\left(t-\frac{h_1}{2h}\right)\right|^{q-1} \right] xR_n^0(x,t)\,dt$$
$$\left. + \int_{\theta-h_1/2h}^{1} \left[\left| S_n\left(t+\frac{h_1}{2h}\right)\right|^{q-1} x^2 - \left| S_n\left(t-\frac{h_1}{2h}\right)\right|^{q-1} x \right] R_n^0(x,t)\,dt \right\}. \tag{3.24}$$

Comparison of (3.22)–(3.24) with (2.57)–(2.60) leads to the validity of (3.21).

Using (2.24), (2.25), (2.40), (2.41), and Lemma 7, we compute $|U_{n+2}(-1,\theta)|$ and $|\tilde{U}_n(-1,\theta)|$. Substitution of $x = -1$ in (3.17) gives us

$$|U_{n+2}(-1,\theta)| = \left| \int_0^\theta |S_n(t)|^{q-1}\,dt \times \int_{-h_1/2h}^{h_1/2h} \sum_{j=0}^{n+1} (-1)^j \sum_{l=0}^{n} (-1)^{n-l} \mu_l^{(n)} \varphi_n((l+z+2-j-t)h_+)\,dz \right|$$
$$= \frac{h_1}{h^2} \int_0^1 |S_n(t)|^q\,dt, \tag{3.25}$$

$$|\tilde{U}_n(-1,\theta)| = \frac{1}{h}\int_0^1 |S_n(t)|^q\,dt. \tag{3.26}$$

It follows from Lemma 7 and (3.25) and (3.26) that

$$U_{n+2}(-1,\theta) \neq 0, \qquad \tilde{U}_n(-1,\theta) \neq 0. \tag{3.27}$$

By (3.20), (3.21), (3.27), and Lemmas 6, 7, and 10, all the conditions in Theorem A (stated in the Introduction) are satisfied for the polynomials $U_{n+2}(x,\theta)/x$ with $0 \leqslant \theta \leqslant 1 - h_1/2h$, $U_{n+2}(x,\theta)$ with $1 - h_1/2h < \theta < 1$, $\tilde{U}_n(x,\theta)$ with $0 < \theta < 1$, and $\tilde{U}_n(x,0)/x$ with $\theta = 0$, and, consequently, the difference equations (3.13) and (3.15) have a unique solution $Z = \{Z_m\}_{-\infty}^{\infty} \in l_p$ expressible by the formula

$$Z_m = \sum_{s=-\infty}^{\infty} d_{-s-m} \Delta^h_{\mathscr{L}_n} y_s, \tag{3.28}$$

where

$$\sum_{s=-\infty}^{\infty} d_s x^s = \begin{cases} h_1\left(h^2 U_{n+2}(x,\Theta)\right)^{-1}, & h_1 > 0, \\ \left(h\tilde{U}_n(x,\Theta)\right)^{-1}, & h_1 = 0, \end{cases}$$

and the solution satisfies the estimate

$$\|Z\|_{l_p} \leqslant \begin{cases} h_1 \|\Delta^h_{\mathscr{L}_n} y\|_{l_p} \left(h^2 |U_{n+2}(-1,\Theta)|\right)^{-1}, & h_1 > 0, \\ \|\Delta^h_{\mathscr{L}_n} y\|_{l_p} \left(h|\tilde{U}_n(-1,\Theta)|\right)^{-1}, & h_1 = 0. \end{cases} \tag{3.29}$$

Setting $Z = \{Z_m\}_{-\infty}^{\infty}$ in (3.10), we get that for any $f \in F_{h,h_1,p}(y) \cap \mathfrak{M}(\mathscr{L}_n, \Theta, S_n)$ of the form

$$f(x) = \sum_{j=1}^{n} c_j v_j(x) + \int_0^x \varphi_n(x-t) u(t)\, dt \tag{3.30}$$

a function u of the form (3.10) is uniquely determined. To complete the proof of the theorem it remains for us to show that the coefficients c_j $(j = 1, \ldots, n)$ are also uniquely determined. We constructed a function f of the form (3.30) by setting

$$\Delta^h_{\mathscr{L}_n} y_m = \begin{cases} \dfrac{1}{h_1} \Delta^h_{\mathscr{L}_n} \displaystyle\int_{-h_1/2}^{h_1/2} f(mh+t)\, dt, & h_1 > 0, \\ \Delta^h_{\mathscr{L}_n} f(mh), & h_1 = 0 \end{cases}$$

$$(m = 0, \pm 1, \pm 2, \ldots). \tag{3.31}$$

Our problem is to find coefficients c_j $(j = 1, \ldots, n)$ such that f satisfies not (3.31) but the interpolation conditions

$$y_m = \begin{cases} \dfrac{1}{h_1} \displaystyle\int_{-h_1/2}^{h_1/2} f(mh+t)\, dt, & h_1 > 0, \\ f(mh), & h_1 = 0 \end{cases}$$

$$(m = 0, \pm 1, \pm 2, \ldots). \tag{3.32}$$

Substituting (3.30) in (3.32) and assuming the interpolation conditions (3.32) to be satisfied for $m = 0, 1, \ldots, n-1$, we get a system of n linear equations for determining the coefficients c_j $(j = 1, 2, \ldots, n)$. It is known [12] that for $h_1 = 0$ and $0 < h < h_0$ the principal determinant W of this system is nonzero. It is not hard to verify that for $0 < h_1 \leqslant h$ and $0 < h < h_0$ the principal determinant of the system differs from W only by a nonzero factor. Consequently, the coefficients c_j ($j = 1,\dots,n$) are uniquely determined by the interpolation conditions (3.32), which are valid for $m = 0, 1,\dots,n-1$. It follows from (3.32) with $m = 0,1,\dots,n-1$ and from (3.31) that the conditions (3.32) hold for any integer m. Thus, a function f in $F_{h,h_1,p}(y) \cap \mathfrak{M}(\mathscr{L}_n, \theta, S_n)$ and of the form (3.30) exists and is uniquely determined.

THEOREM 2. *Suppose that* $0 < h < h_0 = \pi/\max_{1\leqslant s\leqslant k}\alpha_s$, $0 \leqslant h_1 \leqslant h$, $1 \leqslant p \leqslant \infty$ *and* $1/p + 1/q = 1$. *Then*

$$A(\mathscr{L}_n, p, h, h_1) = \left(\|S_n\|_{L_q[0,1]}\right)^{-1}.$$

PROOF. To prove Theorem 2 we estimate $A(\mathscr{L}_n, p, h, h_1)$ from above and below and then show that these estimates coincide. Consider first the case $1 < p \leqslant \infty$.

It follows from Theorem 1 with $1 < p \leqslant \infty$, $0 < h < h_0$ and $0 \leqslant h_1 \leqslant h$ that for any sequence $y = \{y_m\}_{-\infty}^{\infty} \in Y_{h,p}$ there exists an $f \in F_{h,h_1,p}(y)$ of the form

$$f(x) = \sum_{j=1}^{n} c_j v_j(x) + \int_0^x \varphi_n(x-t)u(t)\,dt,$$

where $c_1,\dots,c_n$ are constants, and

$$Z_m = \sum_{s=-\infty}^{\infty} d_{-s-m}\Delta^h_{\mathscr{L}_n} y_s,$$

$$\sum_{s=-\infty}^{\infty} d_s x^s = \begin{cases} h_1\left(h^2 U_{n+2}(x,\Theta)\right)^{-1}, & h_1 > 0,\\ \left(h\tilde{U}_n(x,\Theta)\right)^{-1}, & h_1 = 0.\end{cases}$$

Using (3.29), we get for $\mathscr{L}_n(D)f$ that

$$\|\mathscr{L}_n(D)f\|_{L_p(R)} = \left\{\sum_{m=-\infty}^{\infty}\int_0^1 |S_n(t)|^{(q-1)p}\,dt|Z_m|^p\right\}^{1/p}$$

$$= \left\{\int_0^1 |S_n(t)|^q\,dt\right\}^{1/p}\|Z\|_{l_p} \leqslant \left(\int_0^1 |S_n(t)|^q\,dt\right)^{1/p-1} = \left(\|S_n\|_{L_q[0,1]}\right)^{-1}. \tag{3.33}$$

For $1 < p \leqslant \infty$, $0 < h < h_0$ and $0 \leqslant h_1 \leqslant h$ this gives us the following upper estimate for $A(\mathscr{L}_n, p, h, h_1)$:

$$A(\mathscr{L}_n, p, h, h_1) \leqslant \left(\|S_n\|_{L_q[0,1]}\right)^{-1}. \tag{3.34}$$

We now get a lower estimate for $A(\mathscr{L}_n, p, h, h_1)$ when $1 < p \leqslant \infty$, $0 < h < h_0$ and $0 \leqslant h_1 \leqslant h$. Let $N > n+1$. Consider an arbitrary sequence $y^* = \{y^*_m\}_{-\infty}^{\infty}$ satisfying the condition

$$\Delta^h_{\mathscr{L}_n} y^*_m = \begin{cases} (-1)^m (2N+1)^{-1/p}, & |m| \leqslant N,\\ 0, & |m| > N.\end{cases} \tag{3.35}$$

There are infinitely many such sequences. Indeed, for arbitrary values of $y^*_0, y^*_1,\dots,y^*_{n-1}$ we can compute $y^*_n, y^*_{n-1},\dots$ and $y^*_{-1}, y^*_{-2},\dots$ successively from (3.35). It is not hard to verify that $y^* \in Y_{h,p}$. We show that for any $f \in F_{h,h_1,p}(y^*)$ with $0 < h < h_0$, $0 \leqslant h_1 \leqslant h$ and $1 < p \leqslant \infty$

$$\|\mathscr{L}_n(D)f\|_{L_p(\mathbf{R})} \geqslant \left(\|S_n\|_{L_q[0,1]}\right)^{-1}. \tag{3.36}$$

For $0 \leqslant t \leqslant 1$ let

$$B_1((-N-1+s+t)h) = (-1)^{s+N} \sum_{j=0}^{s} (-1)^j a_{j,n}(t,h,h_1),$$

$$B_2((N+s-1+t)h) = (-1)^{s+N} \sum_{j=s}^{n+1} (-1)^j a_{j,n}(t,h,h_1)$$

$$(s = 0,1,\ldots,n+1), \tag{3.37}$$

where $a_{j,n}(t,h,h_1)$ is defined by (2.40) with $r = n$. Using (3.8), (3.9), (3.35), (3.37), and (2.41), we get

$$\begin{aligned}
(2N+1)^{1/q} &= \sum_{m=-N}^{N} (-1)^m \Delta^h_{\mathscr{L}_n} y^*_m \\
&= \sum_{m=-N}^{N} (-1)^m \sum_{j=0}^{n+1} \int_0^1 \mathscr{L}_n(D) f((m-1+j+t)h) a_{j,n}(t,h,h_1)\, dt \\
&= \sum_{s=-N}^{-N+n+1} \int_0^1 (-1)^s \mathscr{L}_n(D) f((s-1+t)h) \sum_{j=0}^{s+N} (-1)^j a_{j,n}(t,h,h_1)\, dt \\
&\quad + \sum_{s=-N+n+20}^{N-1} \int_0^1 (-1)^s \mathscr{L}_n(D) f((s-1+t)h) \sum_{j=0}^{n+1} (-1)^j a_{j,n}(t,h,h_1)\, dt \\
&\quad + \sum_{s=N}^{N+n+1} \int_0^1 (-1)^s \mathscr{L}_n(D) f((s-1+t)h) \sum_{j=s-N}^{n+1} (-1)^j a_{j,n}(t,h,h_1)\, dt \\
&= \int_{-N-1}^{-N+n+1} \mathscr{L}_n(D) f(th) B_1(th)\, dt - \int_{-N+n+1}^{N-1} \mathscr{L}_n(D) f(th) S_n(t)\, dt \\
&\quad + \int_{N-1}^{N-n+1} \mathscr{L}_n(D) f(th) B_2(th)\, dt.
\end{aligned} \tag{3.38}$$

Since the functions $B_1(th)$ and $B_2(th)$ are bounded and $\mathscr{L}_n(D)f \in L_p(\mathbf{R})$, the first and third integrals in (3.38) tend to zero as $N \to \infty$. Using this and Hölder's inequality, we get that for $N \to \infty$

$$\begin{aligned}
1 &\leqslant (2N+1)^{-1/q} \left\{ \left[\int_{-N+n+1}^{N-1} |\mathscr{L}_n(D) f(th)|^p\, dt \right]^{1/p} \left[\int_{-N+n+1}^{N-1} |S_n(t)|^q\, dt \right]^{1/q} + O(1) \right\} \\
&\leqslant (2N+1)^{-1/q} \|\mathscr{L}_n(D) f\|_{L_p(\mathbf{R})} \left\{ (2N-n-2)^{1/q} \|S_n\|_{L_q[0,1]} + O(1) \right\}.
\end{aligned}$$

Inequality (3.36) follows from this for $N \to \infty$ and $1 < p \leqslant \infty$. A lower estimate $A(\mathscr{L}_n, p, h, h_1)$ when $1 < p \leqslant \infty$, $0 < h < h_0$ and $0 \leqslant h_1 \leqslant h$ is obtained from (3.36), and it coincides with the upper estimate (3.34). Thus, the proof of Theorem 2 is complete for $1 < p \leqslant \infty$.

Consider the case $p = 1$. We show that for any $\delta \to 0$ and any sequence $y = \{y_m\}_{-\infty}^{\infty} \in Y_{h,1}$ there exists for $0 < h < h_0$ and $0 \leqslant h_1 \leqslant h$ a function $f_\delta \in F_{h,h_1,1}(y)$ such that

$$\|\mathscr{L}_n(D) f_\delta\|_{L_1(\mathbf{R})} \overset{\delta \to 0}{\to} \frac{1}{\max_{0 \leqslant t \leqslant 1} |S_n(t)|}. \tag{3.39}$$

In Lemma 8 we already observed that $S_n(t)$ has only one extremum point on $(\theta, \theta+1)$, and, moreover, $S_n(\theta) = S_n(\theta+1) = 0$. Therefore, by (2.47) with $r = n$, $|S_n(t)|$ has only one maximum point on $[0,1)$; denote it by ε. The function $f_\delta \in F_{h,h_1,1}(y)$ will be constructed so that

$$\mathscr{L}_n(D)f_\delta(t) = \begin{cases} Z_m/\delta_1, & (m-1+\varepsilon)h \leqslant t \leqslant (m-1+\varepsilon+\delta_1)h, \\ 0, & (m-1+\varepsilon+\delta_1)h < t < (m+\varepsilon)h \end{cases}$$

$$(m = 0, \pm 1, \pm 2, \dots), \quad (3.40)$$

where $\delta_1 = \delta_1(\delta) \to^{\delta\to 0} 0$ is not hard to choose so that the characteristic polynomials $U_{n+2,\delta_1}(x,\varepsilon)$ and $\tilde{U}_{n,\delta_1}(x,\varepsilon)$ of the difference equations for determining Z_m obtained by substituting (3.40) into (3.8) and (3.9) satisfy all the conditions of Lemma 6. Moreover,

$$\left|U_{n+2,\delta_1}(-1,\varepsilon)\right| = \frac{h_1}{h^2\delta_1}\int_\varepsilon^{\varepsilon+\delta_1} |S_n(t)|dt \overset{\delta_1\to 0}{\to} \frac{h_1}{h^2}|S_n(\varepsilon)|,$$

$$\left|\tilde{U}_{n,\delta_1}(-1,\varepsilon)\right| = \frac{1}{h\delta_1}\int_\varepsilon^{\varepsilon+\delta_1} |S_n(t)\,dt| \overset{\delta_1\to 0}{\to} \overset{\delta_1\to 0}{\to} \frac{1}{h}|S_n(\varepsilon)|.$$

If we now apply Theorem A to the difference equations in Z_m, we get (3.39). We next construct the function f_δ as in Theorem 1 on the basis of (3.4) and conditions (3.31) and (3.32). From (3.39) we get the following upper estimate for $A(\mathscr{L}_n, 1, h, h_1)$ when $0 < h < h_0$ and $0 \leqslant h_1 \leqslant h$:

$$A(\mathscr{L}_n, 1, h, h_1) \leqslant 1/\max_{0\leqslant t\leqslant 1} |S_n(t)|. \quad (3.41)$$

We now obtain an upper estimate for $A(\mathscr{L}_n, 1, h, h_1)$. Consider an arbitrary sequence $y^* = \{y_m^*)_{-\infty}^{\infty}$ satisfying the condition

$$\Delta^h_{\mathscr{L}_n} y^*_m = \begin{cases} 1, & m = 0, \\ 0, & m \neq 0. \end{cases} \quad (3.42)$$

Clearly, $y^* \in Y_{h,1}$. For any $f \in F_{h,h_1,1}(y^*)$ the use of (3.8), (3.9), and (3.42) gives us that for $N > n+1$

$$\begin{aligned} 1 &= \sum_{m=-N}^{N} (-1)^m \Delta^h_{\mathscr{L}_n} y^*_m \\ &= \sum_{m=-N}^{n+1} (-1)^m \sum_{j=0}^{n+1} \int_0^1 \mathscr{L}_n(D) f((m-1+j+t)h) a_{j,n}(t,h,h_1)\,dt \\ &= \int_{-N-1}^{-N+n+1} \mathscr{L}_n(D)f(th)B_1(th)\,dt - \int_{-N+n+1}^{N-1} \mathscr{L}_n(D)f(th)S_n(t)\,dt \\ &\quad + \int_{N-1}^{N-n+1} \mathscr{L}_n(D)f(th)B_2(th)\,dt. \end{aligned} \quad (3.43)$$

Since $\mathscr{L}_n(D)f \in L_1(\mathbf{R})$ and since $B_1(th)$ and $B_1(th)$ are bounded, the first and third integrals in (3.43) tend to 0 as $N \to \infty$, and, consequently, for $N \to \infty$ we get from (3.43) that

$$\|\mathscr{L}_n(D)f\|_{L_1(\mathbf{R})} \geqslant 1/\max_{0\leqslant t\leqslant 1} |S_n(t)|. \quad (3.44)$$

A lower estimate coinciding with the upper estimate (3.41) is obtained for $A(\mathscr{L}_n, 1, h, h_1)$ from (3.44).

The proof of Theorem 2 is complete.

In Theorem 2 the quantity $A(\mathscr{L}_n, p, h, h_1)$ was computed for $1 \leqslant p \leqslant \infty$, $0 < h < h_0$ and $0 \leqslant h_1 \leqslant h$. Moreover, it follows from the properties of $S_n(t)$ that under the indicated conditions $A(\mathscr{L}_n, p, h, h_1) < \infty$. We investigate the case $h = h_0$.

THEOREM 3. *Suppose that* $1 \leqslant p \leqslant \infty$, $h = h_0 = \pi/\max_{1 \leqslant s \leqslant k} \alpha_s$, $0 \leqslant h_1 \leqslant h$, *and* $\alpha_s = \max_{1 \leqslant s \leqslant k} \alpha_s$ $(1 \leqslant s^* \leqslant k)$. *If* $\gamma_{s^*} = 0$, *then*

$$A(\mathscr{L}_n, p, h, h_1) = \infty.$$

PROOF. To prove Theorem 3 it suffices to establish that if h_0, $\gamma_{s^*} = 0$ and $0 \leqslant t \leqslant 1$, then

$$S_n(t) \equiv 0. \tag{3.45}$$

The author proved in [12] that under these conditions

$$R_n^0(-1, t) \equiv 0. \tag{3.46}$$

The identity

$$R_{n-1}(-1, t) = 0 \tag{3.47}$$

is established similarly under the indicated conditions. From (3.46), (3.47), (2.24), (2.38), and (2.48) with $r = n$ we get (3.45).

Theorem 3 is proved.

§4. Computation of $B(\mathscr{L}_n, \omega, h)$

In this section we obtain a solution of problem 1 in another special case and compute the quantity $B(L_n, \omega, h)$ when $0 < h < h_0 = \pi/\max_{1 \leqslant s \leqslant k} \alpha_s$ and $\omega = \omega(\delta)$ is a convex modulus of continuity. Let us first formulate some auxiliary results.

DEFINITION 3. A function $\varphi(x) \in C[a, b]$ is said to be a *simple function on* $[a, b]$ if $\varphi(a) = \varphi(b) = 0$, $|\varphi(x)| > 0$ for any $x \in (a, b)$, and the equation $|\varphi(x)| = d$ has exactly two roots on $[a, b]$ for each $d (0 < d < \max_{a \leqslant x \leqslant b} |\varphi(x)|)$.

Let $\omega = \omega(\delta)$ $(\delta \geqslant 0)$ be a given modulus of continuity, and $H^\omega[a, b]$ the space of functions $f \in C[a, b]$ such that $\omega(\delta, f) \leqslant \omega(\delta)$ for any $\delta > 0$. We present the statement of a lemma of Korniĭchuk and Stechkin that is needed in what follows.

LEMMA 11 ([21], §3.1). *Let* $\psi(t)$ *be an integrable function on* $[a, b]$ *such that*:

1) *the integral* $\Psi(x) = \int_a^x \psi(t)\, dt$ *is a simple function on* $[a, b]$, *and*

2) *on* (a, b) *the function* $\psi(t)$ *vanishes at most on a set of measure zero.*

Then the following assertions are true:

1) $$M_\omega = \sup_{f \in H^\omega[a, b]} \left| \int_a^b \psi(t) f(t)\, dt \right| \leqslant \int_a^c |\psi(t)| \omega(\rho(t) - t)\, dt$$
$$= \int_c^b |\psi(t)| \omega(t - \rho^{-1}(t))\, dt, \tag{4.1}$$

where the function $\rho(t)$ *is defined for* $a \leqslant t \leqslant c$ *by*

$$\Psi(t) = \Psi(\rho(t)) \qquad (a \leqslant t \leqslant c, c \leqslant \rho(t) \leqslant b),$$

and $c = \rho(c)$ *and* $\rho^{-1}(t)$ *is the function inverse to* $\rho(t)$.

2) *If* $\omega = \omega(t)$ *is a convex modulus of continuity, then equality is attained in* (4.1), *and the supremum is realized by the functions of the form* $K \pm f_0(t) \in H^\omega[a, b]$, *where* K *is an arbitrary constant, and*

$$f_0(t) = \begin{cases} -\int_t^c \omega'(\rho(x) - x)\, dx, & a \leqslant t \leqslant c, \\ \int_c^t \omega'(x - \rho^{-1}(x))\, dx, & c \leqslant t \leqslant b. \end{cases}$$

Suppose that $0 < h < h_0 = \pi/\max_{1 \leqslant s \leqslant k} \alpha_s$. Consider the functions $P_n(t)$ and $P^0_{n+1}(t)$ defined by (2.38) with $r = n$. It follows from Lemma 1, Corollary 1, and (2.38) that $P_n(t)$ and $P^0_{n+1}(t)$ have unique zeros α and β on $[0, 1)$ when $0 < h < h_0$, and they are simple and distinct. From (2.20) with $x = -1$ and $r = n$, and from (2.17) and (2.38) we get that for any $t \in \mathbf{R}$

$$\begin{gathered} P^{0'}_{n+1}(t) = 2hP_n(t), \\ P^0_{n+1}(t+1) = -P^0_{n+1}(t), \qquad P_n(t+1) = -P_n(t). \end{gathered} \tag{4.2}$$

This and Corollary 1 give us that for $0 < h < h_0$ the function $P^0_{n+1}(t)$ is simple (in the sense of Definition 3) on each interval $[\beta + m, \beta + m + 1]$ $(m = 0, \pm 1, \pm 2, \ldots)$. Let

$$\tilde{\alpha} = \begin{cases} \alpha, & \beta < \alpha, \\ \alpha + 1, & \beta > \alpha. \end{cases} \tag{4.3}$$

For $\beta h \leqslant t \leqslant \tilde{\alpha} h$ we define $\rho(t)$ with the help of

$$P^0_{n+1}(t/h) = P^0_{n+1}(\rho(t)/h) \qquad (\tilde{\alpha} h \leqslant \rho(t) \leqslant (\beta + 1)h). \tag{4.4}$$

For $\beta h \leqslant t \leqslant (\beta + 1)/h$ consider the function

$$G(t) = K + f_0(t), \tag{4.5}$$

where

$$f_0(t) = \left\{ \begin{array}{ll} -\int_t^{\tilde{\alpha} h} \omega'(\rho(x) - x)\, dx, & \beta h \leqslant t \leqslant \tilde{\alpha} h, \\ \int_{\tilde{\alpha} h}^t \omega'(x - \rho^{-1}(x))\, dx, & \tilde{\alpha} h \leqslant t \leqslant (\beta + 1)h, \end{array} \right\} \tag{4.6}$$

$\omega = \omega(\delta)$ is a given convex modulus of continuity, and the constant K is determined by

$$G((\beta + 1)h) = -G(\beta h). \tag{4.7}$$

It follows from Lemma 11 that the function $G \in H^\omega[\beta h, (\beta + 1)h]$ has exactly one root on $[\beta h, (\beta + 1)h)$, and this root is simple. Denote it by γh. The equality (4.7) enables us to extend G to the whole number line by means of the equality

$$G((t + 1)h) = -G(th). \tag{4.8}$$

From this we get that for any $m = 0, \pm 1, \pm 2, \ldots$

$$G((\gamma + m)h) = 0. \tag{4.9}$$

Denote by $\mathfrak{N}(\mathscr{L}_n, \gamma, G)$ the space of functions that are generalized $\mathscr{L}$-splines with uniform nodes $(l + \gamma)h$ $(l = 0, \pm 1, \pm 2, \ldots)$ corresponding to the operator $\mathscr{L}_n = \mathscr{L}_n(D)$, the parameter γ, and the function $g(t) = |G(t)|$ (see Definition 1).

THEOREM 4. *Suppose that* $0 < h < h_0 = \pi/\max_{1 \leqslant s \leqslant k} \alpha_s$, *and* $\omega = \omega(\delta)$ $(\delta \geqslant 0)$ *is a given convex modulus of continuity. For any sequence* $y = \{y_m\}_{-\infty}^{\infty} \in Y_{h,\infty}$ *there exists a unique function* $f \in F_{h,0,\infty}(y) \cap \mathfrak{N}(\mathscr{L}_n, \gamma, G)$, *and it satisfies the inequality*

$$\omega(\delta, \mathscr{L}_n(D)f) \leqslant \left(h \int_\beta^{\tilde{\alpha}} |P_n(t)| \omega(\rho(th) - th)\, dt \right)^{-1} \omega(\delta) \tag{4.10}$$

for any $0 \leqslant \delta \leqslant h$.

PROOF. Equality (3.9) gives us that for any $f \in F_{h,0,\infty}(y)$

$$\Delta^h_{\mathscr{L}_n} y_m = h \int_{j=0}^n \mathscr{L}_n(D) f((m-1+j+t)h) \times \sum_{l=0}^n (-1)^{n-l} \mu_l^{(n)} \varphi_n((l+1-j-t)h_+)\, dt. \tag{4.11}$$

Any function $f \in N(\mathscr{L}_n, \gamma, G)$ can be regarded as a solution of the equation

$$\mathscr{L}_n(D) f = u, \tag{4.12}$$

where the function u is of the form

$$u(t) = Z_m |G(t)|, \qquad (m-1+\gamma)h \leqslant t < (m+\gamma)h \qquad (m = 0, \pm 1, \pm 2, \ldots) \tag{4.13}$$

and $Z = \{Z_m\}_{-\infty}^{\infty}$ is a sequence of real numbers. Since the general solution of (4.12) has the form

$$f(x) = \sum_{j=1}^n c_j v_j(x) + \int_0^x \varphi_n(x-t) u(t)\, dt, \tag{4.14}$$

where the functions v_j are defined by (3.2), it follows from (4.13) that to prove Theorem 4 we must choose for any $y = \{y_m\}_{-\infty}^{\infty} \in Y_{h,\infty}$ coefficients c_j $(j = 1, \ldots, n)$ and a sequence $Z = \{Z_m\}_{-\infty}^{\infty} \in l_\infty$ of real numbers such that the function $f \in N(\mathscr{L}_n, \gamma, G)$ is in the class $F_{h,0,\infty}(y)$, and show that this choice can be made uniquely.

Replacing $\mathscr{L}_n(D) f$ in (4.11) by a function u of the form (4.13), we get the following difference relation for the sequence $Z = \{Z_m\}_{-\infty}^{\infty}$:

$$\Delta^h_{\mathscr{L}_n} y_m = h \sum_{j=0}^n b_j Z_{m+j} \qquad (m = 0, \pm 1, \pm 2, \ldots), \tag{4.15}$$

where

$$b = 0 = \int_0^{\tilde{\gamma}} |G(th)| \sum_{l=0}^n (-1)^{n-l} \mu_l^{(n)} \varphi_n((l-t)h_+)\, dt,$$

$$b_j = \int_0^{\tilde{\gamma}} |G(th)| \sum_{l=0}^n (-1)^{n-1} \mu_l^{(n)} \varphi_n((l-j-t)h_+)\, dt + \int_{\tilde{\gamma}}^1 |G(th)| \sum_{l=0}^n (-1)^{n-1} \mu_l^{(n)} \varphi_n((l+1-j-t)h_+)\, dt \qquad (j = 1, \ldots, n-1),$$

$$b_n = \int_{\tilde{\gamma}}^1 |G(th)| \sum_{l=0}^n (-1)^{n-l} \mu_l^{(n)} \varphi_n((l+1-n-t)h_+)\, dt \tag{4.16}$$

and

$$\tilde{\gamma} = \begin{cases} \gamma, & 0 \leqslant \gamma < 1, \\ \gamma - 1, & 1 \leqslant \gamma < 2. \end{cases} \tag{4.17}$$

By (2.25), the characteristic polynomial $U_n(x, \tilde{\gamma})$ for the difference equation (4.15) has the form

$$U_n(x,\tilde{\gamma}) = \int_0^{\tilde{\gamma}} |G(th)| \sum_{j=0}^{n-1} x^j \sum_{l=j+1}^{n} (-1)^{n-l} \mu_l^{(n)} \varphi_n((l-j-t)h)\, dt$$

$$+ \int_{\tilde{\gamma}}^{1} |G(th)| x \sum_{j=0}^{n-1} x^j \sum_{l=j+1}^{n} (-1)^{n-l} \mu_l^{(n)} \varphi_n((l-j-t)h)\, dt$$

$$= \int_0^{\tilde{\gamma}} |G(th)| R_{n-1}(x,t)\, dt + \int_{\tilde{\gamma}}^{1} |G(th)| x R_{n-1}(x,t)\, dt. \tag{4.18}$$

We compute the value $|U_n(x, \tilde{\gamma})|$ for $x = -1$. We get from (4.2) that

$$\int_{\beta}^{\beta+1} P_n(t)\, dt = \frac{1}{2h}\left[P_{n+1}^0(\beta+1) - P_{n+1}^0(\beta)\right] = 0. \tag{4.19}$$

By (4.18), (4.19), (4.2), (4.5), (2.24), and (2.38),

$$|U_n(-1,\tilde{\gamma})| = \int_0^{\tilde{\gamma}} |G(th)| R_{n-1}(-1,t)\, dt - \int_{\tilde{\gamma}}^{1} |G(th)| R_{n-1}(-1,t)\, dt\, |$$

$$= \left| \int_0^{\tilde{\gamma}} G(th) P_n(t)\, dt + \int_{\tilde{\gamma}}^{1} G(th) P_n(t)\, dt \right|$$

$$= \left| \int_{\beta}^{\beta+1} G(th) P_n(t)\, dt \right| = \left| \int_{\beta}^{\beta+1} f_0(th) P_n(t)\, dt \right|. \tag{4.20}$$

In view of (4.6) and the fact that $\tilde{\alpha}$ is the simple unique zero of $P_n(t)$ on $[\beta, \beta+1]$, (4.20) gives us that

$$U_n(-1,\tilde{\gamma}) \neq 0. \tag{4.21}$$

We consider separately the case $\tilde{\gamma} = 0$. From (4.16) with $\tilde{\gamma} = 0$ it follows that $b_0 = 0$. Therefore, the difference equation for Z_m has the form

$$\frac{1}{h} \Delta_{\mathscr{L}_n}^h y_m = \sum_{j=1}^{n} b_j Z_{m+j} = \sum_{j'=0}^{n-1} b_{j'+1} Z_{m+j'+1} = \sum_{j=0}^{n-1} \tilde{b}_j \tilde{Z}_{m+j}, \tag{4.22}$$

where $\tilde{b}_j = b_{j+1}$ and $\tilde{Z}_m = Z_{m+1}$. The characteristic polynomial of the difference equation (4.22) has the form

$$\tilde{U}_{n-1}(x,0) = U_n(x,0)/x. \tag{4.23}$$

Note that the characteristic polynomial $U_n(x, \tilde{\gamma})$ of the difference equation (4.15) satisfies all the conditions of Lemma 6, and, consequently, all the zeros of $U_n(x, \tilde{\gamma})$ for $0 < \tilde{\gamma} < 1$ and those of $\tilde{U}_{n-1}(x, 0)$ are negative and simple, and, moreover, we have from (4.21) that $U_n(-1, \tilde{\gamma}) \neq 0$. By this and Theorem A, the difference equations (4.15) and (4.21) have unique bounded solutions

$$Z_m = \frac{1}{h} \sum_{s=-\infty}^{\infty} d_{-s-m} \Delta_{\mathscr{L}_n}^h y_s \qquad (m = 0, \pm 1, \pm 2, \ldots), \tag{4.24}$$

where

$$\sum_{s=-\infty}^{\infty} d_s x^3 = \begin{cases} (U_n(x,\tilde{\gamma}))^{-1}, & 0 < \tilde{\gamma} < 1, \\ (\tilde{U}_{n-1}(x,0))^{-1}, & \tilde{\gamma} = 0, \end{cases}$$

and they satisfy

$$\sup_m |Z_m| \leqslant \left(h|U_n(-1,\tilde{\gamma})|\right)^{-1} \sup_m \left|\Delta^h_{\mathscr{L}_n} y_m\right|. \tag{4.25}$$

Substituting (4.24) in (4.13), we get that for any $f \in N(\mathscr{L}_n, \gamma, G)$ of the form (4.14) a function u of the form (4.13) is uniquely determined. To prove Theorem 4 we need to show that the coefficients c_j $(j = 1,\dots,n)$ in (4.14) are also uniquely determined. We construct a function $f \in \mathfrak{N}(\mathscr{L}_n, \gamma, G)$ by setting

$$\Delta^h_{\mathscr{L}_n} y_m = \Delta^h_{\mathscr{L}_n} f(mh) \qquad (m = 0, \pm 1, \pm 2, \dots). \tag{4.26}$$

Our problem is to find coefficients c_j $(j = 1,\dots,n)$ such that a function $f \in \mathfrak{N}(\mathscr{L}_n, \gamma, G)$ of the form (4.14) satisfies not (4.26) but the interpolation conditions

$$y_m = f(mh) \qquad (m = 0, \pm 1, \pm 2, \dots) \tag{4.27}$$

for it to belong to the class $F_{h,0,\infty}(y)$. Substituting (4.14) into (4.27) and assuming the interpolation conditions (4.27) to be satisfied for $m = 0,\dots,n-1$, we obtain a system of n linear equations for determining the coefficients c_j $(j = 1,\dots,n)$. The principal determinant W of this system coincides witht the determinant in Theorem 1 for $h_1 = 0$ and, consequently, is nonzero for $0 < h < h_0$ by [12]. Therefore, the coefficients c_j $(j = 1,\dots,n)$ are uniquely determined by the interpolation conditions (4.27), satisfied for $m = 0, 1,\dots,n-1$. But the conditions (4.27) hold for any integer m if they hold for $m = 0, 1,\dots,n-1$ and (4.26) holds. Thus, a function $f \in F_{h,0,\infty}(y) \cap \mathfrak{N}(\mathscr{L}_n, \gamma, G)$ of the form (4.14) is uniquely determined. To prove Theorem 4 it remains to prove that f satisfies (4.10).

It follows from the construction of $G(t)$ and from Lemma 11 and (4.13) and (4.25) that for $0 \leqslant \delta \leqslant h$

$$\omega(\delta, \mathscr{L}_n(D)f) \leqslant \left(h|U_n(-1,\tilde{\gamma})|\right)^{-1} \omega(\delta, G) \leqslant \left(h|U_n(-1,\tilde{\gamma})|\right)^{-1} \omega(\delta), \tag{4.28}$$

while for $\delta > h$

$$\omega(\delta, \mathscr{L}_n(D)f) = \omega(h, \mathscr{L}_n(D)f). \tag{4.29}$$

Using (4.20) and Lemma 11, we get (4.10) from (4.28).

The proof of Theorem 4 is complete.

THEOREM 5. *Suppose that* $0 < h < h_0 = \pi/\max_{1\leqslant s\leqslant k} \alpha_s$ *and* $\omega = \omega(\delta)$ *is a given convex modulus of continuity. Then*

$$B(\mathscr{L}_n, \omega, h) = \left(h \int_\beta^{\tilde{\alpha}} |P_n(t)| \omega(\rho(th) - th)\, dt\right)^{-1}.$$

PROOF. It follows from Theorem 4 that for $0 < h < h_0$ and any sequence $y = \{y_m\}_{-\infty}^{\infty} \in Y_{h,\infty}$ there exists a function $f \in F_{h,0,\infty}(y) \cap \mathfrak{N}(\mathscr{L}_n, \gamma, G)$ such that for $0 \leqslant \delta \leqslant h$

$$\omega(\delta, \mathscr{L}_n(D)f) \leqslant \left(h \int_\beta^{\tilde{\alpha}} |P_n(t)| \omega(\rho(th) - th)\, dt\right)^{-1} \omega(\delta), \tag{4.30}$$

while for $\delta > h$

$$\omega(\delta, \mathscr{L}_n(D)f) = \omega(h, \mathscr{L}_n(D)f). \tag{4.31}$$

From (4.30) and (4.31) with $0 < h < h_0$ we get for $B(\mathscr{L}_n, \omega, h)$ the upper estimate

$$B(\mathscr{L}_n, \omega, h) \leqslant \left(h \int_\beta^{\tilde{\alpha}} |P_n(t)| \omega(\rho(th) - th)\, dt \right)^{-1}. \tag{4.32}$$

Let us now obtain a lower estimate for $B(\mathscr{L}_n, \omega, h)$. Choose a sequence $y^* = \{y_m^*\}_{-\infty}^{\infty}$ as follows:

$$y_m^* = (-1)^{m+n} \left(\sum_{s=0}^{n} \mu_s^{(n)} \right)^{-1}. \tag{4.33}$$

It follows from (1.2) and (1.3) that $\sum_{s=0}^{n} \mu_s^{(n)} \neq 0$ for $0 < h < h_0$. Substituting y_m^* in (1.3), we get that

$$\Delta_{\mathscr{L}_n}^h y_m^* = (-1)^m, \tag{4.34}$$

and, consequently, $y^* \in Y_{h,\infty}$. By (4.32) and the definition of $B(\mathscr{L}_n, \omega, h)$,

$$\left(h \int_\beta^{\tilde{\alpha}} |P_n(t)| \omega(\rho(th) - th)\, dt \right)^{-1}$$
$$\geqslant B(\mathscr{L}_n, \omega, h) \geqslant \inf_{f \in F_{h,0,\infty}(y^*)} \sup_{\delta > 0} \frac{\omega(\delta, \mathscr{L}_n(D) f)}{\omega(\delta)}. \tag{4.35}$$

We show that the infimum in the last inequality can be taken over a narrower set, namely, over the functions $f \in F_{h,0,\infty}(y^*)$ such that

$$\sup_x |\mathscr{L}_n(D) f(x)| \leqslant \left(h \int_\beta^{\tilde{\alpha}} |P_n(t)| \omega(\rho(th) - th)\, dt \right)^{-1} \omega[(n+2)h]. \tag{4.36}$$

Assume the opposite. Suppose that at some point x_0

$$|\mathscr{L}_n(D) f(x_0)| > \left(h \int_\beta^{\tilde{\alpha}} |P_n(t)| \omega(\rho(th) - th)\, dt \right)^{-1} \omega[(n+2)h]. \tag{4.37}$$

We consider the sequence of functions $\{f_s(x)\}_{s=0}^{n-k}$: $f_0(x) = f(x)$,

$$f_1(x) = (D - \beta_1) f_0(x), \ldots, f_{n-2k}(x) = (D - \beta_{n-2k}) f_{n-2k}(x),$$
$$f_{n-2k+1}(x) = \left(D^2 - 2\gamma_1 D + \gamma_1^2 + \alpha_1^2\right) f_{n-2k}(x), \ldots, f_{n-k}(x)$$
$$= \left(D^2 - 2\gamma_k D + \gamma_k^2 + \alpha_k^2\right) f_{n-k-1}(x).$$

Since $f \in F_{h,0,\infty}(y^*)$, f vanishes at least once between the points mh and $(m+1)h$ $(m = 0, \pm 1, \pm 2 \ldots)$. From this, using Rolle's theorem, (2.21), (2.22), and the functions $\{f_s(x)\}_{s=0}^{n-k}$, we get that on each closed interval of length $(n+2)h$ the function $\mathscr{L}_n(D) f$ vanishes at least once. Consequently, there exists a point x_1 such that $\mathscr{L}_n(D) f(x_1) = 0$ and $|x_1 - x_0| \leqslant (n+2)h$. Hence,

$$\omega[(n+2)h, \mathscr{L}_n(D) f] \geqslant |\mathscr{L}_n(D) f(x_0)|$$
$$> \left(h \int_\beta^{\tilde{\alpha}} |P_n(t)| \omega(\rho(th) - th)\, dt \right)^{-1} \omega[(n+2)h],$$

which contradicts (4.30) and (4.31). Therefore, the infimum in (4.35) can be taken over the functions $f \in F_{h,0,\infty}(y^*)$ satisfying (4.36). We show that for the set of such functions

$$\sup_{\delta > 0} \frac{\omega(\delta, \mathscr{L}_n(D) f)}{\omega(\delta)} \geqslant \left(h \int_\beta^{\tilde{\alpha}} |P_n(t)| \omega(\rho(th) - th)\, dt \right)^{-1}. \tag{4.38}$$

Let $N > n + 1$. For any $f \in F_{h,0,\infty}(y^*)$ satisfying (4.36) we get by (4.11), (4.34), and (2.40) with $r = n$ and $h_1 = 0$ that

$$\begin{aligned} N + 1 &= \sum_{m=0}^{N} (-1)^m \Delta^h_{\mathscr{L}_n} y^*_m \\ &= \sum_{m=0}^{N} (-1)^m \int_0^1 \sum_{j=0}^{n} \mathscr{L}_n(D) f((m-1+j+t)h) a_{j,n}(t,h,0)\, dt \\ &= \sum_{j=0}^{n} \int_0^1 (-1)^j a_{j,n}(t,h,0) \sum_{p=j}^{j+N} (-1)^p \mathscr{L}_n(D) f((p-1+t)h)\, dt \\ &= \sum_{p=0}^{n-1} \int_0^1 (-1)^p \mathscr{L}_n(D) f((p-1+t)h) \sum_{j=0}^{p} (-1)^j a_{j,n}(t,h,0)\, dt \\ &\quad + \sum_{p=n}^{N} \int_0^1 (-1)^p \mathscr{L}_n(D) f((p-1+t)h) \sum_{j=0}^{n} (-1)^j a_{j,n}(t,h,0)\, dt \\ &\quad + \sum_{p=N+1}^{N+n} \int_0^1 (-1)^p \mathscr{L}_n(D) f((p-1+t)h) \sum_{j=p-N}^{n} (-1)^j a_{j,n}(t,h,0)\, dt. \end{aligned} \tag{4.39}$$

By (4.39), (4.36), and (2.39)–(2.42),

$$\frac{N+1}{h} \leqslant C_1 + \int_0^1 \sum_{p=n}^{N} (-1)^p \mathscr{L}_n(D) f((p-1+t)h) P_n(t)\, dt, \tag{4.40}$$

where C_1 is a constant independent of N. By (4.40) and (2.44) with $r = n$,

$$\begin{aligned} \frac{N+1}{h} &\leqslant C_1 + C_2 + \int_1^{\beta+1} \sum_{p=n}^{N-1} (-1)^{p+1} \mathscr{L}_n(D) f((p+t)h) P_n(t)\, dt \\ &\quad - \int_\beta^1 \sum_{p=n}^{N-1} (-1)^{p+1} \mathscr{L}_n(D) f((p+t)h) P_n(t)\, dt = C_1 + C_2 \\ &\quad + \int_\beta^{\beta+1} \sum_{p=n}^{N-1} (-1)^{p+1} \mathscr{L}_n(D) f((p+t)h) P_n(t)\, dt = C_1 + C_2 \\ &\quad + \sum_{p=n}^{N-1} (-1)^{p+1} \int_{\beta+p}^{\beta+1+p} \mathscr{L}_n(D) f(th) P_n(t)\, dt, \end{aligned} \tag{4.41}$$

where C_2 is a constant independent of N.

From (4.41) and Lemma 11,

$$\begin{aligned} \frac{N+1}{h} &\leqslant C_1 + C_2 \\ &\quad + \frac{N-1-n}{h} \left(\int_{\beta h}^{\tilde{\alpha} h} \left| P_n\left(\frac{t}{h}\right) \right| \omega(\rho(t) - t)\, dt \right) \sup_{\delta > 0} \frac{\omega(\delta, \mathscr{L}_n(D) f)}{\omega(\delta)}. \end{aligned} \tag{4.42}$$

This gives us (4.38) for $N \to \infty$. A lower estimate for $B(\mathscr{L}_n, \omega, h)$ coinciding with the upper estimate (4.32) follows from (4.38) and (4.35).

This proves Theorem 5.

THEOREM 6. *Let* $\omega = \omega(\delta)$ *be a given convex modulus of continuity, and let*

$$h = h_0 = \pi / \max_{1 \leqslant s \leqslant k} \alpha_s \quad \textit{and} \quad \sigma_{s^*} = \max_{1 \leqslant s \leqslant k} \alpha_s \qquad (1 \leqslant s^* \leqslant k).$$

If $\gamma_{s^*} = 0$, *then*

$$B(\mathscr{L}_n, \omega, h) = \infty.$$

The proof is analogous to that of Theorem 3.

§5. A lower estimate for $d_{2m-1}(W^{\mathscr{L}_n}H^\omega, C)$

In this section we establish a lower estimate for the width $d_{2m-1}(W^{\mathscr{L}_n}H^\omega, C)$ in the space $C = C_{2\pi}$ of continuous 2π-periodic functions when $m \geqslant 2\max_{1 \leqslant s \leqslant k} \alpha_s$ and $\omega = \omega(\delta)$ is given convex modulus of continuity. The following theorem of Tikhomirov [21] will be needed.

THEOREM B. *In any Banach space* X *the set* $F = S \cap R_{s+1}$ *satisfies* $d_s(F, X) = 1$, *where* S *is the closed unit ball and* R_{s+1} *is an arbitrary* $(s+1)$*-dimensional subspace of* X.

As before, $N(\mathscr{L}_n, \gamma, G)$ denotes the space of generalized $\mathscr{L}$-splines with uniform nodes $(l+\gamma)h$ $(0 < h < h_0 = \pi/\max_{1 \leqslant s < k} \alpha_s;\ l = 0, \pm 1, \pm 2, \dots)$ corresponding to the operator $\mathscr{L}_n = \mathscr{L}_n(D)$, the parameter γ, and the function $|G(t)|$ defined in §4. In the course of this section $h = \pi/m$, $m \geqslant 2\max_{1 \leqslant s \leqslant k} \alpha_s$. Let $\mathfrak{N}_m(\mathscr{L}_n, \gamma, G) = \mathfrak{N}(\mathscr{L}_n, \gamma, G) \cap C_{2\pi}$. We compute the dimension of the space $N_m(\mathscr{L}_m, \gamma, G)$.

Suppose that the homogeneous linear differential equation $\mathscr{L}_n(D)f = 0$ has exactly μ $(0 \leqslant \mu \leqslant 2k+1)$ linearly independent 2π-periodic solutions: $v_1, \dots, v_\mu$. It is not hard to verify that

$$\mu + 1 \leqslant 4 \max_{1 \leqslant s \leqslant k} \alpha_s. \tag{5.1}$$

The next section assertion is a corollary to Lemma 1 in [22].

LEMMA 12. *A linear differential equation* $\mathscr{L}_n(D)f = u$ *with constant real coefficients, where the function* u *is real-valued,* 2π*-periodic, and integrable over a period, has a* 2π*-periodic solution if and only if*

$$\int_0^{2\pi} v_l(t)u(t)\,dt = 0 \qquad (l = 1, 2, \dots, \mu). \tag{5.2}$$

By using Lemma 12 it is not hard to establish that for $m < (\mu+1)/2$ the space $N_m(\mathscr{L}_n, \gamma, G)$ contains only the subspace of 2π-periodic solutions of differentials equation $\mathscr{L}_n(D)f = 0$, and, consequently, for $m < (\mu+1)/2$

$$\dim \mathfrak{N}_m(\mathscr{L}_n, \gamma, G) = \mu. \tag{5.3}$$

On the other hand, as a consequence of results of Schumaker (Theorem 2.1 in [23]) we get for $N_m(\mathscr{L}_n, \gamma, G)$ that for $m_*(\mu+1)/2$

$$\dim \mathfrak{N}_m(\mathscr{L}_n, \gamma, G) = 2m. \tag{5.4}$$

THEOREM 7. *Suppose that $m \geqslant 2\max_{1\leqslant s\leqslant k}\alpha_s$ and $\omega = \omega(\delta)$ $(0 \leqslant \delta \leqslant \pi)$ is a given convex modulus of continuity. Then for any $f \in N_m(\mathscr{L}_n, \gamma, G)$*

$$\omega(\delta, \mathscr{L}_n(D)f) \leqslant \left(\frac{\pi}{m}\int_\beta^{\tilde\alpha}|P_n(t)|\omega\left(\rho\left(\frac{t\pi}{m}\right)-\frac{t\pi}{m}\right)dt\right)^{-1} \times \sum_{s=0}^{n}\mu_s\left(\mathscr{L}_n, \frac{\pi}{m}\right)\|f\|_{C_{2\pi}}\omega(\delta). \tag{5.5}$$

PROOF. From Theorem 4 with $h = \pi/m \leqslant h_0/2 = \pi/2\max_{1\leqslant s\leqslant k}\alpha_s$ we have that for any $f \in \mathfrak{N}_m(\mathscr{L}_n, \gamma, G)$

$$\omega(\delta, \mathscr{L}_n(D)f) \leqslant \left(\frac{\pi}{m}\int_\beta^{\tilde\alpha}|P_n(t)|\omega\left(\rho\left(\frac{t\pi}{m}\right)-\frac{t\pi}{m}\right)dt\right)^{-1}\sup_s\left|\Delta_{\mathscr{L}_n}^{\pi/m}f\left(\frac{s\pi}{m}\right)\right|\omega(\delta). \tag{5.6}$$

By (1.2) and (1.3) with $m \geqslant 2\max_{1\leqslant s\leqslant k}\alpha_s$,

$$\mu_s^{(n)} = \mu_s(\mathscr{L}_n, \pi/m) \geqslant 0 \qquad (s = 0, 1, \ldots, n), \tag{5.7}$$

and

$$\mu_n^{(n)} = \mu_n(\mathscr{L}_n, \pi/m) = 1. \tag{5.8}$$

From this and (1.3) we get

$$\sup_s\left|\Delta_{\mathscr{L}_n}^{\pi/m}f\left(\frac{s\pi}{m}\right)\right| \leqslant \|f\|_{C_{2\pi}}\sum_{s=0}^{n}\mu_s\left(\mathscr{L}_n, \frac{\pi}{m}\right). \tag{5.9}$$

Inequality (5.5) follows from (5.6) and (5.9). Theorem 7 is proved.

THEOREM 8. *Suppose that $m \geqslant 2\max_{1\leqslant s\leqslant k}\alpha_s$ and $\omega = \omega(\delta)$ $(0 \leqslant \delta \leqslant \pi)$ is a convex modulus of continuity. Then*

$$d_{2m-1}(W^{\mathscr{L}_n}H^\omega, C) \geqslant \left(\frac{\pi}{m}\int_\beta^{\tilde\alpha}|P_n(t)|\omega\left(\rho\left(\frac{t\pi}{m}\right)-\frac{t\pi}{m}\right)dt\right)\left(\sum_{s=0}^{n}\mu_s\left(\mathscr{L}_n, \frac{\pi}{m}\right)\right)^{-1}. \tag{5.10}$$

PROOF. For any $f \in \mathfrak{N}_m(\mathscr{L}_n, \gamma, G)$ such that

$$\omega(\delta, \mathscr{L}_n(D)f) \leqslant \omega(\delta). \tag{5.12}$$

By (5.4), (5.11), and (5.12), the $2m$-dimensional ball of radius r belongs to the class $W^{\mathscr{L}_n}H^\omega$. From this, Theorem B gives us (5.10).
Theorem 8 is proved.

COROLLARY 4. *Under the conditions of Theorem 8*

$$d_{2m-1}(W^{\mathscr{L}_n}H^\omega, C) \geqslant \int_\beta^{\tilde\alpha}\left|\sum_{s=-\infty}^{\infty}\frac{e^{i(2s+1)\pi(1-t)}}{P_n((2s+1)mi)}\right|\omega\left(\rho\left(\frac{t\pi}{m}\right)-\frac{t\pi}{m}\right)dt, \tag{5.13}$$

where $p_n(\lambda)$ is the characteristic polynomial for the operator $\mathscr{L}_n = \mathscr{L}_n(D)$.

PROOF. Let us expand $P_n(t)$ in a Fourier series. Computing the coefficients of the Fourier series with the help of (2.9), (2.10), (2.18), (2.19), and (2.38), we get that

$$P_n(t) = \frac{(-1)^n}{h}\left(\sum_{s=0}^{n}\mu_s(\mathscr{L}_n, h)\right)\sum_{s=-\infty}^{\infty}\frac{e^{i(2s+1)\pi(1-t)}}{p_n((2s+1)\pi i/h)}. \tag{5.14}$$

From (5.10) and (5.14) with $h = \pi/m$ we have (5.13).

REMARK 3. For the operator $\mathscr{L}_n = D^n$ the lower estimate (5.13) for the width was obtained by Lorentz [15] and Subbotin [6] and, as shown by Korniĭchuk [16], [17], is sharp.

BIBLIOGRAPHY

[1] V. S. Ryaben'kiĭ, *On stability of finite-difference schemes and on application of the finite-difference method to the solution of Cauchy's problem for a system of partial differential equations*, Candidate's Dissertation, Moscow State Univ., Moscow, 1952. (Russian)

[2] V. S. Ryaben'kiĭ and A. F. Filippov, *On stability of difference equations*, GITTL, Moscow, 1956; German transl., VEB Deutscher Verlag Wiss., Berlin, 1960.

[3] S. L. Sobolev, *Lectures on the theory of cubature formulas*, Part 2, Novosibirsk. Gos. Univ., Novosibirsk, 1965. (Russian)

[4] Yu. N. Subbotin, *On the relations between finite differences and the corresponding derivatives*, Trudy Mat. Inst. Steklov. **78** (1965), 24–42; English transl. in Proc. Steklov Inst. Math. **78** (1965).

[5] ______, *Functional interpolation in the mean with smallest nth derivative*, Trudy Mat. Inst. Stekov. **88** (1967), 30–60; English transl. in Proc. Steklov Inst. Math. **88** (1967).

[6] ______, *Approximation by spline functions and estimates of widths*, Trudy Mat. Inst. Steklov. **109** (1971), 35–60; English transl. in Proc. Steklov Inst. Math. **109** (1971).

[7] ______, *Extremal problems of functional interpolation and splines for interpolation in the mean*, Trudy Mat. Inst. Steklov. **138** (1975), 118–173; English transl. in Proc. Steklov Inst. Math. **138** (1975).

[8] ______, *Extremal and approximation properties of splines*, Theory of Approximation of Functions (Proc. Internat. Conf., Kaluga, 1975; S. B. Stechkin and S. A. Telyakovskiĭ, editors), "Nauka", Moscow, 1977, pp. 341–345.

[9] I. J. Schoenberg, *Cardinal interpolation and spline functions*, J. Approximation Theory **2** (1969), 167–206.

[10] A. Sharma and I. Tsimbalario [J. Tzimbalario], *Some linear differential operators and generalized differences*, Mat. Zametki **21** (1977), 161–172; English transl. in Math. Notes **21** (1977).

[11] V. T. Shevaldin, *Extremal interpolation with smallest value of the norm of a linear differential operator*, Mat. Zametki **27** (1980), 721–740; English transl. in Math. Notes **27** (1980).

[12] ______, *On a problem of extremal interpolation*, Mat. Zametki **29** (1981), 603–622; English transl. in Math. Notes **29** (1981).

[13] A. Kolmogoroff [A. N. Kolmogorov], *Über die beste Annäherung von Funktioneneiner gegebenen Funktionenklasse*, Ann. of Math. (2) **37** (1936), 107–110.

[14] V. M. Tikhomirov, *Some questions in approximation theory*, Izdat. Moskov. Gos. Univ., Moscow, 1976. (Russian)

[15] G. G. Lorentz, *Approximation of functions*, Holt, Rinehart and Winston, New York, 1966.

[16] N. P. Korneĭchuk, *Exact values of best approximations and widths of certain classes of functions*, Dokl. Akad. Nauk SSSR **150** (1963), 1218–1220; English transl. in Soviet Math. Dokl. **4** (1963).

[17] ______, *Inequalities for differentiable periodic functions and best approximation of one class of functions by another*, Izv. Akad. Nauk SSSR Ser. Mat. **36** (1972), 423–434; English transl. in Math. USSR Izv. **6** (1972).

[18] V. I. Ruban, *Even widths of the classes $W^{(r)}H_\omega$ in the space $C_{2\pi}$*, Mat. Zametki **15** (1974), 387–392; English transl. in Math. Notes **15** (1974).

[19] Charles A. Micchelli, *Cardinal $\mathscr{L}$-splines*, Studies in Spline Functions and Approximation Theory (S. Karlin et al., editors), Academic Press, 1976, pp. 203–250.

[20] I. J. Schoenberg, *On Micchelli's theory of cardinal $\mathscr{L}$-splines*, Studies in Spline Functions and Approximation Theory (S. Karlin et al., editors), Academic Press, 1976, pp. 251–276.

[21] V. M. Tikhomirov, *Widths of sets in function spaces and the theory of best approximations*, Uspekhi Mat. Nauk **15** (1960), no. 3 (93), 81–120; English transl. in Russian Math. Surveys **15** (1960).

[22] V. A. Yakubovich and V. M. Starzhinskiĭ, *Linear differential equations with periodic coefficients, and their applications*, "Nauka", Moscow, 1972; English transl., Vols. 1, 2, Israel Program Sci. Transls., Jerusalem, and Wiley, New York, 1975.

[23] Larry L. Schumaker, *Toward a constructive theory of generalized spline functions*, Spline Functions (Proc. Internat. Sympos., Karlsruhe, 1975), Lecture Notes in Math., Vol. 501, Springer-Verlag, 1976, pp. 265–331.

Translated by H. H. McFADEN

ABCDEFGHIJ–AMS–898765